Dive into Deep Learning

Deep learning has revolutionized pattern recognition, introducing tools that power a wide range of technologies in such diverse fields as computer vision, natural language processing, and automatic speech recognition. Applying deep learning requires you to simultaneously understand how to cast a problem, the basic mathematics of modeling, the algorithms for fitting your models to data, and the engineering techniques to implement it all.

This book is a comprehensive resource that makes deep learning approachable, while still providing sufficient technical depth to enable engineers, scientists, and students to use deep learning in their own work. No previous background in machine learning or deep learning is required—every concept is explained from scratch and the appendix provides a refresher on the mathematics needed. Runnable code is featured throughout, allowing you to develop your own intuition by putting key ideas into practice.

Aston Zhang is Senior Scientist at Amazon Web Services.

Zachary C. Lipton is Assistant Professor of Machine Learning and Operations Research at Carnegie Mellon University.

Mu Li is Senior Principal Scientist at Amazon Web Services.

Alexander J. Smola is VP/Distinguished Scientist for Machine Learning at Amazon Web Services.

"In less than a decade, the AI revolution has swept from research labs to broad industries to every corner of our daily life. *Dive into Deep Learning* is an excellent text on deep learning and deserves attention from anyone who wants to learn why deep learning has ignited the AI revolution: the most powerful technology force of our time."

— Jensen Huang, Founder and CEO, NVIDIA

"This is a timely, fascinating book, providing not only a comprehensive overview of deep learning principles but also detailed algorithms with hands-on programming code, and moreover, a state-of-the-art introduction to deep learning in computer vision and natural language processing. Dive into this book if you want to dive into deep learning!"

— Jiawei Han, Michael Aiken Chair Professor, University of Illinois at Urbana–Champaign

"This is a highly welcome addition to the machine learning literature, with a focus on hands-on experience implemented via the integration of Jupyter notebooks. Students of deep learning should find this invaluable to become proficient in this field."

— Bernhard Schölkopf, Director, Max Planck Institute for Intelligent Systems

"*Dive into Deep Learning* strikes an excellent balance between hands-on learning and in-depth explanation. I've used it in my deep learning course and recommend it to anyone who wants to develop a thorough and practical understanding of deep learning."

— Colin Raffel, Assistant Professor, University of North Carolina, Chapel Hill

Dive into Deep Learning

ASTON ZHANG

Amazon Web Services

ZACHARY C. LIPTON

Carnegie Mellon University, Pennsylvania

MU LI

Amazon Web Services

ALEXANDER J. SMOLA

Amazon Web Services

CAMBRIDGE
UNIVERSITY PRESS

Shaftesbury Road, Cambridge CB2 8EA, United Kingdom

One Liberty Plaza, 20th Floor, New York, NY 10006, USA

477 Williamstown Road, Port Melbourne, VIC 3207, Australia

314–321, 3rd Floor, Plot 3, Splendor Forum, Jasola District Centre, New Delhi – 110025, India

103 Penang Road, #05–06/07, Visioncrest Commercial, Singapore 238467

Cambridge University Press is part of Cambridge University Press & Assessment,
a department of the University of Cambridge.

We share the University's mission to contribute to society through the pursuit of
education, learning and research at the highest international levels of excellence.

www.cambridge.org
Information on this title: www.cambridge.org/9781009389433

DOI: 10.1017/9781009389426

First published 2024

A catalogue record for this publication is available from the British Library

A Cataloging-in-Publication data record for this book is available from the Library of Congress

ISBN 978-1-009-38943-3 Paperback

Cambridge University Press & Assessment has no responsibility for the persistence
or accuracy of URLs for external or third-party internet websites referred to in this
publication and does not guarantee that any content on such websites is, or will
remain, accurate or appropriate.

Contents

Preface

Just a few years ago, there were no legions of deep learning scientists developing intelligent products and services at major companies and startups. When we entered the field, machine learning did not command headlines in daily newspapers. Our parents had no idea what machine learning was, let alone why we might prefer it to a career in medicine or law. Machine learning was a blue skies academic discipline whose industrial significance was limited to a narrow set of real-world applications, including speech recognition and computer vision. Moreover, many of these applications required so much domain knowledge that they were often regarded as entirely separate areas for which machine learning was one small component. At that time, neural networks—the predecessors of the deep learning methods that we focus on in this book—were generally regarded as outmoded.

Yet in just a few years, deep learning has taken the world by surprise, driving rapid progress in such diverse fields as computer vision, natural language processing, automatic speech recognition, reinforcement learning, and biomedical informatics. Moreover, the success of deep learning in so many tasks of practical interest has even catalyzed developments in theoretical machine learning and statistics. With these advances in hand, we can now build cars that drive themselves with more autonomy than ever before (though less autonomy than some companies might have you believe), dialogue systems that debug code by asking clarifying questions, and software agents beating the best human players in the world at board games such as Go, a feat once thought to be decades away. Already, these tools exert ever-wider influence on industry and society, changing the way movies are made, diseases are diagnosed, and playing a growing role in basic sciences—from astrophysics, to climate modeling, to weather prediction, to biomedicine.

About This Book

This book represents our attempt to make deep learning approachable, teaching you the *concepts*, the *context*, and the *code*.

One Medium Combining Code, Math, and HTML

For any computing technology to reach its full impact, it must be well understood, well documented, and supported by mature, well-maintained tools. The key ideas should be clearly distilled, minimizing the onboarding time needed to bring new practitioners up to date. Mature libraries should automate common tasks, and exemplar code should make it easy for practitioners to modify, apply, and extend common applications to suit their needs.

As an example, take dynamic web applications. Despite a large number of companies, such as Amazon, developing successful database-driven web applications in the 1990s, the potential of this technology to aid creative entrepreneurs was realized to a far greater degree only in the past ten years, owing in part to the development of powerful, well-documented frameworks.

Testing the potential of deep learning presents unique challenges because any single application brings together various disciplines. Applying deep learning requires simultaneously understanding (i) the motivations for casting a problem in a particular way; (ii) the mathematical form of a given model; (iii) the optimization algorithms for fitting the models to data; (iv) the statistical principles that tell us when we should expect our models to generalize to unseen data and practical methods for certifying that they have, in fact, generalized; and (v) the engineering techniques required to train models efficiently, navigating the pitfalls of numerical computing and getting the most out of available hardware. Teaching the critical thinking skills required to formulate problems, the mathematics to solve them, and the software tools to implement those solutions all in one place presents formidable challenges. Our goal in this book is to present a unified resource to bring would-be practitioners up to speed.

When we started this book project, there were no resources that simultaneously (i) remained up to date; (ii) covered the breadth of modern machine learning practices with sufficient technical depth; and (iii) interleaved exposition of the quality one expects of a textbook with the clean runnable code that one expects of a hands-on tutorial. We found plenty of code examples illustrating how to use a given deep learning framework (e.g., how to do basic numerical computing with matrices in TensorFlow) or for implementing particular techniques (e.g., code snippets for LeNet, AlexNet, ResNet, etc.) scattered across various blog posts and GitHub repositories. However, these examples typically focused on *how* to implement a given approach, but left out the discussion of *why* certain algorithmic decisions are made. While some interactive resources have popped up sporadically to address a particular topic, e.g., the engaging blog posts published on the website Distill[1], or personal blogs, they only covered selected topics in deep learning, and often lacked associated code. On the other hand, while several deep learning textbooks have emerged—e.g., Goodfellow *et al.* (2016), which offers a comprehensive survey on the basics of deep learning—these resources do not marry the descriptions to realizations of the concepts in code, sometimes leaving readers clueless as to how to implement them. Moreover, too many resources are hidden behind the paywalls of commercial course providers.

We set out to create a resource that could (i) be freely available for everyone; (ii) offer sufficient technical depth to provide a starting point on the path to actually becoming an applied machine learning scientist; (iii) include runnable code, showing readers *how* to solve problems in practice; (iv) allow for rapid updates, both by us and also by the community at large; and (v) be complemented by a forum[2] for interactive discussion of technical details and to answer questions.

These goals were often in conflict. Equations, theorems, and citations are best managed and laid out in LaTeX. Code is best described in Python. And webpages are native in HTML and JavaScript. Furthermore, we want the content to be accessible both as executable code, as a physical book, as a downloadable PDF, and on the Internet as a website. No workflows seemed suited to these demands, so we decided to assemble our own (Section A.6). We settled on GitHub to share the source and to facilitate community contributions; Jupyter notebooks for mixing code, equations and text; Sphinx as a rendering engine; and Discourse as a discussion platform. While our system is not perfect, these choices strike a compromise among the competing concerns. We believe that *Dive into Deep Learning* might be the first book published using such an integrated workflow.

Learning by Doing

Many textbooks present concepts in succession, covering each in exhaustive detail. For example, the excellent textbook of Bishop (2006) teaches each topic so thoroughly that getting to the chapter on linear regression requires a nontrivial amount of work. While experts love this book precisely for its thoroughness, for true beginners, this property limits its usefulness as an introductory text.

In this book, we teach most concepts *just in time*. In other words, you will learn concepts at the very moment that they are needed to accomplish some practical end. While we take some time at the outset to teach fundamental preliminaries, like linear algebra and probability, we want you to taste the satisfaction of training your first model before worrying about more esoteric concepts.

Aside from a few preliminary notebooks that provide a crash course in the basic mathematical background, each subsequent chapter both introduces a reasonable number of new concepts and provides several self-contained working examples, using real datasets. This presented an organizational challenge. Some models might logically be grouped together in a single notebook. And some ideas might be best taught by executing several models in succession. By contrast, there is a big advantage to adhering to a policy of *one working example, one notebook*: This makes it as easy as possible for you to start your own research projects by leveraging our code. Just copy a notebook and start modifying it.

Throughout, we interleave the runnable code with background material as needed. In general, we err on the side of making tools available before explaining them fully (often filling in the background later). For instance, we might use *stochastic gradient descent* before explaining why it is useful or offering some intuition for why it works. This helps to give practitioners the necessary ammunition to solve problems quickly, at the expense of requiring the reader to trust us with some curatorial decisions.

This book teaches deep learning concepts from scratch. Sometimes, we delve into fine details about models that would typically be hidden from users by modern deep learning frameworks. This comes up especially in the basic tutorials, where we want you to understand everything that happens in a given layer or optimizer. In these cases, we often present two versions of the example: one where we implement everything from scratch, relying only on NumPy-like functionality and automatic differentiation, and a more prac-

tical example, where we write succinct code using the high-level APIs of deep learning
frameworks. After explaining how some component works, we rely on the high-level API
in subsequent tutorials.

Content and Structure

The book can be divided into roughly three parts, dealing with preliminaries, deep learning
techniques, and advanced topics focused on real systems and applications (Fig. 1).

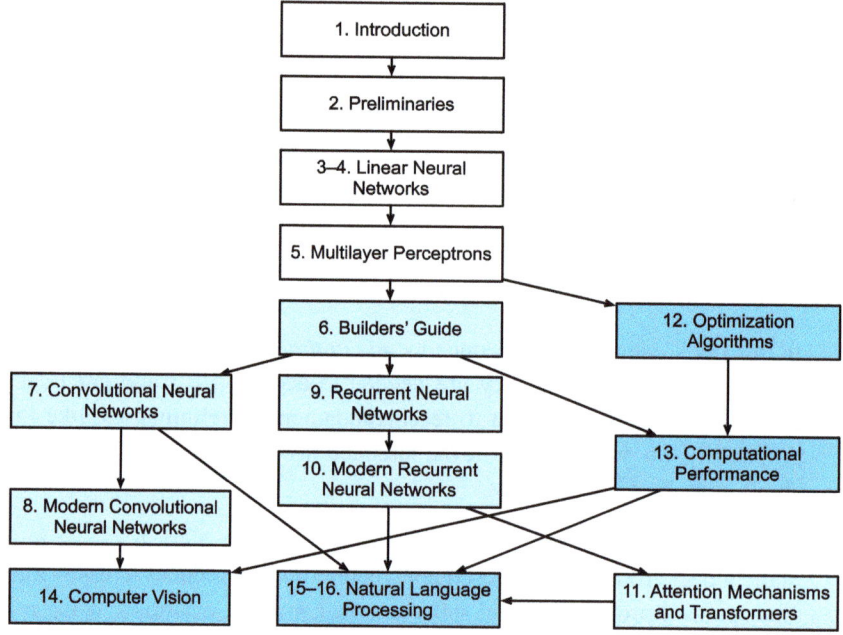

Fig. 1   Book structure.

- **Part 1: Basics and Preliminaries**. Chapter 1 is an introduction to deep learning. Then,
  in Chapter 2, we quickly bring you up to speed on the prerequisites required for hands-
  on deep learning, such as how to store and manipulate data, and how to apply various numerical operations based on elementary concepts from linear algebra, calculus,
  and probability. Chapter 3 and Chapter 5 cover the most fundamental concepts and
  techniques in deep learning, including regression and classification; linear models;
  multilayer perceptrons; and overfitting and regularization.

- **Part 2: Modern Deep Learning Techniques**. Chapter 6 describes the key computational components of deep learning systems and lays the groundwork for our subsequent implementations of more complex models. Next, Chapter 7 and Chapter 8
  present convolutional neural networks (CNNs), powerful tools that form the backbone of most modern computer vision systems. Similarly, Chapter 9 and Chapter 10
  introduce recurrent neural networks (RNNs), models that exploit sequential (e.g., temporal) structure in data and are commonly used for natural language processing and
  time series prediction. In Chapter 11, we describe a relatively new class of models,

based on so-called *attention mechanisms*, that has displaced RNNs as the dominant architecture for most natural language processing tasks. These sections will bring you up to speed on the most powerful and general tools that are widely used by deep learning practitioners.

- **Part 3: Scalability, Efficiency, and Applications** (available online[3]). In Chapter 12, we discuss several common optimization algorithms used to train deep learning models. Next, in Chapter 13, we examine several key factors that influence the computational performance of deep learning code. Then, in Chapter 14, we illustrate major applications of deep learning in computer vision. Finally, in Chapter 15 and Chapter 16, we demonstrate how to pretrain language representation models and apply them to natural language processing tasks.

Code

Most sections of this book feature executable code. We believe that some intuitions are best developed via trial and error, tweaking the code in small ways and observing the results. Ideally, an elegant mathematical theory might tell us precisely how to tweak our code to achieve a desired result. However, deep learning practitioners today must often tread where no solid theory provides guidance. Despite our best attempts, formal explanations for the efficacy of various techniques are still lacking, for a variety of reasons: the mathematics to characterize these models can be so difficult; the explanation likely depends on properties of the data that currently lack clear definitions; and serious inquiry on these topics has only recently kicked into high gear. We are hopeful that as the theory of deep learning progresses, each future edition of this book will provide insights that eclipse those presently available.

To avoid unnecessary repetition, we capture some of our most frequently imported and used functions and classes in the d2l package. Throughout, we mark blocks of code (such as functions, classes, or collection of import statements) with #@save to indicate that they will be accessed later via the d2l package. We offer a detailed overview of these classes and functions in Section A.7. The d2l package is lightweight and only requires the following dependencies:

```
#@save
import collections
import hashlib
import inspect
import math
import os
import random
import re
import shutil
import sys
import tarfile
import time
import zipfile
from collections import defaultdict
import pandas as pd
```

(continues on next page)

(continued from previous page)

```
import requests
from IPython import display
from matplotlib import pyplot as plt
from matplotlib_inline import backend_inline

d2l = sys.modules[__name__]
```

Most of the code in this book is based on PyTorch, a popular open-source framework that has been enthusiastically embraced by the deep learning research community. All of the code in this book has passed tests under the latest stable version of PyTorch. However, due to the rapid development of deep learning, some code *in the print edition* may not work properly in future versions of PyTorch. We plan to keep the online version up to date. In case you encounter any problems, please consult *Installation* (page xxvi) to update your code and runtime environment. Below lists dependencies in our PyTorch implementation.

```
#@save
import numpy as np
import torch
import torchvision
from PIL import Image
from scipy.spatial import distance_matrix
from torch import nn
from torch.nn import functional as F
from torchvision import transforms
```

Target Audience

This book is for students (undergraduate or graduate), engineers, and researchers, who seek a solid grasp of the practical techniques of deep learning. Because we explain every concept from scratch, no previous background in deep learning or machine learning is required. Fully explaining the methods of deep learning requires some mathematics and programming, but we will only assume that you enter with some basics, including modest amounts of linear algebra, calculus, probability, and Python programming. Just in case you have forgotten anything, the online Appendix[4] provides a refresher on most of the mathematics you will find in this book. Usually, we will prioritize intuition and ideas over mathematical rigor. If you would like to extend these foundations beyond the prerequisites to understand our book, we happily recommend some other terrific resources: *Linear Analysis* by Bollobás (1999) covers linear algebra and functional analysis in great depth. *All of Statistics* (Wasserman, 2013) provides a marvelous introduction to statistics. Joe Blitzstein's books[5] and courses[6] on probability and inference are pedagogical gems. And if you have not used Python before, you may want to peruse this Python tutorial[7].

Notebooks, Website, GitHub, and Forum

All our notebooks can be downloaded from the D2L.ai website[8] and from GitHub[9]. Associated with this book we have launched a discussion forum at discuss.d2l.ai[10]. Whenever you have questions on any section of the book, you can find a link to the associated discussion page at the end of each notebook.

Acknowledgments

We are indebted to the hundreds of contributors for both the English and the Chinese drafts. They helped improve the content and offered valuable feedback. This book was originally implemented with MXNet as the primary framework. We thank Anirudh Dagar and Yuan Tang for adapting a majority part of earlier MXNet code into PyTorch and TensorFlow implementations, respectively. Since July 2021, we have redesigned and reimplemented this book in PyTorch, MXNet, and TensorFlow, choosing PyTorch as the primary framework. We thank Anirudh Dagar for adapting a majority part of more recent PyTorch code into JAX implementations. We thank Gaosheng Wu, Liujun Hu, Ge Zhang, and Jiehang Xie from Baidu for adapting a majority part of more recent PyTorch code into PaddlePaddle implementations in the Chinese draft. We thank Shuai Zhang for integrating the LaTeX style from the press into the PDF building.

On GitHub, we thank every contributor of this English draft for making it better for everyone. Their GitHub IDs or names are (in no particular order): alxnorden, avinashingit, bowen0701, brettkoonce, Chaitanya Prakash Bapat, cryptonaut, Davide Fiocco, edgarroman, gkutiel, John Mitro, Liang Pu, Rahul Agarwal, Mohamed Ali Jamaoui, Michael (Stu) Stewart, Mike Müller, NRauschmayr, Prakhar Srivastav, sad-, sfermigier, Sheng Zha, sundeepteki, topecongiro, tpdi, vermicelli, Vishaal Kapoor, Vishwesh Ravi Shrimali, YaYaB, Yuhong Chen, Evgeniy Smirnov, lgov, Simon Corston-Oliver, Igor Dzreyev, Ha Nguyen, pmuens, Andrei Lukovenko, senorcinco, vfdev-5, dsweet, Mohammad Mahdi Rahimi, Abhishek Gupta, uwsd, DomKM, Lisa Oakley, Bowen Li, Aarush Ahuja, Prasanth Buddareddygari, brianhendee, mani2106, mtn, lkevinzc, caojilin, Lakshya, Fiete Lüer, Surbhi Vijayvargeeya, Muhyun Kim, dennismalmgren, adursun, Anirudh Dagar, liqingnz, Pedro Larroy, lgov, ati-ozgur, Jun Wu, Matthias Blume, Lin Yuan, geogunow, Josh Gardner, Maximilian Böther, Rakib Islam, Leonard Lausen, Abhinav Upadhyay, rongruosong, Steve Sedlmeyer, Ruslan Baratov, Rafael Schlatter, liusy182, Giannis Pappas, ati-ozgur, qbaza, dchoi77, Adam Gerson, Phuc Le, Mark Atwood, christabella, vn09, Haibin Lin, jjangga0214, RichyChen, noelo, hansent, Giel Dops, dvincent1337, WhiteD3vil, Peter Kulits, codypenta, joseppinilla, ahmaurya, karolszk, heytitle, Peter Goetz, rigtorp, Tiep Vu, sfilip, mlxd, Kale-ab Tessera, Sanjar Adilov, MatteoFerrara, hsneto, Katarzyna Biesialska, Gregory Bruss, Duy–Thanh Doan, paulaurel, graytowne, Duc Pham, sl7423, Jaedong Hwang, Yida Wang, cys4, clhm, Jean Kaddour, austinmw, trebeljahr, tbaums, Cuong V. Nguyen, pavelkomarov, vzlamal, NotAnotherSystem, J-Arun-Mani, jancio, eldarkurtic, the-great-shazbot, doctorcolossus, gducharme, cclauss, Daniel-Mietchen, hoonose, biagiom, abhinavsp0730, jonathanhrandall, ysraell, Nodar Okroshiashvili, UgurKap, Jiyang Kang, StevenJokes, Tomer Kaftan, liweiwp, netyster, ypandya, NishantTharani, heiligerl, SportsTHU, Hoa Nguyen, manuel-arno-korfmann-webentwicklung, aterzis-personal, nxby,

Xiaoting He, Josiah Yoder, mathresearch, mzz2017, jroberayalas, iluu, ghejc, BSharmi, vkramdev, simonwardjones, LakshKD, TalNeoran, djliden, Nikhil95, Oren Barkan, guoweis, haozhu233, pratikhack, Yue Ying, tayfununal, steinsag, charleybeller, Andrew Lumsdaine, Jiekui Zhang, Deepak Pathak, Florian Donhauser, Tim Gates, Adriaan Tijsseling, Ron Medina, Gaurav Saha, Murat Semerci, Lei Mao, Levi McClenny, Joshua Broyde, jake221, jonbally, zyhazwraith, Brian Pulfer, Nick Tomasino, Lefan Zhang, Hongshen Yang, Vinney Cavallo, yuntai, Yuanxiang Zhu, amarazov, pasricha, Ben Greenawald, Shivam Upadhyay, Quanshangze Du, Biswajit Sahoo, Parthe Pandit, Ishan Kumar, HomunculusK, Lane Schwartz, varadgunjal, Jason Wiener, Armin Gholampoor, Shreshtha13, eigenarnav, Hyeonggyu Kim, EmilyOng, Bálint Mucsányi, Chase DuBois, Juntian Tao, Wenxiang Xu, Lifu Huang, filevich, quake2005, nils-werner, Yiming Li, Marsel Khisamutdinov, Francesco "Fuma" Fumagalli, Peilin Sun, Vincent Gurgul, qingfengtommy, Janmey Shukla, Mo Shan, Kaan Sancak, regob, AlexSauer, Gopalakrishna Ramachandra, Tobias Uelwer, Chao Wang, Tian Cao, Nicolas Corthorn, akash5474, kxxt, zxydi1992, Jacob Britton, Shuangchi He, zhmou, krahets, Jie-Han Chen, Atishay Garg, Marcel Flygare, adtygan, Nik Vaessen, bolded, Louis Schlessinger, Balaji Varatharajan, atgctg, Kaixin Li, Victor Barbaros, Riccardo Musto, Elizabeth Ho, azimjonn, Guilherme Miotto, Alessandro Finamore, Joji Joseph, Anthony Biel, Zeming Zhao, shjustinbaek, gab-chen, nantekoto, Yutaro Nishiyama, Oren Amsalem, Tian-MaoMao, Amin Allahyar, Gijs van Tulder, Mikhail Berkov, iamorphen, Matthew Caseres, Andrew Walsh, pggPL, RohanKarthikeyan, Ryan Choi, and Likun Lei.

We thank Amazon Web Services, especially Wen-Ming Ye, George Karypis, Swami Sivasubramanian, Peter DeSantis, Adam Selipsky, and Andrew Jassy for their generous support in writing this book. Without the available time, resources, discussions with colleagues, and continuous encouragement, this book would not have happened. During the preparation of the book for publication, Cambridge University Press has offered excellent support. We thank our commissioning editor David Tranah for his help and professionalism.

Summary

Deep learning has revolutionized pattern recognition, introducing technology that now powers a wide range of technologies, in such diverse fields as computer vision, natural language processing, and automatic speech recognition. To successfully apply deep learning, you must understand how to cast a problem, the basic mathematics of modeling, the algorithms for fitting your models to data, and the engineering techniques to implement it all. This book presents a comprehensive resource, including prose, figures, mathematics, and code, all in one place.

Exercises

1. Register an account on the discussion forum of this book discuss.d2l.ai[11].

2. Install Python on your computer.

3. Follow the links at the bottom of the section to the forum, where you will be able to seek out help and discuss the book and find answers to your questions by engaging the authors and broader community.

Discussions[12].

Installation

In order to get up and running, we will need an environment for running Python, the Jupyter Notebook, the relevant libraries, and the code needed to run the book itself.

Installing Miniconda

 Your simplest option is to install Miniconda[13]. Note that the Python 3.x version is required. You can skip the following steps if your machine already has conda installed.

Visit the Miniconda website and determine the appropriate version for your system based on your Python 3.x version and machine architecture. Suppose that your Python version is 3.9 (our tested version). If you are using macOS, you would download the bash script whose name contains the strings "MacOSX", navigate to the download location, and execute the installation as follows (taking Intel Macs as an example):

```
# The file name is subject to changes
sh Miniconda3-py39_4.12.0-MacOSX-x86_64.sh -b
```

A Linux user would download the file whose name contains the strings "Linux" and execute the following at the download location:

```
# The file name is subject to changes
sh Miniconda3-py39_4.12.0-Linux-x86_64.sh -b
```

 On Windows, download and install Miniconda by following the online instructions[14]. Windows users should search for cmd to open the Command Prompt (command-line interpreter) for running commands.

Next, initialize the shell so we can run conda directly.

```
~/miniconda3/bin/conda init
```

Then close and reopen your current shell. You should be able to create a new environment as follows:

```
conda create --name d2l python=3.9 -y
```

Now we can activate the d2l environment:

```
conda activate d2l
```

Installing the Deep Learning Framework and the d2l **Package**

Before installing any deep learning framework, please first check whether or not you have proper GPUs on your machine (the GPUs that power the display on a standard laptop are not relevant for our purposes). For example, if your computer has NVIDIA GPUs and has installed CUDA [15], then you are all set. If your machine does not house any GPU, there is no need to worry just yet. Your CPU provides more than enough horsepower to get you through the first few chapters. Just remember that you will want to access GPUs before running larger models.

You can install PyTorch (the specified versions are tested at the time of writing) with either CPU or GPU support as follows:

```
pip install torch==2.0.0 torchvision==0.15.1
```

Our next step is to install the d2l package that we developed in order to encapsulate frequently used functions and classes found throughout this book:

```
pip install d2l==1.0.3
```

Downloading and Running the Code

Next, you will want to download the notebooks so that you can run each of the book's code blocks. Simply click on the "Notebooks" tab at the top of any HTML page on the D2L.ai website [16] to download the code and then unzip it. Alternatively, you can fetch the notebooks from the command line as follows:

```
mkdir d2l-en && cd d2l-en
curl https://d2l.ai/d2l-en-1.0.3.zip -o d2l-en.zip
unzip d2l-en.zip && rm d2l-en.zip
cd pytorch
```

If you do not already have `unzip` installed, first run `sudo apt-get install unzip`. Now we can start the Jupyter Notebook server by running:

```
jupyter notebook
```

At this point, you can open http://localhost:8888 (it may have already opened automatically) in your web browser. Then we can run the code for each section of the book. Whenever you open a new command line window, you will need to execute `conda activate d2l` to activate the runtime environment before running the D2L notebooks, or updating your packages (either the deep learning framework or the d2l package). To exit the environment, run `conda deactivate`.

Discussions[17].

17

Notation

Throughout this book, we adhere to the following notational conventions. Note that some of these symbols are placeholders, while others refer to specific objects. As a general rule of thumb, the indefinite article "a" often indicates that the symbol is a placeholder and that similarly formatted symbols can denote other objects of the same type. For example, "x: a scalar" means that lowercased letters generally represent scalar values, but "$\mathbb{Z}$: the set of integers" refers specifically to the symbol $\mathbb{Z}$.

Numerical Objects

- x: a scalar

- $\mathbf{x}$: a vector

- $\mathbf{X}$: a matrix

- X: a general tensor

- $\mathbf{I}$: the identity matrix (of some given dimension), i.e., a square matrix with 1 on all diagonal entries and 0 on all off-diagonals

- x_i, $[\mathbf{x}]_i$: the ith element of vector $\mathbf{x}$

- x_{ij}, $x_{i,j}$, $[\mathbf{X}]_{ij}$, $[\mathbf{X}]_{i,j}$: the element of matrix $\mathbf{X}$ at row i and column j.

Set Theory

- $\mathcal{X}$: a set

- $\mathbb{Z}$: the set of integers

- $\mathbb{Z}^+$: the set of positive integers

- $\mathbb{R}$: the set of real numbers

- $\mathbb{R}^n$: the set of n-dimensional vectors of real numbers

- $\mathbb{R}^{a \times b}$: The set of matrices of real numbers with a rows and b columns

- $|\mathcal{X}|$: cardinality (number of elements) of set $\mathcal{X}$

- $\mathcal{A} \cup \mathcal{B}$: union of sets $\mathcal{A}$ and $\mathcal{B}$

- $\mathcal{A} \cap \mathcal{B}$: intersection of sets $\mathcal{A}$ and $\mathcal{B}$

- $\mathcal{A} \setminus \mathcal{B}$: set subtraction of $\mathcal{B}$ from $\mathcal{A}$ (contains only those elements of $\mathcal{A}$ that do not belong to $\mathcal{B}$)

Functions and Operators

- $f(\cdot)$: a function

- $\log(\cdot)$: the natural logarithm (base e)

- $\log_2(\cdot)$: logarithm to base 2

- $\exp(\cdot)$: the exponential function

- $\mathbf{1}(\cdot)$: the indicator function; evaluates to 1 if the boolean argument is true, and 0 otherwise

- $\mathbf{1}_\mathcal{X}(z)$: the set-membership indicator function; evaluates to 1 if the element z belongs to the set $\mathcal{X}$ and 0 otherwise

- $(\cdot)^\top$: transpose of a vector or a matrix

- $\mathbf{X}^{-1}$: inverse of matrix $\mathbf{X}$

- $\odot$: Hadamard (elementwise) product

- $[\cdot, \cdot]$: concatenation

- $\|\cdot\|_p$: ℓ_p norm

- $\|\cdot\|$: ℓ_2 norm

- $\langle \mathbf{x}, \mathbf{y} \rangle$: inner (dot) product of vectors $\mathbf{x}$ and $\mathbf{y}$

- $\sum$: summation over a collection of elements

- $\prod$: product over a collection of elements

- $\stackrel{\text{def}}{=}$: an equality asserted as a definition of the symbol on the left-hand side

Calculus

- $\frac{dy}{dx}$: derivative of y with respect to x

- $\frac{\partial y}{\partial x}$: partial derivative of y with respect to x

- $\nabla_{\mathbf{x}} y$: gradient of y with respect to $\mathbf{x}$

- $\int_a^b f(x)\, dx$: definite integral of f from a to b with respect to x

- $\int f(x)\, dx$: indefinite integral of f with respect to x

Probability and Information Theory

- X: a random variable

- P: a probability distribution

- $X \sim P$: the random variable X follows distribution P

- $P(X = x)$: the probability assigned to the event where random variable X takes value x

- $P(X \mid Y)$: the conditional probability distribution of X given Y

- $p(\cdot)$: a probability density function (PDF) associated with distribution P

- $E[X]$: expectation of a random variable X

- $X \perp Y$: random variables X and Y are independent

- $X \perp Y \mid Z$: random variables X and Y are conditionally independent given Z

- σ_X: standard deviation of random variable X

- $\mathrm{Var}(X)$: variance of random variable X, equal to σ_X^2

- $\mathrm{Cov}(X, Y)$: covariance of random variables X and Y

- $\rho(X, Y)$: the Pearson correlation coefficient between X and Y, equals $\frac{\mathrm{Cov}(X,Y)}{\sigma_X \sigma_Y}$

- $H(X)$: entropy of random variable X

- $D_{\mathrm{KL}}(P \| Q)$: the KL-divergence (or relative entropy) from distribution Q to distribution P

Discussions[18].

18

Introduction

Until recently, nearly every computer program that you might have interacted with during an ordinary day was coded up as a rigid set of rules specifying precisely how it should behave. Say that we wanted to write an application to manage an e-commerce platform. After huddling around a whiteboard for a few hours to ponder the problem, we might settle on the broad strokes of a working solution, for example: (i) users interact with the application through an interface running in a web browser or mobile application; (ii) our application interacts with a commercial-grade database engine to keep track of each user's state and maintain records of historical transactions; and (iii) at the heart of our application, the *business logic* (you might say, the *brains*) of our application spells out a set of rules that map every conceivable circumstance to the corresponding action that our program should take.

To build the brains of our application, we might enumerate all the common events that our program should handle. For example, whenever a customer clicks to add an item to their shopping cart, our program should add an entry to the shopping cart database table, associating that user's ID with the requested product's ID. We might then attempt to step through every possible corner case, testing the appropriateness of our rules and making any necessary modifications. What happens if a user initiates a purchase with an empty cart? While few developers ever get it completely right the first time (it might take some test runs to work out the kinks), for the most part we can write such programs and confidently launch them *before* ever seeing a real customer. Our ability to manually design automated systems that drive functioning products and systems, often in novel situations, is a remarkable cognitive feat. And when you are able to devise solutions that work 100% of the time, you typically should not be worrying about machine learning.

Fortunately for the growing community of machine learning scientists, many tasks that we would like to automate do not bend so easily to human ingenuity. Imagine huddling around the whiteboard with the smartest minds you know, but this time you are tackling one of the following problems:

- Write a program that predicts tomorrow's weather given geographic information, satellite images, and a trailing window of past weather.

- Write a program that takes in a factoid question, expressed in free-form text, and answers it correctly.

- Write a program that, given an image, identifies every person depicted in it and draws outlines around each.

- Write a program that presents users with products that they are likely to enjoy but unlikely, in the natural course of browsing, to encounter.

For these problems, even elite programmers would struggle to code up solutions from scratch. The reasons can vary. Sometimes the program that we are looking for follows a pattern that changes over time, so there is no fixed right answer! In such cases, any successful solution must adapt gracefully to a changing world. At other times, the relationship (say between pixels, and abstract categories) may be too complicated, requiring thousands or millions of computations and following unknown principles. In the case of image recognition, the precise steps required to perform the task lie beyond our conscious understanding, even though our subconscious cognitive processes execute the task effortlessly.

Machine learning is the study of algorithms that can learn from experience. As a machine learning algorithm accumulates more experience, typically in the form of observational data or interactions with an environment, its performance improves. Contrast this with our deterministic e-commerce platform, which follows the same business logic, no matter how much experience accrues, until the developers themselves learn and decide that it is time to update the software. In this book, we will teach you the fundamentals of machine learning, focusing in particular on *deep learning*, a powerful set of techniques driving innovations in areas as diverse as computer vision, natural language processing, healthcare, and genomics.

1.1 A Motivating Example

Before beginning writing, the authors of this book, like much of the work force, had to become caffeinated. We hopped in the car and started driving. Using an iPhone, Alex called out "Hey Siri", awakening the phone's voice recognition system. Then Mu commanded "directions to Blue Bottle coffee shop". The phone quickly displayed the transcription of his command. It also recognized that we were asking for directions and launched the Maps application (app) to fulfill our request. Once launched, the Maps app identified a number of routes. Next to each route, the phone displayed a predicted transit time. While this story was fabricated for pedagogical convenience, it demonstrates that in the span of just a few seconds, our everyday interactions with a smart phone can engage several machine learning models.

Imagine just writing a program to respond to a *wake word* such as "Alexa", "OK Google", and "Hey Siri". Try coding it up in a room by yourself with nothing but a computer and a code editor, as illustrated in Fig. 1.1.1. How would you write such a program from first principles? Think about it... the problem is hard. Every second, the microphone will collect roughly 44,000 samples. Each sample is a measurement of the amplitude of the sound wave. What rule could map reliably from a snippet of raw audio to confident predictions {yes, no} about whether the snippet contains the wake word? If you are stuck, do not worry.

We do not know how to write such a program from scratch either. That is why we use machine learning.

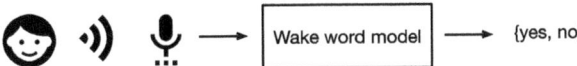

Fig. 1.1.1 Identify a wake word.

Here is the trick. Often, even when we do not know how to tell a computer explicitly how to map from inputs to outputs, we are nonetheless capable of performing the cognitive feat ourselves. In other words, even if you do not know how to program a computer to recognize the word "Alexa", you yourself are able to recognize it. Armed with this ability, we can collect a huge *dataset* containing examples of audio snippets and associated labels, indicating which snippets contain the wake word. In the currently dominant approach to machine learning, we do not attempt to design a system *explicitly* to recognize wake words. Instead, we define a flexible program whose behavior is determined by a number of *parameters*. Then we use the dataset to determine the best possible parameter values, i.e., those that improve the performance of our program with respect to a chosen performance measure.

You can think of the parameters as knobs that we can turn, manipulating the behavior of the program. Once the parameters are fixed, we call the program a *model*. The set of all distinct programs (input–output mappings) that we can produce just by manipulating the parameters is called a *family* of models. And the "meta-program" that uses our dataset to choose the parameters is called a *learning algorithm*.

Before we can go ahead and engage the learning algorithm, we have to define the problem precisely, pinning down the exact nature of the inputs and outputs, and choosing an appropriate model family. In this case, our model receives a snippet of audio as *input*, and the model generates a selection among {yes, no} as *output*. If all goes according to plan the model's guesses will typically be correct as to whether the snippet contains the wake word.

If we choose the right family of models, there should exist one setting of the knobs such that the model fires "yes" every time it hears the word "Alexa". Because the exact choice of the wake word is arbitrary, we will probably need a model family sufficiently rich that, via another setting of the knobs, it could fire "yes" only upon hearing the word "Apricot". We expect that the same model family should be suitable for "Alexa" recognition and "Apricot" recognition because they seem, intuitively, to be similar tasks. However, we might need a different family of models entirely if we want to deal with fundamentally different inputs or outputs, say if we wanted to map from images to captions, or from English sentences to Chinese sentences.

As you might guess, if we just set all of the knobs randomly, it is unlikely that our model will recognize "Alexa", "Apricot", or any other English word. In machine learning, the *learning* is the process by which we discover the right setting of the knobs for coercing the desired behavior from our model. In other words, we *train* our model with data. As shown in Fig. 1.1.2, the training process usually looks like the following:

1. Start off with a randomly initialized model that cannot do anything useful.

2. Grab some of your data (e.g., audio snippets and corresponding {yes, no} labels).

3. Tweak the knobs to make the model perform better as assessed on those examples.

4. Repeat Steps 2 and 3 until the model is awesome.

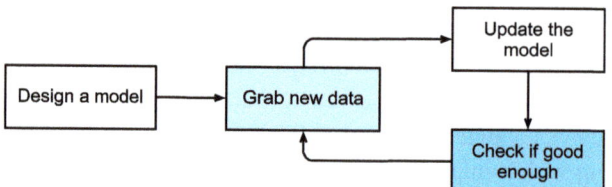

Fig. 1.1.2   A typical training process.

To summarize, rather than code up a wake word recognizer, we code up a program that can *learn* to recognize wake words, if presented with a large labeled dataset. You can think of this act of determining a program's behavior by presenting it with a dataset as *programming with data*. That is to say, we can "program" a cat detector by providing our machine learning system with many examples of cats and dogs. This way the detector will eventually learn to emit a very large positive number if it is a cat, a very large negative number if it is a dog, and something closer to zero if it is not sure. This barely scratches the surface of what machine learning can do. Deep learning, which we will explain in greater detail later, is just one among many popular methods for solving machine learning problems.

1.2 Key Components

In our wake word example, we described a dataset consisting of audio snippets and binary labels, and we gave a hand-wavy sense of how we might train a model to approximate a mapping from snippets to classifications. This sort of problem, where we try to predict a designated unknown label based on known inputs given a dataset consisting of examples for which the labels are known, is called *supervised learning*. This is just one among many kinds of machine learning problems. Before we explore other varieties, we would like to shed more light on some core components that will follow us around, no matter what kind of machine learning problem we tackle:

1. The *data* that we can learn from.

2. A *model* of how to transform the data.

3. An *objective function* that quantifies how well (or badly) the model is doing.

4. An *algorithm* to adjust the model's parameters to optimize the objective function.

1.2.1 Data

It might go without saying that you cannot do data science without data. We could lose hundreds of pages pondering what precisely data *is*, but for now, we will focus on the key properties of the datasets that we will be concerned with. Generally, we are concerned with a collection of examples. In order to work with data usefully, we typically need to come up with a suitable numerical representation. Each *example* (or *data point*, *data instance*, *sample*) typically consists of a set of attributes called *features* (sometimes called *covariates* or *inputs*), based on which the model must make its predictions. In supervised learning problems, our goal is to predict the value of a special attribute, called the *label* (or *target*), that is not part of the model's input.

If we were working with image data, each example might consist of an individual photograph (the features) and a number indicating the category to which the photograph belongs (the label). The photograph would be represented numerically as three grids of numerical values representing the brightness of red, green, and blue light at each pixel location. For example, a 200×200 pixel color photograph would consist of $200 \times 200 \times 3 = 120000$ numerical values.

Alternatively, we might work with electronic health record data and tackle the task of predicting the likelihood that a given patient will survive the next 30 days. Here, our features might consist of a collection of readily available attributes and frequently recorded measurements, including age, vital signs, comorbidities, current medications, and recent procedures. The label available for training would be a binary value indicating whether each patient in the historical data survived within the 30-day window.

In such cases, when every example is characterized by the same number of numerical features, we say that the inputs are fixed-length vectors and we call the (constant) length of the vectors the *dimensionality* of the data. As you might imagine, fixed-length inputs can be convenient, giving us one less complication to worry about. However, not all data can easily be represented as *fixed-length* vectors. While we might expect microscope images to come from standard equipment, we cannot expect images mined from the Internet all to have the same resolution or shape. For images, we might consider cropping them to a standard size, but that strategy only gets us so far. We risk losing information in the cropped-out portions. Moreover, text data resists fixed-length representations even more stubbornly. Consider the customer reviews left on e-commerce sites such as Amazon, IMDb, and TripAdvisor. Some are short: "it stinks!". Others ramble for pages. One major advantage of deep learning over traditional methods is the comparative grace with which modern models can handle *varying-length* data.

Generally, the more data we have, the easier our job becomes. When we have more data, we can train more powerful models and rely less heavily on preconceived assumptions. The regime change from (comparatively) small to big data is a major contributor to the success of modern deep learning. To drive the point home, many of the most exciting models in deep learning do not work without large datasets. Some others might work in the small data regime, but are no better than traditional approaches.

Finally, it is not enough to have lots of data and to process it cleverly. We need the *right* data. If the data is full of mistakes, or if the chosen features are not predictive of the target quantity of interest, learning is going to fail. The situation is captured well by the cliché: *garbage in, garbage out.* Moreover, poor predictive performance is not the only potential consequence. In sensitive applications of machine learning, like predictive policing, resume screening, and risk models used for lending, we must be especially alert to the consequences of garbage data. One commonly occurring failure mode concerns datasets where some groups of people are unrepresented in the training data. Imagine applying a skin cancer recognition system that had never seen black skin before. Failure can also occur when the data does not only under-represent some groups but reflects societal prejudices. For example, if past hiring decisions are used to train a predictive model that will be used to screen resumes then machine learning models could inadvertently capture and automate historical injustices. Note that this can all happen without the data scientist actively conspiring, or even being aware.

1.2.2 Models

Most machine learning involves transforming the data in some sense. We might want to build a system that ingests photos and predicts smiley-ness. Alternatively, we might want to ingest a set of sensor readings and predict how normal vs. anomalous the readings are. By *model*, we denote the computational machinery for ingesting data of one type, and spitting out predictions of a possibly different type. In particular, we are interested in *statistical models* that can be estimated from data. While simple models are perfectly capable of addressing appropriately simple problems, the problems that we focus on in this book stretch the limits of classical methods. Deep learning is differentiated from classical approaches principally by the set of powerful models that it focuses on. These models consist of many successive transformations of the data that are chained together top to bottom, thus the name *deep learning*. On our way to discussing deep models, we will also discuss some more traditional methods.

1.2.3 Objective Functions

Earlier, we introduced machine learning as learning from experience. By *learning* here, we mean improving at some task over time. But who is to say what constitutes an improvement? You might imagine that we could propose updating our model, and some people might disagree on whether our proposal constituted an improvement or not.

In order to develop a formal mathematical system of learning machines, we need to have formal measures of how good (or bad) our models are. In machine learning, and optimization more generally, we call these *objective functions*. By convention, we usually define objective functions so that lower is better. This is merely a convention. You can take any function for which higher is better, and turn it into a new function that is qualitatively identical but for which lower is better by flipping the sign. Because we choose lower to be better, these functions are sometimes called *loss functions*.

When trying to predict numerical values, the most common loss function is *squared error*, i.e., the square of the difference between the prediction and the ground truth target. For classification, the most common objective is to minimize error rate, i.e., the fraction of examples on which our predictions disagree with the ground truth. Some objectives (e.g., squared error) are easy to optimize, while others (e.g., error rate) are difficult to optimize directly, owing to non-differentiability or other complications. In these cases, it is common instead to optimize a *surrogate objective*.

During optimization, we think of the loss as a function of the model's parameters, and treat the training dataset as a constant. We learn the best values of our model's parameters by minimizing the loss incurred on a set consisting of some number of examples collected for training. However, doing well on the training data does not guarantee that we will do well on unseen data. So we will typically want to split the available data into two partitions: the *training dataset* (or *training set*), for learning model parameters; and the *test dataset* (or *test set*), which is held out for evaluation. At the end of the day, we typically report how our models perform on both partitions. You could think of training performance as analogous to the scores that a student achieves on the practice exams used to prepare for some real final exam. Even if the results are encouraging, that does not guarantee success on the final exam. Over the course of studying, the student might begin to memorize the practice questions, appearing to master the topic but faltering when faced with previously unseen questions on the actual final exam. When a model performs well on the training set but fails to generalize to unseen data, we say that it is *overfitting* to the training data.

1.2.4 Optimization Algorithms

Once we have got some data source and representation, a model, and a well-defined objective function, we need an algorithm capable of searching for the best possible parameters for minimizing the loss function. Popular optimization algorithms for deep learning are based on an approach called *gradient descent*. In brief, at each step, this method checks to see, for each parameter, how that training set loss would change if you perturbed that parameter by just a small amount. It would then update the parameter in the direction that lowers the loss.

1.3 Kinds of Machine Learning Problems

The wake word problem in our motivating example is just one among many that machine learning can tackle. To motivate the reader further and provide us with some common language that will follow us throughout the book, we now provide a broad overview of the landscape of machine learning problems.

1.3.1 Supervised Learning

Supervised learning describes tasks where we are given a dataset containing both features and labels and asked to produce a model that predicts the labels when given input features. Each feature–label pair is called an example. Sometimes, when the context is clear, we may use the term *examples* to refer to a collection of inputs, even when the corresponding labels are unknown. The supervision comes into play because, for choosing the parameters, we (the supervisors) provide the model with a dataset consisting of labeled examples. In probabilistic terms, we typically are interested in estimating the conditional probability of a label given input features. While it is just one among several paradigms, supervised learning accounts for the majority of successful applications of machine learning in industry. Partly that is because many important tasks can be described crisply as estimating the probability of something unknown given a particular set of available data:

- Predict cancer vs. not cancer, given a computer tomography image.

- Predict the correct translation in French, given a sentence in English.

- Predict the price of a stock next month based on this month's financial reporting data.

While all supervised learning problems are captured by the simple description "predicting the labels given input features", supervised learning itself can take diverse forms and require tons of modeling decisions, depending on (among other considerations) the type, size, and quantity of the inputs and outputs. For example, we use different models for processing sequences of arbitrary lengths and fixed-length vector representations. We will visit many of these problems in depth throughout this book.

Informally, the learning process looks something like the following. First, grab a big collection of examples for which the features are known and select from them a random subset, acquiring the ground truth labels for each. Sometimes these labels might be available data that have already been collected (e.g., did a patient die within the following year?) and other times we might need to employ human annotators to label the data, (e.g., assigning images to categories). Together, these inputs and corresponding labels comprise the training set. We feed the training dataset into a supervised learning algorithm, a function that takes as input a dataset and outputs another function: the learned model. Finally, we can feed previously unseen inputs to the learned model, using its outputs as predictions of the corresponding label. The full process is drawn in Fig. 1.3.1.

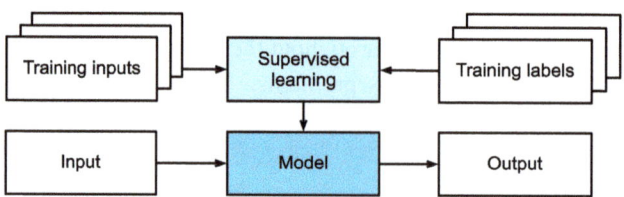

Fig. 1.3.1   Supervised learning.

Regression

Perhaps the simplest supervised learning task to wrap your head around is *regression*. Consider, for example, a set of data harvested from a database of home sales. We might construct a table, in which each row corresponds to a different house, and each column corresponds to some relevant attribute, such as the square footage of a house, the number of bedrooms, the number of bathrooms, and the number of minutes (walking) to the center of town. In this dataset, each example would be a specific house, and the corresponding feature vector would be one row in the table. If you live in New York or San Francisco, and you are not the CEO of Amazon, Google, Microsoft, or Facebook, the (sq. footage, no. of bedrooms, no. of bathrooms, walking distance) feature vector for your home might look something like: $[600, 1, 1, 60]$. However, if you live in Pittsburgh, it might look more like $[3000, 4, 3, 10]$. Fixed-length feature vectors like this are essential for most classic machine learning algorithms.

What makes a problem a regression is actually the form of the target. Say that you are in the market for a new home. You might want to estimate the fair market value of a house, given some features such as above. The data here might consist of historical home listings and the labels might be the observed sales prices. When labels take on arbitrary numerical values (even within some interval), we call this a *regression* problem. The goal is to produce a model whose predictions closely approximate the actual label values.

Lots of practical problems are easily described as regression problems. Predicting the rating that a user will assign to a movie can be thought of as a regression problem and if you designed a great algorithm to accomplish this feat in 2009, you might have won the 1-million-dollar Netflix prize[19]. Predicting the length of stay for patients in the hospital is also a regression problem. A good rule of thumb is that any *how much?* or *how many?* problem is likely to be regression. For example:

- How many hours will this surgery take?

- How much rainfall will this town have in the next six hours?

Even if you have never worked with machine learning before, you have probably worked through a regression problem informally. Imagine, for example, that you had your drains repaired and that your contractor spent 3 hours removing gunk from your sewage pipes. Then they sent you a bill of 350 dollars. Now imagine that your friend hired the same contractor for 2 hours and received a bill of 250 dollars. If someone then asked you how much to expect on their upcoming gunk-removal invoice you might make some reasonable assumptions, such as more hours worked costs more dollars. You might also assume that there is some base charge and that the contractor then charges per hour. If these assumptions held true, then given these two data examples, you could already identify the contractor's pricing structure: 100 dollars per hour plus 50 dollars to show up at your house. If you followed that much, then you already understand the high-level idea behind *linear* regression.

In this case, we could produce the parameters that exactly matched the contractor's prices. Sometimes this is not possible, e.g., if some of the variation arises from factors beyond your two features. In these cases, we will try to learn models that minimize the distance between our predictions and the observed values. In most of our chapters, we will focus on minimizing the squared error loss function. As we will see later, this loss corresponds to the assumption that our data were corrupted by Gaussian noise.

Classification

While regression models are great for addressing *how many?* questions, lots of problems do not fit comfortably in this template. Consider, for example, a bank that wants to develop a check scanning feature for its mobile app. Ideally, the customer would simply snap a photo of a check and the app would automatically recognize the text from the image. Assuming that we had some ability to segment out image patches corresponding to each handwritten character, then the primary remaining task would be to determine which character among some known set is depicted in each image patch. These kinds of *which one?* problems are called *classification* and require a different set of tools from those used for regression, although many techniques will carry over.

In *classification*, we want our model to look at features, e.g., the pixel values in an image, and then predict to which *category* (sometimes called a *class*) among some discrete set of options, an example belongs. For handwritten digits, we might have ten classes, corresponding to the digits 0 through 9. The simplest form of classification is when there are only two classes, a problem which we call *binary classification*. For example, our dataset could consist of images of animals and our labels might be the classes {cat, dog}. Whereas in regression we sought a regressor to output a numerical value, in classification we seek a classifier, whose output is the predicted class assignment.

For reasons that we will get into as the book gets more technical, it can be difficult to optimize a model that can only output a *firm* categorical assignment, e.g., either "cat" or "dog". In these cases, it is usually much easier to express our model in the language of probabilities. Given features of an example, our model assigns a probability to each possible class. Returning to our animal classification example where the classes are {cat, dog}, a classifier might see an image and output the probability that the image is a cat as 0.9. We can interpret this number by saying that the classifier is 90% sure that the image depicts a cat. The magnitude of the probability for the predicted class conveys a notion of uncertainty. It is not the only one available and we will discuss others in chapters dealing with more advanced topics.

When we have more than two possible classes, we call the problem *multiclass classification*. Common examples include handwritten character recognition {0, 1, 2, ...9, a, b, c, ...}. While we attacked regression problems by trying to minimize the squared error loss function, the common loss function for classification problems is called *cross-entropy*, whose name will be demystified when we introduce information theory in later chapters.

Note that the most likely class is not necessarily the one that you are going to use for your decision. Assume that you find a beautiful mushroom in your backyard as shown in Fig. 1.3.2.

Now, assume that you built a classifier and trained it to predict whether a mushroom is poi-

Fig. 1.3.2   Death cap—do not eat!

sonous based on a photograph. Say our poison-detection classifier outputs that the probability that Fig. 1.3.2 shows a death cap is 0.2. In other words, the classifier is 80% sure that our mushroom is not a death cap. Still, you would have to be a fool to eat it. That is because the certain benefit of a delicious dinner is not worth a 20% risk of dying from it. In other words, the effect of the uncertain risk outweighs the benefit by far. Thus, in order to make a decision about whether to eat the mushroom, we need to compute the expected detriment associated with each action which depends both on the likely outcomes and the benefits or harms associated with each. In this case, the detriment incurred by eating the mushroom might be $0.2 \times \infty + 0.8 \times 0 = \infty$, whereas the loss of discarding it is $0.2 \times 0 + 0.8 \times 1 = 0.8$. Our caution was justified: as any mycologist would tell us, the mushroom in Fig. 1.3.2 is actually a death cap.

Classification can get much more complicated than just binary or multiclass classification. For instance, there are some variants of classification addressing hierarchically structured classes. In such cases not all errors are equal—if we must err, we might prefer to misclassify to a related class rather than a distant class. Usually, this is referred to as *hierarchical classification*. For inspiration, you might think of Linnaeus[20], who organized fauna in a hierarchy.

20

In the case of animal classification, it might not be so bad to mistake a poodle for a schnauzer, but our model would pay a huge penalty if it confused a poodle with a dinosaur. Which hierarchy is relevant might depend on how you plan to use the model. For example, rattlesnakes and garter snakes might be close on the phylogenetic tree, but mistaking a rattler for a garter could have fatal consequences.

Tagging

Some classification problems fit neatly into the binary or multiclass classification setups. For example, we could train a normal binary classifier to distinguish cats from dogs. Given the current state of computer vision, we can do this easily, with off-the-shelf tools. Nonetheless, no matter how accurate our model gets, we might find ourselves in trouble when the classifier encounters an image of the *Town Musicians of Bremen*, a popular German fairy tale featuring four animals (Fig. 1.3.3).

Fig. 1.3.3   A donkey, a dog, a cat, and a rooster.

As you can see, the photo features a cat, a rooster, a dog, and a donkey, with some trees in the background. If we anticipate encountering such images, multiclass classification might not be the right problem formulation. Instead, we might want to give the model the option of saying the image depicts a cat, a dog, a donkey, *and* a rooster.

The problem of learning to predict classes that are not mutually exclusive is called *multilabel classification*. Auto-tagging problems are typically best described in terms of multilabel classification. Think of the tags people might apply to posts on a technical blog, e.g., "machine learning", "technology", "gadgets", "programming languages", "Linux", "cloud computing", "AWS". A typical article might have 5–10 tags applied. Typically, tags will exhibit some correlation structure. Posts about "cloud computing" are likely to mention "AWS" and posts about "machine learning" are likely to mention "GPUs".

Sometimes such tagging problems draw on enormous label sets. The National Library of Medicine employs many professional annotators who associate each article to be indexed in PubMed with a set of tags drawn from the Medical Subject Headings (MeSH) ontology, a collection of roughly 28,000 tags. Correctly tagging articles is important because it allows researchers to conduct exhaustive reviews of the literature. This is a time-consuming process and typically there is a one-year lag between archiving and tagging. Machine learning can provide provisional tags until each article has a proper manual review. Indeed, for several years, the BioASQ organization has hosted competitions[21] for this task.

[21]

Search

In the field of information retrieval, we often impose ranks on sets of items. Take web search for example. The goal is less to determine *whether* a particular page is relevant for a query, but rather which, among a set of relevant results, should be shown most prominently to a particular user. One way of doing this might be to first assign a score to every element in the set and then to retrieve the top-rated elements. PageRank[22], the original secret sauce behind the Google search engine, was an early example of such a scoring system. Weirdly, the scoring provided by PageRank did not depend on the actual query. Instead, they relied on a simple relevance filter to identify the set of relevant candidates and then used PageRank to prioritize the more authoritative pages. Nowadays, search engines use machine learning and behavioral models to obtain query-dependent relevance scores. There are entire academic conferences devoted to this subject.

[22]

Recommender Systems

Recommender systems are another problem setting that is related to search and ranking. The problems are similar insofar as the goal is to display a set of items relevant to the user. The main difference is the emphasis on *personalization* to specific users in the context of recommender systems. For instance, for movie recommendations, the results page for a science fiction fan and the results page for a connoisseur of Peter Sellers comedies might differ significantly. Similar problems pop up in other recommendation settings, e.g., for retail products, music, and news recommendation.

In some cases, customers provide explicit feedback, communicating how much they liked a particular product (e.g., the product ratings and reviews on Amazon, IMDb, or Goodreads). In other cases, they provide implicit feedback, e.g., by skipping titles on a playlist, which might indicate dissatisfaction or maybe just indicate that the song was inappropriate in context. In the simplest formulations, these systems are trained to estimate some score, such as an expected star rating or the probability that a given user will purchase a particular item.

Given such a model, for any given user, we could retrieve the set of objects with the largest scores, which could then be recommended to the user. Production systems are considerably more advanced and take detailed user activity and item characteristics into account when computing such scores. Fig. 1.3.4 displays the deep learning books recommended by Amazon based on personalization algorithms tuned to capture Aston's preferences.

Despite their tremendous economic value, recommender systems naively built on top of predictive models suffer some serious conceptual flaws. To start, we only observe *censored*

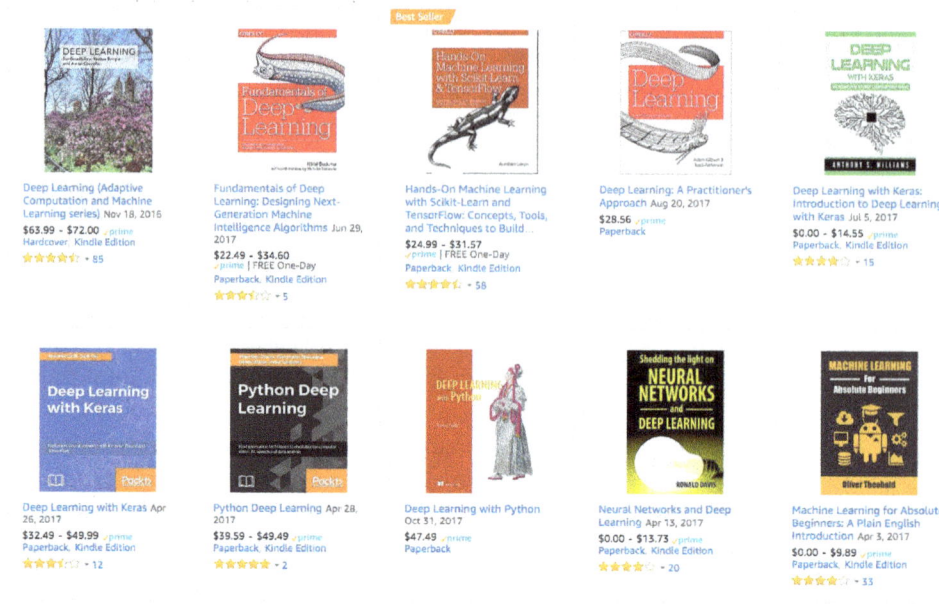

Fig. 1.3.4   Deep learning books recommended by Amazon.

feedback: users preferentially rate movies that they feel strongly about. For example, on a five-point scale, you might notice that items receive many one- and five-star ratings but that there are conspicuously few three-star ratings. Moreover, current purchase habits are often a result of the recommendation algorithm currently in place, but learning algorithms do not always take this detail into account. Thus it is possible for feedback loops to form where a recommender system preferentially pushes an item that is then taken to be better (due to greater purchases) and in turn is recommended even more frequently. Many of these problems—about how to deal with censoring, incentives, and feedback loops—are important open research questions.

Sequence Learning

So far, we have looked at problems where we have some fixed number of inputs and produce a fixed number of outputs. For example, we considered predicting house prices given a fixed set of features: square footage, number of bedrooms, number of bathrooms, and the transit time to downtown. We also discussed mapping from an image (of fixed dimension) to the predicted probabilities that it belongs to each among a fixed number of classes and predicting star ratings associated with purchases based on the user ID and product ID alone. In these cases, once our model is trained, after each test example is fed into our model, it is immediately forgotten. We assumed that successive observations were independent and thus there was no need to hold on to this context.

But how should we deal with video snippets? In this case, each snippet might consist of a different number of frames. And our guess of what is going on in each frame might be much stronger if we take into account the previous or succeeding frames. The same goes for language. For example, one popular deep learning problem is machine translation: the task of ingesting sentences in some source language and predicting their translations in another language.

Such problems also occur in medicine. We might want a model to monitor patients in the intensive care unit and to fire off alerts whenever their risk of dying in the next 24 hours exceeds some threshold. Here, we would not throw away everything that we know about the patient history every hour, because we might not want to make predictions based only on the most recent measurements.

Questions like these are among the most exciting applications of machine learning and they are instances of *sequence learning*. They require a model either to ingest sequences of inputs or to emit sequences of outputs (or both). Specifically, *sequence-to-sequence learning* considers problems where both inputs and outputs consist of variable-length sequences. Examples include machine translation and speech-to-text transcription. While it is impossible to consider all types of sequence transformations, the following special cases are worth mentioning.

Tagging and Parsing. This involves annotating a text sequence with attributes. Here, the inputs and outputs are *aligned*, i.e., they are of the same number and occur in a corresponding order. For instance, in *part-of-speech (PoS) tagging*, we annotate every word in a sentence with the corresponding part of speech, i.e., "noun" or "direct object". Alternatively, we might want to know which groups of contiguous words refer to named entities, like *people*, *places*, or *organizations*. In the cartoonishly simple example below, we might just want to indicate whether or not any word in the sentence is part of a named entity (tagged as "Ent").

```
Tom has dinner in Washington with Sally
Ent  -   -      -  Ent        -    Ent
```

Automatic Speech Recognition. With speech recognition, the input sequence is an audio recording of a speaker (Fig. 1.3.5), and the output is a transcript of what the speaker said. The challenge is that there are many more audio frames (sound is typically sampled at 8kHz or 16kHz) than text, i.e., there is no 1:1 correspondence between audio and text, since thousands of samples may correspond to a single spoken word. These are sequence-to-sequence learning problems, where the output is much shorter than the input. While humans are remarkably good at recognizing speech, even from low-quality audio, getting computers to perform the same feat is a formidable challenge.

Text to Speech. This is the inverse of automatic speech recognition. Here, the input is text and the output is an audio file. In this case, the output is much longer than the input.

Machine Translation. Unlike the case of speech recognition, where corresponding inputs and outputs occur in the same order, in machine translation, unaligned data poses a new challenge. Here the input and output sequences can have different lengths, and the corresponding regions of the respective sequences may appear in a different order. Consider the

Fig. 1.3.5 -D-e-e-p- L-ea-r-ni-ng- in an audio recording.

following illustrative example of the peculiar tendency of Germans to place the verbs at the end of sentences:

```
German:              Haben Sie sich schon dieses grossartige Lehrwerk angeschaut?
English:             Have you already looked at this excellent textbook?
Wrong alignment:     Have you yourself already this excellent textbook looked at?
```

Many related problems pop up in other learning tasks. For instance, determining the order in which a user reads a webpage is a two-dimensional layout analysis problem. Dialogue problems exhibit all kinds of additional complications, where determining what to say next requires taking into account real-world knowledge and the prior state of the conversation across long temporal distances. Such topics are active areas of research.

1.3.2 Unsupervised and Self-Supervised Learning

The previous examples focused on supervised learning, where we feed the model a giant dataset containing both the features and corresponding label values. You could think of the supervised learner as having an extremely specialized job and an extremely dictatorial boss. The boss stands over the learner's shoulder and tells them exactly what to do in every situation until they learn to map from situations to actions. Working for such a boss sounds pretty lame. On the other hand, pleasing such a boss is pretty easy. You just recognize the pattern as quickly as possible and imitate the boss's actions.

Considering the opposite situation, it could be frustrating to work for a boss who has no idea what they want you to do. However, if you plan to be a data scientist, you had better get used to it. The boss might just hand you a giant dump of data and tell you to *do some data science with it!* This sounds vague because it is vague. We call this class of problems *unsupervised learning*, and the type and number of questions we can ask is limited only by our creativity. We will address unsupervised learning techniques in later chapters. To whet your appetite for now, we describe a few of the following questions you might ask.

- Can we find a small number of prototypes that accurately summarize the data? Given a set of photos, can we group them into landscape photos, pictures of dogs, babies, cats, and mountain peaks? Likewise, given a collection of users' browsing activities, can we group them into users with similar behavior? This problem is typically known as *clustering*.

- Can we find a small number of parameters that accurately capture the relevant properties of the data? The trajectories of a ball are well described by velocity, diameter, and mass of the ball. Tailors have developed a small number of parameters that describe human body shape fairly accurately for the purpose of fitting clothes. These problems are referred to as *subspace estimation*. If the dependence is linear, it is called *principal component analysis*.

- Is there a representation of (arbitrarily structured) objects in Euclidean space such that symbolic properties can be well matched? This can be used to describe entities and their relations, such as "Rome" − "Italy" + "France" = "Paris".

- Is there a description of the root causes of much of the data that we observe? For instance, if we have demographic data about house prices, pollution, crime, location, education, and salaries, can we discover how they are related simply based on empirical data? The fields concerned with *causality* and *probabilistic graphical models* tackle such questions.

- Another important and exciting recent development in unsupervised learning is the advent of *deep generative models*. These models estimate the density of the data, either explicitly or *implicitly*. Once trained, we can use a generative model either to score examples according to how likely they are, or to sample synthetic examples from the learned distribution. Early deep learning breakthroughs in generative modeling came with the invention of *variational autoencoders* (Kingma and Welling, 2014, Rezende *et al.*, 2014) and continued with the development of *generative adversarial networks* (Goodfellow *et al.*, 2014). More recent advances include normalizing flows (Dinh *et al.*, 2014, Dinh *et al.*, 2017) and diffusion models (Ho *et al.*, 2020, Sohl-Dickstein *et al.*, 2015, Song and Ermon, 2019, Song *et al.*, 2021).

A further development in unsupervised learning has been the rise of *self-supervised learning*, techniques that leverage some aspect of the unlabeled data to provide supervision. For text, we can train models to "fill in the blanks" by predicting randomly masked words using their surrounding words (contexts) in big corpora without any labeling effort (Devlin *et al.*, 2018)! For images, we may train models to tell the relative position between two cropped regions of the same image (Doersch *et al.*, 2015), to predict an occluded part of an image based on the remaining portions of the image, or to predict whether two examples are perturbed versions of the same underlying image. Self-supervised models often learn representations that are subsequently leveraged by fine-tuning the resulting models on some downstream task of interest.

1.3.3 Interacting with an Environment

So far, we have not discussed where data actually comes from, or what actually happens when a machine learning model generates an output. That is because supervised learning and unsupervised learning do not address these issues in a very sophisticated way. In each case, we grab a big pile of data upfront, then set our pattern recognition machines in motion without ever interacting with the environment again. Because all the learning takes place after the algorithm is disconnected from the environment, this is sometimes called *offline learning*. For example, supervised learning assumes the simple interaction pattern depicted in Fig. 1.3.6.

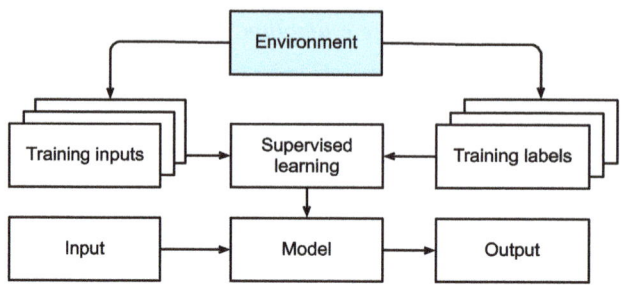

Fig. 1.3.6   Collecting data for supervised learning from an environment.

This simplicity of offline learning has its charms. The upside is that we can worry about pattern recognition in isolation, with no concern about complications arising from interactions with a dynamic environment. But this problem formulation is limiting. If you grew up reading Asimov's Robot novels, then you probably picture artificially intelligent agents capable not only of making predictions, but also of taking actions in the world. We want to think about intelligent *agents*, not just predictive models. This means that we need to think about choosing *actions*, not just making predictions. In contrast to mere predictions, actions actually impact the environment. If we want to train an intelligent agent, we must account for the way its actions might impact the future observations of the agent, and so offline learning is inappropriate.

Considering the interaction with an environment opens a whole set of new modeling questions. The following are just a few examples.

- Does the environment remember what we did previously?

- Does the environment want to help us, e.g., a user reading text into a speech recognizer?

- Does the environment want to beat us, e.g., spammers adapting their emails to evade spam filters?

- Does the environment have shifting dynamics? For example, would future data always resemble the past or would the patterns change over time, either naturally or in response to our automated tools?

These questions raise the problem of *distribution shift*, where training and test data are different. An example of this, that many of us may have met, is when taking exams written by a lecturer, while the homework was composed by their teaching assistants. Next, we briefly describe reinforcement learning, a rich framework for posing learning problems in which an agent interacts with an environment.

1.3.4 Reinforcement Learning

If you are interested in using machine learning to develop an agent that interacts with an environment and takes actions, then you are probably going to wind up focusing on *re-*

inforcement learning. This might include applications to robotics, to dialogue systems, and even to developing artificial intelligence (AI) for video games. *Deep reinforcement learning*, which applies deep learning to reinforcement learning problems, has surged in popularity. The breakthrough deep Q-network, that beat humans at Atari games using only the visual input (Mnih *et al.*, 2015), and the AlphaGo program, which dethroned the world champion at the board game Go (Silver *et al.*, 2016), are two prominent examples.

Reinforcement learning gives a very general statement of a problem in which an agent interacts with an environment over a series of time steps. At each time step, the agent receives some *observation* from the environment and must choose an *action* that is subsequently transmitted back to the environment via some mechanism (sometimes called an *actuator*), when, after each loop, the agent receives a reward from the environment. This process is illustrated in Fig. 1.3.7. The agent then receives a subsequent observation, and chooses a subsequent action, and so on. The behavior of a reinforcement learning agent is governed by a *policy*. In brief, a *policy* is just a function that maps from observations of the environment to actions. The goal of reinforcement learning is to produce good policies.

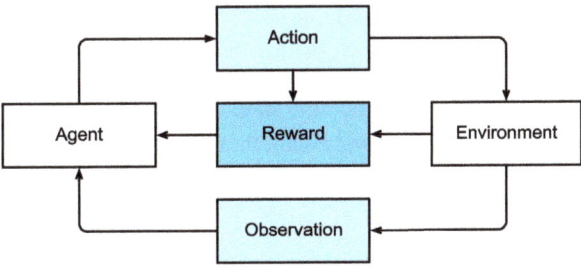

Fig. 1.3.7   The interaction between reinforcement learning and an environment.

It is hard to overstate the generality of the reinforcement learning framework. For example, supervised learning can be recast as reinforcement learning. Say we had a classification problem. We could create a reinforcement learning agent with one action corresponding to each class. We could then create an environment which gave a reward that was exactly equal to the loss function from the original supervised learning problem.

Further, reinforcement learning can also address many problems that supervised learning cannot. For example, in supervised learning, we always expect that the training input comes associated with the correct label. But in reinforcement learning, we do not assume that, for each observation the environment tells us the optimal action. In general, we just get some reward. Moreover, the environment may not even tell us which actions led to the reward.

Consider the game of chess. The only real reward signal comes at the end of the game when we either win, earning a reward of, say, 1, or when we lose, receiving a reward of, say, −1. So reinforcement learners must deal with the *credit assignment* problem: determining which actions to credit or blame for an outcome. The same goes for an employee who gets a promotion on October 11. That promotion likely reflects a number of well-chosen actions over the previous year. Getting promoted in the future requires figuring out which actions along the way led to the earlier promotions.

Reinforcement learners may also have to deal with the problem of partial observability. That is, the current observation might not tell you everything about your current state. Say your cleaning robot found itself trapped in one of many identical closets in your house. Rescuing the robot involves inferring its precise location which might require considering earlier observations prior to it entering the closet.

Finally, at any given point, reinforcement learners might know of one good policy, but there might be many other better policies that the agent has never tried. The reinforcement learner must constantly choose whether to *exploit* the best (currently) known strategy as a policy, or to *explore* the space of strategies, potentially giving up some short-term reward in exchange for knowledge.

The general reinforcement learning problem has a very general setting. Actions affect subsequent observations. Rewards are only observed when they correspond to the chosen actions. The environment may be either fully or partially observed. Accounting for all this complexity at once may be asking too much. Moreover, not every practical problem exhibits all this complexity. As a result, researchers have studied a number of special cases of reinforcement learning problems.

When the environment is fully observed, we call the reinforcement learning problem a *Markov decision process*. When the state does not depend on the previous actions, we call it a *contextual bandit problem*. When there is no state, just a set of available actions with initially unknown rewards, we have the classic *multi-armed bandit problem*.

1.4 Roots

We have just reviewed a small subset of problems that machine learning can address. For a diverse set of machine learning problems, deep learning provides powerful tools for their solution. Although many deep learning methods are recent inventions, the core ideas behind learning from data have been studied for centuries. In fact, humans have held the desire to analyze data and to predict future outcomes for ages, and it is this desire that is at the root of much of natural science and mathematics. Two examples are the Bernoulli distribution, named after Jacob Bernoulli (1655–1705)[23] , and the Gaussian distribution discovered by Carl Friedrich Gauss (1777–1855)[24]. Gauss invented, for instance, the least mean squares algorithm, which is still used today for a multitude of problems from insurance calculations to medical diagnostics. Such tools enhanced the experimental approach in the natural sciences—for instance, Ohm's law relating current and voltage in a resistor is perfectly described by a linear model.

Even in the middle ages, mathematicians had a keen intuition of estimates. For instance, the geometry book of Jacob Köbel (1460–1533)[25] illustrates averaging the length of 16 adult men's feet to estimate the typical foot length in the population (Fig. 1.4.1).

As a group of individuals exited a church, 16 adult men were asked to line up in a row

Fig. 1.4.1   Estimating the length of a foot.

and have their feet measured. The sum of these measurements was then divided by 16 to obtain an estimate for what now is called one foot. This "algorithm" was later improved to deal with misshapen feet; The two men with the shortest and longest feet were sent away, averaging only over the remainder. This is among the earliest examples of a trimmed mean estimate.

Statistics really took off with the availability and collection of data. One of its pioneers, Ronald Fisher (1890–1962)[26], contributed significantly to its theory and also its applications in genetics. Many of his algorithms (such as linear discriminant analysis) and concepts (such as the Fisher information matrix) still hold a prominent place in the foundations of modern statistics. Even his data resources had a lasting impact. The Iris dataset that Fisher released in 1936 is still sometimes used to demonstrate machine learning algorithms. Fisher was also a proponent of eugenics, which should remind us that the morally dubious use of data science has as long and enduring a history as its productive use in industry and the natural sciences.

Other influences for machine learning came from the information theory of Claude Shannon (1916–2001)[27] and the theory of computation that was proposed by Alan Turing (1912–1954)[28]. Turing posed the question "can machines think?" in his famous paper *Computing Machinery and Intelligence* (Turing, 1950). Describing what is now known as the Turing test, he proposed that a machine can be considered *intelligent* if it is difficult for a human evaluator to distinguish between the replies from a machine and those of a human, based purely on textual interactions.

Further influences came from neuroscience and psychology. After all, humans clearly exhibit intelligent behavior. Many scholars have asked whether one could explain and possibly reverse engineer this capacity. One of the first biologically inspired algorithms was formulated by Donald Hebb (1904–1985) [29]. In his groundbreaking book *The Organization of Behavior* (Hebb, 1949), he posited that neurons learn by positive reinforcement. This became known as the Hebbian learning rule. These ideas inspired later work, such as Rosenblatt's perceptron learning algorithm, and laid the foundations of many stochastic gradient descent algorithms that underpin deep learning today: reinforce desirable behavior and diminish undesirable behavior to obtain good settings of the parameters in a neural network.

Biological inspiration is what gave *neural networks* their name. For over a century (dating back to the models of Alexander Bain, 1873, and James Sherrington, 1890), researchers have tried to assemble computational circuits that resemble networks of interacting neurons. Over time, the interpretation of biology has become less literal, but the name stuck. At its heart lie a few key principles that can be found in most networks today:

- The alternation of linear and nonlinear processing units, often referred to as *layers*.

- The use of the chain rule (also known as *backpropagation*) for adjusting parameters in the entire network at once.

After initial rapid progress, research in neural networks languished from around 1995 until 2005. This was mainly due to two reasons. First, training a network is computationally very expensive. While random-access memory was plentiful at the end of the past century, computational power was scarce. Second, datasets were relatively small. In fact, Fisher's Iris dataset from 1936 was still a popular tool for testing the efficacy of algorithms. The MNIST dataset with its 60,000 handwritten digits was considered huge.

Given the scarcity of data and computation, strong statistical tools such as kernel methods, decision trees, and graphical models proved empirically superior in many applications. Moreover, unlike neural networks, they did not require weeks to train and provided predictable results with strong theoretical guarantees.

1.5 The Road to Deep Learning

Much of this changed with the availability of massive amounts of data, thanks to the World Wide Web, the advent of companies serving hundreds of millions of users online, a dissemination of low-cost, high-quality sensors, inexpensive data storage (Kryder's law), and cheap computation (Moore's law). In particular, the landscape of computation in deep learning was revolutionized by advances in GPUs that were originally engineered for com-

Decade	Dataset	Memory	Floating point calculations per second
1970	100 (Iris)	1 KB	100 KF (Intel 8080)
1980	1 K (house prices in Boston)	100 KB	1 MF (Intel 80186)
1990	10 K (optical character recognition)	10 MB	10 MF (Intel 80486)
2000	10 M (web pages)	100 MB	1 GF (Intel Core)
2010	10 G (advertising)	1 GB	1 TF (NVIDIA C2050)
2020	1 T (social network)	100 GB	1 PF (NVIDIA DGX-2)

Table 1.5.1 Dataset vs. computer memory and computational power

puter gaming. Suddenly algorithms and models that seemed computationally infeasible were within reach. This is best illustrated in Table 1.5.1.

Note that random-access memory has not kept pace with the growth in data. At the same time, increases in computational power have outpaced the growth in datasets. This means that statistical models need to become more memory efficient, and so they are free to spend more computer cycles optimizing parameters, thanks to the increased compute budget. Consequently, the sweet spot in machine learning and statistics moved from (generalized) linear models and kernel methods to deep neural networks. This is also one of the reasons why many of the mainstays of deep learning, such as multilayer perceptrons (McCulloch and Pitts, 1943), convolutional neural networks (LeCun *et al.*, 1998), long short-term memory (Hochreiter and Schmidhuber, 1997), and Q-Learning (Watkins and Dayan, 1992), were essentially "rediscovered" in the past decade, after lying comparatively dormant for considerable time.

The recent progress in statistical models, applications, and algorithms has sometimes been likened to the Cambrian explosion: a moment of rapid progress in the evolution of species. Indeed, the state of the art is not just a mere consequence of available resources applied to decades-old algorithms. Note that the list of ideas below barely scratches the surface of what has helped researchers achieve tremendous progress over the past decade.

- Novel methods for capacity control, such as *dropout* (Srivastava *et al.*, 2014), have helped to mitigate overfitting. Here, noise is injected (Bishop, 1995) throughout the neural network during training.

- *Attention mechanisms* solved a second problem that had plagued statistics for over a century: how to increase the memory and complexity of a system without increasing the number of learnable parameters. Researchers found an elegant solution by using what can only be viewed as a *learnable pointer structure* (Bahdanau *et al.*, 2014). Rather than having to remember an entire text sequence, e.g., for machine translation in a fixed-dimensional representation, all that needed to be stored was a pointer to the intermediate state of the translation process. This allowed for significantly increased accuracy for long sequences, since the model no longer needed to remember the entire sequence before commencing the generation of a new one.

- Built solely on attention mechanisms, the *Transformer* architecture (Vaswani *et al.*, 2017) has demonstrated superior *scaling* behavior: it performs better with an increase in dataset size, model size, and amount of training compute (Kaplan *et al.*, 2020). This architecture has demonstrated compelling success in a wide range of areas, such as natural language processing (Devlin *et al.*, 2018, Brown *et al.*, 2020), computer vision (Dosovitskiy *et al.*, 2021, Liu *et al.*, 2021), speech recognition (Gulati *et al.*, 2020), reinforcement learning (Chen *et al.*, 2021), and graph neural networks (Dwivedi and Bresson, 2020). For example, a single Transformer pretrained on modalities as diverse as text, images, joint torques, and button presses can play Atari, caption images, chat, and control a robot (Reed *et al.*, 2022).

- Modeling probabilities of text sequences, *language models* can predict text given other text. Scaling up the data, model, and compute has unlocked a growing number of capabilities of language models to perform desired tasks via human-like text generation based on input text (Brown *et al.*, 2020, OpenAI, 2023, Rae *et al.*, 2021, Chowdhery *et al.*, 2022, Hoffmann *et al.*, 2022, Anil *et al.*, 2023, Touvron *et al.*, 2023a, 2023b). For instance, aligning language models with human intent (Ouyang *et al.*, 2022), OpenAI's ChatGPT[30] allows users to interact with it in a conversational way to solve problems, such as code debugging and creative writing.

30

- Multi-stage designs, e.g., via the memory networks (Sukhbaatar *et al.*, 2015) and the neural programmer-interpreter (Reed and De Freitas, 2015) permitted statistical modelers to describe iterative approaches to reasoning. These tools allow for an internal state of the deep neural network to be modified repeatedly, thus carrying out subsequent steps in a chain of reasoning, just as a processor can modify memory for a computation.

- A key development in *deep generative model* was the invention of *generative adversarial networks* (Goodfellow *et al.*, 2014). Traditionally, statistical methods for density estimation and generative models focused on finding proper probability distributions and (often approximate) algorithms for sampling from them. As a result, these algorithms were largely limited by the lack of flexibility inherent in the statistical models. The crucial innovation in generative adversarial networks was to replace the sampler by an arbitrary algorithm with differentiable parameters. These are then adjusted in such a way that the discriminator (effectively a two-sample test) cannot distinguish fake from real data. Through the ability to use arbitrary algorithms to generate data, density estimation was opened up to a wide variety of techniques. Examples of galloping zebras (Zhu *et al.*, 2017) and of fake celebrity faces (Karras *et al.*, 2017) are each testimony to this progress. Even amateur doodlers can produce photorealistic images just based on sketches describing the layout of a scene (Park *et al.*, 2019).

- Furthermore, while the diffusion process gradually adds random noise to data samples, *diffusion models* (Ho *et al.*, 2020, Sohl-Dickstein *et al.*, 2015) learn the denoising process to gradually construct data samples from random noise, reversing the diffusion process. They have started to replace generative adversarial networks in more recent deep generative models, such as in DALL-E 2 (Ramesh *et al.*, 2022) and Imagen (Saharia *et al.*, 2022) for creative art and image generation based on text descriptions.

- In many cases, a single GPU is insufficient for processing the large amounts of data available for training. Over the past decade the ability to build parallel and distributed training algorithms has improved significantly. One of the key challenges in designing scalable algorithms is that the workhorse of deep learning optimization, stochastic gradient descent, relies on relatively small minibatches of data to be processed. At the same time, small batches limit the efficiency of GPUs. Hence, training on 1,024 GPUs with a minibatch size of, say, 32 images per batch amounts to an aggregate minibatch of about 32,000 images. Work, first by Li (2017) and subsequently by You *et al.* (2017) and Jia *et al.* (2018) pushed the size up to 64,000 observations, reducing training time for the ResNet-50 model on the ImageNet dataset to less than 7 minutes. By comparison, training times were initially of the order of days.

- The ability to parallelize computation has also contributed to progress in *reinforcement learning*. This has led to significant progress in computers achieving superhuman performance on tasks like Go, Atari games, Starcraft, and in physics simulations (e.g., using MuJoCo) where environment simulators are available. See, e.g., Silver *et al.* (2016) for a description of such achievements in AlphaGo. In a nutshell, reinforcement learning works best if plenty of (state, action, reward) tuples are available. Simulation provides such an avenue.

- Deep learning frameworks played a crucial role in disseminating ideas. The first generation of open-source frameworks for neural network modeling consisted of Caffe[31], Torch[32], and Theano[33]. Many seminal papers were written using them. These have now been superseded by TensorFlow[34] (often used via its high-level API Keras[35]), CNTK[36], Caffe 2[37], and Apache MXNet[38]. The third generation of frameworks consists of so-called *imperative* tools for deep learning, a trend that was arguably ignited by Chainer[39], which used a syntax similar to Python NumPy to describe models. This idea was adopted by both PyTorch[40], the Gluon API[41] of MXNet, and JAX[42].

The division of labor between system researchers building better tools and statistical modelers building better neural networks has greatly simplified things. For instance, training a linear logistic regression model used to be a nontrivial homework problem, worthy to give to new machine learning Ph.D. students at Carnegie Mellon University in 2014. By now, this task can be accomplished with under 10 lines of code, putting it firmly within the reach of any programmer.

1.6 Success Stories

Artificial intelligence has a long history of delivering results that would be difficult to accomplish otherwise. For instance, mail sorting systems using optical character recognition have been deployed since the 1990s. This is, after all, the source of the famous MNIST dataset of handwritten digits. The same applies to reading checks for bank deposits and scoring creditworthiness of applicants. Financial transactions are checked for fraud automatically. This forms the backbone of many e-commerce payment systems, such as PayPal, Stripe, AliPay, WeChat, Apple, Visa, and MasterCard. Computer programs for chess have been competitive for decades. Machine learning feeds search, recommendation, personalization, and ranking on the Internet. In other words, machine learning is pervasive, albeit often hidden from sight.

It is only recently that AI has been in the limelight, mostly due to solutions to problems that were considered intractable previously and that are directly related to consumers. Many of such advances are attributed to deep learning.

- Intelligent assistants, such as Apple's Siri, Amazon's Alexa, and Google's assistant, are able to respond to spoken requests with a reasonable degree of accuracy. This includes menial jobs, like turning on light switches, and more complex tasks, such as arranging barber's appointments and offering phone support dialog. This is likely the most noticeable sign that AI is affecting our lives.

- A key ingredient in digital assistants is their ability to recognize speech accurately. The accuracy of such systems has gradually increased to the point of achieving parity with humans for certain applications (Xiong *et al.*, 2018).

- Object recognition has likewise come a long way. Identifying the object in a picture was a fairly challenging task in 2010. On the ImageNet benchmark researchers from NEC Labs and University of Illinois at Urbana-Champaign achieved a top-five error rate of 28% (Lin *et al.*, 2010). By 2017, this error rate was reduced to 2.25% (Hu *et al.*, 2018). Similarly, stunning results have been achieved for identifying birdsong and for diagnosing skin cancer.

- Prowess in games used to provide a measuring stick for human ability. Starting from TD-Gammon, a program for playing backgammon using temporal difference reinforcement learning, algorithmic and computational progress has led to algorithms for a wide range of applications. Compared with backgammon, chess has a much more complex state space and set of actions. DeepBlue beat Garry Kasparov using massive parallelism, special-purpose hardware and efficient search through the game tree (Campbell *et al.*, 2002). Go is more difficult still, due to its huge state space. AlphaGo reached human parity in 2015, using deep learning combined with Monte Carlo tree sampling (Silver *et al.*, 2016). The challenge in Poker was that the state space is large and only partially observed (we do not know the opponents' cards). Libratus exceeded human performance in Poker using efficiently structured strategies (Brown and Sandholm, 2017).

- Another indication of progress in AI is the advent of self-driving vehicles. While full autonomy is not yet within reach, excellent progress has been made in this direction, with companies such as Tesla, NVIDIA, and Waymo shipping products that enable partial autonomy. What makes full autonomy so challenging is that proper driving requires the ability to perceive, to reason and to incorporate rules into a system. At present, deep learning is used primarily in the visual aspect of these problems. The rest is heavily tuned by engineers.

This barely scratches the surface of significant applications of machine learning. For instance, robotics, logistics, computational biology, particle physics, and astronomy owe some of their most impressive recent advances at least in parts to machine learning, which is thus becoming a ubiquitous tool for engineers and scientists.

Frequently, questions about a coming AI apocalypse and the plausibility of a *singularity* have been raised in non-technical articles. The fear is that somehow machine learning systems will become sentient and make decisions, independently of their programmers, that directly impact the lives of humans. To some extent, AI already affects the livelihood of humans in direct ways: creditworthiness is assessed automatically, autopilots mostly navigate vehicles, decisions about whether to grant bail use statistical data as input. More frivolously, we can ask Alexa to switch on the coffee machine.

Fortunately, we are far from a sentient AI system that could deliberately manipulate its human creators. First, AI systems are engineered, trained, and deployed in a specific, goal-oriented manner. While their behavior might give the illusion of general intelligence, it is a combination of rules, heuristics and statistical models that underlie the design. Second, at present, there are simply no tools for *artificial general intelligence* that are able to improve themselves, reason about themselves, and that are able to modify, extend, and improve their own architecture while trying to solve general tasks.

A much more pressing concern is how AI is being used in our daily lives. It is likely that many routine tasks, currently fulfilled by humans, can and will be automated. Farm robots will likely reduce the costs for organic farmers but they will also automate harvesting operations. This phase of the industrial revolution may have profound consequences for large swaths of society, since menial jobs provide much employment in many countries. Furthermore, statistical models, when applied without care, can lead to racial, gender, or age bias and raise reasonable concerns about procedural fairness if automated to drive consequential decisions. It is important to ensure that these algorithms are used with care. With what we know today, this strikes us as a much more pressing concern than the potential of malevolent superintelligence for destroying humanity.

1.7 The Essence of Deep Learning

Thus far, we have talked in broad terms about machine learning. Deep learning is the subset of machine learning concerned with models based on many-layered neural networks. It is *deep* in precisely the sense that its models learn many *layers* of transformations. While this might sound narrow, deep learning has given rise to a dizzying array of models, techniques, problem formulations, and applications. Many intuitions have been developed to explain the benefits of depth. Arguably, all machine learning has many layers of computation, the first consisting of feature processing steps. What differentiates deep learning is that the operations learned at each of the many layers of representations are learned jointly from data.

The problems that we have discussed so far, such as learning from the raw audio signal, the raw pixel values of images, or mapping between sentences of arbitrary lengths and their counterparts in foreign languages, are those where deep learning excels and traditional methods falter. It turns out that these many-layered models are capable of addressing low-level perceptual data in a way that previous tools could not. Arguably the most significant commonality in deep learning methods is *end-to-end training*. That is, rather than assembling a system based on components that are individually tuned, one builds the system and then tunes their performance jointly. For instance, in computer vision scientists used to separate the process of *feature engineering* from the process of building machine learning models. The Canny edge detector (Canny, 1987) and Lowe's SIFT feature extractor (Lowe, 2004) reigned supreme for over a decade as algorithms for mapping images into feature vectors. In bygone days, the crucial part of applying machine learning to these problems consisted of coming up with manually-engineered ways of transforming the data into some form amenable to shallow models. Unfortunately, there is only so much that humans can accomplish by ingenuity in comparison with a consistent evaluation over millions of choices carried out automatically by an algorithm. When deep learning took over, these feature extractors were replaced by automatically tuned filters that yielded superior accuracy.

Thus, one key advantage of deep learning is that it replaces not only the shallow models at the end of traditional learning pipelines, but also the labor-intensive process of feature engineering. Moreover, by replacing much of the domain-specific preprocessing, deep learning has eliminated many of the boundaries that previously separated computer vision, speech recognition, natural language processing, medical informatics, and other application areas, thereby offering a unified set of tools for tackling diverse problems.

Beyond end-to-end training, we are experiencing a transition from parametric statistical descriptions to fully nonparametric models. When data is scarce, one needs to rely on simplifying assumptions about reality in order to obtain useful models. When data is abundant, these can be replaced by nonparametric models that better fit the data. To some extent, this mirrors the progress that physics experienced in the middle of the previous century with the availability of computers. Rather than solving by hand parametric approximations of how electrons behave, one can now resort to numerical simulations of the associated partial differential equations. This has led to much more accurate models, albeit often at the expense of interpretation.

Another difference from previous work is the acceptance of suboptimal solutions, dealing with nonconvex nonlinear optimization problems, and the willingness to try things before proving them. This new-found empiricism in dealing with statistical problems, combined with a rapid influx of talent has led to rapid progress in the development of practical algorithms, albeit in many cases at the expense of modifying and re-inventing tools that existed for decades.

In the end, the deep learning community prides itself on sharing tools across academic and corporate boundaries, releasing many excellent libraries, statistical models, and trained networks as open source. It is in this spirit that the notebooks forming this book are freely available for distribution and use. We have worked hard to lower the barriers of access for anyone wishing to learn about deep learning and we hope that our readers will benefit from this.

1.8 Summary

Machine learning studies how computer systems can leverage experience (often data) to improve performance at specific tasks. It combines ideas from statistics, data mining, and optimization. Often, it is used as a means of implementing AI solutions. As a class of machine learning, representational learning focuses on how to automatically find the appropriate way to represent data. Considered as multi-level representation learning through learning many layers of transformations, deep learning replaces not only the shallow models at the end of traditional machine learning pipelines, but also the labor-intensive process of feature engineering. Much of the recent progress in deep learning has been triggered by an abundance of data arising from cheap sensors and Internet-scale applications, and by significant progress in computation, mostly through GPUs. Furthermore, the availability of efficient deep learning frameworks has made design and implementation of whole system optimization significantly easier, and this is a key component in obtaining high performance.

1.9 Exercises

1. Which parts of code that you are currently writing could be "learned", i.e., improved by learning and automatically determining design choices that are made in your code? Does your code include heuristic design choices? What data might you need to learn the desired behavior?

2. Which problems that you encounter have many examples for their solution, yet no specific way for automating them? These may be prime candidates for using deep learning.

3. Describe the relationships between algorithms, data, and computation. How do characteristics of the data and the current available computational resources influence the appropriateness of various algorithms?

4. Name some settings where end-to-end training is not currently the default approach but where it might be useful.

Discussions[43].

43

2   Preliminaries

To prepare for your dive into deep learning, you will need a few survival skills: (i) techniques for storing and manipulating data; (ii) libraries for ingesting and preprocessing data from a variety of sources; (iii) knowledge of the basic linear algebraic operations that we apply to high-dimensional data elements; (iv) just enough calculus to determine which direction to adjust each parameter in order to decrease the loss function; (v) the ability to automatically compute derivatives so that you can forget much of the calculus you just learned; (vi) some basic fluency in probability, our primary language for reasoning under uncertainty; and (vii) some aptitude for finding answers in the official documentation when you get stuck.

In short, this chapter provides a rapid introduction to the basics that you will need to follow *most* of the technical content in this book.

2.1 Data Manipulation

In order to get anything done, we need some way to store and manipulate data. Generally, there are two important things we need to do with data: (i) acquire them; and (ii) process them once they are inside the computer. There is no point in acquiring data without some way to store it, so to start, let's get our hands dirty with n-dimensional arrays, which we also call *tensors*. If you already know the NumPy scientific computing package, this will be a breeze. For all modern deep learning frameworks, the *tensor class* (ndarray in MXNet, Tensor in PyTorch and TensorFlow) resembles NumPy's ndarray, with a few killer features added. First, the tensor class supports automatic differentiation. Second, it leverages GPUs to accelerate numerical computation, whereas NumPy only runs on CPUs. These properties make neural networks both easy to code and fast to run.

2.1.1 Getting Started

To start, we import the PyTorch library. Note that the package name is torch.

```
import torch
```

A tensor represents a (possibly multidimensional) array of numerical values. In the one-dimensional case, i.e., when only one axis is needed for the data, a tensor is called a *vector*.

With two axes, a tensor is called a *matrix*. With $k > 2$ axes, we drop the specialized names and just refer to the object as a kth-*order tensor*.

PyTorch provides a variety of functions for creating new tensors prepopulated with values. For example, by invoking arange(n), we can create a vector of evenly spaced values, starting at 0 (included) and ending at n (not included). By default, the interval size is 1. Unless otherwise specified, new tensors are stored in main memory and designated for CPU-based computation.

```
x = torch.arange(12, dtype=torch.float32)
x
```

```
tensor([ 0.,  1.,  2.,  3.,  4.,  5.,  6.,  7.,  8.,  9., 10., 11.])
```

Each of these values is called an *element* of the tensor. The tensor x contains 12 elements. We can inspect the total number of elements in a tensor via its numel method.

```
x.numel()
```

```
12
```

We can access a tensor's *shape* (the length along each axis) by inspecting its shape attribute. Because we are dealing with a vector here, the shape contains just a single element and is identical to the size.

```
x.shape
```

```
torch.Size([12])
```

We can change the shape of a tensor without altering its size or values, by invoking reshape. For example, we can transform our vector x whose shape is (12,) to a matrix X with shape (3, 4). This new tensor retains all elements but reconfigures them into a matrix. Notice that the elements of our vector are laid out one row at a time and thus x[3] == X[0, 3].

```
X = x.reshape(3, 4)
X
```

```
tensor([[ 0.,  1.,  2.,  3.],
        [ 4.,  5.,  6.,  7.],
        [ 8.,  9., 10., 11.]])
```

Note that specifying every shape component to reshape is redundant. Because we already know our tensor's size, we can work out one component of the shape given the rest. For example, given a tensor of size n and target shape (h, w), we know that $w = n/h$. To automatically infer one component of the shape, we can place a -1 for the shape component that should be inferred automatically. In our case, instead of calling x.reshape(3, 4), we could have equivalently called x.reshape(-1, 4) or x.reshape(3, -1).

Practitioners often need to work with tensors initialized to contain all 0s or 1s. We can construct a tensor with all elements set to 0 and a shape of (2, 3, 4) via the zeros function.

```
torch.zeros((2, 3, 4))
```

```
tensor([[[0., 0., 0., 0.],
         [0., 0., 0., 0.],
         [0., 0., 0., 0.]],

        [[0., 0., 0., 0.],
         [0., 0., 0., 0.],
         [0., 0., 0., 0.]]])
```

Similarly, we can create a tensor with all 1s by invoking ones.

```
torch.ones((2, 3, 4))
```

```
tensor([[[1., 1., 1., 1.],
         [1., 1., 1., 1.],
         [1., 1., 1., 1.]],

        [[1., 1., 1., 1.],
         [1., 1., 1., 1.],
         [1., 1., 1., 1.]]])
```

We often wish to sample each element randomly (and independently) from a given probability distribution. For example, the parameters of neural networks are often initialized randomly. The following snippet creates a tensor with elements drawn from a standard Gaussian (normal) distribution with mean 0 and standard deviation 1.

```
torch.randn(3, 4)
```

```
tensor([[ 0.4191, -1.5045,  0.5484,  0.6978],
        [-0.3852,  0.6876,  0.2580,  0.8139],
        [ 0.0716, -1.3135, -1.1745,  0.2992]])
```

Finally, we can construct tensors by supplying the exact values for each element by supplying (possibly nested) Python list(s) containing numerical literals. Here, we construct a matrix with a list of lists, where the outermost list corresponds to axis 0, and the inner list corresponds to axis 1.

```
torch.tensor([[2, 1, 4, 3], [1, 2, 3, 4], [4, 3, 2, 1]])
```

```
tensor([[2, 1, 4, 3],
        [1, 2, 3, 4],
        [4, 3, 2, 1]])
```

2.1.2 Indexing and Slicing

As with Python lists, we can access tensor elements by indexing (starting with 0). To access an element based on its position relative to the end of the list, we can use negative indexing. Finally, we can access whole ranges of indices via slicing (e.g., X[start:stop]), where the returned value includes the first index (start) *but not the last* (stop). Finally, when only one index (or slice) is specified for a kth-order tensor, it is applied along axis 0. Thus, in the following code, [-1] selects the last row and [1:3] selects the second and third rows.

```
X[-1], X[1:3]
```

```
(tensor([ 8.,  9., 10., 11.]),
 tensor([[ 4.,  5.,  6.,  7.],
         [ 8.,  9., 10., 11.]]))
```

Beyond reading them, we can also *write* elements of a matrix by specifying indices.

```
X[1, 2] = 17
X
```

```
tensor([[ 0.,  1.,  2.,  3.],
        [ 4.,  5., 17.,  7.],
        [ 8.,  9., 10., 11.]])
```

If we want to assign multiple elements the same value, we apply the indexing on the left-hand side of the assignment operation. For instance, [:2, :] accesses the first and second rows, where : takes all the elements along axis 1 (column). While we discussed indexing for matrices, this also works for vectors and for tensors of more than two dimensions.

```
X[:2, :] = 12
X
```

```
tensor([[12., 12., 12., 12.],
        [12., 12., 12., 12.],
        [ 8.,  9., 10., 11.]])
```

2.1.3 Operations

Now that we know how to construct tensors and how to read from and write to their elements, we can begin to manipulate them with various mathematical operations. Among the most useful of these are the *elementwise* operations. These apply a standard scalar operation to each element of a tensor. For functions that take two tensors as inputs, elementwise operations apply some standard binary operator on each pair of corresponding elements. We can create an elementwise function from any function that maps from a scalar to a scalar.

In mathematical notation, we denote such *unary* scalar operators (taking one input) by the signature $f : \mathbb{R} \to \mathbb{R}$. This just means that the function maps from any real number onto some other real number. Most standard operators, including unary ones like e^x, can be applied elementwise.

```
torch.exp(x)
```

```
tensor([162754.7969, 162754.7969, 162754.7969, 162754.7969, 162754.7969,
        162754.7969, 162754.7969, 162754.7969,   2980.9580,    8103.0840,
         22026.4648,   59874.1406])
```

Likewise, we denote *binary* scalar operators, which map pairs of real numbers to a (single) real number via the signature $f : \mathbb{R}, \mathbb{R} \to \mathbb{R}$. Given any two vectors **u** and **v** *of the same shape*, and a binary operator f, we can produce a vector $\mathbf{c} = F(\mathbf{u}, \mathbf{v})$ by setting $c_i \leftarrow f(u_i, v_i)$ for all i, where c_i, u_i, and v_i are the ith elements of vectors **c**, **u**, and **v**. Here, we produced the vector-valued $F : \mathbb{R}^d, \mathbb{R}^d \to \mathbb{R}^d$ by *lifting* the scalar function to an elementwise vector operation. The common standard arithmetic operators for addition (+), subtraction (-), multiplication (*), division (/), and exponentiation (**) have all been *lifted* to elementwise operations for identically-shaped tensors of arbitrary shape.

```
x = torch.tensor([1.0, 2, 4, 8])
y = torch.tensor([2, 2, 2, 2])
x + y, x - y, x * y, x / y, x ** y
```

```
(tensor([ 3.,   4.,   6.,  10.]),
 tensor([-1.,   0.,   2.,   6.]),
 tensor([ 2.,   4.,   8.,  16.]),
 tensor([0.5000, 1.0000, 2.0000, 4.0000]),
 tensor([ 1.,   4.,  16.,  64.]))
```

In addition to elementwise computations, we can also perform linear algebraic operations, such as dot products and matrix multiplications. We will elaborate on these in Section 2.3.

We can also *concatenate* multiple tensors, stacking them end-to-end to form a larger one. We just need to provide a list of tensors and tell the system along which axis to concatenate. The example below shows what happens when we concatenate two matrices along rows (axis 0) instead of columns (axis 1). We can see that the first output's axis-0 length (6) is the sum of the two input tensors' axis-0 lengths (3 + 3); while the second output's axis-1 length (8) is the sum of the two input tensors' axis-1 lengths (4 + 4).

```
X = torch.arange(12, dtype=torch.float32).reshape((3,4))
Y = torch.tensor([[2.0, 1, 4, 3], [1, 2, 3, 4], [4, 3, 2, 1]])
torch.cat((X, Y), dim=0), torch.cat((X, Y), dim=1)
```

```
(tensor([[ 0.,  1.,  2.,  3.],
         [ 4.,  5.,  6.,  7.],
         [ 8.,  9., 10., 11.],
         [ 2.,  1.,  4.,  3.],
         [ 1.,  2.,  3.,  4.],
         [ 4.,  3.,  2.,  1.]]),
 tensor([[ 0.,  1.,  2.,  3.,  2.,  1.,  4.,  3.],
         [ 4.,  5.,  6.,  7.,  1.,  2.,  3.,  4.],
         [ 8.,  9., 10., 11.,  4.,  3.,  2.,  1.]]))
```

Sometimes, we want to construct a binary tensor via *logical statements*. Take X == Y as an example. For each position i, j, if X[i, j] and Y[i, j] are equal, then the corresponding entry in the result takes value 1, otherwise it takes value 0.

```
X == Y
```

```
tensor([[False,  True, False,  True],
        [False, False, False, False],
        [False, False, False, False]])
```

Summing all the elements in the tensor yields a tensor with only one element.

```
X.sum()
```

```
tensor(66.)
```

2.1.4 Broadcasting

By now, you know how to perform elementwise binary operations on two tensors of the same shape. Under certain conditions, even when shapes differ, we can still perform elementwise binary operations by invoking the *broadcasting mechanism*. Broadcasting works according to the following two-step procedure: (i) expand one or both arrays by copying elements along axes with length 1 so that after this transformation, the two tensors have the same shape; (ii) perform an elementwise operation on the resulting arrays.

```
a = torch.arange(3).reshape((3, 1))
b = torch.arange(2).reshape((1, 2))
a, b
```

```
(tensor([[0],
         [1],
         [2]]),
 tensor([[0, 1]]))
```

Since a and b are 3×1 and 1×2 matrices, respectively, their shapes do not match up. Broadcasting produces a larger 3×2 matrix by replicating matrix a along the columns and matrix b along the rows before adding them elementwise.

```
a + b
```

```
tensor([[0, 1],
        [1, 2],
        [2, 3]])
```

2.1.5 Saving Memory

Running operations can cause new memory to be allocated to host results. For example, if we write Y = X + Y, we dereference the tensor that Y used to point to and instead point Y at the newly allocated memory. We can demonstrate this issue with Python's id() function, which gives us the exact address of the referenced object in memory. Note that after we run Y = Y + X, id(Y) points to a different location. That is because Python first evaluates Y + X, allocating new memory for the result and then points Y to this new location in memory.

```
before = id(Y)
Y = Y + X
id(Y) == before
```

```
False
```

This might be undesirable for two reasons. First, we do not want to run around allocating memory unnecessarily all the time. In machine learning, we often have hundreds of megabytes of parameters and update all of them multiple times per second. Whenever possible, we want to perform these updates *in place*. Second, we might point at the same parameters from multiple variables. If we do not update in place, we must be careful to update all of these references, lest we spring a memory leak or inadvertently refer to stale parameters.

Fortunately, performing in-place operations is easy. We can assign the result of an operation to a previously allocated array Y by using slice notation: Y[:] = <expression>. To illustrate this concept, we overwrite the values of tensor Z, after initializing it, using zeros_like, to have the same shape as Y.

```
Z = torch.zeros_like(Y)
print('id(Z):', id(Z))
Z[:] = X + Y
print('id(Z):', id(Z))
```

```
id(Z): 140651747691872
id(Z): 140651747691872
```

If the value of X is not reused in subsequent computations, we can also use X[:] = X + Y
or X += Y to reduce the memory overhead of the operation.

```
before = id(X)
X += Y
id(X) == before
```

```
True
```

2.1.6 Conversion to Other Python Objects

Converting to a NumPy tensor (ndarray), or vice versa, is easy. The torch tensor and
NumPy array will share their underlying memory, and changing one through an in-place
operation will also change the other.

```
A = X.numpy()
B = torch.from_numpy(A)
type(A), type(B)
```

```
(numpy.ndarray, torch.Tensor)
```

To convert a size-1 tensor to a Python scalar, we can invoke the item function or Python's
built-in functions.

```
a = torch.tensor([3.5])
a, a.item(), float(a), int(a)
```

```
(tensor([3.5000]), 3.5, 3.5, 3)
```

2.1.7 Summary

The tensor class is the main interface for storing and manipulating data in deep learning libraries. Tensors provide a variety of functionalities including construction routines; indexing and slicing; basic mathematics operations; broadcasting; memory-efficient assignment;
and conversion to and from other Python objects.

2.1.8 Exercises

1. Run the code in this section. Change the conditional statement X == Y to X < Y or X > Y, and then see what kind of tensor you can get.

2. Replace the two tensors that operate by element in the broadcasting mechanism with other shapes, e.g., 3-dimensional tensors. Is the result the same as expected?

Discussions[44].

2.2 Data Preprocessing

So far, we have been working with synthetic data that arrived in ready-made tensors. However, to apply deep learning in the wild we must extract messy data stored in arbitrary formats, and preprocess it to suit our needs. Fortunately, the *pandas* library[45] can do much of the heavy lifting. This section, while no substitute for a proper *pandas* tutorial[46], will give you a crash course on some of the most common routines.

2.2.1 Reading the Dataset

Comma-separated values (CSV) files are ubiquitous for the storing of tabular (spreadsheet-like) data. In them, each line corresponds to one record and consists of several (comma-separated) fields, e.g., "Albert Einstein,March 14 1879,Ulm,Federal polytechnic school,field of gravitational physics". To demonstrate how to load CSV files with pandas, we create a CSV file below ../data/house_tiny.csv. This file represents a dataset of homes, where each row corresponds to a distinct home and the columns correspond to the number of rooms (NumRooms), the roof type (RoofType), and the price (Price).

```
import os

os.makedirs(os.path.join('..', 'data'), exist_ok=True)
data_file = os.path.join('..', 'data', 'house_tiny.csv')
with open(data_file, 'w') as f:
    f.write('''NumRooms,RoofType,Price
NA,NA,127500
2,NA,106000
4,Slate,178100
NA,NA,140000''')
```

Now let's import pandas and load the dataset with read_csv.

```
import pandas as pd

data = pd.read_csv(data_file)
print(data)
```

```
   NumRooms RoofType   Price
0       NaN      NaN  127500
1       2.0      NaN  106000
2       4.0    Slate  178100
3       NaN      NaN  140000
```

2.2.2 Data Preparation

In supervised learning, we train models to predict a designated *target* value, given some set of *input* values. Our first step in processing the dataset is to separate out columns corresponding to input versus target values. We can select columns either by name or via integer-location based indexing (iloc).

You might have noticed that pandas replaced all CSV entries with value NA with a special NaN (*not a number*) value. This can also happen whenever an entry is empty, e.g., "3,,,270000". These are called *missing values* and they are the "bed bugs" of data science, a persistent menace that you will confront throughout your career. Depending upon the context, missing values might be handled either via *imputation* or *deletion*. Imputation replaces missing values with estimates of their values while deletion simply discards either those rows or those columns that contain missing values.

Here are some common imputation heuristics. For categorical input fields, we can treat NaN as a category. Since the RoofType column takes values Slate and NaN, pandas can convert this column into two columns RoofType_Slate and RoofType_nan. A row whose roof type is Slate will set values of RoofType_Slate and RoofType_nan to 1 and 0, respectively. The converse holds for a row with a missing RoofType value.

```
inputs, targets = data.iloc[:, 0:2], data.iloc[:, 2]
inputs = pd.get_dummies(inputs, dummy_na=True)
print(inputs)
```

```
   NumRooms  RoofType_Slate  RoofType_nan
0       NaN           False          True
1       2.0           False          True
2       4.0            True         False
3       NaN           False          True
```

For missing numerical values, one common heuristic is to replace the NaN entries with the mean value of the corresponding column.

```
inputs = inputs.fillna(inputs.mean())
print(inputs)
```

```
   NumRooms  RoofType_Slate  RoofType_nan
0       3.0           False          True
1       2.0           False          True
```

(continues on next page)

(continued from previous page)

```
2       4.0         True        False
3       3.0         False       True
```

2.2.3 Conversion to the Tensor Format

Now that all the entries in `inputs` and `targets` are numerical, we can load them into a tensor (recall Section 2.1).

```python
import torch

X = torch.tensor(inputs.to_numpy(dtype=float))
y = torch.tensor(targets.to_numpy(dtype=float))
X, y
```

```
(tensor([[3., 0., 1.],
         [2., 0., 1.],
         [4., 1., 0.],
         [3., 0., 1.]], dtype=torch.float64),
 tensor([127500., 106000., 178100., 140000.], dtype=torch.float64))
```

2.2.4 Discussion

You now know how to partition data columns, impute missing variables, and load pandas data into tensors. In Section 5.7, you will pick up some more data processing skills. While this crash course kept things simple, data processing can get hairy. For example, rather than arriving in a single CSV file, our dataset might be spread across multiple files extracted from a relational database. For instance, in an e-commerce application, customer addresses might live in one table and purchase data in another. Moreover, practitioners face myriad data types beyond categorical and numeric, for example, text strings, images, audio data, and point clouds. Oftentimes, advanced tools and efficient algorithms are required in order to prevent data processing from becoming the biggest bottleneck in the machine learning pipeline. These problems will arise when we get to computer vision and natural language processing. Finally, we must pay attention to data quality. Real-world datasets are often plagued by outliers, faulty measurements from sensors, and recording errors, which must be addressed before feeding the data into any model. Data visualization tools, for example seaborn[47], Bokeh[48], or matplotlib[49], can help you to manually inspect the data and develop intuitions about the type of problems you may need to address.

2.2.5 Exercises

1. Try loading datasets, e.g., Abalone from the UCI Machine Learning Repository[50] and inspect their properties. What fraction of them has missing values? What fraction of the variables is numerical, categorical, or text?

2. Try indexing and selecting data columns by name rather than by column number. The pandas documentation on indexing[51] has further details on how to do this.

3. How large a dataset do you think you could load this way? What might be the limitations? Hint: consider the time to read the data, representation, processing, and memory footprint. Try this out on your laptop. What happens if you try it out on a server?

4. How would you deal with data that has a very large number of categories? What if the category labels are all unique? Should you include the latter?

5. What alternatives to pandas can you think of? How about loading NumPy tensors from a file[52]? Check out Pillow[53], the Python Imaging Library.

Discussions[54].

2.3 Linear Algebra

By now, we can load datasets into tensors and manipulate these tensors with basic mathematical operations. To start building sophisticated models, we will also need a few tools from linear algebra. This section offers a gentle introduction to the most essential concepts, starting from scalar arithmetic and ramping up to matrix multiplication.

```
import torch
```

2.3.1 Scalars

Most everyday mathematics consists of manipulating numbers one at a time. Formally, we call these values *scalars*. For example, the temperature in Palo Alto is a balmy 72 degrees Fahrenheit. If you wanted to convert the temperature to Celsius you would evaluate the expression $c = \frac{5}{9}(f - 32)$, setting f to 72. In this equation, the values 5, 9, and 32 are constant scalars. The variables c and f in general represent unknown scalars.

We denote scalars by ordinary lower-cased letters (e.g., x, y, and z) and the space of all (continuous) *real-valued* scalars by $\mathbb{R}$. For expedience, we will skip past rigorous definitions of *spaces*: just remember that the expression $x \in \mathbb{R}$ is a formal way to say that x is a real-valued scalar. The symbol $\in$ (pronounced "in") denotes membership in a set. For example, $x, y \in \{0, 1\}$ indicates that x and y are variables that can only take values 0 or 1.

Scalars are implemented as tensors that contain only one element. Below, we assign two scalars and perform the familiar addition, multiplication, division, and exponentiation operations.

```
x = torch.tensor(3.0)
y = torch.tensor(2.0)

x + y, x * y, x / y, x**y
```

```
(tensor(5.), tensor(6.), tensor(1.5000), tensor(9.))
```

2.3.2 Vectors

For current purposes, you can think of a vector as a fixed-length array of scalars. As with their code counterparts, we call these scalars the *elements* of the vector (synonyms include *entries* and *components*). When vectors represent examples from real-world datasets, their values hold some real-world significance. For example, if we were training a model to predict the risk of a loan defaulting, we might associate each applicant with a vector whose components correspond to quantities like their income, length of employment, or number of previous defaults. If we were studying the risk of heart attack, each vector might represent a patient and its components might correspond to their most recent vital signs, cholesterol levels, minutes of exercise per day, etc. We denote vectors by bold lowercase letters, (e.g., **x**, **y**, and **z**).

Vectors are implemented as 1st-order tensors. In general, such tensors can have arbitrary lengths, subject to memory limitations. Caution: in Python, as in most programming languages, vector indices start at 0, also known as *zero-based indexing*, whereas in linear algebra subscripts begin at 1 (one-based indexing).

```
x = torch.arange(3)
x
```

```
tensor([0, 1, 2])
```

We can refer to an element of a vector by using a subscript. For example, x_2 denotes the second element of **x**. Since x_2 is a scalar, we do not bold it. By default, we visualize vectors by stacking their elements vertically.

$$\mathbf{x} = \begin{bmatrix} x_1 \\ \vdots \\ x_n \end{bmatrix}. \tag{2.3.1}$$

Here $x_1, \ldots, x_n$ are elements of the vector. Later on, we will distinguish between such *column vectors* and *row vectors* whose elements are stacked horizontally. Recall that we access a tensor's elements via indexing.

```
x[2]
```

```
tensor(2)
```

To indicate that a vector contains n elements, we write $\mathbf{x} \in \mathbb{R}^n$. Formally, we call n the *dimensionality* of the vector. In code, this corresponds to the tensor's length, accessible via Python's built-in `len` function.

```
len(x)
```

```
3
```

We can also access the length via the shape attribute. The shape is a tuple that indicates a tensor's length along each axis. Tensors with just one axis have shapes with just one element.

```
x.shape
```

```
torch.Size([3])
```

Oftentimes, the word "dimension" gets overloaded to mean both the number of axes and the length along a particular axis. To avoid this confusion, we use *order* to refer to the number of axes and *dimensionality* exclusively to refer to the number of components.

2.3.3 Matrices

Just as scalars are 0th-order tensors and vectors are 1st-order tensors, matrices are 2nd-order tensors. We denote matrices by bold capital letters (e.g., $\mathbf{X}$, $\mathbf{Y}$, and $\mathbf{Z}$), and represent them in code by tensors with two axes. The expression $\mathbf{A} \in \mathbb{R}^{m \times n}$ indicates that a matrix $\mathbf{A}$ contains $m \times n$ real-valued scalars, arranged as m rows and n columns. When $m = n$, we say that a matrix is *square*. Visually, we can illustrate any matrix as a table. To refer to an individual element, we subscript both the row and column indices, e.g., a_{ij} is the value that belongs to $\mathbf{A}$'s i^{th} row and j^{th} column:

$$\mathbf{A} = \begin{bmatrix} a_{11} & a_{12} & \cdots & a_{1n} \\ a_{21} & a_{22} & \cdots & a_{2n} \\ \vdots & \vdots & \ddots & \vdots \\ a_{m1} & a_{m2} & \cdots & a_{mn} \end{bmatrix}. \tag{2.3.2}$$

In code, we represent a matrix $\mathbf{A} \in \mathbb{R}^{m \times n}$ by a 2nd-order tensor with shape (m, n). We can convert any appropriately sized $m \times n$ tensor into an $m \times n$ matrix by passing the desired shape to reshape:

```
A = torch.arange(6).reshape(3, 2)
A
```

```
tensor([[0, 1],
        [2, 3],
        [4, 5]])
```

Sometimes we want to flip the axes. When we exchange a matrix's rows and columns, the result is called its *transpose*. Formally, we signify a matrix $\mathbf{A}$'s transpose by $\mathbf{A}^{\top}$ and if

$\mathbf{B} = \mathbf{A}^\top$, then $b_{ij} = a_{ji}$ for all i and j. Thus, the transpose of an $m \times n$ matrix is an $n \times m$ matrix:

$$\mathbf{A}^\top = \begin{bmatrix} a_{11} & a_{21} & \dots & a_{m1} \\ a_{12} & a_{22} & \dots & a_{m2} \\ \vdots & \vdots & \ddots & \vdots \\ a_{1n} & a_{2n} & \dots & a_{mn} \end{bmatrix}. \tag{2.3.3}$$

In code, we can access any matrix's transpose as follows:

```
A.T
```

```
tensor([[0, 2, 4],
        [1, 3, 5]])
```

Symmetric matrices are the subset of square matrices that are equal to their own transposes: $\mathbf{A} = \mathbf{A}^\top$. The following matrix is symmetric:

```
A = torch.tensor([[1, 2, 3], [2, 0, 4], [3, 4, 5]])
A == A.T
```

```
tensor([[True, True, True],
        [True, True, True],
        [True, True, True]])
```

Matrices are useful for representing datasets. Typically, rows correspond to individual records and columns correspond to distinct attributes.

2.3.4 Tensors

While you can go far in your machine learning journey with only scalars, vectors, and matrices, eventually you may need to work with higher-order tensors. Tensors give us a generic way of describing extensions to n^{th}-order arrays. We call software objects of the *tensor class* "tensors" precisely because they too can have arbitrary numbers of axes. While it may be confusing to use the word *tensor* for both the mathematical object and its realization in code, our meaning should usually be clear from context. We denote general tensors by capital letters with a special font face (e.g., X, Y, and Z) and their indexing mechanism (e.g., x_{ijk} and $[\mathsf{X}]_{1,2i-1,3}$) follows naturally from that of matrices.

Tensors will become more important when we start working with images. Each image arrives as a 3rd-order tensor with axes corresponding to the height, width, and *channel*. At each spatial location, the intensities of each color (red, green, and blue) are stacked along the channel. Furthermore, a collection of images is represented in code by a 4th-order tensor, where distinct images are indexed along the first axis. Higher-order tensors are constructed, as were vectors and matrices, by growing the number of shape components.

```
torch.arange(24).reshape(2, 3, 4)
```

```
tensor([[[ 0,  1,  2,  3],
         [ 4,  5,  6,  7],
         [ 8,  9, 10, 11]],

        [[12, 13, 14, 15],
         [16, 17, 18, 19],
         [20, 21, 22, 23]]])
```

2.3.5 Basic Properties of Tensor Arithmetic

Scalars, vectors, matrices, and higher-order tensors all have some handy properties. For example, elementwise operations produce outputs that have the same shape as their operands.

```
A = torch.arange(6, dtype=torch.float32).reshape(2, 3)
B = A.clone()  # Assign a copy of A to B by allocating new memory
A, A + B
```

```
(tensor([[0., 1., 2.],
         [3., 4., 5.]]),
 tensor([[ 0.,  2.,  4.],
         [ 6.,  8., 10.]]))
```

The elementwise product of two matrices is called their *Hadamard product* (denoted $\odot$). We can spell out the entries of the Hadamard product of two matrices $\mathbf{A}, \mathbf{B} \in \mathbb{R}^{m \times n}$:

$$\mathbf{A} \odot \mathbf{B} = \begin{bmatrix} a_{11}b_{11} & a_{12}b_{12} & \dots & a_{1n}b_{1n} \\ a_{21}b_{21} & a_{22}b_{22} & \dots & a_{2n}b_{2n} \\ \vdots & \vdots & \ddots & \vdots \\ a_{m1}b_{m1} & a_{m2}b_{m2} & \dots & a_{mn}b_{mn} \end{bmatrix}. \tag{2.3.4}$$

```
A * B
```

```
tensor([[ 0.,  1.,  4.],
        [ 9., 16., 25.]])
```

Adding or multiplying a scalar and a tensor produces a result with the same shape as the original tensor. Here, each element of the tensor is added to (or multiplied by) the scalar.

```
a = 2
X = torch.arange(24).reshape(2, 3, 4)
a + X, (a * X).shape
```

```
(tensor([[[ 2,   3,   4,   5],
          [ 6,   7,   8,   9],
          [10,  11,  12,  13]],

         [[14,  15,  16,  17],
          [18,  19,  20,  21],
          [22,  23,  24,  25]]]),
 torch.Size([2, 3, 4]))
```

2.3.6 Reduction

Often, we wish to calculate the sum of a tensor's elements. To express the sum of the elements in a vector $\mathbf{x}$ of length n, we write $\sum_{i=1}^{n} x_i$. There is a simple function for it:

```
x = torch.arange(3, dtype=torch.float32)
x, x.sum()
```

```
(tensor([0., 1., 2.]), tensor(3.))
```

To express sums over the elements of tensors of arbitrary shape, we simply sum over all its axes. For example, the sum of the elements of an $m \times n$ matrix $\mathbf{A}$ could be written $\sum_{i=1}^{m} \sum_{j=1}^{n} a_{ij}$.

```
A.shape, A.sum()
```

```
(torch.Size([2, 3]), tensor(15.))
```

By default, invoking the sum function *reduces* a tensor along all of its axes, eventually producing a scalar. Our libraries also allow us to specify the axes along which the tensor should be reduced. To sum over all elements along the rows (axis 0), we specify axis=0 in sum. Since the input matrix reduces along axis 0 to generate the output vector, this axis is missing from the shape of the output.

```
A.shape, A.sum(axis=0).shape
```

```
(torch.Size([2, 3]), torch.Size([3]))
```

Specifying axis=1 will reduce the column dimension (axis 1) by summing up elements of all the columns.

```
A.shape, A.sum(axis=1).shape
```

```
(torch.Size([2, 3]), torch.Size([2]))
```

Reducing a matrix along both rows and columns via summation is equivalent to summing up all the elements of the matrix.

```
A.sum(axis=[0, 1]) == A.sum()   # Same as A.sum()
```

```
tensor(True)
```

A related quantity is the *mean*, also called the *average*. We calculate the mean by dividing the sum by the total number of elements. Because computing the mean is so common, it gets a dedicated library function that works analogously to sum.

```
A.mean(), A.sum() / A.numel()
```

```
(tensor(2.5000), tensor(2.5000))
```

Likewise, the function for calculating the mean can also reduce a tensor along specific axes.

```
A.mean(axis=0), A.sum(axis=0) / A.shape[0]
```

```
(tensor([1.5000, 2.5000, 3.5000]), tensor([1.5000, 2.5000, 3.5000]))
```

2.3.7 Non-Reduction Sum

Sometimes it can be useful to keep the number of axes unchanged when invoking the function for calculating the sum or mean. This matters when we want to use the broadcast mechanism.

```
sum_A = A.sum(axis=1, keepdims=True)
sum_A, sum_A.shape
```

```
(tensor([[ 3.],
         [12.]]),
 torch.Size([2, 1]))
```

For instance, since sum_A keeps its two axes after summing each row, we can divide A by sum_A with broadcasting to create a matrix where each row sums up to 1.

```
A / sum_A
```

```
tensor([[0.0000, 0.3333, 0.6667],
        [0.2500, 0.3333, 0.4167]])
```

If we want to calculate the cumulative sum of elements of A along some axis, say `axis=0` (row by row), we can call the `cumsum` function. By design, this function does not reduce the input tensor along any axis.

```
A.cumsum(axis=0)
```

```
tensor([[0., 1., 2.],
        [3., 5., 7.]])
```

2.3.8 Dot Products

So far, we have only performed elementwise operations, sums, and averages. And if this was all we could do, linear algebra would not deserve its own section. Fortunately, this is where things get more interesting. One of the most fundamental operations is the dot product. Given two vectors $\mathbf{x}, \mathbf{y} \in \mathbb{R}^d$, their *dot product* $\mathbf{x}^\top \mathbf{y}$ (also known as *inner product*, $\langle \mathbf{x}, \mathbf{y} \rangle$) is a sum over the products of the elements at the same position: $\mathbf{x}^\top \mathbf{y} = \sum_{i=1}^{d} x_i y_i$.

```
y = torch.ones(3, dtype = torch.float32)
x, y, torch.dot(x, y)
```

```
(tensor([0., 1., 2.]), tensor([1., 1., 1.]), tensor(3.))
```

Equivalently, we can calculate the dot product of two vectors by performing an elementwise multiplication followed by a sum:

```
torch.sum(x * y)
```

```
tensor(3.)
```

Dot products are useful in a wide range of contexts. For example, given some set of values, denoted by a vector $\mathbf{x} \in \mathbb{R}^n$, and a set of weights, denoted by $\mathbf{w} \in \mathbb{R}^n$, the weighted sum of the values in $\mathbf{x}$ according to the weights $\mathbf{w}$ could be expressed as the dot product $\mathbf{x}^\top \mathbf{w}$. When the weights are nonnegative and sum to 1, i.e., $\left(\sum_{i=1}^{n} w_i = 1 \right)$, the dot product expresses a *weighted average*. After normalizing two vectors to have unit length, the dot products express the cosine of the angle between them. Later in this section, we will formally introduce this notion of *length*.

2.3.9 Matrix–Vector Products

Now that we know how to calculate dot products, we can begin to understand the *product* between an $m \times n$ matrix $\mathbf{A}$ and an n-dimensional vector $\mathbf{x}$. To start off, we visualize our matrix in terms of its row vectors

$$
\mathbf{A} = \begin{bmatrix} \mathbf{a}_1^\top \\ \mathbf{a}_2^\top \\ \vdots \\ \mathbf{a}_m^\top \end{bmatrix},
\tag{2.3.5}
$$

where each $\mathbf{a}_i^\top \in \mathbb{R}^n$ is a row vector representing the ith row of the matrix $\mathbf{A}$.

The matrix–vector product $\mathbf{A}\mathbf{x}$ is simply a column vector of length m, whose i^{th} element is the dot product $\mathbf{a}_i^\top \mathbf{x}$:

$$
\mathbf{A}\mathbf{x} = \begin{bmatrix} \mathbf{a}_1^\top \\ \mathbf{a}_2^\top \\ \vdots \\ \mathbf{a}_m^\top \end{bmatrix} \mathbf{x} = \begin{bmatrix} \mathbf{a}_1^\top \mathbf{x} \\ \mathbf{a}_2^\top \mathbf{x} \\ \vdots \\ \mathbf{a}_m^\top \mathbf{x} \end{bmatrix}.
\tag{2.3.6}
$$

We can think of multiplication with a matrix $\mathbf{A} \in \mathbb{R}^{m \times n}$ as a transformation that projects vectors from $\mathbb{R}^n$ to $\mathbb{R}^m$. These transformations are remarkably useful. For example, we can represent rotations as multiplications by certain square matrices. Matrix–vector products also describe the key calculation involved in computing the outputs of each layer in a neural network given the outputs from the previous layer.

To express a matrix–vector product in code, we use the `mv` function. Note that the column dimension of A (its length along axis 1) must be the same as the dimension of x (its length). Python has a convenience operator @ that can execute both matrix–vector and matrix–matrix products (depending on its arguments). Thus we can write A@x.

```
A.shape, x.shape, torch.mv(A, x), A@x
```

```
(torch.Size([2, 3]), torch.Size([3]), tensor([ 5., 14.]), tensor([ 5., 14.]))
```

2.3.10 Matrix–Matrix Multiplication

Once you have gotten the hang of dot products and matrix–vector products, then *matrix–matrix multiplication* should be straightforward.

Say that we have two matrices $\mathbf{A} \in \mathbb{R}^{n \times k}$ and $\mathbf{B} \in \mathbb{R}^{k \times m}$:

$$
\mathbf{A} = \begin{bmatrix} a_{11} & a_{12} & \cdots & a_{1k} \\ a_{21} & a_{22} & \cdots & a_{2k} \\ \vdots & \vdots & \ddots & \vdots \\ a_{n1} & a_{n2} & \cdots & a_{nk} \end{bmatrix}, \quad \mathbf{B} = \begin{bmatrix} b_{11} & b_{12} & \cdots & b_{1m} \\ b_{21} & b_{22} & \cdots & b_{2m} \\ \vdots & \vdots & \ddots & \vdots \\ b_{k1} & b_{k2} & \cdots & b_{km} \end{bmatrix}.
\tag{2.3.7}
$$

Let $\mathbf{a}_i^\top \in \mathbb{R}^k$ denote the row vector representing the ith row of the matrix $\mathbf{A}$ and let $\mathbf{b}_j \in \mathbb{R}^k$

denote the column vector from the jth column of the matrix $\mathbf{B}$:

$$\mathbf{A} = \begin{bmatrix} \mathbf{a}_1^\top \\ \mathbf{a}_2^\top \\ \vdots \\ \mathbf{a}_n^\top \end{bmatrix}, \quad \mathbf{B} = \begin{bmatrix} \mathbf{b}_1 & \mathbf{b}_2 & \cdots & \mathbf{b}_m \end{bmatrix}. \tag{2.3.8}$$

To form the matrix product $\mathbf{C} \in \mathbb{R}^{n \times m}$, we simply compute each element c_{ij} as the dot product between the ith row of $\mathbf{A}$ and the jth column of $\mathbf{B}$, i.e., $\mathbf{a}_i^\top \mathbf{b}_j$:

$$\mathbf{C} = \mathbf{AB} = \begin{bmatrix} \mathbf{a}_1^\top \\ \mathbf{a}_2^\top \\ \vdots \\ \mathbf{a}_n^\top \end{bmatrix} \begin{bmatrix} \mathbf{b}_1 & \mathbf{b}_2 & \cdots & \mathbf{b}_m \end{bmatrix} = \begin{bmatrix} \mathbf{a}_1^\top \mathbf{b}_1 & \mathbf{a}_1^\top \mathbf{b}_2 & \cdots & \mathbf{a}_1^\top \mathbf{b}_m \\ \mathbf{a}_2^\top \mathbf{b}_1 & \mathbf{a}_2^\top \mathbf{b}_2 & \cdots & \mathbf{a}_2^\top \mathbf{b}_m \\ \vdots & \vdots & \ddots & \vdots \\ \mathbf{a}_n^\top \mathbf{b}_1 & \mathbf{a}_n^\top \mathbf{b}_2 & \cdots & \mathbf{a}_n^\top \mathbf{b}_m \end{bmatrix}. \tag{2.3.9}$$

We can think of the matrix–matrix multiplication $\mathbf{AB}$ as performing m matrix–vector products or $m \times n$ dot products and stitching the results together to form an $n \times m$ matrix. In the following snippet, we perform matrix multiplication on A and B. Here, A is a matrix with two rows and three columns, and B is a matrix with three rows and four columns. After multiplication, we obtain a matrix with two rows and four columns.

```
B = torch.ones(3, 4)
torch.mm(A, B), A@B
```

```
(tensor([[ 3.,  3.,  3.,  3.],
         [12., 12., 12., 12.]]),
 tensor([[ 3.,  3.,  3.,  3.],
         [12., 12., 12., 12.]]))
```

The term *matrix–matrix multiplication* is often simplified to *matrix multiplication*, and should not be confused with the Hadamard product.

2.3.11 Norms

Some of the most useful operators in linear algebra are *norms*. Informally, the norm of a vector tells us how *big* it is. For instance, the ℓ_2 norm measures the (Euclidean) length of a vector. Here, we are employing a notion of *size* that concerns the magnitude of a vector's components (not its dimensionality).

A norm is a function $\| \cdot \|$ that maps a vector to a scalar and satisfies the following three properties:

1. Given any vector $\mathbf{x}$, if we scale (all elements of) the vector by a scalar $\alpha \in \mathbb{R}$, its norm scales accordingly:

$$\|\alpha \mathbf{x}\| = |\alpha| \|\mathbf{x}\|. \tag{2.3.10}$$

2. For any vectors $\mathbf{x}$ and $\mathbf{y}$: norms satisfy the triangle inequality:

$$\|\mathbf{x} + \mathbf{y}\| \leq \|\mathbf{x}\| + \|\mathbf{y}\|. \tag{2.3.11}$$

3. The norm of a vector is nonnegative and it only vanishes if the vector is zero:

$$\|\mathbf{x}\| > 0 \text{ for all } \mathbf{x} \neq 0. \tag{2.3.12}$$

Many functions are valid norms and different norms encode different notions of size. The Euclidean norm that we all learned in elementary school geometry when calculating the hypotenuse of a right triangle is the square root of the sum of squares of a vector's elements. Formally, this is called the ℓ_2 *norm* and expressed as

$$\|\mathbf{x}\|_2 = \sqrt{\sum_{i=1}^{n} x_i^2}. \tag{2.3.13}$$

The method norm calculates the ℓ_2 norm.

```
u = torch.tensor([3.0, -4.0])
torch.norm(u)
```

```
tensor(5.)
```

The ℓ_1 norm is also common and the associated measure is called the Manhattan distance. By definition, the ℓ_1 norm sums the absolute values of a vector's elements:

$$\|\mathbf{x}\|_1 = \sum_{i=1}^{n} |x_i|. \tag{2.3.14}$$

Compared to the ℓ_2 norm, it is less sensitive to outliers. To compute the ℓ_1 norm, we compose the absolute value with the sum operation.

```
torch.abs(u).sum()
```

```
tensor(7.)
```

Both the ℓ_2 and ℓ_1 norms are special cases of the more general ℓ_p *norms*:

$$\|\mathbf{x}\|_p = \left(\sum_{i=1}^{n} |x_i|^p \right)^{1/p}. \tag{2.3.15}$$

In the case of matrices, matters are more complicated. After all, matrices can be viewed both as collections of individual entries *and* as objects that operate on vectors and transform them into other vectors. For instance, we can ask by how much longer the matrix–vector product $\mathbf{Xv}$ could be relative to $\mathbf{v}$. This line of thought leads to what is called the *spectral* norm. For now, we introduce the *Frobenius norm*, which is much easier to compute and defined as the square root of the sum of the squares of a matrix's elements:

$$\|\mathbf{X}\|_F = \sqrt{\sum_{i=1}^{m} \sum_{j=1}^{n} x_{ij}^2}. \tag{2.3.16}$$

The Frobenius norm behaves as if it were an ℓ_2 norm of a matrix-shaped vector. Invoking the following function will calculate the Frobenius norm of a matrix.

```
torch.norm(torch.ones((4, 9)))
```

```
tensor(6.)
```

While we do not want to get too far ahead of ourselves, we already can plant some intuition about why these concepts are useful. In deep learning, we are often trying to solve optimization problems: *maximize* the probability assigned to observed data; *maximize* the revenue associated with a recommender model; *minimize* the distance between predictions and the ground truth observations; *minimize* the distance between representations of photos of the same person while *maximizing* the distance between representations of photos of different people. These distances, which constitute the objectives of deep learning algorithms, are often expressed as norms.

2.3.12 Discussion

In this section, we have reviewed all the linear algebra that you will need to understand a significant chunk of modern deep learning. There is a lot more to linear algebra, though, and much of it is useful for machine learning. For example, matrices can be decomposed into factors, and these decompositions can reveal low-dimensional structure in real-world datasets. There are entire subfields of machine learning that focus on using matrix decompositions and their generalizations to high-order tensors to discover structure in datasets and solve prediction problems. But this book focuses on deep learning. And we believe you will be more inclined to learn more mathematics once you have gotten your hands dirty applying machine learning to real datasets. So while we reserve the right to introduce more mathematics later on, we wrap up this section here.

If you are eager to learn more linear algebra, there are many excellent books and online resources. For a more advanced crash course, consider checking out Strang (1993), Kolter (2008), and Petersen and Pedersen (2008).

To recap:

- Scalars, vectors, matrices, and tensors are the basic mathematical objects used in linear algebra and have zero, one, two, and an arbitrary number of axes, respectively.

- Tensors can be sliced or reduced along specified axes via indexing, or operations such as sum and mean, respectively.

- Elementwise products are called Hadamard products. By contrast, dot products, matrix–vector products, and matrix–matrix products are not elementwise operations and in general return objects having shapes that are different from the the operands.

- Compared to Hadamard products, matrix–matrix products take considerably longer to compute (cubic rather than quadratic time).

- Norms capture various notions of the magnitude of a vector (or matrix), and are commonly applied to the difference of two vectors to measure their distance apart.

- Common vector norms include the ℓ_1 and ℓ_2 norms, and common matrix norms include the *spectral* and *Frobenius* norms.

2.3.13 Exercises

1. Prove that the transpose of the transpose of a matrix is the matrix itself: $(\mathbf{A}^\top)^\top = \mathbf{A}$.

2. Given two matrices $\mathbf{A}$ and $\mathbf{B}$, show that sum and transposition commute: $\mathbf{A}^\top + \mathbf{B}^\top = (\mathbf{A} + \mathbf{B})^\top$.

3. Given any square matrix $\mathbf{A}$, is $\mathbf{A} + \mathbf{A}^\top$ always symmetric? Can you prove the result by using only the results of the previous two exercises?

4. We defined the tensor X of shape (2, 3, 4) in this section. What is the output of len(X)? Write your answer without implementing any code, then check your answer using code.

5. For a tensor X of arbitrary shape, does len(X) always correspond to the length of a certain axis of X? What is that axis?

6. Run A / A.sum(axis=1) and see what happens. Can you analyze the results?

7. When traveling between two points in downtown Manhattan, what is the distance that you need to cover in terms of the coordinates, i.e., in terms of avenues and streets? Can you travel diagonally?

8. Consider a tensor of shape (2, 3, 4). What are the shapes of the summation outputs along axes 0, 1, and 2?

9. Feed a tensor with three or more axes to the linalg.norm function and observe its output. What does this function compute for tensors of arbitrary shape?

10. Consider three large matrices, say $\mathbf{A} \in \mathbb{R}^{2^{10} \times 2^{16}}$, $\mathbf{B} \in \mathbb{R}^{2^{16} \times 2^5}$ and $\mathbf{C} \in \mathbb{R}^{2^5 \times 2^{14}}$, initialized with Gaussian random variables. You want to compute the product $\mathbf{ABC}$. Is there any difference in memory footprint and speed, depending on whether you compute $(\mathbf{AB})\mathbf{C}$ or $\mathbf{A}(\mathbf{BC})$? Why?

11. Consider three large matrices, say $\mathbf{A} \in \mathbb{R}^{2^{10} \times 2^{16}}$, $\mathbf{B} \in \mathbb{R}^{2^{16} \times 2^5}$ and $\mathbf{C} \in \mathbb{R}^{2^5 \times 2^{16}}$. Is there any difference in speed depending on whether you compute $\mathbf{AB}$ or $\mathbf{AC}^\top$? Why? What changes if you initialize $\mathbf{C} = \mathbf{B}^\top$ without cloning memory? Why?

12. Consider three matrices, say $\mathbf{A}, \mathbf{B}, \mathbf{C} \in \mathbb{R}^{100 \times 200}$. Construct a tensor with three axes by

stacking $[\mathbf{A}, \mathbf{B}, \mathbf{C}]$. What is the dimensionality? Slice out the second coordinate of the third axis to recover $\mathbf{B}$. Check that your answer is correct.

Discussions[55].

2.4 Calculus

For a long time, how to calculate the area of a circle remained a mystery. Then, in Ancient Greece, the mathematician Archimedes came up with the clever idea to inscribe a series of polygons with increasing numbers of vertices on the inside of a circle (Fig. 2.4.1). For a polygon with n vertices, we obtain n triangles. The height of each triangle approaches the radius r as we partition the circle more finely. At the same time, its base approaches $2\pi r/n$, since the ratio between arc and secant approaches 1 for a large number of vertices. Thus, the area of the polygon approaches $n \cdot r \cdot \frac{1}{2}(2\pi r/n) = \pi r^2$.

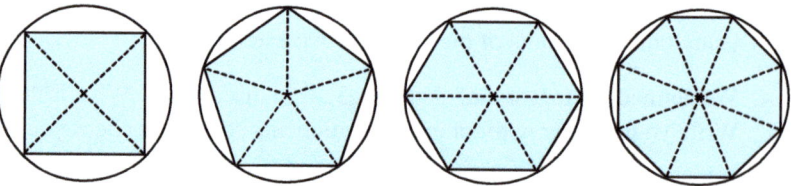

Fig. 2.4.1   Finding the area of a circle as a limit procedure.

This limiting procedure is at the root of both *differential calculus* and *integral calculus*. The former can tell us how to increase or decrease a function's value by manipulating its arguments. This comes in handy for the *optimization problems* that we face in deep learning, where we repeatedly update our parameters in order to decrease the loss function. Optimization addresses how to fit our models to training data, and calculus is its key prerequisite. However, do not forget that our ultimate goal is to perform well on *previously unseen* data. That problem is called *generalization* and will be a key focus of other chapters.

```
%matplotlib inline
import numpy as np
from matplotlib_inline import backend_inline
from d2l import torch as d2l
```

2.4.1 Derivatives and Differentiation

Put simply, a *derivative* is the rate of change in a function with respect to changes in its arguments. Derivatives can tell us how rapidly a loss function would increase or decrease were we to *increase* or *decrease* each parameter by an infinitesimally small amount. Formally, for functions $f : \mathbb{R} \to \mathbb{R}$, that map from scalars to scalars, the *derivative* of f at a point x is defined as

$$f'(x) = \lim_{h \to 0} \frac{f(x+h) - f(x)}{h}. \qquad (2.4.1)$$

This term on the right hand side is called a *limit* and it tells us what happens to the value of an expression as a specified variable approaches a particular value. This limit tells us what the ratio between a perturbation h and the change in the function value $f(x+h) - f(x)$ converges to as we shrink its size to zero.

When $f'(x)$ exists, f is said to be *differentiable* at x; and when $f'(x)$ exists for all x on a set, e.g., the interval $[a, b]$, we say that f is differentiable on this set. Not all functions are differentiable, including many that we wish to optimize, such as accuracy and the area under the receiving operating characteristic (AUC). However, because computing the derivative of the loss is a crucial step in nearly all algorithms for training deep neural networks, we often optimize a differentiable *surrogate* instead.

We can interpret the derivative $f'(x)$ as the *instantaneous* rate of change of $f(x)$ with respect to x. Let's develop some intuition with an example. Define $u = f(x) = 3x^2 - 4x$.

```
def f(x):
    return 3 * x ** 2 - 4 * x
```

Setting $x = 1$, we see that $\frac{f(x+h)-f(x)}{h}$ approaches 2 as h approaches 0. While this experiment lacks the rigor of a mathematical proof, we can quickly see that indeed $f'(1) = 2$.

```
for h in 10.0**np.arange(-1, -6, -1):
    print(f'h={h:.5f}, numerical limit={(f(1+h)-f(1))/h:.5f}')
```

```
h=0.10000, numerical limit=2.30000
h=0.01000, numerical limit=2.03000
h=0.00100, numerical limit=2.00300
h=0.00010, numerical limit=2.00030
h=0.00001, numerical limit=2.00003
```

There are several equivalent notational conventions for derivatives. Given $y = f(x)$, the following expressions are equivalent:

$$f'(x) = y' = \frac{dy}{dx} = \frac{df}{dx} = \frac{d}{dx} f(x) = Df(x) = D_x f(x), \qquad (2.4.2)$$

where the symbols $\frac{d}{dx}$ and D are *differentiation operators*. Below, we present the derivatives of some common functions:

$$\frac{d}{dx}C = 0 \qquad \text{for any constant } C$$

$$\frac{d}{dx}x^n = nx^{n-1} \quad \text{for } n \neq 0$$

$$\frac{d}{dx}e^x = e^x$$

$$\frac{d}{dx}\ln x = x^{-1}.$$

(2.4.3)

Functions composed from differentiable functions are often themselves differentiable. The following rules come in handy for working with compositions of any differentiable functions f and g, and constant C.

$$\frac{d}{dx}[Cf(x)] = C\frac{d}{dx}f(x) \qquad\qquad \text{Constant multiple rule}$$

$$\frac{d}{dx}[f(x) + g(x)] = \frac{d}{dx}f(x) + \frac{d}{dx}g(x) \qquad \text{Sum rule}$$

$$\frac{d}{dx}[f(x)g(x)] = f(x)\frac{d}{dx}g(x) + g(x)\frac{d}{dx}f(x) \quad \text{Product rule}$$

$$\frac{d}{dx}\frac{f(x)}{g(x)} = \frac{g(x)\frac{d}{dx}f(x) - f(x)\frac{d}{dx}g(x)}{g^2(x)} \qquad \text{Quotient rule}$$

(2.4.4)

Using this, we can apply the rules to find the derivative of $3x^2 - 4x$ via

$$\frac{d}{dx}[3x^2 - 4x] = 3\frac{d}{dx}x^2 - 4\frac{d}{dx}x = 6x - 4. \tag{2.4.5}$$

Plugging in $x = 1$ shows that, indeed, the derivative equals 2 at this location. Note that derivatives tell us the *slope* of a function at a particular location.

2.4.2 Visualization Utilities

We can visualize the slopes of functions using the matplotlib library. We need to define a few functions. As its name indicates, use_svg_display tells matplotlib to output graphics in SVG format for crisper images. The comment #@save is a special modifier that allows us to save any function, class, or other code block to the d2l package so that we can invoke it later without repeating the code, e.g., via d2l.use_svg_display().

```
def use_svg_display():  #@save
    """Use the svg format to display a plot in Jupyter."""
    backend_inline.set_matplotlib_formats('svg')
```

Conveniently, we can set figure sizes with set_figsize. Since the import statement from matplotlib import pyplot as plt was marked via #@save in the d2l package, we can call d2l.plt.

```
def set_figsize(figsize=(3.5, 2.5)):  #@save
    """Set the figure size for matplotlib."""
```

(continues on next page)

(continued from previous page)

```
    use_svg_display()
    d2l.plt.rcParams['figure.figsize'] = figsize
```

The `set_axes` function can associate axes with properties, including labels, ranges, and scales.

```
#@save
def set_axes(axes, xlabel, ylabel, xlim, ylim, xscale, yscale, legend):
    """Set the axes for matplotlib."""
    axes.set_xlabel(xlabel), axes.set_ylabel(ylabel)
    axes.set_xscale(xscale), axes.set_yscale(yscale)
    axes.set_xlim(xlim),     axes.set_ylim(ylim)
    if legend:
        axes.legend(legend)
    axes.grid()
```

With these three functions, we can define a `plot` function to overlay multiple curves. Much of the code here is just ensuring that the sizes and shapes of inputs match.

```
#@save
def plot(X, Y=None, xlabel=None, ylabel=None, legend=[], xlim=None,
         ylim=None, xscale='linear', yscale='linear',
         fmts=('-', 'm--', 'g-.', 'r:'), figsize=(3.5, 2.5), axes=None):
    """Plot data points."""

    def has_one_axis(X):  # True if X (tensor or list) has 1 axis
        return (hasattr(X, "ndim") and X.ndim == 1 or isinstance(X, list)
                and not hasattr(X[0], "__len__"))

    if has_one_axis(X): X = [X]
    if Y is None:
        X, Y = [[]] * len(X), X
    elif has_one_axis(Y):
        Y = [Y]
    if len(X) != len(Y):
        X = X * len(Y)

    set_figsize(figsize)
    if axes is None:
        axes = d2l.plt.gca()
    axes.cla()
    for x, y, fmt in zip(X, Y, fmts):
        axes.plot(x,y,fmt) if len(x) else axes.plot(y,fmt)
    set_axes(axes, xlabel, ylabel, xlim, ylim, xscale, yscale, legend)
```

Now we can plot the function $u = f(x)$ and its tangent line $y = 2x - 3$ at $x = 1$, where the coefficient 2 is the slope of the tangent line.

```
x = np.arange(0, 3, 0.1)
plot(x, [f(x), 2 * x - 3], 'x', 'f(x)', legend=['f(x)', 'Tangent line (x=1)'])
```

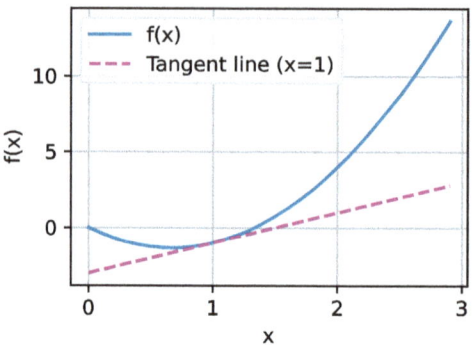

2.4.3 Partial Derivatives and Gradients

Thus far, we have been differentiating functions of just one variable. In deep learning, we also need to work with functions of *many* variables. We briefly introduce notions of the derivative that apply to such *multivariate* functions.

Let $y = f(x_1, x_2, \ldots, x_n)$ be a function with n variables. The *partial derivative* of y with respect to its ith parameter x_i is

$$\frac{\partial y}{\partial x_i} = \lim_{h \to 0} \frac{f(x_1, \ldots, x_{i-1}, x_i + h, x_{i+1}, \ldots, x_n) - f(x_1, \ldots, x_i, \ldots, x_n)}{h}. \tag{2.4.6}$$

To calculate $\frac{\partial y}{\partial x_i}$, we can treat $x_1, \ldots, x_{i-1}, x_{i+1}, \ldots, x_n$ as constants and calculate the derivative of y with respect to x_i. The following notational conventions for partial derivatives are all common and all mean the same thing:

$$\frac{\partial y}{\partial x_i} = \frac{\partial f}{\partial x_i} = \partial_{x_i} f = \partial_i f = f_{x_i} = f_i = D_i f = D_{x_i} f. \tag{2.4.7}$$

We can concatenate partial derivatives of a multivariate function with respect to all its variables to obtain a vector that is called the *gradient* of the function. Suppose that the input of function $f : \mathbb{R}^n \to \mathbb{R}$ is an n-dimensional vector $\mathbf{x} = [x_1, x_2, \ldots, x_n]^\top$ and the output is a scalar. The gradient of the function f with respect to $\mathbf{x}$ is a vector of n partial derivatives:

$$\nabla_{\mathbf{x}} f(\mathbf{x}) = \left[\partial_{x_1} f(\mathbf{x}), \partial_{x_2} f(\mathbf{x}), \ldots \partial_{x_n} f(\mathbf{x}) \right]^\top. \tag{2.4.8}$$

When there is no ambiguity, $\nabla_{\mathbf{x}} f(\mathbf{x})$ is typically replaced by $\nabla f(\mathbf{x})$. The following rules come in handy for differentiating multivariate functions:

- For all $\mathbf{A} \in \mathbb{R}^{m \times n}$ we have $\nabla_{\mathbf{x}} \mathbf{A} \mathbf{x} = \mathbf{A}^\top$ and $\nabla_{\mathbf{x}} \mathbf{x}^\top \mathbf{A} = \mathbf{A}$.

- For square matrices $\mathbf{A} \in \mathbb{R}^{n \times n}$ we have that $\nabla_{\mathbf{x}} \mathbf{x}^\top \mathbf{A} \mathbf{x} = (\mathbf{A} + \mathbf{A}^\top) \mathbf{x}$ and in particular $\nabla_{\mathbf{x}} \|\mathbf{x}\|^2 = \nabla_{\mathbf{x}} \mathbf{x}^\top \mathbf{x} = 2\mathbf{x}$.

Similarly, for any matrix $\mathbf{X}$, we have $\nabla_{\mathbf{X}} \|\mathbf{X}\|_{\text{F}}^2 = 2\mathbf{X}$.

2.4.4 Chain Rule

In deep learning, the gradients of concern are often difficult to calculate because we are working with deeply nested functions (of functions (of functions…)). Fortunately, the *chain rule* takes care of this. Returning to functions of a single variable, suppose that $y = f(g(x))$ and that the underlying functions $y = f(u)$ and $u = g(x)$ are both differentiable. The chain rule states that

$$\frac{dy}{dx} = \frac{dy}{du}\frac{du}{dx}. \tag{2.4.9}$$

Turning back to multivariate functions, suppose that $y = f(\mathbf{u})$ has variables $u_1, u_2, \ldots, u_m$, where each $u_i = g_i(\mathbf{x})$ has variables $x_1, x_2, \ldots, x_n$, i.e., $\mathbf{u} = g(\mathbf{x})$. Then the chain rule states that

$$\frac{\partial y}{\partial x_i} = \frac{\partial y}{\partial u_1}\frac{\partial u_1}{\partial x_i} + \frac{\partial y}{\partial u_2}\frac{\partial u_2}{\partial x_i} + \ldots + \frac{\partial y}{\partial u_m}\frac{\partial u_m}{\partial x_i} \quad \text{and so} \quad \nabla_{\mathbf{x}}y = \mathbf{A}\nabla_{\mathbf{u}}y, \tag{2.4.10}$$

where $\mathbf{A} \in \mathbb{R}^{n \times m}$ is a *matrix* that contains the derivative of vector $\mathbf{u}$ with respect to vector $\mathbf{x}$. Thus, evaluating the gradient requires computing a vector–matrix product. This is one of the key reasons why linear algebra is such an integral building block in building deep learning systems.

2.4.5 Discussion

While we have just scratched the surface of a deep topic, a number of concepts already come into focus: first, the composition rules for differentiation can be applied routinely, enabling us to compute gradients *automatically*. This task requires no creativity and thus we can focus our cognitive powers elsewhere. Second, computing the derivatives of vector-valued functions requires us to multiply matrices as we trace the dependency graph of variables from output to input. In particular, this graph is traversed in a *forward* direction when we evaluate a function and in a *backwards* direction when we compute gradients. Later chapters will formally introduce backpropagation, a computational procedure for applying the chain rule.

From the viewpoint of optimization, gradients allow us to determine how to move the parameters of a model in order to lower the loss, and each step of the optimization algorithms used throughout this book will require calculating the gradient.

2.4.6 Exercises

1. So far we took the rules for derivatives for granted. Using the definition and limits prove the properties for (i) $f(x) = c$, (ii) $f(x) = x^n$, (iii) $f(x) = e^x$ and (iv) $f(x) = \log x$.

2. In the same vein, prove the product, sum, and quotient rule from first principles.

3. Prove that the constant multiple rule follows as a special case of the product rule.

4. Calculate the derivative of $f(x) = x^x$.

5. What does it mean that $f'(x) = 0$ for some x? Give an example of a function f and a location x for which this might hold.

6. Plot the function $y = f(x) = x^3 - \frac{1}{x}$ and plot its tangent line at $x = 1$.

7. Find the gradient of the function $f(\mathbf{x}) = 3x_1^2 + 5e^{x_2}$.

8. What is the gradient of the function $f(\mathbf{x}) = \|\mathbf{x}\|_2$? What happens for $\mathbf{x} = \mathbf{0}$?

9. Can you write out the chain rule for the case where $u = f(x, y, z)$ and $x = x(a, b)$, $y = y(a, b)$, and $z = z(a, b)$?

10. Given a function $f(x)$ that is invertible, compute the derivative of its inverse $f^{-1}(x)$. Here we have that $f^{-1}(f(x)) = x$ and conversely $f(f^{-1}(y)) = y$. Hint: use these properties in your derivation.

 Discussions[56].

2.5 Automatic Differentiation

Recall from Section 2.4 that calculating derivatives is the crucial step in all the optimization algorithms that we will use to train deep networks. While the calculations are straightforward, working them out by hand can be tedious and error-prone, and these issues only grow as our models become more complex.

Fortunately all modern deep learning frameworks take this work off our plates by offering *automatic differentiation* (often shortened to *autograd*). As we pass data through each successive function, the framework builds a *computational graph* that tracks how each value depends on others. To calculate derivatives, automatic differentiation works backwards through this graph applying the chain rule. The computational algorithm for applying the chain rule in this fashion is called *backpropagation*.

While autograd libraries have become a hot concern over the past decade, they have a long history. In fact the earliest references to autograd date back over half of a century (Wengert, 1964). The core ideas behind modern backpropagation date to a PhD thesis from 1980 (Speelpenning, 1980) and were further developed in the late 1980s (Griewank, 1989). While backpropagation has become the default method for computing gradients, it is not the only option. For instance, the Julia programming language employs forward propagation (Revels *et al.*, 2016). Before exploring methods, let's first master the autograd package.

```
import torch
```

2.5.1 A Simple Function

Let's assume that we are interested in differentiating the function $y = 2\mathbf{x}^\top\mathbf{x}$ with respect to the column vector $\mathbf{x}$. To start, we assign x an initial value.

```
x = torch.arange(4.0)
x
```

```
tensor([0., 1., 2., 3.])
```

Before we calculate the gradient of y with respect to $\mathbf{x}$, we need a place to store it. In general, we avoid allocating new memory every time we take a derivative because deep learning requires successively computing derivatives with respect to the same parameters a great many times, and we might risk running out of memory. Note that the gradient of a scalar-valued function with respect to a vector $\mathbf{x}$ is vector-valued with the same shape as $\mathbf{x}$.

```
# Can also create x = torch.arange(4.0, requires_grad=True)
x.requires_grad_(True)
x.grad  # The gradient is None by default
```

We now calculate our function of x and assign the result to y.

```
y = 2 * torch.dot(x, x)
y
```

```
tensor(28., grad_fn=<MulBackward0>)
```

We can now take the gradient of y with respect to x by calling its backward method. Next, we can access the gradient via x's grad attribute.

```
y.backward()
x.grad
```

```
tensor([ 0.,  4.,  8., 12.])
```

We already know that the gradient of the function $y = 2\mathbf{x}^\top\mathbf{x}$ with respect to $\mathbf{x}$ should be $4\mathbf{x}$. We can now verify that the automatic gradient computation and the expected result are identical.

```
x.grad == 4 * x
```

```
tensor([True, True, True, True])
```

Now let's calculate another function of x and take its gradient. Note that PyTorch does not automatically reset the gradient buffer when we record a new gradient. Instead, the new gradient is added to the already-stored gradient. This behavior comes in handy when we want to optimize the sum of multiple objective functions. To reset the gradient buffer, we can call x.grad.zero_() as follows:

```
x.grad.zero_()   # Reset the gradient
y = x.sum()
y.backward()
x.grad
```

```
tensor([1., 1., 1., 1.])
```

2.5.2 Backward for Non-Scalar Variables

When y is a vector, the most natural representation of the derivative of y with respect to a vector x is a matrix called the *Jacobian* that contains the partial derivatives of each component of y with respect to each component of x. Likewise, for higher-order y and x, the result of differentiation could be an even higher-order tensor.

While Jacobians do show up in some advanced machine learning techniques, more commonly we want to sum up the gradients of each component of y with respect to the full vector x, yielding a vector of the same shape as x. For example, we often have a vector representing the value of our loss function calculated separately for each example among a *batch* of training examples. Here, we just want to sum up the gradients computed individually for each example.

Because deep learning frameworks vary in how they interpret gradients of non-scalar tensors, PyTorch takes some steps to avoid confusion. Invoking backward on a non-scalar elicits an error unless we tell PyTorch how to reduce the object to a scalar. More formally, we need to provide some vector $\mathbf{v}$ such that backward will compute $\mathbf{v}^\top \partial_{\mathbf{x}} \mathbf{y}$ rather than $\partial_{\mathbf{x}} \mathbf{y}$. This next part may be confusing, but for reasons that will become clear later, this argument (representing $\mathbf{v}$) is named gradient. For a more detailed description, see Yang Zhang's Medium post[57].

57

```
x.grad.zero_()
y = x * x
y.backward(gradient=torch.ones(len(y)))   # Faster: y.sum().backward()
x.grad
```

```
tensor([0., 2., 4., 6.])
```

2.5.3 Detaching Computation

Sometimes, we wish to move some calculations outside of the recorded computational graph. For example, say that we use the input to create some auxiliary intermediate terms for which we do not want to compute a gradient. In this case, we need to *detach* the respective computational graph from the final result. The following toy example makes this clearer: suppose we have z = x * y and y = x * x but we want to focus on the *direct* influence of x on z rather than the influence conveyed via y. In this case, we can create a new variable u that takes the same value as y but whose *provenance* (how it was created) has been wiped out. Thus u has no ancestors in the graph and gradients do not flow through u to x. For example, taking the gradient of z = x * u will yield the result u, (not 3 * x * x as you might have expected since z = x * x * x).

```
x.grad.zero_()
y = x * x
u = y.detach()
z = u * x

z.sum().backward()
x.grad == u
```

```
tensor([True, True, True, True])
```

Note that while this procedure detaches y's ancestors from the graph leading to z, the computational graph leading to y persists and thus we can calculate the gradient of y with respect to x.

```
x.grad.zero_()
y.sum().backward()
x.grad == 2 * x
```

```
tensor([True, True, True, True])
```

2.5.4 Gradients and Python Control Flow

So far we reviewed cases where the path from input to output was well defined via a function such as z = x * x * x. Programming offers us a lot more freedom in how we compute results. For instance, we can make them depend on auxiliary variables or condition choices on intermediate results. One benefit of using automatic differentiation is that even if building the computational graph of a function required passing through a maze of Python control flow (e.g., conditionals, loops, and arbitrary function calls), we can still calculate the gradient of the resulting variable. To illustrate this, consider the following code snippet where the number of iterations of the while loop and the evaluation of the if statement both depend on the value of the input a.

```
def f(a):
    b = a * 2
    while b.norm() < 1000:
        b = b * 2
    if b.sum() > 0:
        c = b
    else:
        c = 100 * b
    return c
```

Below, we call this function, passing in a random value, as input. Since the input is a random variable, we do not know what form the computational graph will take. However, whenever we execute f(a) on a specific input, we realize a specific computational graph and can subsequently run backward.

```
a = torch.randn(size=(), requires_grad=True)
d = f(a)
d.backward()
```

Even though our function f is, for demonstration purposes, a bit contrived, its dependence on the input is quite simple: it is a *linear* function of a with piecewise defined scale. As such, f(a) / a is a vector of constant entries and, moreover, f(a) / a needs to match the gradient of f(a) with respect to a.

```
a.grad == d / a
```

```
tensor(True)
```

Dynamic control flow is very common in deep learning. For instance, when processing text, the computational graph depends on the length of the input. In these cases, automatic differentiation becomes vital for statistical modeling since it is impossible to compute the gradient *a priori*.

2.5.5 Discussion

You have now gotten a taste of the power of automatic differentiation. The development of libraries for calculating derivatives both automatically and efficiently has been a massive productivity booster for deep learning practitioners, liberating them so they can focus on less menial. Moreover, autograd lets us design massive models for which pen and paper gradient computations would be prohibitively time consuming. Interestingly, while we use autograd to *optimize* models (in a statistical sense) the *optimization* of autograd libraries themselves (in a computational sense) is a rich subject of vital interest to framework designers. Here, tools from compilers and graph manipulation are leveraged to compute results in the most expedient and memory-efficient manner.

For now, try to remember these basics: (i) attach gradients to those variables with respect to which we desire derivatives; (ii) record the computation of the target value; (iii) execute the backpropagation function; and (iv) access the resulting gradient.

2.5.6 Exercises

1. Why is the second derivative much more expensive to compute than the first derivative?

2. After running the function for backpropagation, immediately run it again and see what happens. Investigate.

3. In the control flow example where we calculate the derivative of d with respect to a,

what would happen if we changed the variable a to a random vector or a matrix? At this point, the result of the calculation f(a) is no longer a scalar. What happens to the result? How do we analyze this?

4. Let $f(x) = \sin(x)$. Plot the graph of f and of its derivative f'. Do not exploit the fact that $f'(x) = \cos(x)$ but rather use automatic differentiation to get the result.

5. Let $f(x) = ((\log x^2) \cdot \sin x) + x^{-1}$. Write out a dependency graph tracing results from x to $f(x)$.

6. Use the chain rule to compute the derivative $\frac{df}{dx}$ of the aforementioned function, placing each term on the dependency graph that you constructed previously.

7. Given the graph and the intermediate derivative results, you have a number of options when computing the gradient. Evaluate the result once starting from x to f and once from f tracing back to x. The path from x to f is commonly known as *forward differentiation*, whereas the path from f to x is known as backward differentiation.

8. When might you want to use forward, and when backward, differentiation? Hint: consider the amount of intermediate data needed, the ability to parallelize steps, and the size of matrices and vectors involved.

 Discussions[58].

2.6 Probability and Statistics

One way or another, machine learning is all about uncertainty. In supervised learning, we want to predict something unknown (the *target*) given something known (the *features*). Depending on our objective, we might attempt to predict the most likely value of the target. Or we might predict the value with the smallest expected distance from the target. And sometimes we wish not only to predict a specific value but to *quantify our uncertainty*. For example, given some features describing a patient, we might want to know *how likely* they are to suffer a heart attack in the next year. In unsupervised learning, we often care about uncertainty. To determine whether a set of measurements are anomalous, it helps to know how likely one is to observe values in a population of interest. Furthermore, in reinforcement learning, we wish to develop agents that act intelligently in various environments. This requires reasoning about how an environment might be expected to change and what rewards one might expect to encounter in response to each of the available actions.

Probability is the mathematical field concerned with reasoning under uncertainty. Given a probabilistic model of some process, we can reason about the likelihood of various events. The use of probabilities to describe the frequencies of repeatable events (like coin tosses) is fairly uncontroversial. In fact, *frequentist* scholars adhere to an interpretation of probability that applies *only* to such repeatable events. By contrast *Bayesian* scholars use the language of probability more broadly to formalize reasoning under uncertainty. Bayesian probability is characterized by two unique features: (i) assigning degrees of belief to non-repeatable events, e.g., what is the *probability* that a dam will collapse?; and (ii) subjectivity. While Bayesian probability provides unambiguous rules for how one should update their beliefs in light of new evidence, it allows for different individuals to start off with different *prior* beliefs. *Statistics* helps us to reason backwards, starting off with collection and organization of data and backing out to what inferences we might draw about the process that generated the data. Whenever we analyze a dataset, hunting for patterns that we hope might characterize a broader population, we are employing statistical thinking. Many courses, majors, theses, careers, departments, companies, and institutions have been devoted to the study of probability and statistics. While this section only scratches the surface, we will provide the foundation that you need to begin building models.

```
%matplotlib inline
import random
import torch
from torch.distributions.multinomial import Multinomial
from d2l import torch as d2l
```

2.6.1 A Simple Example: Tossing Coins

Imagine that we plan to toss a coin and want to quantify how likely we are to see heads (vs. tails). If the coin is *fair*, then both outcomes (heads and tails), are equally likely. Moreover if we plan to toss the coin n times then the fraction of heads that we *expect* to see should exactly match the *expected* fraction of tails. One intuitive way to see this is by symmetry: for every possible outcome with n_h heads and $n_t = (n - n_h)$ tails, there is an equally likely outcome with n_t heads and n_h tails. Note that this is only possible if on average we expect to see $1/2$ of tosses come up heads and $1/2$ come up tails. Of course, if you conduct this experiment many times with $n = 1000000$ tosses each, you might never see a trial where $n_h = n_t$ exactly.

Formally, the quantity $1/2$ is called a *probability* and here it captures the certainty with which any given toss will come up heads. Probabilities assign scores between 0 and 1 to outcomes of interest, called *events*. Here the event of interest is heads and we denote the corresponding probability $P(\text{heads})$. A probability of 1 indicates absolute certainty (imagine a trick coin where both sides were heads) and a probability of 0 indicates impossibility (e.g., if both sides were tails). The frequencies n_h/n and n_t/n are not probabilities but rather *statistics*. Probabilities are *theoretical* quantities that underly the data generating process. Here, the probability $1/2$ is a property of the coin itself. By contrast, statistics are *empirical* quantities that are computed as functions of the observed data. Our interests in probabilistic and statistical quantities are inextricably intertwined. We often design special statistics called *estimators* that, given a dataset, produce *estimates* of model parameters such as probabilities. Moreover, when those estimators satisfy a nice property called *consistency*, our estimates will converge to the corresponding probability. In turn, these inferred probabilities tell about the likely statistical properties of data from the same population that we might encounter in the future.

Suppose that we stumbled upon a real coin for which we did not know the true $P(\text{heads})$.

To investigate this quantity with statistical methods, we need to (i) collect some data; and (ii) design an estimator. Data acquisition here is easy; we can toss the coin many times and record all the outcomes. Formally, drawing realizations from some underlying random process is called *sampling*. As you might have guessed, one natural estimator is the ratio of the number of observed *heads* to the total number of tosses.

Now, suppose that the coin was in fact fair, i.e., $P(\text{heads}) = 0.5$. To simulate tosses of a fair coin, we can invoke any random number generator. There are some easy ways to draw samples of an event with probability 0.5. For example Python's `random.random` yields numbers in the interval $[0, 1]$ where the probability of lying in any sub-interval $[a, b] \subset [0, 1]$ is equal to $b - a$. Thus we can get out 0 and 1 with probability 0.5 each by testing whether the returned float number is greater than 0.5:

```
num_tosses = 100
heads = sum([random.random() > 0.5 for _ in range(num_tosses)])
tails = num_tosses - heads
print("heads, tails: ", [heads, tails])
```

```
heads, tails:  [51, 49]
```

More generally, we can simulate multiple draws from any variable with a finite number of possible outcomes (like the toss of a coin or roll of a die) by calling the multinomial function, setting the first argument to the number of draws and the second as a list of probabilities associated with each of the possible outcomes. To simulate ten tosses of a fair coin, we assign probability vector [0.5, 0.5], interpreting index 0 as heads and index 1 as tails. The function returns a vector with length equal to the number of possible outcomes (here, 2), where the first component tells us the number of occurrences of heads and the second component tells us the number of occurrences of tails.

```
fair_probs = torch.tensor([0.5, 0.5])
Multinomial(100, fair_probs).sample()
```

```
tensor([51., 49.])
```

Each time you run this sampling process, you will receive a new random value that may differ from the previous outcome. Dividing by the number of tosses gives us the *frequency* of each outcome in our data. Note that these frequencies, just like the probabilities that they are intended to estimate, sum to 1.

```
Multinomial(100, fair_probs).sample() / 100
```

```
tensor([0.5500, 0.4500])
```

Here, even though our simulated coin is fair (we ourselves set the probabilities [0.5, 0. 5]), the counts of heads and tails may not be identical. That is because we only drew a relatively small number of samples. If we did not implement the simulation ourselves, and only saw the outcome, how would we know if the coin were slightly unfair or if the possible deviation from 1/2 was just an artifact of the small sample size? Let's see what happens when we simulate 10,000 tosses.

```
counts = Multinomial(10000, fair_probs).sample()
counts / 10000
```

```
tensor([0.4911, 0.5089])
```

In general, for averages of repeated events (like coin tosses), as the number of repetitions grows, our estimates are guaranteed to converge to the true underlying probabilities. The mathematical formulation of this phenomenon is called the *law of large numbers* and the *central limit theorem* tells us that in many situations, as the sample size n grows, these errors should go down at a rate of $(1/\sqrt{n})$. Let's get some more intuition by studying how our estimate evolves as we grow the number of tosses from 1 to 10,000.

```
counts = Multinomial(1, fair_probs).sample((10000,))
cum_counts = counts.cumsum(dim=0)
estimates = cum_counts / cum_counts.sum(dim=1, keepdims=True)
estimates = estimates.numpy()

d2l.set_figsize((4.5, 3.5))
d2l.plt.plot(estimates[:, 0], label=("P(coin=heads)"))
d2l.plt.plot(estimates[:, 1], label=("P(coin=tails)"))
d2l.plt.axhline(y=0.5, color='black', linestyle='dashed')
d2l.plt.gca().set_xlabel('Samples')
d2l.plt.gca().set_ylabel('Estimated probability')
d2l.plt.legend();
```

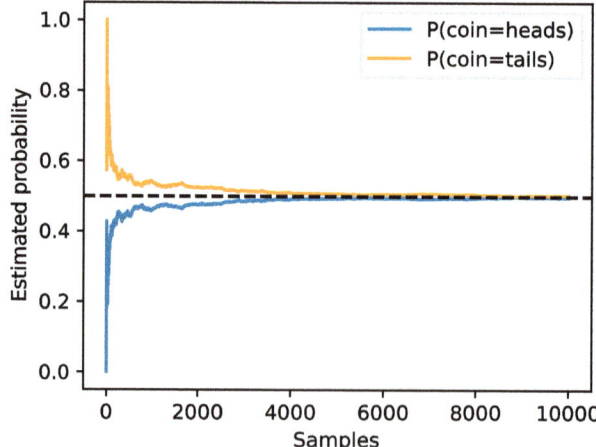

Each solid curve corresponds to one of the two values of the coin and gives our estimated probability that the coin turns up that value after each group of experiments. The dashed black line gives the true underlying probability. As we get more data by conducting more experiments, the curves converge towards the true probability. You might already begin to see the shape of some of the more advanced questions that preoccupy statisticians: How quickly does this convergence happen? If we had already tested many coins manufactured at the same plant, how might we incorporate this information?

2.6.2 A More Formal Treatment

We have already gotten pretty far: posing a probabilistic model, generating synthetic data, running a statistical estimator, empirically assessing convergence, and reporting error metrics (checking the deviation). However, to go much further, we will need to be more precise.

When dealing with randomness, we denote the set of possible outcomes S and call it the *sample space* or *outcome space*. Here, each element is a distinct possible *outcome*. In the case of rolling a single coin, $S = \{\text{heads}, \text{tails}\}$. For a single die, $S = \{1, 2, 3, 4, 5, 6\}$. When flipping two coins, the possible outcomes are $\{(\text{heads}, \text{heads}), (\text{heads}, \text{tails}), (\text{tails}, \text{heads}), (\text{tails}, \text{tails})\}$. *Events* are subsets of the sample space. For instance, the event "the first coin toss comes up heads" corresponds to the set $\{(\text{heads}, \text{heads}), (\text{heads}, \text{tails})\}$. Whenever the outcome z of a random experiment satisfies $z \in \mathcal{A}$, then event $\mathcal{A}$ has occurred. For a single roll of a die, we could define the events "seeing a 5" ($\mathcal{A} = \{5\}$) and "seeing an odd number" ($\mathcal{B} = \{1, 3, 5\}$). In this case, if the die came up 5, we would say that both $\mathcal{A}$ and $\mathcal{B}$ occurred. On the other hand, if $z = 3$, then $\mathcal{A}$ did not occur but $\mathcal{B}$ did.

A *probability* function maps events onto real values $P : \mathcal{A} \subseteq S \rightarrow [0, 1]$. The probability, denoted $P(\mathcal{A})$, of an event $\mathcal{A}$ in the given sample space S, has the following properties:

- The probability of any event $\mathcal{A}$ is a nonnegative real number, i.e., $P(\mathcal{A}) \geq 0$;

- The probability of the entire sample space is 1, i.e., $P(S) = 1$;

- For any countable sequence of events $\mathcal{A}_1, \mathcal{A}_2, \ldots$ that are *mutually exclusive* (i.e., $\mathcal{A}_i \cap \mathcal{A}_j = \emptyset$ for all $i \neq j$), the probability that any of them happens is equal to the sum of their individual probabilities, i.e., $P(\bigcup_{i=1}^{\infty} \mathcal{A}_i) = \sum_{i=1}^{\infty} P(\mathcal{A}_i)$.

These axioms of probability theory, proposed by Kolmogorov (1933), can be applied to rapidly derive a number of important consequences. For instance, it follows immediately that the probability of any event $\mathcal{A}$ *or* its complement $\mathcal{A}'$ occurring is 1 (because $\mathcal{A} \cup \mathcal{A}' = S$). We can also prove that $P(\emptyset) = 0$ because $1 = P(S \cup S') = P(S \cup \emptyset) = P(S) + P(\emptyset) = 1 + P(\emptyset)$. Consequently, the probability of any event $\mathcal{A}$ *and* its complement $\mathcal{A}'$ occurring simultaneously is $P(\mathcal{A} \cap \mathcal{A}') = 0$. Informally, this tells us that impossible events have zero probability of occurring.

2.6.3 Random Variables

When we spoke about events like the roll of a die coming up odds or the first coin toss coming up heads, we were invoking the idea of a *random variable*. Formally, random variables are mappings from an underlying sample space to a set of (possibly many) values. You might wonder how a random variable is different from the sample space, since both are collections of outcomes. Importantly, random variables can be much coarser than the raw sample space. We can define a binary random variable like "greater than 0.5" even when the underlying sample space is infinite, e.g., points on the line segment between 0 and 1. Additionally, multiple random variables can share the same underlying sample space. For example "whether my home alarm goes off" and "whether my house was burgled" are both binary random variables that share an underlying sample space. Consequently, knowing the value taken by one random variable can tell us something about the likely value of another random variable. Knowing that the alarm went off, we might suspect that the house was likely burgled.

Every value taken by a random variable corresponds to a subset of the underlying sample space. Thus the occurrence where the random variable X takes value v, denoted by $X = v$, is an *event* and $P(X = v)$ denotes its probability. Sometimes this notation can get clunky, and we can abuse notation when the context is clear. For example, we might use $P(X)$ to refer broadly to the *distribution* of X, i.e., the function that tells us the probability that X takes any given value. Other times we write expressions like $P(X, Y) = P(X)P(Y)$, as a shorthand to express a statement that is true for all of the values that the random variables X and Y can take, i.e., for all i, j it holds that $P(X = i \text{ and } Y = j) = P(X = i)P(Y = j)$. Other times, we abuse notation by writing $P(v)$ when the random variable is clear from the context. Since an event in probability theory is a set of outcomes from the sample space, we can specify a range of values for a random variable to take. For example, $P(1 \leq X \leq 3)$ denotes the probability of the event $\{1 \leq X \leq 3\}$.

Note that there is a subtle difference between *discrete* random variables, like flips of a coin or tosses of a die, and *continuous* ones, like the weight and the height of a person sampled at random from the population. In this case we seldom really care about someone's exact height. Moreover, if we took precise enough measurements, we would find that no two people on the planet have the exact same height. In fact, with fine enough measurements, you would never have the same height when you wake up and when you go to sleep. There is little point in asking about the exact probability that someone is 1.801392782910287192 meters tall. Instead, we typically care more about being able to say whether someone's height falls into a given interval, say between 1.79 and 1.81 meters. In these cases we work with probability *densities*. The height of exactly 1.80 meters has no probability, but nonzero density. To work out the probability assigned to an interval, we must take an *integral* of the density over that interval.

2.6.4 Multiple Random Variables

You might have noticed that we could not even make it through the previous section without making statements involving interactions among multiple random variables (recall that $P(X, Y) = P(X)P(Y)$). Most of machine learning is concerned with such relationships. Here, the sample space would be the population of interest, say customers who transact with a business, photographs on the Internet, or proteins known to biologists. Each random variable would represent the (unknown) value of a different attribute. Whenever we sample an individual from the population, we observe a realization of each of the random variables. Because the values taken by random variables correspond to subsets of the sample space that could be overlapping, partially overlapping, or entirely disjoint, knowing the value taken by one random variable can cause us to update our beliefs about which values of another random variable are likely. If a patient walks into a hospital and we observe that they are having trouble breathing and have lost their sense of smell, then we believe that they are more likely to have COVID-19 than we might if they had no trouble breathing and a perfectly ordinary sense of smell.

When working with multiple random variables, we can construct events corresponding to every combination of values that the variables can jointly take. The probability function that assigns probabilities to each of these combinations (e.g. $A = a$ and $B = b$) is called the *joint probability* function and simply returns the probability assigned to the intersection of the corresponding subsets of the sample space. The *joint probability* assigned to the event where random variables A and B take values a and b, respectively, is denoted $P(A = a, B = b)$, where the comma indicates "and". Note that for any values a and b, it follows that

$$P(A = a, B = b) \le P(A = a) \text{ and } P(A = a, B = b) \le P(B = b), \qquad (2.6.1)$$

since for $A = a$ and $B = b$ to happen, $A = a$ has to happen *and* $B = b$ also has to happen. Interestingly, the joint probability tells us all that we can know about these random variables in a probabilistic sense, and can be used to derive many other useful quantities, including recovering the individual distributions $P(A)$ and $P(B)$. To recover $P(A = a)$ we simply sum up $P(A = a, B = v)$ over all values v that the random variable B can take: $P(A = a) = \sum_v P(A = a, B = v)$.

The ratio $\frac{P(A=a,B=b)}{P(A=a)} \le 1$ turns out to be extremely important. It is called the *conditional probability*, and is denoted via the "|" symbol:

$$P(B = b \mid A = a) = P(A = a, B = b)/P(A = a). \qquad (2.6.2)$$

It tells us the new probability associated with the event $B = b$, once we condition on the fact $A = a$ took place. We can think of this conditional probability as restricting attention only to the subset of the sample space associated with $A = a$ and then renormalizing so that all probabilities sum to 1. Conditional probabilities are in fact just ordinary probabilities and thus respect all of the axioms, as long as we condition all terms on the same event and thus restrict attention to the same sample space. For instance, for disjoint events $\mathcal{B}$ and $\mathcal{B}'$, we have that $P(\mathcal{B} \cup \mathcal{B}' \mid A = a) = P(\mathcal{B} \mid A = a) + P(\mathcal{B}' \mid A = a)$.

Using the definition of conditional probabilities, we can derive the famous result called *Bayes' theorem*. By construction, we have that $P(A, B) = P(B \mid A)P(A)$ and $P(A, B) = P(A \mid B)P(B)$. Combining both equations yields $P(B \mid A)P(A) = P(A \mid B)P(B)$ and hence

$$P(A \mid B) = \frac{P(B \mid A)P(A)}{P(B)}. \qquad (2.6.3)$$

This simple equation has profound implications because it allows us to reverse the order of conditioning. If we know how to estimate $P(B \mid A)$, $P(A)$, and $P(B)$, then we can estimate

$P(A \mid B)$. We often find it easier to estimate one term directly but not the other and Bayes' theorem can come to the rescue here. For instance, if we know the prevalence of symptoms for a given disease, and the overall prevalences of the disease and symptoms, respectively, we can determine how likely someone is to have the disease based on their symptoms. In some cases we might not have direct access to $P(B)$, such as the prevalence of symptoms. In this case a simplified version of Bayes' theorem comes in handy:

$$P(A \mid B) \propto P(B \mid A)P(A). \tag{2.6.4}$$

Since we know that $P(A \mid B)$ must be normalized to 1, i.e., $\sum_a P(A = a \mid B) = 1$, we can use it to compute

$$P(A \mid B) = \frac{P(B \mid A)P(A)}{\sum_a P(B \mid A = a)P(A = a)}. \tag{2.6.5}$$

In Bayesian statistics, we think of an observer as possessing some (subjective) prior beliefs about the plausibility of the available hypotheses encoded in the *prior* $P(H)$, and a *likelihood function* that says how likely one is to observe any value of the collected evidence for each of the hypotheses in the class $P(E \mid H)$. Bayes' theorem is then interpreted as telling us how to update the initial *prior* $P(H)$ in light of the available evidence E to produce *posterior* beliefs $P(H \mid E) = \frac{P(E|H)P(H)}{P(E)}$. Informally, this can be stated as "posterior equals prior times likelihood, divided by the evidence". Now, because the evidence $P(E)$ is the same for all hypotheses, we can get away with simply normalizing over the hypotheses.

Note that $\sum_a P(A = a \mid B) = 1$ also allows us to *marginalize* over random variables. That is, we can drop variables from a joint distribution such as $P(A, B)$. After all, we have that

$$\sum_a P(B \mid A = a)P(A = a) = \sum_a P(B, A = a) = P(B). \tag{2.6.6}$$

Independence is another fundamentally important concept that forms the backbone of many important ideas in statistics. In short, two variables are *independent* if conditioning on the value of A does not cause any change to the probability distribution associated with B and vice versa. More formally, independence, denoted $A \perp B$, requires that $P(A \mid B) = P(A)$ and, consequently, that $P(A, B) = P(A \mid B)P(B) = P(A)P(B)$. Independence is often an appropriate assumption. For example, if the random variable A represents the outcome from tossing one fair coin and the random variable B represents the outcome from tossing another, then knowing whether A came up heads should not influence the probability of B coming up heads.

Independence is especially useful when it holds among the successive draws of our data from some underlying distribution (allowing us to make strong statistical conclusions) or when it holds among various variables in our data, allowing us to work with simpler models that encode this independence structure. On the other hand, estimating the dependencies among random variables is often the very aim of learning. We care to estimate the probability of disease given symptoms specifically because we believe that diseases and symptoms are *not* independent.

Note that because conditional probabilities are proper probabilities, the concepts of independence and dependence also apply to them. Two random variables A and B are *conditionally independent* given a third variable C if and only if $P(A, B \mid C) = P(A \mid C)P(B \mid C)$. Interestingly, two variables can be independent in general but become dependent when conditioning on a third. This often occurs when the two random variables A and B correspond to causes of some third variable C. For example, broken bones and lung cancer might be independent in the general population but if we condition on being in the hospital then we might find that broken bones are negatively correlated with lung cancer. That is because the broken bone *explains away* why some person is in the hospital and thus lowers the probability that they are hospitalized because of having lung cancer.

And conversely, two dependent random variables can become independent upon conditioning on a third. This often happens when two otherwise unrelated events have a common cause. Shoe size and reading level are highly correlated among elementary school students, but this correlation disappears if we condition on age.

2.6.5 An Example

Let's put our skills to the test. Assume that a doctor administers an HIV test to a patient. This test is fairly accurate and fails only with 1% probability if the patient is healthy but reported as diseased, i.e., healthy patients test positive in 1% of cases. Moreover, it never fails to detect HIV if the patient actually has it. We use $D_1 \in \{0, 1\}$ to indicate the diagnosis (0 if negative and 1 if positive) and $H \in \{0, 1\}$ to denote the HIV status.

Conditional probability	$H = 1$	$H = 0$
$P(D_1 = 1 \mid H)$	1	0.01
$P(D_1 = 0 \mid H)$	0	0.99

Note that the column sums are all 1 (but the row sums do not), since they are conditional probabilities. Let's compute the probability of the patient having HIV if the test comes back positive, i.e., $P(H = 1 \mid D_1 = 1)$. Intuitively this is going to depend on how common the disease is, since it affects the number of false alarms. Assume that the population is fairly free of the disease, e.g., $P(H = 1) = 0.0015$. To apply Bayes' theorem, we need to apply marginalization to determine

$$
\begin{aligned}
P(D_1 = 1) &= P(D_1 = 1, H = 0) + P(D_1 = 1, H = 1) \\
&= P(D_1 = 1 \mid H = 0)P(H = 0) + P(D_1 = 1 \mid H = 1)P(H = 1) \qquad (2.6.7) \\
&= 0.011485.
\end{aligned}
$$

This leads us to

$$
P(H = 1 \mid D_1 = 1) = \frac{P(D_1 = 1 \mid H = 1)P(H = 1)}{P(D_1 = 1)} = 0.1306. \qquad (2.6.8)
$$

In other words, there is only a 13.06% chance that the patient actually has HIV, despite the test being pretty accurate. As we can see, probability can be counterintuitive. What should a patient do upon receiving such terrifying news? Likely, the patient would ask the physician to administer another test to get clarity. The second test has different characteristics and it is not as good as the first one.

Conditional probability	$H = 1$	$H = 0$
$P(D_2 = 1 \mid H)$	0.98	0.03
$P(D_2 = 0 \mid H)$	0.02	0.97

Unfortunately, the second test comes back positive, too. Let's calculate the requisite probabilities to invoke Bayes' theorem by assuming conditional independence:

$$P(D_1 = 1, D_2 = 1 \mid H = 0) = P(D_1 = 1 \mid H = 0)P(D_2 = 1 \mid H = 0) = \quad 0.0003,$$
$$P(D_1 = 1, D_2 = 1 \mid H = 1) = P(D_1 = 1 \mid H = 1)P(D_2 = 1 \mid H = 1) = \quad 0.98.$$

$$(2.6.9)$$

Now we can apply marginalization to obtain the probability that both tests come back positive:

$$P(D_1 = 1, D_2 = 1)$$
$$= P(D_1 = 1, D_2 = 1, H = 0) + P(D_1 = 1, D_2 = 1, H = 1)$$
$$= P(D_1 = 1, D_2 = 1 \mid H = 0)P(H = 0) + P(D_1 = 1, D_2 = 1 \mid H = 1)P(H = 1)$$
$$= 0.00176955.$$

$$(2.6.10)$$

Finally, the probability of the patient having HIV given that both tests are positive is

$$P(H = 1 \mid D_1 = 1, D_2 = 1) = \frac{P(D_1 = 1, D_2 = 1 \mid H = 1)P(H = 1)}{P(D_1 = 1, D_2 = 1)} = 0.8307. \quad (2.6.11)$$

That is, the second test allowed us to gain much higher confidence that not all is well. Despite the second test being considerably less accurate than the first one, it still significantly improved our estimate. The assumption of both tests being conditionally independent of each other was crucial for our ability to generate a more accurate estimate. Take the extreme case where we run the same test twice. In this situation we would expect the same outcome both times, hence no additional insight is gained from running the same test again. The astute reader might have noticed that the diagnosis behaved like a classifier hiding in plain sight where our ability to decide whether a patient is healthy increases as we obtain more features (test outcomes).

2.6.6 Expectations

Often, making decisions requires not just looking at the probabilities assigned to individual events but composing them together into useful aggregates that can provide us with guidance. For example, when random variables take continuous scalar values, we often care about knowing what value to expect *on average*. This quantity is formally called an *expectation*. If we are making investments, the first quantity of interest might be the return we can expect, averaging over all the possible outcomes (and weighting by the appropriate probabilities). For instance, say that with 50% probability, an investment might fail altogether, with 40% probability it might provide a 2× return, and with 10% probability it might provide a 10× return 10×. To calculate the expected return, we sum over all returns, multiplying each by the probability that they will occur. This yields the expectation $0.5 \cdot 0 + 0.4 \cdot 2 + 0.1 \cdot 10 = 1.8$. Hence the expected return is 1.8×.

In general, the *expectation* (or average) of the random variable X is defined as

$$E[X] = E_{x \sim P}[x] = \sum_x xP(X = x). \qquad (2.6.12)$$

Likewise, for densities we obtain $E[X] = \int x \, dp(x)$. Sometimes we are interested in the expected value of some function of x. We can calculate these expectations as

$$E_{x \sim P}[f(x)] = \sum_x f(x)P(x) \text{ and } E_{x \sim P}[f(x)] = \int f(x)p(x) \, dx \qquad (2.6.13)$$

for discrete probabilities and densities, respectively. Returning to the investment example from above, f might be the *utility* (happiness) associated with the return. Behavior economists have long noted that people associate greater disutility with losing money than the utility gained from earning one dollar relative to their baseline. Moreover, the value of money tends to be sub-linear. Possessing 100k dollars versus zero dollars can make the difference between paying the rent, eating well, and enjoying quality healthcare versus suffering through homelessness. On the other hand, the gains due to possessing 200k versus 100k are less dramatic. Reasoning like this motivates the cliché that "the utility of money is logarithmic".

If the utility associated with a total loss were -1, and the utilities associated with returns of 1, 2, and 10 were 1, 2 and 4, respectively, then the expected happiness of investing would be $0.5 \cdot (-1) + 0.4 \cdot 2 + 0.1 \cdot 4 = 0.7$ (an expected loss of utility of 30%). If indeed this were your utility function, you might be best off keeping the money in the bank.

For financial decisions, we might also want to measure how *risky* an investment is. Here, we care not just about the expected value but how much the actual values tend to *vary* relative to this value. Note that we cannot just take the expectation of the difference between the actual and expected values. This is because the expectation of a difference is the difference of the expectations, i.e., $E[X - E[X]] = E[X] - E[E[X]] = 0$. However, we can look at the expectation of any non-negative function of this difference. The *variance* of a random variable is calculated by looking at the expected value of the *squared* differences:

$$\text{Var}[X] = E\left[(X - E[X])^2\right] = E[X^2] - E[X]^2. \qquad (2.6.14)$$

Here the equality follows by expanding $(X - E[X])^2 = X^2 - 2XE[X] + E[X]^2$ and taking expectations for each term. The square root of the variance is another useful quantity called the *standard deviation*. While this and the variance convey the same information (either can be calculated from the other), the standard deviation has the nice property that it is expressed in the same units as the original quantity represented by the random variable.

Lastly, the variance of a function of a random variable is defined analogously as

$$\text{Var}_{x \sim P}[f(x)] = E_{x \sim P}[f^2(x)] - E_{x \sim P}[f(x)]^2. \qquad (2.6.15)$$

Returning to our investment example, we can now compute the variance of the investment. It is given by $0.5 \cdot 0 + 0.4 \cdot 2^2 + 0.1 \cdot 10^2 - 1.8^2 = 8.36$. For all intents and purposes this is a risky investment. Note that by mathematical convention mean and variance are often referenced as μ and σ^2. This is particularly the case whenever we use it to parametrize a Gaussian distribution.

In the same way as we introduced expectations and variance for *scalar* random variables, we can do so for vector-valued ones. Expectations are easy, since we can apply them elementwise. For instance, $\boldsymbol{\mu} \stackrel{\text{def}}{=} E_{\mathbf{x} \sim P}[\mathbf{x}]$ has coordinates $\mu_i = E_{\mathbf{x} \sim P}[x_i]$. *Covariances* are more complicated. We define them by taking expectations of the *outer product* of the difference between random variables and their mean:

$$\boldsymbol{\Sigma} \stackrel{\text{def}}{=} \text{Cov}_{\mathbf{x} \sim P}[\mathbf{x}] = E_{\mathbf{x} \sim P}\left[(\mathbf{x} - \boldsymbol{\mu})(\mathbf{x} - \boldsymbol{\mu})^\top\right]. \tag{2.6.16}$$

This matrix $\boldsymbol{\Sigma}$ is referred to as the covariance matrix. An easy way to see its effect is to consider some vector $\mathbf{v}$ of the same size as $\mathbf{x}$. It follows that

$$\mathbf{v}^\top \boldsymbol{\Sigma} \mathbf{v} = E_{\mathbf{x} \sim P}\left[\mathbf{v}^\top(\mathbf{x} - \boldsymbol{\mu})(\mathbf{x} - \boldsymbol{\mu})^\top \mathbf{v}\right] = \text{Var}_{x \sim P}[\mathbf{v}^\top \mathbf{x}]. \tag{2.6.17}$$

As such, $\boldsymbol{\Sigma}$ allows us to compute the variance for any linear function of $\mathbf{x}$ by a simple matrix multiplication. The off-diagonal elements tell us how correlated the coordinates are: a value of 0 means no correlation, where a larger positive value means that they are more strongly correlated.

2.6.7 Discussion

In machine learning, there are many things to be uncertain about! We can be uncertain about the value of a label given an input. We can be uncertain about the estimated value of a parameter. We can even be uncertain about whether data arriving at deployment is even from the same distribution as the training data.

By *aleatoric uncertainty*, we mean uncertainty that is intrinsic to the problem, and due to genuine randomness unaccounted for by the observed variables. By *epistemic uncertainty*, we mean uncertainty over a model's parameters, the sort of uncertainty that we can hope to reduce by collecting more data. We might have epistemic uncertainty concerning the probability that a coin turns up heads, but even once we know this probability, we are left with aleatoric uncertainty about the outcome of any future toss. No matter how long we watch someone tossing a fair coin, we will never be more or less than 50% certain that the next toss will come up heads. These terms come from mechanical modeling, (see e.g., Der Kiureghian and Ditlevsen (2009) for a review on this aspect of uncertainty quantification[59]). It is worth noting, however, that these terms constitute a slight abuse of language. The term *epistemic* refers to anything concerning *knowledge* and thus, in the philosophical sense, all uncertainty is epistemic.

59

We saw that sampling data from some unknown probability distribution can provide us with information that can be used to estimate the parameters of the data generating distribution. That said, the rate at which this is possible can be quite slow. In our coin tossing example (and many others) we can do no better than to design estimators that converge at a rate of

$1/\sqrt{n}$, where n is the sample size (e.g., the number of tosses). This means that by going from 10 to 1000 observations (usually a very achievable task) we see a tenfold reduction of uncertainty, whereas the next 1000 observations help comparatively little, offering only a 1.41 times reduction. This is a persistent feature of machine learning: while there are often easy gains, it takes a very large amount of data, and often with it an enormous amount of computation, to make further gains. For an empirical review of this fact for large scale language models see Revels *et al.* (2016).

We also sharpened our language and tools for statistical modeling. In the process of that we learned about conditional probabilities and about one of the most important equations in statistics—Bayes' theorem. It is an effective tool for decoupling information conveyed by data through a likelihood term $P(B \mid A)$ that addresses how well observations B match a choice of parameters A, and a prior probability $P(A)$ which governs how plausible a particular choice of A was in the first place. In particular, we saw how this rule can be applied to assign probabilities to diagnoses, based on the efficacy of the test *and* the prevalence of the disease itself (i.e., our prior).

Lastly, we introduced a first set of nontrivial questions about the effect of a specific probability distribution, namely expectations and variances. While there are many more than just linear and quadratic expectations for a probability distribution, these two already provide a good deal of knowledge about the possible behavior of the distribution. For instance, Chebyshev's inequality [60] states that $P(|X - \mu| \geq k\sigma) \leq 1/k^2$, where μ is the expectation, σ^2 is the variance of the distribution, and $k > 1$ is a confidence parameter of our choosing. It tells us that draws from a distribution lie with at least 50% probability within a $[-\sqrt{2}\sigma, \sqrt{2}\sigma]$ interval centered on the expectation.

60

2.6.8 Exercises

1. Give an example where observing more data can reduce the amount of uncertainty about the outcome to an arbitrarily low level.

2. Give an example where observing more data will only reduce the amount of uncertainty up to a point and then no further. Explain why this is the case and where you expect this point to occur.

3. We empirically demonstrated convergence to the mean for the toss of a coin. Calculate the variance of the estimate of the probability that we see a head after drawing n samples.

    1. How does the variance scale with the number of observations?

    2. Use Chebyshev's inequality to bound the deviation from the expectation.

    3. How does it relate to the central limit theorem?

4. Assume that we draw m samples x_i from a probability distribution with zero mean and unit variance. Compute the averages $z_m \stackrel{\text{def}}{=} m^{-1} \sum_{i=1}^{m} x_i$. Can we apply Chebyshev's inequality for every z_m independently? Why not?

5. Given two events with probability $P(\mathcal{A})$ and $P(\mathcal{B})$, compute upper and lower bounds on $P(\mathcal{A} \cup \mathcal{B})$ and $P(\mathcal{A} \cap \mathcal{B})$. Hint: graph the situation using a Venn diagram[61].

61

6. Assume that we have a sequence of random variables, say A, B, and C, where B only depends on A, and C only depends on B, can you simplify the joint probability $P(A, B, C)$? Hint: this is a Markov chain[62].

62

7. In Section 2.6.5, assume that the outcomes of the two tests are not independent. In particular assume that either test on its own has a false positive rate of 10% and a false negative rate of 1%. That is, assume that $P(D = 1 \mid H = 0) = 0.1$ and that $P(D = 0 \mid H = 1) = 0.01$. Moreover, assume that for $H = 1$ (infected) the test outcomes are conditionally independent, i.e., that $P(D_1, D_2 \mid H = 1) = P(D_1 \mid H = 1)P(D_2 \mid H = 1)$ but that for healthy patients the outcomes are coupled via $P(D_1 = D_2 = 1 \mid H = 0) = 0.02$.

   1. Work out the joint probability table for D_1 and D_2, given $H = 0$ based on the information you have so far.

   2. Derive the probability that the patient is diseased ($H = 1$) after one test returns positive. You can assume the same baseline probability $P(H = 1) = 0.0015$ as before.

   3. Derive the probability that the patient is diseased ($H = 1$) after both tests return positive.

8. Assume that you are an asset manager for an investment bank and you have a choice of stocks s_i to invest in. Your portfolio needs to add up to 1 with weights α_i for each stock. The stocks have an average return $\boldsymbol{\mu} = E_{\mathbf{s} \sim P}[\mathbf{s}]$ and covariance $\boldsymbol{\Sigma} = \text{Cov}_{\mathbf{s} \sim P}[\mathbf{s}]$.

   1. Compute the expected return for a given portfolio $\boldsymbol{\alpha}$.

   2. If you wanted to maximize the return of the portfolio, how should you choose your investment?

   3. Compute the *variance* of the portfolio.

   4. Formulate an optimization problem of maximizing the return while keeping the variance constrained to an upper bound. This is the Nobel-Prize winning Markovitz portfolio[63] (Mangram, 2013). To solve it you will need a quadratic programming solver, something way beyond the scope of this book.

63

Discussions[64].

64

65

2.7 Documentation

66

While we cannot possibly introduce every single PyTorch function and class (and the information might become outdated quickly), the API documentation[65] and extra tutorials[66] and examples provide such documentation. This section provides some guidance for how to explore the PyTorch API.

```
import torch
```

2.7.1 Functions and Classes in a Module

To know which functions and classes can be called in a module, we invoke the `dir` function. For instance, we can query all properties in the module for generating random numbers:

```
print(dir(torch.distributions))
```

```
['AbsTransform', 'AffineTransform', 'Bernoulli', 'Beta', 'Binomial',
↪'CatTransform', 'Categorical', 'Cauchy', 'Chi2', 'ComposeTransform',
↪'ContinuousBernoulli', 'CorrCholeskyTransform',
↪'CumulativeDistributionTransform', 'Dirichlet', 'Distribution', 'ExpTransform
↪', 'Exponential', 'ExponentialFamily', 'FisherSnedecor', 'Gamma', 'Geometric
↪', 'Gumbel', 'HalfCauchy', 'HalfNormal', 'Independent', 'IndependentTransform
↪', 'Kumaraswamy', 'LKJCholesky', 'Laplace', 'LogNormal', 'LogisticNormal',
↪'LowRankMultivariateNormal', 'LowerCholeskyTransform', 'MixtureSameFamily',
↪'Multinomial', 'MultivariateNormal', 'NegativeBinomial', 'Normal',
↪'OneHotCategorical', 'OneHotCategoricalStraightThrough', 'Pareto', 'Poisson',
↪ 'PositiveDefiniteTransform', 'PowerTransform', 'RelaxedBernoulli',
↪'RelaxedOneHotCategorical', 'ReshapeTransform', 'SigmoidTransform',
↪'SoftmaxTransform', 'SoftplusTransform', 'StackTransform',
↪'StickBreakingTransform', 'StudentT', 'TanhTransform', 'Transform',
↪'TransformedDistribution', 'Uniform', 'VonMises', 'Weibull', 'Wishart', '__
↪all__', '__builtins__', '__cached__', '__doc__', '__file__', '__loader__', '_
↪_name__', '__package__', '__path__', '__spec__', 'bernoulli', 'beta',
↪'biject_to', 'binomial', 'categorical', 'cauchy', 'chi2', 'constraint_
↪registry', 'constraints', 'continuous_bernoulli', 'dirichlet', 'distribution
↪', 'exp_family', 'exponential', 'fishersnedecor', 'gamma', 'geometric',
↪'gumbel', 'half_cauchy', 'half_normal', 'identity_transform', 'independent',
↪'kl', 'kl_divergence', 'kumaraswamy', 'laplace', 'lkj_cholesky', 'log_normal
↪', 'logistic_normal', 'lowrank_multivariate_normal', 'mixture_same_family',
↪'multinomial', 'multivariate_normal', 'negative_binomial', 'normal', 'one_
↪hot_categorical', 'pareto', 'poisson', 'register_kl', 'relaxed_bernoulli',
↪'relaxed_categorical', 'studentT', 'transform_to', 'transformed_distribution
↪', 'transforms', 'uniform', 'utils', 'von_mises', 'weibull', 'wishart']
```

Generally, we can ignore functions that start and end with __ (special objects in Python) or functions that start with a single _(usually internal functions). Based on the remaining function or attribute names, we might hazard a guess that this module offers various methods for generating random numbers, including sampling from the uniform distribution (`uniform`), normal distribution (`normal`), and multinomial distribution (`multinomial`).

2.7.2 Specific Functions and Classes

For specific instructions on how to use a given function or class, we can invoke the `help` function. As an example, let's explore the usage instructions for tensors' ones function.

```
help(torch.ones)
```

```
Help on built-in function ones in module torch:

ones(...)
    ones(*size, *, out=None, dtype=None, layout=torch.strided,
     device=None, requires_grad=False) -> Tensor

    Returns a tensor filled with the scalar value 1, with the shape

    defined by the variable argument size.

    Args:
        size (int...): a sequence of integers defining the shape of
        the output tensor.
            Can be a variable number of arguments or a collection like
            a list or tuple.

    Keyword arguments:
        out (Tensor, optional): the output tensor.
        dtype (torch.dtype, optional): the desired data type of
        returned tensor.
            Default: if None, uses a global default
              (see torch.set_default_tensor_type()).
        layout (torch.layout, optional): the desired layout of
        returned Tensor.
            Default: torch.strided.
        device (torch.device, optional): the desired device of returned
        tensor.
            Default: if None, uses the current device for the default
        tensor type
            (see torch.set_default_tensor_type()). device will be the
            CPU for CPU tensor types and the current CUDA device for
            CUDA tensor types.
        requires_grad (bool, optional): If autograd should record
        operations on the returned tensor. Default: False.

    Example::

        >>> torch.ones(2, 3)
        tensor([[ 1.,   1.,   1.],
                [ 1.,   1.,   1.]])

        >>> torch.ones(5)
```

```
tensor([ 1.,   1.,   1.,   1.,   1.])
```

From the documentation, we can see that the ones function creates a new tensor with the specified shape and sets all the elements to the value of 1. Whenever possible, you should run a quick test to confirm your interpretation:

```
torch.ones(4)
```

```
tensor([1., 1., 1., 1.])
```

In the Jupyter notebook, we can use ? to display the document in another window. For example, list? will create content that is almost identical to help(list), displaying it in a new browser window. In addition, if we use two question marks, such as list??, the Python code implementing the function will also be displayed.

The official documentation provides plenty of descriptions and examples that are beyond this book. We emphasize important use cases that will get you started quickly with practical problems, rather than completeness of coverage. We also encourage you to study the source code of the libraries to see examples of high-quality implementations of production code. By doing this you will become a better engineer in addition to becoming a better scientist.

Discussions[67].

67

3 Linear Neural Networks for Regression

Before we worry about making our neural networks deep, it will be helpful to implement some shallow ones, for which the inputs connect directly to the outputs. This will prove important for a few reasons. First, rather than getting distracted by complicated architectures, we can focus on the basics of neural network training, including parametrizing the output layer, handling data, specifying a loss function, and training the model. Second, this class of shallow networks happens to comprise the set of linear models, which subsumes many classical methods of statistical prediction, including linear and softmax regression. Understanding these classical tools is pivotal because they are widely used in many contexts and we will often need to use them as baselines when justifying the use of fancier architectures. This chapter will focus narrowly on linear regression and the next one will extend our modeling repertoire by developing linear neural networks for classification.

3.1 Linear Regression

Regression problems pop up whenever we want to predict a numerical value. Common examples include predicting prices (of homes, stocks, etc.), predicting the length of stay (for patients in the hospital), forecasting demand (for retail sales), among numerous others. Not every prediction problem is one of classical regression. Later on, we will introduce classification problems, where the goal is to predict membership among a set of categories.

As a running example, suppose that we wish to estimate the prices of houses (in dollars) based on their area (in square feet) and age (in years). To develop a model for predicting house prices, we need to get our hands on data, including the sales price, area, and age for each home. In the terminology of machine learning, the dataset is called a *training dataset* or *training set*, and each row (containing the data corresponding to one sale) is called an *example* (or *data point*, *instance*, *sample*). The thing we are trying to predict (price) is called a *label* (or *target*). The variables (age and area) upon which the predictions are based are called *features* (or *covariates*).

```
%matplotlib inline
import math
import time
import numpy as np
```

(continues on next page)

(continued from previous page)

```
import torch
from d2l import torch as d2l
```

3.1.1 Basics

Linear regression is both the simplest and most popular among the standard tools for tackling regression problems. Dating back to the dawn of the 19th century (Gauss, 1809, Legendre, 1805), linear regression flows from a few simple assumptions. First, we assume that the relationship between features $\mathbf{x}$ and target y is approximately linear, i.e., that the conditional mean $E[Y \mid X = \mathbf{x}]$ can be expressed as a weighted sum of the features $\mathbf{x}$. This setup allows that the target value may still deviate from its expected value on account of observation noise. Next, we can impose the assumption that any such noise is well behaved, following a Gaussian distribution. Typically, we will use n to denote the number of examples in our dataset. We use superscripts to enumerate samples and targets, and subscripts to index coordinates. More concretely, $\mathbf{x}^{(i)}$ denotes the ith sample and $x_j^{(i)}$ denotes its jth coordinate.

Model

At the heart of every solution is a model that describes how features can be transformed into an estimate of the target. The assumption of linearity means that the expected value of the target (price) can be expressed as a weighted sum of the features (area and age):

$$\text{price} = w_{\text{area}} \cdot \text{area} + w_{\text{age}} \cdot \text{age} + b. \tag{3.1.1}$$

Here w_{area} and w_{age} are called *weights*, and b is called a *bias* (or *offset* or *intercept*). The weights determine the influence of each feature on our prediction. The bias determines the value of the estimate when all features are zero. Even though we will never see any newly-built homes with precisely zero area, we still need the bias because it allows us to express all linear functions of our features (rather than restricting us to lines that pass through the origin). Strictly speaking, (3.1.1) is an *affine transformation* of input features, which is characterized by a *linear transformation* of features via a weighted sum, combined with a *translation* via the added bias. Given a dataset, our goal is to choose the weights $\mathbf{w}$ and the bias b that, on average, make our model's predictions fit the true prices observed in the data as closely as possible.

In disciplines where it is common to focus on datasets with just a few features, explicitly expressing models long-form, as in (3.1.1), is common. In machine learning, we usually work with high-dimensional datasets, where it is more convenient to employ compact linear algebra notation. When our inputs consist of d features, we can assign each an index (between 1 and d) and express our prediction $\hat{y}$ (in general the "hat" symbol denotes an estimate) as

$$\hat{y} = w_1 x_1 + \cdots + w_d x_d + b. \tag{3.1.2}$$

Collecting all features into a vector $\mathbf{x} \in \mathbb{R}^d$ and all weights into a vector $\mathbf{w} \in \mathbb{R}^d$, we can express our model compactly via the dot product between $\mathbf{w}$ and $\mathbf{x}$:

$$\hat{y} = \mathbf{w}^\top \mathbf{x} + b. \tag{3.1.3}$$

In (3.1.3), the vector $\mathbf{x}$ corresponds to the features of a single example. We will often find it convenient to refer to features of our entire dataset of n examples via the *design matrix* $\mathbf{X} \in \mathbb{R}^{n \times d}$. Here, $\mathbf{X}$ contains one row for every example and one column for every feature. For a collection of features $\mathbf{X}$, the predictions $\hat{\mathbf{y}} \in \mathbb{R}^n$ can be expressed via the matrix–vector product:

$$\hat{\mathbf{y}} = \mathbf{X}\mathbf{w} + b, \tag{3.1.4}$$

where broadcasting (Section 2.1.4) is applied during the summation. Given features of a training dataset $\mathbf{X}$ and corresponding (known) labels $\mathbf{y}$, the goal of linear regression is to find the weight vector $\mathbf{w}$ and the bias term b such that, given features of a new data example sampled from the same distribution as $\mathbf{X}$, the new example's label will (in expectation) be predicted with the smallest error.

Even if we believe that the best model for predicting y given $\mathbf{x}$ is linear, we would not expect to find a real-world dataset of n examples where $y^{(i)}$ exactly equals $\mathbf{w}^\top \mathbf{x}^{(i)} + b$ for all $1 \leq i \leq n$. For example, whatever instruments we use to observe the features $\mathbf{X}$ and labels $\mathbf{y}$, there might be a small amount of measurement error. Thus, even when we are confident that the underlying relationship is linear, we will incorporate a noise term to account for such errors.

Before we can go about searching for the best *parameters* (or *model parameters*) $\mathbf{w}$ and b, we will need two more things: (i) a measure of the quality of some given model; and (ii) a procedure for updating the model to improve its quality.

Loss Function

Naturally, fitting our model to the data requires that we agree on some measure of *fitness* (or, equivalently, of *unfitness*). *Loss functions* quantify the distance between the *real* and *predicted* values of the target. The loss will usually be a nonnegative number where smaller values are better and perfect predictions incur a loss of 0. For regression problems, the most common loss function is the squared error. When our prediction for an example i is $\hat{y}^{(i)}$ and the corresponding true label is $y^{(i)}$, the *squared error* is given by:

$$l^{(i)}(\mathbf{w}, b) = \frac{1}{2}\left(\hat{y}^{(i)} - y^{(i)}\right)^2. \tag{3.1.5}$$

The constant $\frac{1}{2}$ makes no real difference but proves to be notationally convenient, since it cancels out when we take the derivative of the loss. Because the training dataset is given to us, and thus is out of our control, the empirical error is only a function of the model parameters. In Fig. 3.1.1, we visualize the fit of a linear regression model in a problem with one-dimensional inputs.

Note that large differences between estimates $\hat{y}^{(i)}$ and targets $y^{(i)}$ lead to even larger contributions to the loss, due to its quadratic form (this quadraticity can be a double-edge sword;

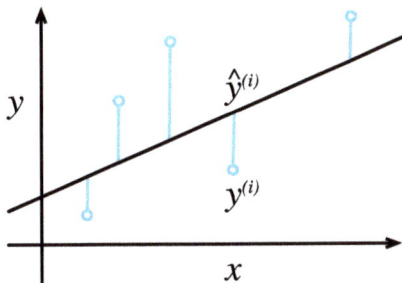

Fig. 3.1.1 Fitting a linear regression model to one-dimensional data.

while it encourages the model to avoid large errors it can also lead to excessive sensitivity to anomalous data). To measure the quality of a model on the entire dataset of n examples, we simply average (or equivalently, sum) the losses on the training set:

$$L(\mathbf{w}, b) = \frac{1}{n} \sum_{i=1}^{n} l^{(i)}(\mathbf{w}, b) = \frac{1}{n} \sum_{i=1}^{n} \frac{1}{2} \left(\mathbf{w}^\top \mathbf{x}^{(i)} + b - y^{(i)} \right)^2. \qquad (3.1.6)$$

When training the model, we seek parameters $(\mathbf{w}^*, b^*)$ that minimize the total loss across all training examples:

$$\mathbf{w}^*, b^* = \operatorname*{argmin}_{\mathbf{w}, b} \ L(\mathbf{w}, b). \qquad (3.1.7)$$

Analytic Solution

Unlike most of the models that we will cover, linear regression presents us with a surprisingly easy optimization problem. In particular, we can find the optimal parameters (as assessed on the training data) analytically by applying a simple formula as follows. First, we can subsume the bias b into the parameter $\mathbf{w}$ by appending a column to the design matrix consisting of all 1s. Then our prediction problem is to minimize $\|\mathbf{y} - \mathbf{X}\mathbf{w}\|^2$. As long as the design matrix $\mathbf{X}$ has full rank (no feature is linearly dependent on the others), then there will be just one critical point on the loss surface and it corresponds to the minimum of the loss over the entire domain. Taking the derivative of the loss with respect to $\mathbf{w}$ and setting it equal to zero yields:

$$\partial_{\mathbf{w}} \|\mathbf{y} - \mathbf{X}\mathbf{w}\|^2 = 2\mathbf{X}^\top(\mathbf{X}\mathbf{w} - \mathbf{y}) = 0 \text{ and hence } \mathbf{X}^\top \mathbf{y} = \mathbf{X}^\top \mathbf{X}\mathbf{w}. \qquad (3.1.8)$$

Solving for $\mathbf{w}$ provides us with the optimal solution for the optimization problem. Note that this solution

$$\mathbf{w}^* = (\mathbf{X}^\top \mathbf{X})^{-1} \mathbf{X}^\top \mathbf{y} \qquad (3.1.9)$$

will only be unique when the matrix $\mathbf{X}^\top \mathbf{X}$ is invertible, i.e., when the columns of the design matrix are linearly independent (Golub and Van Loan, 1996).

While simple problems like linear regression may admit analytic solutions, you should not get used to such good fortune. Although analytic solutions allow for nice mathematical analysis, the requirement of an analytic solution is so restrictive that it would exclude almost all exciting aspects of deep learning.

Minibatch Stochastic Gradient Descent

Fortunately, even in cases where we cannot solve the models analytically, we can still often train models effectively in practice. Moreover, for many tasks, those hard-to-optimize models turn out to be so much better that figuring out how to train them ends up being well worth the trouble.

The key technique for optimizing nearly every deep learning model, and which we will call upon throughout this book, consists of iteratively reducing the error by updating the parameters in the direction that incrementally lowers the loss function. This algorithm is called *gradient descent*.

The most naive application of gradient descent consists of taking the derivative of the loss function, which is an average of the losses computed on every single example in the dataset. In practice, this can be extremely slow: we must pass over the entire dataset before making a single update, even if the update steps might be very powerful (Liu and Nocedal, 1989). Even worse, if there is a lot of redundancy in the training data, the benefit of a full update is limited.

The other extreme is to consider only a single example at a time and to take update steps based on one observation at a time. The resulting algorithm, *stochastic gradient descent* (SGD) can be an effective strategy (Bottou, 2010), even for large datasets. Unfortunately, SGD has drawbacks, both computational and statistical. One problem arises from the fact that processors are a lot faster multiplying and adding numbers than they are at moving data from main memory to processor cache. It is up to an order of magnitude more efficient to perform a matrix–vector multiplication than a corresponding number of vector–vector operations. This means that it can take a lot longer to process one sample at a time compared to a full batch. A second problem is that some of the layers, such as batch normalization (to be described in Section 8.5), only work well when we have access to more than one observation at a time.

The solution to both problems is to pick an intermediate strategy: rather than taking a full batch or only a single sample at a time, we take a *minibatch* of observations (Li *et al.*, 2014). The specific choice of the size of the said minibatch depends on many factors, such as the amount of memory, the number of accelerators, the choice of layers, and the total dataset size. Despite all that, a number between 32 and 256, preferably a multiple of a large power of 2, is a good start. This leads us to *minibatch stochastic gradient descent*.

In its most basic form, in each iteration t, we first randomly sample a minibatch $\mathcal{B}_t$ consisting of a fixed number $|\mathcal{B}|$ of training examples. We then compute the derivative (gradient) of the average loss on the minibatch with respect to the model parameters. Finally, we multiply the gradient by a predetermined small positive value η, called the *learning rate*, and subtract the resulting term from the current parameter values. We can express the update as follows:

$$(\mathbf{w}, b) \leftarrow (\mathbf{w}, b) - \frac{\eta}{|\mathcal{B}|} \sum_{i \in \mathcal{B}_t} \partial_{(\mathbf{w},b)} l^{(i)}(\mathbf{w}, b). \qquad (3.1.10)$$

In summary, minibatch SGD proceeds as follows: (i) initialize the values of the model parameters, typically at random; (ii) iteratively sample random minibatches from the data, updating the parameters in the direction of the negative gradient. For quadratic losses and affine transformations, this has a closed-form expansion:

$$
\begin{aligned}
\mathbf{w} &\leftarrow \mathbf{w} - \frac{\eta}{|\mathcal{B}|} \sum_{i \in \mathcal{B}_t} \partial_{\mathbf{w}} l^{(i)}(\mathbf{w}, b) &&= \mathbf{w} - \frac{\eta}{|\mathcal{B}|} \sum_{i \in \mathcal{B}_t} \mathbf{x}^{(i)} \left(\mathbf{w}^\top \mathbf{x}^{(i)} + b - y^{(i)} \right) \\
b &\leftarrow b - \frac{\eta}{|\mathcal{B}|} \sum_{i \in \mathcal{B}_t} \partial_b l^{(i)}(\mathbf{w}, b) &&= b - \frac{\eta}{|\mathcal{B}|} \sum_{i \in \mathcal{B}_t} \left(\mathbf{w}^\top \mathbf{x}^{(i)} + b - y^{(i)} \right).
\end{aligned}
\qquad (3.1.11)
$$

Since we pick a minibatch $\mathcal{B}$ we need to normalize by its size $|\mathcal{B}|$. Frequently minibatch size and learning rate are user-defined. Such tunable parameters that are not updated in the training loop are called *hyperparameters*. They can be tuned automatically by a number of techniques, such as Bayesian optimization (Frazier, 2018). In the end, the quality of the solution is typically assessed on a separate *validation dataset* (or *validation set*).

After training for some predetermined number of iterations (or until some other stopping criterion is met), we record the estimated model parameters, denoted $\hat{\mathbf{w}}, \hat{b}$. Note that even if our function is truly linear and noiseless, these parameters will not be the exact minimizers of the loss, nor even deterministic. Although the algorithm converges slowly towards the minimizers it typically will not find them exactly in a finite number of steps. Moreover, the minibatches $\mathcal{B}$ used for updating the parameters are chosen at random. This breaks determinism.

Linear regression happens to be a learning problem with a global minimum (whenever $\mathbf{X}$ is full rank, or equivalently, whenever $\mathbf{X}^\top \mathbf{X}$ is invertible). However, the loss surfaces for deep networks contain many saddle points and minima. Fortunately, we typically do not care about finding an exact set of parameters but merely any set of parameters that leads to accurate predictions (and thus low loss). In practice, deep learning practitioners seldom struggle to find parameters that minimize the loss *on training sets* (Frankle and Carbin, 2018, Izmailov *et al.*, 2018). The more formidable task is to find parameters that lead to accurate predictions on previously unseen data, a challenge called *generalization*. We return to these topics throughout the book.

Predictions

Given the model $\hat{\mathbf{w}}^\top \mathbf{x} + \hat{b}$, we can now make *predictions* for a new example, e.g., predicting the sales price of a previously unseen house given its area x_1 and age x_2. Deep learning practitioners have taken to calling the prediction phase *inference* but this is a bit of a misnomer—*inference* refers broadly to any conclusion reached on the basis of evidence, including both the values of the parameters and the likely label for an unseen instance. If anything, in the statistics literature *inference* more often denotes parameter inference and this overloading of terminology creates unnecessary confusion when deep learning practitioners talk to statisticians. In the following we will stick to *prediction* whenever possible.

3.1.2 Vectorization for Speed

When training our models, we typically want to process whole minibatches of examples simultaneously. Doing this efficiently requires that we vectorize the calculations and leverage fast linear algebra libraries rather than writing costly for-loops in Python.

To see why this matters so much, let's consider two methods for adding vectors. To start, we instantiate two 10,000-dimensional vectors containing all 1s. In the first method, we loop over the vectors with a Python for-loop. In the second, we rely on a single call to +.

```
n = 10000
a = torch.ones(n)
b = torch.ones(n)
```

Now we can benchmark the workloads. First, we add them, one coordinate at a time, using a for-loop.

```
c = torch.zeros(n)
t = time.time()
for i in range(n):
    c[i] = a[i] + b[i]
f'{time.time() - t:.5f} sec'
```

```
'0.18559 sec'
```

Alternatively, we rely on the reloaded + operator to compute the elementwise sum.

```
t = time.time()
d = a + b
f'{time.time() - t:.5f} sec'
```

```
'0.00313 sec'
```

The second method is dramatically faster than the first. Vectorizing code often yields order-of-magnitude speedups. Moreover, we push more of the mathematics to the library so we do not have to write as many calculations ourselves, reducing the potential for errors and increasing portability of the code.

3.1.3 The Normal Distribution and Squared Loss

So far we have given a fairly functional motivation of the squared loss objective: the optimal parameters return the conditional expectation $E[Y \mid X]$ whenever the underlying pattern is truly linear, and the loss assigns large penalties for outliers. We can also provide a more formal motivation for the squared loss objective by making probabilistic assumptions about the distribution of noise.

Linear regression was invented at the turn of the 19th century. While it has long been debated whether Gauss or Legendre first thought up the idea, it was Gauss who also discovered the normal distribution (also called the *Gaussian*). It turns out that the normal distribution and linear regression with squared loss share a deeper connection than common parentage.

To begin, recall that a normal distribution with mean μ and variance σ^2 (standard deviation σ) is given as

$$p(x) = \frac{1}{\sqrt{2\pi\sigma^2}} \exp\left(-\frac{1}{2\sigma^2}(x - \mu)^2\right). \tag{3.1.12}$$

Below we define a function to compute the normal distribution.

```
def normal(x, mu, sigma):
    p = 1 / math.sqrt(2 * math.pi * sigma**2)
    return p * np.exp(-0.5 * (x - mu)**2 / sigma**2)
```

We can now visualize the normal distributions.

```
# Use NumPy again for visualization
x = np.arange(-7, 7, 0.01)

# Mean and standard deviation pairs
params = [(0, 1), (0, 2), (3, 1)]
d2l.plot(x, [normal(x, mu, sigma) for mu, sigma in params], xlabel='x',
         ylabel='p(x)', figsize=(4.5, 2.5),
         legend=[f'mean {mu}, std {sigma}' for mu, sigma in params])
```

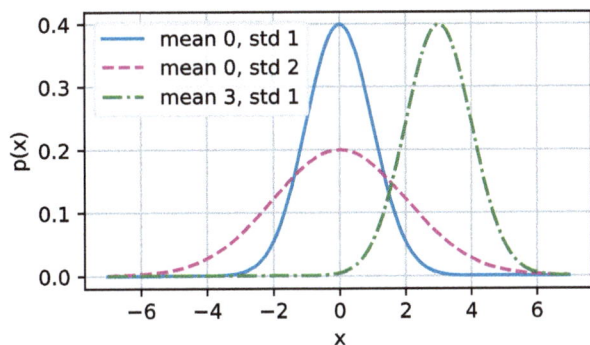

Note that changing the mean corresponds to a shift along the *x*-axis, and increasing the variance spreads the distribution out, lowering its peak.

One way to motivate linear regression with squared loss is to assume that observations arise from noisy measurements, where the noise ϵ follows the normal distribution $\mathcal{N}(0, \sigma^2)$:

$$y = \mathbf{w}^\top \mathbf{x} + b + \epsilon \text{ where } \epsilon \sim \mathcal{N}(0, \sigma^2). \tag{3.1.13}$$

Thus, we can now write out the *likelihood* of seeing a particular y for a given $\mathbf{x}$ via

$$P(y \mid \mathbf{x}) = \frac{1}{\sqrt{2\pi\sigma^2}} \exp\left(-\frac{1}{2\sigma^2}(y - \mathbf{w}^\top \mathbf{x} - b)^2\right). \tag{3.1.14}$$

As such, the likelihood factorizes. According to *the principle of maximum likelihood*, the best values of parameters $\mathbf{w}$ and b are those that maximize the *likelihood* of the entire dataset:

$$P(\mathbf{y} \mid \mathbf{X}) = \prod_{i=1}^{n} p(y^{(i)} \mid \mathbf{x}^{(i)}). \tag{3.1.15}$$

The equality follows since all pairs $(\mathbf{x}^{(i)}, y^{(i)})$ were drawn independently of each other. Estimators chosen according to the principle of maximum likelihood are called *maximum likelihood estimators*. While, maximizing the product of many exponential functions, might look difficult, we can simplify things significantly, without changing the objective, by maximizing the logarithm of the likelihood instead. For historical reasons, optimizations are more often expressed as minimization rather than maximization. So, without changing anything, we can *minimize* the *negative log-likelihood*, which we can express as follows:

$$-\log P(\mathbf{y} \mid \mathbf{X}) = \sum_{i=1}^{n} \frac{1}{2} \log(2\pi\sigma^2) + \frac{1}{2\sigma^2}\left(y^{(i)} - \mathbf{w}^\top \mathbf{x}^{(i)} - b\right)^2. \tag{3.1.16}$$

If we assume that σ is fixed, we can ignore the first term, because it does not depend on $\mathbf{w}$ or b. The second term is identical to the squared error loss introduced earlier, except for the multiplicative constant $\frac{1}{\sigma^2}$. Fortunately, the solution does not depend on σ either. It follows that minimizing the mean squared error is equivalent to the maximum likelihood estimation of a linear model under the assumption of additive Gaussian noise.

3.1.4 Linear Regression as a Neural Network

While linear models are not sufficiently rich to express the many complicated networks that we will introduce in this book, (artificial) neural networks are rich enough to subsume linear models as networks in which every feature is represented by an input neuron, all of which are connected directly to the output.

Fig. 3.1.2 depicts linear regression as a neural network. The diagram highlights the connectivity pattern, such as how each input is connected to the output, but not the specific values taken by the weights or biases.

The inputs are $x_1, \ldots, x_d$. We refer to d as the *number of inputs* or the *feature dimensionality* in the input layer. The output of the network is o_1. Because we are just trying to predict a single numerical value, we have only one output neuron. Note that the input values are all *given*. There is just a single *computed* neuron. In summary, we can think of linear regression as a single-layer fully connected neural network. We will encounter networks with far more layers in later chapters.

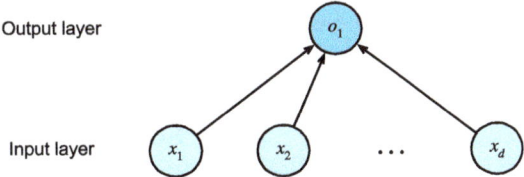

Fig. 3.1.2 Linear regression is a single-layer neural network.

Biology

Because linear regression predates computational neuroscience, it might seem anachronistic to describe linear regression in terms of neural networks. Nonetheless, they were a natural place to start when the cyberneticists and neurophysiologists Warren McCulloch and Walter Pitts began to develop models of artificial neurons. Consider the cartoonish picture of a biological neuron in Fig. 3.1.3, consisting of *dendrites* (input terminals), the *nucleus* (CPU), the *axon* (output wire), and the *axon terminals* (output terminals), enabling connections to other neurons via *synapses*.

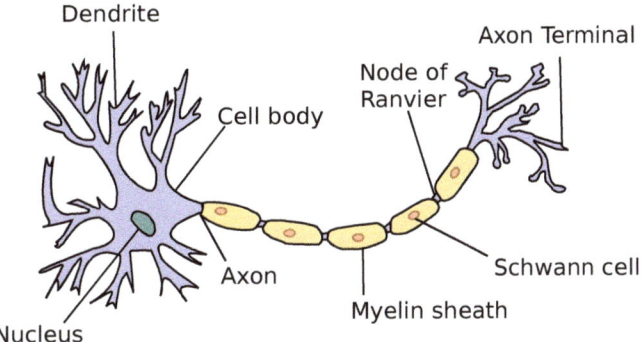

Fig. 3.1.3 The real neuron (source: "Anatomy and Physiology" by the US National Cancer Institute's Surveillance, Epidemiology and End Results (SEER) Program).

Information x_i arriving from other neurons (or environmental sensors) is received in the dendrites. In particular, that information is weighted by *synaptic weights* w_i, determining the effect of the inputs, e.g., activation or inhibition via the product $x_i w_i$. The weighted inputs arriving from multiple sources are aggregated in the nucleus as a weighted sum $y = \sum_i x_i w_i + b$, possibly subject to some nonlinear postprocessing via a function $\sigma(y)$. This information is then sent via the axon to the axon terminals, where it reaches its destination (e.g., an actuator such as a muscle) or it is fed into another neuron via its dendrites.

Certainly, the high-level idea that many such units could be combined, provided they have the correct connectivity and learning algorithm, to produce far more interesting and complex behavior than any one neuron alone could express arises from our study of real biological neural systems. At the same time, most research in deep learning today draws inspiration from a much wider source. We invoke Russell and Norvig (2016) who pointed out that although airplanes might have been *inspired* by birds, ornithology has not been the primary driver of aeronautics innovation for some centuries. Likewise, inspiration in deep learning these days comes in equal or greater measure from mathematics, linguistics, psychology, statistics, computer science, and many other fields.

3.1.5 Summary

In this section, we introduced traditional linear regression, where the parameters of a linear function are chosen to minimize squared loss on the training set. We also motivated this choice of objective both via some practical considerations and through an interpretation of linear regression as maximimum likelihood estimation under an assumption of linearity and Gaussian noise. After discussing both computational considerations and connections to statistics, we showed how such linear models could be expressed as simple neural networks where the inputs are directly wired to the output(s). While we will soon move past linear models altogether, they are sufficient to introduce most of the components that all of our models require: parametric forms, differentiable objectives, optimization via minibatch stochastic gradient descent, and ultimately, evaluation on previously unseen data.

3.1.6 Exercises

1. Assume that we have some data $x_1, \ldots, x_n \in \mathbb{R}$. Our goal is to find a constant b such that $\sum_i (x_i - b)^2$ is minimized.

    1. Find an analytic solution for the optimal value of b.

    2. How does this problem and its solution relate to the normal distribution?

    3. What if we change the loss from $\sum_i (x_i - b)^2$ to $\sum_i |x_i - b|$? Can you find the optimal solution for b?

2. Prove that the affine functions that can be expressed by $\mathbf{x}^\top \mathbf{w} + b$ are equivalent to linear functions on $(\mathbf{x}, 1)$.

3. Assume that you want to find quadratic functions of $\mathbf{x}$, i.e., $f(\mathbf{x}) = b + \sum_i w_i x_i + \sum_{j \leq i} w_{ij} x_i x_j$. How would you formulate this in a deep network?

4. Recall that one of the conditions for the linear regression problem to be solvable was that the design matrix $\mathbf{X}^\top \mathbf{X}$ has full rank.

    1. What happens if this is not the case?

    2. How could you fix it? What happens if you add a small amount of coordinate-wise independent Gaussian noise to all entries of $\mathbf{X}$?

    3. What is the expected value of the design matrix $\mathbf{X}^\top \mathbf{X}$ in this case?

    4. What happens with stochastic gradient descent when $\mathbf{X}^\top \mathbf{X}$ does not have full rank?

5. Assume that the noise model governing the additive noise ϵ is the exponential distribution. That is, $p(\epsilon) = \frac{1}{2} \exp(-|\epsilon|)$.

    1. Write out the negative log-likelihood of the data under the model $-\log P(\mathbf{y} \mid \mathbf{X})$.

    2. Can you find a closed form solution?

3. Suggest a minibatch stochastic gradient descent algorithm to solve this problem. What could possibly go wrong (hint: what happens near the stationary point as we keep on updating the parameters)? Can you fix this?

6. Assume that we want to design a neural network with two layers by composing two linear layers. That is, the output of the first layer becomes the input of the second layer. Why would such a naive composition not work?

7. What happens if you want to use regression for realistic price estimation of houses or stock prices?

   1. Show that the additive Gaussian noise assumption is not appropriate. Hint: can we have negative prices? What about fluctuations?

   2. Why would regression to the logarithm of the price be much better, i.e., $y = \log$ price?

   3. What do you need to worry about when dealing with pennystock, i.e., stock with very low prices? Hint: can you trade at all possible prices? Why is this a bigger problem for cheap stock? For more information review the celebrated Black–Scholes model for option pricing (Black and Scholes, 1973).

8. Suppose we want to use regression to estimate the *number* of apples sold in a grocery store.

   1. What are the problems with a Gaussian additive noise model? Hint: you are selling apples, not oil.

   2. The Poisson distribution[68] captures distributions over counts. It is given by $p(k \mid \lambda) = \lambda^k e^{-\lambda}/k!$. Here λ is the rate function and k is the number of events you see. Prove that λ is the expected value of counts k.

   3. Design a loss function associated with the Poisson distribution.

   4. Design a loss function for estimating $\log \lambda$ instead.

Discussions[69].

3.2 Object-Oriented Design for Implementation

In our introduction to linear regression, we walked through various components including the data, the model, the loss function, and the optimization algorithm. Indeed, linear regression is one of the simplest machine learning models. Training it, however, uses many of the same components that other models in this book require. Therefore, before diving into the implementation details it is worth designing some of the APIs that we use throughout. Treating components in deep learning as objects, we can start by defining classes for these objects and their interactions. This object-oriented design for implementation will greatly streamline the presentation and you might even want to use it in your projects.

70
Inspired by open-source libraries such as PyTorch Lightning [70], at a high level we wish to have three classes: (i) Module contains models, losses, and optimization methods; (ii) DataModule provides data loaders for training and validation; (iii) both classes are combined using the Trainer class, which allows us to train models on a variety of hardware platforms. Most code in this book adapts Module and DataModule. We will touch upon the Trainer class only when we discuss GPUs, CPUs, parallel training, and optimization algorithms.

```
import time
import numpy as np
import torch
from torch import nn
from d2l import torch as d2l
```

3.2.1 Utilities

We need a few utilities to simplify object-oriented programming in Jupyter notebooks. One of the challenges is that class definitions tend to be fairly long blocks of code. Notebook readability demands short code fragments, interspersed with explanations, a requirement incompatible with the style of programming common for Python libraries. The first utility function allows us to register functions as methods in a class *after* the class has been created. In fact, we can do so *even after* we have created instances of the class! It allows us to split the implementation of a class into multiple code blocks.

```
def add_to_class(Class):  #@save
    """Register functions as methods in created class."""
    def wrapper(obj):
        setattr(Class, obj.__name__, obj)
    return wrapper
```

Let's have a quick look at how to use it. We plan to implement a class A with a method do. Instead of having code for both A and do in the same code block, we can first declare the class A and create an instance a.

```
class A:
    def __init__(self):
        self.b = 1

a = A()
```

Next we define the method do as we normally would, but not in class A's scope. Instead, we decorate this method by add_to_class with class A as its argument. In doing so, the method is able to access the member variables of A just as we would expect had it been included as part of A's definition. Let's see what happens when we invoke it for the instance a.

```
@add_to_class(A)
def do(self):
    print('Class attribute "b" is', self.b)

a.do()
```

```
Class attribute "b" is 1
```

The second one is a utility class that saves all arguments in a class's `__init__` method as class attributes. This allows us to extend constructor call signatures implicitly without additional code.

```
class HyperParameters:  #@save
    """The base class of hyperparameters."""
    def save_hyperparameters(self, ignore=[]):
        raise NotImplemented
```

We defer its implementation into `sec_utils`. To use it, we define our class that inherits from `HyperParameters` and calls `save_hyperparameters` in the `__init__` method.

```
# Call the fully implemented HyperParameters class saved in d2l
class B(d2l.HyperParameters):
    def __init__(self, a, b, c):
        self.save_hyperparameters(ignore=['c'])
        print('self.a =', self.a, 'self.b =', self.b)
        print('There is no self.c =', not hasattr(self, 'c'))

b = B(a=1, b=2, c=3)
```

```
self.a = 1 self.b = 2
There is no self.c = True
```

The final utility allows us to plot experiment progress interactively while it is going on. In deference to the much more powerful (and complex) TensorBoard[71] we name it ProgressBoard. The implementation is deferred to `sec_utils`. For now, let's simply see it in action.

71

The draw method plots a point (x, y) in the figure, with `label` specified in the legend. The optional `every_n` smooths the line by only showing $1/n$ points in the figure. Their values are averaged from the n neighbor points in the original figure.

```
class ProgressBoard(d2l.HyperParameters):  #@save
    """The board that plots data points in animation."""
    def __init__(self, xlabel=None, ylabel=None, xlim=None,
                 ylim=None, xscale='linear', yscale='linear',
                 ls=['-', '--', '-.', ':'], colors=['C0', 'C1', 'C2', 'C3'],
                 fig=None, axes=None, figsize=(3.5, 2.5), display=True):
        self.save_hyperparameters()
```

(continues on next page)

(continued from previous page)

```
    def draw(self, x, y, label, every_n=1):
        raise NotImplemented
```

In the following example, we draw `sin` and `cos` with a different smoothness. If you run this code block, you will see the lines grow in animation.

```
board = d2l.ProgressBoard('x')
for x in np.arange(0, 10, 0.1):
    board.draw(x, np.sin(x), 'sin', every_n=2)
    board.draw(x, np.cos(x), 'cos', every_n=10)
```

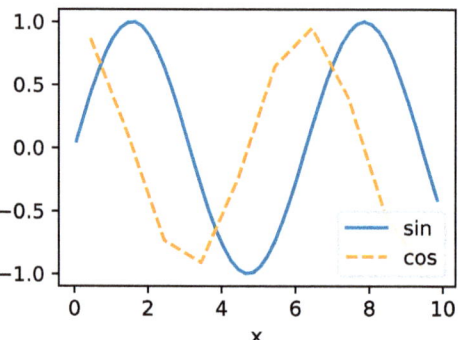

3.2.2 Models

The `Module` class is the base class of all models we will implement. At the very least we need three methods. The first, `__init__`, stores the learnable parameters, the `training_step` method accepts a data batch to return the loss value, and finally, `configure_optimizers` returns the optimization method, or a list of them, that is used to update the learnable parameters. Optionally we can define `validation_step` to report the evaluation measures. Sometimes we put the code for computing the output into a separate `forward` method to make it more reusable.

```
class Module(nn.Module, d2l.HyperParameters):  #@save
    """The base class of models."""
    def __init__(self, plot_train_per_epoch=2, plot_valid_per_epoch=1):
        super().__init__()
        self.save_hyperparameters()
        self.board = ProgressBoard()

    def loss(self, y_hat, y):
        raise NotImplementedError

    def forward(self, X):
        assert hasattr(self, 'net'), 'Neural network is defined'
```

(continues on next page)

(continued from previous page)

```
        return self.net(X)

    def plot(self, key, value, train):
        """Plot a point in animation."""
        assert hasattr(self, 'trainer'), 'Trainer is not inited'
        self.board.xlabel = 'epoch'
        if train:
            x = self.trainer.train_batch_idx / \
                self.trainer.num_train_batches
            n = self.trainer.num_train_batches / \
                self.plot_train_per_epoch
        else:
            x = self.trainer.epoch + 1
            n = self.trainer.num_val_batches / \
                self.plot_valid_per_epoch
        self.board.draw(x, value.to(d2l.cpu()).detach().numpy(),
                        ('train_' if train else 'val_') + key,
                        every_n=int(n))

    def training_step(self, batch):
        l = self.loss(self(*batch[:-1]), batch[-1])
        self.plot('loss', l, train=True)
        return l

    def validation_step(self, batch):
        l = self.loss(self(*batch[:-1]), batch[-1])
        self.plot('loss', l, train=False)

    def configure_optimizers(self):
        raise NotImplementedError
```

You may notice that Module is a subclass of nn.Module, the base class of neural networks in PyTorch. It provides convenient features for handling neural networks. For example, if we define a forward method, such as forward(self, X), then for an instance a we can invoke this method by a(X). This works since it calls the forward method in the built-in __call__ method. You can find more details and examples about nn.Module in Section 6.1.

3.2.3 Data

The DataModule class is the base class for data. Quite frequently the __init__ method is used to prepare the data. This includes downloading and preprocessing if needed. The train_dataloader returns the data loader for the training dataset. A data loader is a (Python) generator that yields a data batch each time it is used. This batch is then fed into the training_step method of Module to compute the loss. There is an optional val_dataloader to return the validation dataset loader. It behaves in the same manner, except that it yields data batches for the validation_step method in Module.

```
class DataModule(d2l.HyperParameters):  #@save
    """The base class of data."""
```

(continues on next page)

(continued from previous page)

```python
def __init__(self, root='../data', num_workers=4):
    self.save_hyperparameters()

def get_dataloader(self, train):
    raise NotImplementedError

def train_dataloader(self):
    return self.get_dataloader(train=True)

def val_dataloader(self):
    return self.get_dataloader(train=False)
```

3.2.4 Training

The `Trainer` class trains the learnable parameters in the `Module` class with data specified in `DataModule`. The key method is `fit`, which accepts two arguments: `model`, an instance of `Module`, and `data`, an instance of `DataModule`. It then iterates over the entire dataset `max_epochs` times to train the model. As before, we will defer the implementation of this method to later chapters.

```python
class Trainer(d2l.HyperParameters):  #@save
    """The base class for training models with data."""
    def __init__(self, max_epochs, num_gpus=0, gradient_clip_val=0):
        self.save_hyperparameters()
        assert num_gpus == 0, 'No GPU support yet'

    def prepare_data(self, data):
        self.train_dataloader = data.train_dataloader()
        self.val_dataloader = data.val_dataloader()
        self.num_train_batches = len(self.train_dataloader)
        self.num_val_batches = (len(self.val_dataloader)
                                if self.val_dataloader is not None else 0)

    def prepare_model(self, model):
        model.trainer = self
        model.board.xlim = [0, self.max_epochs]
        self.model = model

    def fit(self, model, data):
        self.prepare_data(data)
        self.prepare_model(model)
        self.optim = model.configure_optimizers()
        self.epoch = 0
        self.train_batch_idx = 0
        self.val_batch_idx = 0
        for self.epoch in range(self.max_epochs):
            self.fit_epoch()

    def fit_epoch(self):
        raise NotImplementedError
```

3.2.5 Summary

To highlight the object-oriented design for our future deep learning implementation, the above classes simply show how their objects store data and interact with each other. We will keep enriching implementations of these classes, such as via @add_to_class, in the rest of the book. Moreover, these fully implemented classes are saved in the D2L library[72], a *lightweight toolkit* that makes structured modeling for deep learning easy. In particular, it facilitates reusing many components between projects without changing much at all. For instance, we can replace just the optimizer, just the model, just the dataset, etc.; this degree of modularity pays dividends throughout the book in terms of conciseness and simplicity (this is why we added it) and it can do the same for your own projects.

3.2.6 Exercises

1. Locate full implementations of the above classes that are saved in the D2L library[73]. We strongly recommend that you look at the implementation in detail once you have gained some more familiarity with deep learning modeling.

2. Remove the save_hyperparameters statement in the B class. Can you still print self.a and self.b? Optional: if you have dived into the full implementation of the HyperParameters class, can you explain why?

Discussions[74].

3.3 Synthetic Regression Data

Machine learning is all about extracting information from data. So you might wonder, what could we possibly learn from synthetic data? While we might not care intrinsically about the patterns that we ourselves baked into an artificial data generating model, such datasets are nevertheless useful for didactic purposes, helping us to evaluate the properties of our learning algorithms and to confirm that our implementations work as expected. For example, if we create data for which the correct parameters are known *a priori*, then we can check that our model can in fact recover them.

```
%matplotlib inline
import random
import torch
from d2l import torch as d2l
```

3.3.1 Generating the Dataset

For this example, we will work in low dimension for succinctness. The following code snippet generates 1000 examples with 2-dimensional features drawn from a standard normal distribution. The resulting design matrix $\mathbf{X}$ belongs to $\mathbb{R}^{1000\times2}$. We generate each label by applying a *ground truth* linear function, corrupting them via additive noise ϵ, drawn independently and identically for each example:

$$\mathbf{y} = \mathbf{X}\mathbf{w} + b + \epsilon. \tag{3.3.1}$$

For convenience we assume that ϵ is drawn from a normal distribution with mean $\mu = 0$ and standard deviation $\sigma = 0.01$. Note that for object-oriented design we add the code to the __init__ method of a subclass of d2l.DataModule (introduced in Section 3.2.3). It is good practice to allow the setting of any additional hyperparameters. We accomplish this with save_hyperparameters(). The batch_size will be determined later.

```
class SyntheticRegressionData(d2l.DataModule):  #@save
    """Synthetic data for linear regression."""
    def __init__(self, w, b, noise=0.01, num_train=1000, num_val=1000,
                 batch_size=32):
        super().__init__()
        self.save_hyperparameters()
        n = num_train + num_val
        self.X = torch.randn(n, len(w))
        noise = torch.randn(n, 1) * noise
        self.y = torch.matmul(self.X, w.reshape((-1, 1))) + b + noise
```

Below, we set the true parameters to $\mathbf{w} = [2, -3.4]^\top$ and $b = 4.2$. Later, we can check our estimated parameters against these *ground truth* values.

```
data = SyntheticRegressionData(w=torch.tensor([2, -3.4]), b=4.2)
```

Each row in features consists of a vector in $\mathbb{R}^2$ and each row in labels is a scalar. Let's have a look at the first entry.

```
print('features:', data.X[0],'\nlabel:', data.y[0])
```

```
features: tensor([-0.0455, -1.0667])
label: tensor([7.7415])
```

3.3.2 Reading the Dataset

Training machine learning models often requires multiple passes over a dataset, grabbing one minibatch of examples at a time. This data is then used to update the model. To illustrate how this works, we implement the get_dataloader method, registering it in the SyntheticRegressionData class via add_to_class (introduced in Section 3.2.1). It takes a batch size, a matrix of features, and a vector of labels, and generates minibatches of size batch_size. As such, each minibatch consists of a tuple of features and labels. Note that we need to be mindful of whether we're in training or validation mode: in the former, we will want to read the data in random order, whereas for the latter, being able to read data in a pre-defined order may be important for debugging purposes.

```
@d21.add_to_class(SyntheticRegressionData)
def get_dataloader(self, train):
    if train:
        indices = list(range(0, self.num_train))
        # The examples are read in random order
        random.shuffle(indices)
    else:
        indices = list(range(self.num_train, self.num_train+self.num_val))
    for i in range(0, len(indices), self.batch_size):
        batch_indices = torch.tensor(indices[i: i+self.batch_size])
        yield self.X[batch_indices], self.y[batch_indices]
```

To build some intuition, let's inspect the first minibatch of data. Each minibatch of features provides us with both its size and the dimensionality of input features. Likewise, our minibatch of labels will have a matching shape given by batch_size.

```
X, y = next(iter(data.train_dataloader()))
print('X shape:', X.shape, '\ny shape:', y.shape)
```

```
X shape: torch.Size([32, 2])
y shape: torch.Size([32, 1])
```

While seemingly innocuous, the invocation of iter(data.train_dataloader()) illustrates the power of Python's object-oriented design. Note that we added a method to the SyntheticRegressionData class *after* creating the data object. Nonetheless, the object benefits from the *ex post facto* addition of functionality to the class.

Throughout the iteration we obtain distinct minibatches until the entire dataset has been exhausted (try this). While the iteration implemented above is good for didactic purposes, it is inefficient in ways that might get us into trouble with real problems. For example, it requires that we load all the data in memory and that we perform lots of random memory access. The built-in iterators implemented in a deep learning framework are considerably more efficient and they can deal with sources such as data stored in files, data received via a stream, and data generated or processed on the fly. Next let's try to implement the same method using built-in iterators.

3.3.3 Concise Implementation of the Data Loader

Rather than writing our own iterator, we can call the existing API in a framework to load data. As before, we need a dataset with features X and labels y. Beyond that, we set batch_size in the built-in data loader and let it take care of shuffling examples efficiently.

```
@d21.add_to_class(d21.DataModule)  #@save
def get_tensorloader(self, tensors, train, indices=slice(0, None)):
    tensors = tuple(a[indices] for a in tensors)
    dataset = torch.utils.data.TensorDataset(*tensors)
```

(continues on next page)

(continued from previous page)

```
    return torch.utils.data.DataLoader(dataset, self.batch_size,
                                       shuffle=train)
```

```
@d2l.add_to_class(SyntheticRegressionData)  #@save
def get_dataloader(self, train):
    i = slice(0, self.num_train) if train else slice(self.num_train, None)
    return self.get_tensorloader((self.X, self.y), train, i)
```

The new data loader behaves just like the previous one, except that it is more efficient and
has some added functionality.

```
X, y = next(iter(data.train_dataloader()))
print('X shape:', X.shape, '\ny shape:', y.shape)
```

```
X shape: torch.Size([32, 2])
y shape: torch.Size([32, 1])
```

For instance, the data loader provided by the framework API supports the built-in `__len__`
method, so we can query its length, i.e., the number of batches.

```
len(data.train_dataloader())
```

```
32
```

3.3.4 Summary

Data loaders are a convenient way of abstracting out the process of loading and manipulating data. This way the same machine learning *algorithm* is capable of processing many
different types and sources of data without the need for modification. One of the nice things
about data loaders is that they can be composed. For instance, we might be loading images
and then have a postprocessing filter that crops them or modifies them in other ways. As
such, data loaders can be used to describe an entire data processing pipeline.

As for the model itself, the two-dimensional linear model is about the simplest we might
encounter. It lets us test out the accuracy of regression models without worrying about
having insufficient amounts of data or an underdetermined system of equations. We will
put this to good use in the next section.

3.3.5 Exercises

1. What will happen if the number of examples cannot be divided by the batch size. How
   would you change this behavior by specifying a different argument by using the framework's API?

2. Suppose that we want to generate a huge dataset, where both the size of the parameter vector w and the number of examples num_examples are large.

   1. What happens if we cannot hold all data in memory?

   2. How would you shuffle the data if it is held on disk? Your task is to design an *efficient* algorithm that does not require too many random reads or writes. Hint: pseudorandom permutation generators[75] allow you to design a reshuffle without the need to store the permutation table explicitly (Naor and Reingold, 1999).

3. Implement a data generator that produces new data on the fly, every time the iterator is called.

4. How would you design a random data generator that generates *the same* data each time it is called?

Discussions[76].

3.4 Linear Regression Implementation from Scratch

We are now ready to work through a fully functioning implementation of linear regression. In this section, we will implement the entire method from scratch, including (i) the model; (ii) the loss function; (iii) a minibatch stochastic gradient descent optimizer; and (iv) the training function that stitches all of these pieces together. Finally, we will run our synthetic data generator from Section 3.3 and apply our model on the resulting dataset. While modern deep learning frameworks can automate nearly all of this work, implementing things from scratch is the only way to make sure that you really know what you are doing. Moreover, when it is time to customize models, defining our own layers or loss functions, understanding how things work under the hood will prove handy. In this section, we will rely only on tensors and automatic differentiation. Later, we will introduce a more concise implementation, taking advantage of the bells and whistles of deep learning frameworks while retaining the structure of what follows below.

```
%matplotlib inline
import torch
from d2l import torch as d2l
```

3.4.1 Defining the Model

Before we can begin optimizing our model's parameters by minibatch SGD, we need to have some parameters in the first place. In the following we initialize weights by drawing random numbers from a normal distribution with mean 0 and a standard deviation of 0.01. The magic number 0.01 often works well in practice, but you can specify a different value through the argument sigma. Moreover we set the bias to 0. Note that for object-oriented design we add the code to the __init__ method of a subclass of d2l.Module (introduced in Section 3.2.2).

```
class LinearRegressionScratch(d2l.Module):  #@save
    """The linear regression model implemented from scratch."""
    def __init__(self, num_inputs, lr, sigma=0.01):
        super().__init__()
        self.save_hyperparameters()
        self.w = torch.normal(0, sigma, (num_inputs, 1), requires_grad=True)
        self.b = torch.zeros(1, requires_grad=True)
```

Next we must define our model, relating its input and parameters to its output. Using the same notation as (3.1.4) for our linear model we simply take the matrix–vector product of the input features $\mathbf{X}$ and the model weights $\mathbf{w}$, and add the offset b to each example. The product $\mathbf{Xw}$ is a vector and b is a scalar. Because of the broadcasting mechanism (see Section 2.1.4), when we add a vector and a scalar, the scalar is added to each component of the vector. The resulting forward method is registered in the LinearRegressionScratch class via add_to_class (introduced in Section 3.2.1).

```
@d2l.add_to_class(LinearRegressionScratch)  #@save
def forward(self, X):
    return torch.matmul(X, self.w) + self.b
```

3.4.2 Defining the Loss Function

Since updating our model requires taking the gradient of our loss function, we ought to define the loss function first. Here we use the squared loss function in (3.1.5). In the implementation, we need to transform the true value y into the predicted value's shape y_hat. The result returned by the following method will also have the same shape as y_hat. We also return the averaged loss value among all examples in the minibatch.

```
@d2l.add_to_class(LinearRegressionScratch)  #@save
def loss(self, y_hat, y):
    l = (y_hat - y) ** 2 / 2
    return l.mean()
```

3.4.3 Defining the Optimization Algorithm

As discussed in Section 3.1, linear regression has a closed-form solution. However, our goal here is to illustrate how to train more general neural networks, and that requires that we teach you how to use minibatch SGD. Hence we will take this opportunity to introduce your first working example of SGD. At each step, using a minibatch randomly drawn from our dataset, we estimate the gradient of the loss with respect to the parameters. Next, we update the parameters in the direction that may reduce the loss.

The following code applies the update, given a set of parameters, a learning rate lr. Since our loss is computed as an average over the minibatch, we do not need to adjust the learning rate against the batch size. In later chapters we will investigate how learning rates should be adjusted for very large minibatches as they arise in distributed large-scale learning. For now, we can ignore this dependency.

We define our SGD class, a subclass of d2l.HyperParameters (introduced in Section 3.2.1), to have a similar API as the built-in SGD optimizer. We update the parameters in the step method. The zero_grad method sets all gradients to 0, which must be run before a back-propagation step.

```
class SGD(d2l.HyperParameters):  #@save
    """Minibatch stochastic gradient descent."""
    def __init__(self, params, lr):
        self.save_hyperparameters()

    def step(self):
        for param in self.params:
            param -= self.lr * param.grad

    def zero_grad(self):
        for param in self.params:
            if param.grad is not None:
                param.grad.zero_()
```

We next define the configure_optimizers method, which returns an instance of the SGD class.

```
@d2l.add_to_class(LinearRegressionScratch)  #@save
def configure_optimizers(self):
    return SGD([self.w, self.b], self.lr)
```

3.4.4 Training

Now that we have all of the parts in place (parameters, loss function, model, and optimizer), we are ready to implement the main training loop. It is crucial that you understand this code fully since you will employ similar training loops for every other deep learning model covered in this book. In each *epoch*, we iterate through the entire training dataset, passing once through every example (assuming that the number of examples is divisible by the batch size). In each *iteration*, we grab a minibatch of training examples, and compute its loss through the model's training_step method. Then we compute the gradients with respect to each parameter. Finally, we will call the optimization algorithm to update the model parameters. In summary, we will execute the following loop:

- Initialize parameters $(\mathbf{w}, b)$

- Repeat until done

  - Compute gradient $\mathbf{g} \leftarrow \partial_{(\mathbf{w},b)} \frac{1}{|\mathcal{B}|} \sum_{i \in \mathcal{B}} l(\mathbf{x}^{(i)}, y^{(i)}, \mathbf{w}, b)$

  - Update parameters $(\mathbf{w}, b) \leftarrow (\mathbf{w}, b) - \eta \mathbf{g}$

Recall that the synthetic regression dataset that we generated in Section 3.3 does not provide a validation dataset. In most cases, however, we will want a validation dataset to measure our model quality. Here we pass the validation dataloader once in each epoch to measure the model performance. Following our object-oriented design, the `prepare_batch` and `fit_epoch` methods are registered in the `d2l.Trainer` class (introduced in Section 3.2.4).

```python
@d2l.add_to_class(d2l.Trainer)  #@save
def prepare_batch(self, batch):
    return batch
```

```python
@d2l.add_to_class(d2l.Trainer)  #@save
def fit_epoch(self):
    self.model.train()
    for batch in self.train_dataloader:
        loss = self.model.training_step(self.prepare_batch(batch))
        self.optim.zero_grad()
        with torch.no_grad():
            loss.backward()
            if self.gradient_clip_val > 0:  # To be discussed later
                self.clip_gradients(self.gradient_clip_val, self.model)
            self.optim.step()
        self.train_batch_idx += 1
    if self.val_dataloader is None:
        return
    self.model.eval()
    for batch in self.val_dataloader:
        with torch.no_grad():
            self.model.validation_step(self.prepare_batch(batch))
        self.val_batch_idx += 1
```

We are almost ready to train the model, but first we need some training data. Here we use the `SyntheticRegressionData` class and pass in some ground truth parameters. Then we train our model with the learning rate `lr=0.03` and set `max_epochs=3`. Note that in general, both the number of epochs and the learning rate are hyperparameters. In general, setting hyperparameters is tricky and we will usually want to use a three-way split, one set for training, a second for hyperparameter selection, and the third reserved for the final evaluation. We elide these details for now but will revise them later.

```python
model = LinearRegressionScratch(2, lr=0.03)
data = d2l.SyntheticRegressionData(w=torch.tensor([2, -3.4]), b=4.2)
trainer = d2l.Trainer(max_epochs=3)
trainer.fit(model, data)
```

Because we synthesized the dataset ourselves, we know precisely what the true parameters are. Thus, we can evaluate our success in training by comparing the true parameters with those that we learned through our training loop. Indeed they turn out to be very close to each other.

```python
with torch.no_grad():
    print(f'error in estimating w: {data.w - model.w.reshape(data.w.shape)}')
    print(f'error in estimating b: {data.b - model.b}')
```

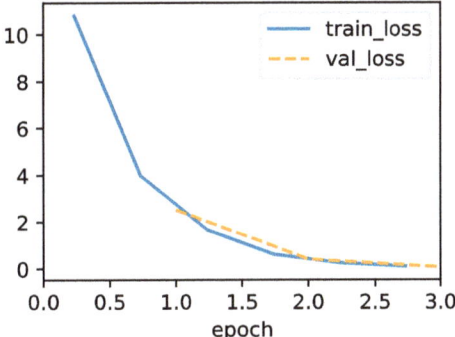

```
error in estimating w: tensor([ 0.0793, -0.2406])
error in estimating b: tensor([0.2664])
```

We should not take the ability to exactly recover the ground truth parameters for granted. In general, for deep models unique solutions for the parameters do not exist, and even for linear models, exactly recovering the parameters is only possible when no feature is linearly dependent on the others. However, in machine learning, we are often less concerned with recovering true underlying parameters, but rather with parameters that lead to highly accurate prediction (Vapnik, 1992). Fortunately, even on difficult optimization problems, stochastic gradient descent can often find remarkably good solutions, owing partly to the fact that, for deep networks, there exist many configurations of the parameters that lead to highly accurate prediction.

3.4.5 Summary

In this section, we took a significant step towards designing deep learning systems by implementing a fully functional neural network model and training loop. In this process, we built a data loader, a model, a loss function, an optimization procedure, and a visualization and monitoring tool. We did this by composing a Python object that contains all relevant components for training a model. While this is not yet a professional-grade implementation it is perfectly functional and code like this could already help you to solve small problems quickly. In the coming sections, we will see how to do this both *more concisely* (avoiding boilerplate code) and *more efficiently* (using our GPUs to their full potential).

3.4.6 Exercises

1. What would happen if we were to initialize the weights to zero. Would the algorithm still work? What if we initialized the parameters with variance 1000 rather than 0.01?

77

2. Assume that you are Georg Simon Ohm [77] trying to come up with a model for resistance that relates voltage and current. Can you use automatic differentiation to learn the parameters of your model?

78

3. Can you use Planck's Law [78] to determine the temperature of an object using spectral

energy density? For reference, the spectral density B of radiation emanating from a black body is $B(\lambda, T) = \frac{2hc^2}{\lambda^5} \cdot \left(\exp \frac{hc}{\lambda kT} - 1\right)^{-1}$. Here λ is the wavelength, T is the temperature, c is the speed of light, h is Planck's constant, and k is the Boltzmann constant. You measure the energy for different wavelengths λ and you now need to fit the spectral density curve to Planck's law.

4. What are the problems you might encounter if you wanted to compute the second derivatives of the loss? How would you fix them?

5. Why is the `reshape` method needed in the `loss` function?

6. Experiment using different learning rates to find out how quickly the loss function value drops. Can you reduce the error by increasing the number of epochs of training?

7. If the number of examples cannot be divided by the batch size, what happens to `data_iter` at the end of an epoch?

8. Try implementing a different loss function, such as the absolute value loss (`y_hat - d2l.reshape(y, y_hat.shape)).abs().sum()`.

    1. Check what happens for regular data.

    2. Check whether there is a difference in behavior if you actively perturb some entries, such as $y_5 = 10000$, of $\mathbf{y}$.

    3. Can you think of a cheap solution for combining the best aspects of squared loss and absolute value loss? Hint: how can you avoid really large gradient values?

9. Why do we need to reshuffle the dataset? Can you design a case where a maliciously constructed dataset would break the optimization algorithm otherwise?

 Discussions[79].

3.5 Concise Implementation of Linear Regression

Deep learning has witnessed a sort of Cambrian explosion over the past decade. The sheer number of techniques, applications and algorithms by far surpasses the progress of previous decades. This is due to a fortuitous combination of multiple factors, one of which is the powerful free tools offered by a number of open-source deep learning frameworks. Theano (Bergstra *et al.*, 2010), DistBelief (Dean *et al.*, 2012), and Caffe (Jia *et al.*, 2014) arguably represent the first generation of such models that found widespread adoption. In contrast to earlier (seminal) works like SN2 (Simulateur Neuristique) (Bottou and Le Cun, 1988), which provided a Lisp-like programming experience, modern frameworks offer automatic differentiation and the convenience of Python. These frameworks allow us to automate and modularize the repetitive work of implementing gradient-based learning algorithms.

In Section 3.4, we relied only on (i) tensors for data storage and linear algebra; and (ii) automatic differentiation for calculating gradients. In practice, because data iterators, loss functions, optimizers, and neural network layers are so common, modern libraries implement these components for us as well. In this section, we will show you how to implement the linear regression model from Section 3.4 concisely by using high-level APIs of deep learning frameworks.

```python
import numpy as np
import torch
from torch import nn
from d2l import torch as d2l
```

3.5.1 Defining the Model

When we implemented linear regression from scratch in Section 3.4, we defined our model parameters explicitly and coded up the calculations to produce output using basic linear algebra operations. You *should* know how to do this. But once your models get more complex, and once you have to do this nearly every day, you will be glad of the assistance. The situation is similar to coding up your own blog from scratch. Doing it once or twice is rewarding and instructive, but you would be a lousy web developer if you spent a month reinventing the wheel.

For standard operations, we can use a framework's predefined layers, which allow us to focus on the layers used to construct the model rather than worrying about their implementation. Recall the architecture of a single-layer network as described in Fig. 3.1.2. The layer is called *fully connected*, since each of its inputs is connected to each of its outputs by means of a matrix–vector multiplication.

In PyTorch, the fully connected layer is defined in `Linear` and `LazyLinear` classes (available since version 1.8.0). The latter allows users to specify *merely* the output dimension, while the former additionally asks for how many inputs go into this layer. Specifying input shapes is inconvenient and may require nontrivial calculations (such as in convolutional layers). Thus, for simplicity, we will use such "lazy" layers whenever we can.

```python
class LinearRegression(d2l.Module):  #@save
    """The linear regression model implemented with high-level APIs."""
    def __init__(self, lr):
        super().__init__()
        self.save_hyperparameters()
        self.net = nn.LazyLinear(1)
        self.net.weight.data.normal_(0, 0.01)
        self.net.bias.data.fill_(0)
```

In the `forward` method we just invoke the built-in `__call__` method of the predefined layers to compute the outputs.

```python
@d2l.add_to_class(LinearRegression)  #@save
```

(continues on next page)

(continued from previous page)

```
def forward(self, X):
    return self.net(X)
```

3.5.2 Defining the Loss Function

The `MSELoss` class computes the mean squared error (without the $1/2$ factor in (3.1.5)). By default, `MSELoss` returns the average loss over examples. It is faster (and easier to use) than implementing our own.

```
@d21.add_to_class(LinearRegression)  #@save
def loss(self, y_hat, y):
    fn = nn.MSELoss()
    return fn(y_hat, y)
```

3.5.3 Defining the Optimization Algorithm

Minibatch SGD is a standard tool for optimizing neural networks and thus PyTorch supports it alongside a number of variations on this algorithm in the `optim` module. When we instantiate an SGD instance, we specify the parameters to optimize over, obtainable from our model via `self.parameters()`, and the learning rate (`self.lr`) required by our optimization algorithm.

```
@d21.add_to_class(LinearRegression)  #@save
def configure_optimizers(self):
    return torch.optim.SGD(self.parameters(), self.lr)
```

3.5.4 Training

You might have noticed that expressing our model through high-level APIs of a deep learning framework requires fewer lines of code. We did not have to allocate parameters individually, define our loss function, or implement minibatch SGD. Once we start working with much more complex models, the advantages of the high-level API will grow considerably.

Now that we have all the basic pieces in place, the training loop itself is the same as the one we implemented from scratch. So we just call the `fit` method (introduced in Section 3.2.4), which relies on the implementation of the `fit_epoch` method in Section 3.4, to train our model.

```
model = LinearRegression(lr=0.03)
data = d21.SyntheticRegressionData(w=torch.tensor([2, -3.4]), b=4.2)
trainer = d21.Trainer(max_epochs=3)
trainer.fit(model, data)
```

Below, we compare the model parameters learned by training on finite data and the actual parameters that generated our dataset. To access parameters, we access the weights and bias

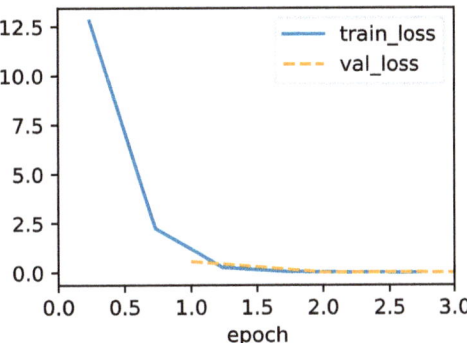

of the layer that we need. As in our implementation from scratch, note that our estimated parameters are close to their true counterparts.

```
@d2l.add_to_class(LinearRegression)  #@save
def get_w_b(self):
    return (self.net.weight.data, self.net.bias.data)
w, b = model.get_w_b()
```

```
print(f'error in estimating w: {data.w - w.reshape(data.w.shape)}')
print(f'error in estimating b: {data.b - b}')
```

```
error in estimating w: tensor([ 0.0035, -0.0086])
error in estimating b: tensor([0.0120])
```

3.5.5 Summary

This section contains the first implementation of a deep network (in this book) to tap into the conveniences afforded by modern deep learning frameworks, such as MXNet (Chen et al., 2015), JAX (Frostig et al., 2018), PyTorch (Paszke et al., 2019), and Tensorflow (Abadi et al., 2016). We used framework defaults for loading data, defining a layer, a loss function, an optimizer and a training loop. Whenever the framework provides all necessary features, it is generally a good idea to use them, since the library implementations of these components tend to be heavily optimized for performance and properly tested for reliability. At the same time, try not to forget that these modules *can* be implemented directly. This is especially important for aspiring researchers who wish to live on the leading edge of model development, where you will be inventing new components that cannot possibly exist in any current library.

In PyTorch, the data module provides tools for data processing, the nn module defines a large number of neural network layers and common loss functions. We can initialize the parameters by replacing their values with methods ending with _. Note that we need to specify the input dimensions of the network. While this is trivial for now, it can have significant knock-on effects when we want to design complex networks with many layers. Careful considerations of how to parametrize these networks is needed to allow portability.

3.5.6 Exercises

1. How would you need to change the learning rate if you replace the aggregate loss over the minibatch with an average over the loss on the minibatch?

2. Review the framework documentation to see which loss functions are provided. In particular, replace the squared loss with Huber's robust loss function. That is, use the loss function

$$l(y, y') = \begin{cases} |y - y'| - \frac{\sigma}{2} & \text{if } |y - y'| > \sigma \\ \frac{1}{2\sigma}(y - y')^2 & \text{otherwise} \end{cases} \tag{3.5.1}$$

3. How do you access the gradient of the weights of the model?

4. What is the effect on the solution if you change the learning rate and the number of epochs? Does it keep on improving?

5. How does the solution change as you vary the amount of data generated?

   1. Plot the estimation error for $\hat{\mathbf{w}} - \mathbf{w}$ and $\hat{b} - b$ as a function of the amount of data. Hint: increase the amount of data logarithmically rather than linearly, i.e., 5, 10, 20, 50, ..., 10,000 rather than 1000, 2000, ..., 10,000.

   2. Why is the suggestion in the hint appropriate?

 Discussions[80].

3.6 Generalization

Consider two college students diligently preparing for their final exam. Commonly, this preparation will consist of practicing and testing their abilities by taking exams administered in previous years. Nonetheless, doing well on past exams is no guarantee that they will excel when it matters. For instance, imagine a student, Extraordinary Ellie, whose preparation consisted entirely of memorizing the answers to previous years' exam questions. Even if Ellie were endowed with an extraordinary memory, and thus could perfectly recall the answer to any *previously seen* question, she might nevertheless freeze when faced with a new (*previously unseen*) question. By comparison, imagine another student, Inductive Irene, with comparably poor memorization skills, but a knack for picking up patterns. Note that if the exam truly consisted of recycled questions from a previous year, Ellie would handily outperform Irene. Even if Irene's inferred patterns yielded 90% accurate predictions, they could never compete with Ellie's 100% recall. However, even if the exam consisted entirely of fresh questions, Irene might maintain her 90% average.

As machine learning scientists, our goal is to discover *patterns*. But how can we be sure that we have truly discovered a *general* pattern and not simply memorized our data? Most of the time, our predictions are only useful if our model discovers such a pattern. We do not want to predict yesterday's stock prices, but tomorrow's. We do not need to recognize already diagnosed diseases for previously seen patients, but rather previously undiagnosed ailments in previously unseen patients. This problem—how to discover patterns that *generalize*—is the fundamental problem of machine learning, and arguably of all of statistics. We might cast this problem as just one slice of a far grander question that engulfs all of science: when are we ever justified in making the leap from particular observations to more general statements?

In real life, we must fit our models using a finite collection of data. The typical scales of that data vary wildly across domains. For many important medical problems, we can only access a few thousand data points. When studying rare diseases, we might be lucky to access hundreds. By contrast, the largest public datasets consisting of labeled photographs, e.g., ImageNet (Deng *et al.*, 2009), contain millions of images. And some unlabeled image collections such as the Flickr YFC100M dataset can be even larger, containing over 100 million images (Thomee *et al.*, 2016). However, even at this extreme scale, the number of available data points remains infinitesimally small compared to the space of all possible images at a megapixel resolution. Whenever we work with finite samples, we must keep in mind the risk that we might fit our training data, only to discover that we failed to discover a generalizable pattern.

The phenomenon of fitting closer to our training data than to the underlying distribution is called *overfitting*, and techniques for combatting overfitting are often called *regularization* methods. While it is no substitute for a proper introduction to statistical learning theory (see Boucheron *et al.* (2005), Vapnik (1998)), we will give you just enough intuition to get going. We will revisit generalization in many chapters throughout the book, exploring both what is known about the principles underlying generalization in various models, and also heuristic techniques that have been found (empirically) to yield improved generalization on tasks of practical interest.

3.6.1 Training Error and Generalization Error

In the standard supervised learning setting, we assume that the training data and the test data are drawn *independently* from *identical* distributions. This is commonly called the *IID assumption*. While this assumption is strong, it is worth noting that, absent any such assumption, we would be dead in the water. Why should we believe that training data sampled from distribution $P(X, Y)$ should tell us how to make predictions on test data generated by a *different distribution* $Q(X, Y)$? Making such leaps turns out to require strong assumptions about how P and Q are related. Later on we will discuss some assumptions that allow for shifts in distribution but first we need to understand the IID case, where $P(\cdot) = Q(\cdot)$.

To begin with, we need to differentiate between the *training error* R_{emp}, which is a *statistic* calculated on the training dataset, and the *generalization error* R, which is an *expectation* taken with respect to the underlying distribution. You can think of the generalization error as what you would see if you applied your model to an infinite stream of additional data examples drawn from the same underlying data distribution. Formally the training error is expressed as a *sum* (with the same notation as Section 3.1):

$$R_{\text{emp}}[\mathbf{X}, \mathbf{y}, f] = \frac{1}{n} \sum_{i=1}^{n} l(\mathbf{x}^{(i)}, y^{(i)}, f(\mathbf{x}^{(i)})), \tag{3.6.1}$$

while the generalization error is expressed as an integral:

$$R[p, f] = E_{(\mathbf{x}, y) \sim P}[l(\mathbf{x}, y, f(\mathbf{x}))] = \int \int l(\mathbf{x}, y, f(\mathbf{x})) p(\mathbf{x}, y) \, d\mathbf{x} dy. \tag{3.6.2}$$

Problematically, we can never calculate the generalization error R exactly. Nobody ever tells us the precise form of the density function $p(\mathbf{x}, y)$. Moreover, we cannot sample an infinite stream of data points. Thus, in practice, we must *estimate* the generalization error by applying our model to an independent test set constituted of a random selection of examples $\mathbf{X}'$ and labels $\mathbf{y}'$ that were withheld from our training set. This consists of applying the same formula that was used for calculating the empirical training error but to a test set $\mathbf{X}', \mathbf{y}'$.

Crucially, when we evaluate our classifier on the test set, we are working with a *fixed* classifier (it does not depend on the sample of the test set), and thus estimating its error is simply the problem of mean estimation. However the same cannot be said for the training set. Note that the model we wind up with depends explicitly on the selection of the training set and thus the training error will in general be a biased estimate of the true error on the underlying population. The central question of generalization is then when should we expect our training error to be close to the population error (and thus the generalization error).

Model Complexity

In classical theory, when we have simple models and abundant data, the training and generalization errors tend to be close. However, when we work with more complex models and/or fewer examples, we expect the training error to go down but the generalization gap to grow. This should not be surprising. Imagine a model class so expressive that for any dataset of n examples, we can find a set of parameters that can perfectly fit arbitrary labels, even if randomly assigned. In this case, even if we fit our training data perfectly, how can we conclude anything about the generalization error? For all we know, our generalization error might be no better than random guessing.

In general, absent any restriction on our model class, we cannot conclude, based on fitting the training data alone, that our model has discovered any generalizable pattern (Vapnik *et al.*, 1994). On the other hand, if our model class was not capable of fitting arbitrary labels, then it must have discovered a pattern. Learning-theoretic ideas about model complexity derived some inspiration from the ideas of Karl Popper, an influential philosopher of science, who formalized the criterion of falsifiability. According to Popper, a theory that can explain any and all observations is not a scientific theory at all! After all, what has it told us about the world if it has not ruled out any possibility? In short, what we want is a hypothesis that *could not* explain any observations we might conceivably make and yet nevertheless happens to be compatible with those observations that we *in fact* make.

Now what precisely constitutes an appropriate notion of model complexity is a complex matter. Often, models with more parameters are able to fit a greater number of arbitrarily assigned labels. However, this is not necessarily true. For instance, kernel methods operate in spaces with infinite numbers of parameters, yet their complexity is controlled by other means (Schölkopf and Smola, 2002). One notion of complexity that often proves useful is the range of values that the parameters can take. Here, a model whose parameters are permitted to take arbitrary values would be more complex. We will revisit this idea in the next section, when we introduce *weight decay*, your first practical regularization technique. Notably, it can be difficult to compare complexity among members of substantially different model classes (say, decision trees vs. neural networks).

At this point, we must stress another important point that we will revisit when introducing deep neural networks. When a model is capable of fitting arbitrary labels, low training error does not necessarily imply low generalization error. *However, it does not necessarily imply high generalization error either!* All we can say with confidence is that low training error alone is not enough to certify low generalization error. Deep neural networks turn out to be just such models: while they generalize well in practice, they are too powerful to allow us to conclude much on the basis of training error alone. In these cases we must rely more heavily on our holdout data to certify generalization after the fact. Error on the holdout data, i.e., validation set, is called the *validation error*.

3.6.2 Underfitting or Overfitting?

When we compare the training and validation errors, we want to be mindful of two common situations. First, we want to watch out for cases when our training error and validation error are both substantial but there is a little gap between them. If the model is unable to reduce the training error, that could mean that our model is too simple (i.e., insufficiently expressive) to capture the pattern that we are trying to model. Moreover, since the *generalization gap* ($R_{emp} - R$) between our training and generalization errors is small, we have reason to believe that we could get away with a more complex model. This phenomenon is known as *underfitting*.

On the other hand, as we discussed above, we want to watch out for the cases when our training error is significantly lower than our validation error, indicating severe *overfitting*. Note that overfitting is not always a bad thing. In deep learning especially, the best predictive models often perform far better on training data than on holdout data. Ultimately, we usually care about driving the generalization error lower, and only care about the gap insofar as it becomes an obstacle to that end. Note that if the training error is zero, then the generalization gap is precisely equal to the generalization error and we can make progress only by reducing the gap.

Polynomial Curve Fitting

To illustrate some classical intuition about overfitting and model complexity, consider the following: given training data consisting of a single feature x and a corresponding real-valued label y, we try to find the polynomial of degree d

$$\hat{y} = \sum_{i=0}^{d} x^i w_i \tag{3.6.3}$$

for estimating the label y. This is just a linear regression problem where our features are given by the powers of x, the model's weights are given by w_i, and the bias is given by w_0 since $x^0 = 1$ for all x. Since this is just a linear regression problem, we can use the squared error as our loss function.

A higher-order polynomial function is more complex than a lower-order polynomial function, since the higher-order polynomial has more parameters and the model function's selection range is wider. Fixing the training dataset, higher-order polynomial functions should always achieve lower (at worst, equal) training error relative to lower-degree polynomials. In fact, whenever each data example has a distinct value of x, a polynomial function with degree equal to the number of data examples can fit the training set perfectly. We compare the relationship between polynomial degree (model complexity) and both underfitting and overfitting in Fig. 3.6.1.

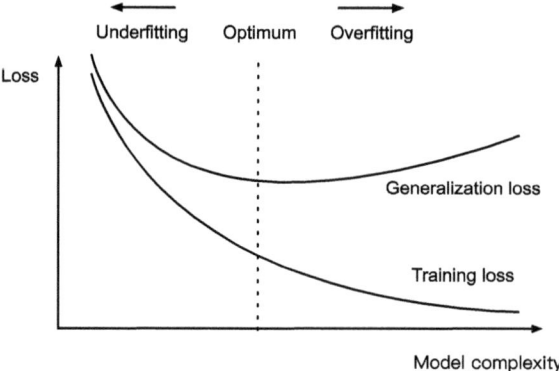

Fig. 3.6.1 Influence of model complexity on underfitting and overfitting.

Dataset Size

As the above bound already indicates, another big consideration to bear in mind is dataset size. Fixing our model, the fewer samples we have in the training dataset, the more likely (and more severely) we are to encounter overfitting. As we increase the amount of training data, the generalization error typically decreases. Moreover, in general, more data never hurts. For a fixed task and data distribution, model complexity should not increase more rapidly than the amount of data. Given more data, we might attempt to fit a more complex model. Absent sufficient data, simpler models may be more difficult to beat. For many

tasks, deep learning only outperforms linear models when many thousands of training examples are available. In part, the current success of deep learning owes considerably to the
abundance of massive datasets arising from Internet companies, cheap storage, connected
devices, and the broad digitization of the economy.

3.6.3 Model Selection

Typically, we select our final model only after evaluating multiple models that differ in various ways (different architectures, training objectives, selected features, data preprocessing,
learning rates, etc.). Choosing among many models is aptly called *model selection*.

In principle, we should not touch our test set until after we have chosen all our hyperparameters. Were we to use the test data in the model selection process, there is a risk that we
might overfit the test data. Then we would be in serious trouble. If we overfit our training
data, there is always the evaluation on test data to keep us honest. But if we overfit the test
data, how would we ever know? See Ong *et al.* (2005) for an example of how this can lead
to absurd results even for models where the complexity can be tightly controlled.

Thus, we should never rely on the test data for model selection. And yet we cannot rely
solely on the training data for model selection either because we cannot estimate the generalization error on the very data that we use to train the model.

In practical applications, the picture gets muddier. While ideally we would only touch the
test data once, to assess the very best model or to compare a small number of models with
each other, real-world test data is seldom discarded after just one use. We can seldom
afford a new test set for each round of experiments. In fact, recycling benchmark data for
decades can have a significant impact on the development of algorithms, e.g., for image

81 classification[81] and optical character recognition[82].

82 The common practice for addressing the problem of *training on the test set* is to split our
data three ways, incorporating a *validation set* in addition to the training and test datasets.
The result is a murky business where the boundaries between validation and test data are
worryingly ambiguous. Unless explicitly stated otherwise, in the experiments in this book
we are really working with what should rightly be called training data and validation data,
with no true test sets. Therefore, the accuracy reported in each experiment of the book is
really the validation accuracy and not a true test set accuracy.

Cross-Validation

When training data is scarce, we might not even be able to afford to hold out enough data to
constitute a proper validation set. One popular solution to this problem is to employ *K-fold
cross-validation*. Here, the original training data is split into K non-overlapping subsets.
Then model training and validation are executed K times, each time training on $K - 1$
subsets and validating on a different subset (the one not used for training in that round).
Finally, the training and validation errors are estimated by averaging over the results from
the K experiments.

3.6.4 Summary

This section explored some of the underpinnings of generalization in machine learning. Some of these ideas become complicated and counterintuitive when we get to deeper models; here, models are capable of overfitting data badly, and the relevant notions of complexity can be both implicit and counterintuitive (e.g., larger architectures with more parameters generalizing better). We leave you with a few rules of thumb:

1. Use validation sets (or *K-fold cross-validation*) for model selection;

2. More complex models often require more data;

3. Relevant notions of complexity include both the number of parameters and the range of values that they are allowed to take;

4. Keeping all else equal, more data almost always leads to better generalization;

5. This entire talk of generalization is all predicated on the IID assumption. If we relax this assumption, allowing for distributions to shift between the train and testing periods, then we cannot say anything about generalization absent a further (perhaps milder) assumption.

3.6.5 Exercises

1. When can you solve the problem of polynomial regression exactly?

2. Give at least five examples where dependent random variables make treating the problem as IID data inadvisable.

3. Can you ever expect to see zero training error? Under which circumstances would you see zero generalization error?

4. Why is K-fold cross-validation very expensive to compute?

5. Why is the K-fold cross-validation error estimate biased?

6. The VC dimension is defined as the maximum number of points that can be classified with arbitrary labels $\{\pm 1\}$ by a function of a class of functions. Why might this not be a good idea for measuring how complex the class of functions is? Hint: consider the magnitude of the functions.

83

7. Your manager gives you a difficult dataset on which your current algorithm does not perform so well. How would you justify to them that you need more data? Hint: you cannot increase the data but you can decrease it.

Discussions[83].

3.7 Weight Decay

Now that we have characterized the problem of overfitting, we can introduce our first *regularization* technique. Recall that we can always mitigate overfitting by collecting more training data. However, that can be costly, time consuming, or entirely out of our control, making it impossible in the short run. For now, we can assume that we already have as much high-quality data as our resources permit and focus the tools at our disposal when the dataset is taken as a given.

Recall that in our polynomial regression example (Section 3.6.2) we could limit our model's capacity by tweaking the degree of the fitted polynomial. Indeed, limiting the number of features is a popular technique for mitigating overfitting. However, simply tossing aside features can be too blunt an instrument. Sticking with the polynomial regression example, consider what might happen with high-dimensional input. The natural extensions of polynomials to multivariate data are called *monomials*, which are simply products of powers of variables. The degree of a monomial is the sum of the powers. For example, $x_1^2 x_2$, and $x_3 x_5^2$ are both monomials of degree 3.

Note that the number of terms with degree d blows up rapidly as d grows larger. Given k variables, the number of monomials of degree d is $\binom{k-1+d}{k-1}$. Even small changes in degree, say from 2 to 3, dramatically increase the complexity of our model. Thus we often need a more fine-grained tool for adjusting function complexity.

```
%matplotlib inline
import torch
from torch import nn
from d2l import torch as d2l
```

3.7.1 Norms and Weight Decay

Rather than directly manipulating the number of parameters, *weight decay* operates by restricting the values that the parameters can take. More commonly called ℓ_2 regularization outside of deep learning circles when optimized by minibatch stochastic gradient descent, weight decay might be the most widely used technique for regularizing parametric machine learning models. The technique is motivated by the basic intuition that among all functions f, the function $f = 0$ (assigning the value 0 to all inputs) is in some sense the *simplest*, and that we can measure the complexity of a function by the distance of its parameters from zero. But how precisely should we measure the distance between a function and zero? There is no single right answer. In fact, entire branches of mathematics, including parts of functional analysis and the theory of Banach spaces, are devoted to addressing such issues.

One simple interpretation might be to measure the complexity of a linear function $f(\mathbf{x}) = \mathbf{w}^\top \mathbf{x}$ by some norm of its weight vector, e.g., $\|\mathbf{w}\|^2$. Recall that we introduced the ℓ_2 norm and ℓ_1 norm, which are special cases of the more general ℓ_p norm, in Section 2.3.11. The most common method for ensuring a small weight vector is to add its norm as a penalty term to the problem of minimizing the loss. Thus we replace our original objective, *minimizing the prediction loss on the training labels*, with new objective, *minimizing the sum of the prediction loss and the penalty term*. Now, if our weight vector grows too large, our learning algorithm might focus on minimizing the weight norm $\|\mathbf{w}\|^2$ rather than minimizing the training error. That is exactly what we want. To illustrate things in code, we revive our previous example from Section 3.1 for linear regression. There, our loss was given by

$$L(\mathbf{w}, b) = \frac{1}{n} \sum_{i=1}^{n} \frac{1}{2} \left(\mathbf{w}^\top \mathbf{x}^{(i)} + b - y^{(i)} \right)^2. \tag{3.7.1}$$

Recall that $\mathbf{x}^{(i)}$ are the features, $y^{(i)}$ is the label for any data example i, and $(\mathbf{w}, b)$ are the weight and bias parameters, respectively. To penalize the size of the weight vector, we must somehow add $\|\mathbf{w}\|^2$ to the loss function, but how should the model trade off the standard loss for this new additive penalty? In practice, we characterize this trade-off via the *regularization constant* λ, a nonnegative hyperparameter that we fit using validation data:

$$L(\mathbf{w}, b) + \frac{\lambda}{2} \|\mathbf{w}\|^2. \tag{3.7.2}$$

For $\lambda = 0$, we recover our original loss function. For $\lambda > 0$, we restrict the size of $\|\mathbf{w}\|$. We divide by 2 by convention: when we take the derivative of a quadratic function, the 2 and $1/2$ cancel out, ensuring that the expression for the update looks nice and simple. The astute reader might wonder why we work with the squared norm and not the standard norm (i.e., the Euclidean distance). We do this for computational convenience. By squaring the ℓ_2 norm, we remove the square root, leaving the sum of squares of each component of the weight vector. This makes the derivative of the penalty easy to compute: the sum of derivatives equals the derivative of the sum.

Moreover, you might ask why we work with the ℓ_2 norm in the first place and not, say, the ℓ_1 norm. In fact, other choices are valid and popular throughout statistics. While ℓ_2-regularized linear models constitute the classic *ridge regression* algorithm, ℓ_1-regularized linear regression is a similarly fundamental method in statistics, popularly known as *lasso regression*. One reason to work with the ℓ_2 norm is that it places an outsize penalty on large components of the weight vector. This biases our learning algorithm towards models that distribute weight evenly across a larger number of features. In practice, this might make them more robust to measurement error in a single variable. By contrast, ℓ_1 penalties lead to models that concentrate weights on a small set of features by clearing the other weights to zero. This gives us an effective method for *feature selection*, which may be desirable for other reasons. For example, if our model only relies on a few features, then we may not need to collect, store, or transmit data for the other (dropped) features.

Using the same notation in (3.1.11), minibatch stochastic gradient descent updates for ℓ_2-regularized regression as follows:

$$\mathbf{w} \leftarrow (1 - \eta\lambda) \, \mathbf{w} - \frac{\eta}{|\mathcal{B}|} \sum_{i \in \mathcal{B}} \mathbf{x}^{(i)} \left(\mathbf{w}^\top \mathbf{x}^{(i)} + b - y^{(i)} \right). \tag{3.7.3}$$

As before, we update **w** based on the amount by which our estimate differs from the observation. However, we also shrink the size of **w** towards zero. That is why the method is sometimes called "weight decay": given the penalty term alone, our optimization algorithm *decays* the weight at each step of training. In contrast to feature selection, weight decay offers us a mechanism for continuously adjusting the complexity of a function. Smaller values of λ correspond to less constrained **w**, whereas larger values of λ constrain **w** more considerably. Whether we include a corresponding bias penalty b^2 can vary across implementations, and may vary across layers of a neural network. Often, we do not regularize the bias term. Besides, although ℓ_2 regularization may not be equivalent to weight decay for other optimization algorithms, the idea of regularization through shrinking the size of weights still holds true.

3.7.2 High-Dimensional Linear Regression

We can illustrate the benefits of weight decay through a simple synthetic example.

First, we generate some data as before:

$$y = 0.05 + \sum_{i=1}^{d} 0.01x_i + \epsilon \text{ where } \epsilon \sim \mathcal{N}(0, 0.01^2). \tag{3.7.4}$$

In this synthetic dataset, our label is given by an underlying linear function of our inputs, corrupted by Gaussian noise with zero mean and standard deviation 0.01. For illustrative purposes, we can make the effects of overfitting pronounced, by increasing the dimensionality of our problem to $d = 200$ and working with a small training set with only 20 examples.

```
class Data(d2l.DataModule):
    def __init__(self, num_train, num_val, num_inputs, batch_size):
        self.save_hyperparameters()
        n = num_train + num_val
        self.X = torch.randn(n, num_inputs)
        noise = torch.randn(n, 1) * 0.01
        w, b = torch.ones((num_inputs, 1)) * 0.01, 0.05
        self.y = torch.matmul(self.X, w) + b + noise

    def get_dataloader(self, train):
        i = slice(0, self.num_train) if train else slice(self.num_train, None)
        return self.get_tensorloader([self.X, self.y], train, i)
```

3.7.3 Implementation from Scratch

Now, let's try implementing weight decay from scratch. Since minibatch stochastic gradient descent is our optimizer, we just need to add the squared ℓ_2 penalty to the original loss function.

Defining ℓ_2 Norm Penalty

Perhaps the most convenient way of implementing this penalty is to square all terms in place and sum them.

```python
def l2_penalty(w):
    return (w ** 2).sum() / 2
```

Defining the Model

In the final model, the linear regression and the squared loss have not changed since Section 3.4, so we will just define a subclass of d2l.LinearRegressionScratch. The only change here is that our loss now includes the penalty term.

```python
class WeightDecayScratch(d2l.LinearRegressionScratch):
    def __init__(self, num_inputs, lambd, lr, sigma=0.01):
        super().__init__(num_inputs, lr, sigma)
        self.save_hyperparameters()

    def loss(self, y_hat, y):
        return (super().loss(y_hat, y) +
                self.lambd * l2_penalty(self.w))
```

The following code fits our model on the training set with 20 examples and evaluates it on the validation set with 100 examples.

```python
data = Data(num_train=20, num_val=100, num_inputs=200, batch_size=5)
trainer = d2l.Trainer(max_epochs=10)

def train_scratch(lambd):
    model = WeightDecayScratch(num_inputs=200, lambd=lambd, lr=0.01)
    model.board.yscale='log'
    trainer.fit(model, data)
    print('L2 norm of w:', float(l2_penalty(model.w)))
```

Training without Regularization

We now run this code with lambd = 0, disabling weight decay. Note that we overfit badly, decreasing the training error but not the validation error—a textbook case of overfitting.

```python
train_scratch(0)
```

```
L2 norm of w: 0.011355810798704624
```

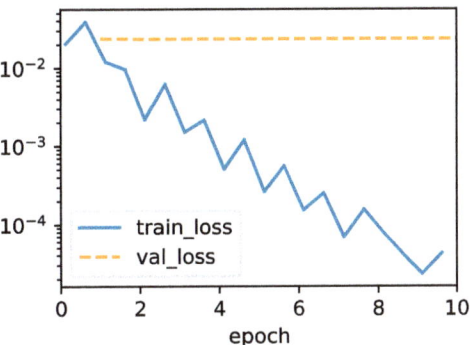

Using Weight Decay

Below, we run with substantial weight decay. Note that the training error increases but the validation error decreases. This is precisely the effect we expect from regularization.

```
train_scratch(3)
```

```
L2 norm of w: 0.0016978129278868437
```

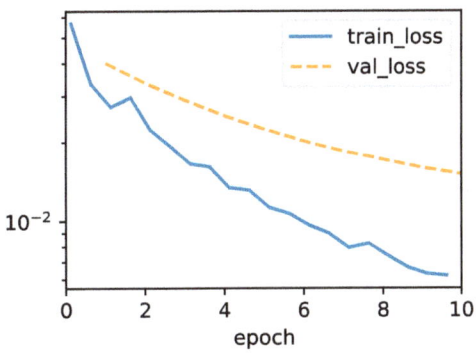

3.7.4 Concise Implementation

Because weight decay is ubiquitous in neural network optimization, the deep learning framework makes it especially convenient, integrating weight decay into the optimization algorithm itself for easy use in combination with any loss function. Moreover, this integration serves a computational benefit, allowing implementation tricks to add weight decay to the algorithm, without any additional computational overhead. Since the weight decay portion of the update depends only on the current value of each parameter, the optimizer must touch each parameter once anyway.

Below, we specify the weight decay hyperparameter directly through `weight_decay` when instantiating our optimizer. By default, PyTorch decays both weights and biases simultaneously, but we can configure the optimizer to handle different parameters according to different policies. Here, we only set `weight_decay` for the weights (the `net.weight` parameters), hence the bias (the `net.bias` parameter) will not decay.

```python
class WeightDecay(d2l.LinearRegression):
    def __init__(self, wd, lr):
        super().__init__(lr)
        self.save_hyperparameters()
        self.wd = wd

    def configure_optimizers(self):
        return torch.optim.SGD([
            {'params': self.net.weight, 'weight_decay': self.wd},
            {'params': self.net.bias}], lr=self.lr)
```

The plot looks similar to that when we implemented weight decay from scratch. However, this version runs faster and is easier to implement, benefits that will become more pronounced as you address larger problems and this work becomes more routine.

```python
model = WeightDecay(wd=3, lr=0.01)
model.board.yscale='log'
trainer.fit(model, data)

print('L2 norm of w:', float(l2_penalty(model.get_w_b()[0])))
```

```
L2 norm of w: 0.014042328111827374
```

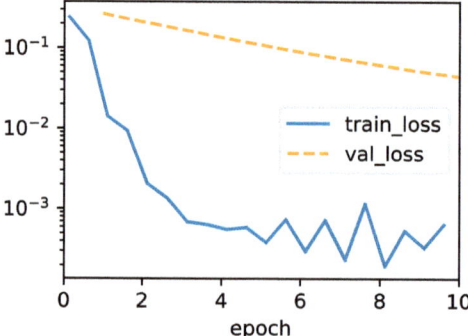

So far, we have touched upon one notion of what constitutes a simple linear function. However, even for simple nonlinear functions, the situation can be much more complex. To see this, the concept of reproducing kernel Hilbert space (RKHS)[84] allows one to apply tools introduced for linear functions in a nonlinear context. Unfortunately, RKHS-based algorithms tend to scale poorly to large, high-dimensional data. In this book we will often adopt the common heuristic whereby weight decay is applied to all layers of a deep network.

3.7.5 Summary

Regularization is a common method for dealing with overfitting. Classical regularization techniques add a penalty term to the loss function (when training) to reduce the complexity of the learned model. One particular choice for keeping the model simple is using an ℓ_2 penalty. This leads to weight decay in the update steps of the minibatch stochastic gradient descent algorithm. In practice, the weight decay functionality is provided in optimizers from deep learning frameworks. Different sets of parameters can have different update behaviors within the same training loop.

3.7.6 Exercises

1. Experiment with the value of λ in the estimation problem in this section. Plot training and validation accuracy as a function of λ. What do you observe?

2. Use a validation set to find the optimal value of λ. Is it really the optimal value? Does this matter?

3. What would the update equations look like if instead of $\|\mathbf{w}\|^2$ we used $\sum_i |w_i|$ as our penalty of choice (ℓ_1 regularization)?

4. We know that $\|\mathbf{w}\|^2 = \mathbf{w}^\top \mathbf{w}$. Can you find a similar equation for matrices (see the Frobenius norm in Section 2.3.11)?

5. Review the relationship between training error and generalization error. In addition to weight decay, increased training, and the use of a model of suitable complexity, what other ways might help us deal with overfitting?

6. In Bayesian statistics we use the product of prior and likelihood to arrive at a posterior via $P(w \mid x) \propto P(x \mid w)P(w)$. How can you identify $P(w)$ with regularization?

Discussions[85].

4 Linear Neural Networks for Classification

Now that you have worked through all of the mechanics you are ready to apply the skills you have learned to broader kinds of tasks. Even as we pivot towards classification, most of the plumbing remains the same: loading the data, passing it through the model, generating output, calculating the loss, taking gradients with respect to weights, and updating the model. However, the precise form of the targets, the parametrization of the output layer, and the choice of loss function will adapt to suit the *classification* setting.

4.1 Softmax Regression

In Section 3.1, we introduced linear regression, working through implementations from scratch in Section 3.4 and again using high-level APIs of a deep learning framework in Section 3.5 to do the heavy lifting.

Regression is the hammer we reach for when we want to answer *how much?* or *how many?* questions. If you want to predict the number of dollars (price) at which a house will be sold, or the number of wins a baseball team might have, or the number of days that a patient will remain hospitalized before being discharged, then you are probably looking for a regression model. However, even within regression models, there are important distinctions. For instance, the price of a house will never be negative and changes might often be *relative* to its baseline price. As such, it might be more effective to regress on the logarithm of the price. Likewise, the number of days a patient spends in hospital is a *discrete nonnegative* random variable. As such, least mean squares might not be an ideal approach either. This sort of time-to-event modeling comes with a host of other complications that are dealt with in a specialized subfield called *survival modeling*.

The point here is not to overwhelm you but just to let you know that there is a lot more to estimation than simply minimizing squared errors. And more broadly, there is a lot more to supervised learning than regression. In this section, we focus on *classification* problems where we put aside *how much?* questions and instead focus on *which category?* questions.

- Does this email belong in the spam folder or the inbox?

- Is this customer more likely to sign up or not to sign up for a subscription service?

- Does this image depict a donkey, a dog, a cat, or a rooster?

- Which movie is Aston most likely to watch next?

- Which section of the book are you going to read next?

Colloquially, machine learning practitioners overload the word *classification* to describe two subtly different problems: (i) those where we are interested only in hard assignments of examples to categories (classes); and (ii) those where we wish to make soft assignments, i.e., to assess the probability that each category applies. The distinction tends to get blurred, in part, because often, even when we only care about hard assignments, we still use models that make soft assignments.

Even more, there are cases where more than one label might be true. For instance, a news article might simultaneously cover the topics of entertainment, business, and space flight, but not the topics of medicine or sports. Thus, categorizing it into one of the above categories on their own would not be very useful. This problem is commonly known as multi-label classification [86]. See Tsoumakas and Katakis (2007) for an overview and Huang *et al.* (2015) for an effective algorithm when tagging images.

4.1.1 Classification

To get our feet wet, let's start with a simple image classification problem. Here, each input consists of a 2×2 grayscale image. We can represent each pixel value with a single scalar, giving us four features x_1, x_2, x_3, x_4. Further, let's assume that each image belongs to one among the categories "cat", "chicken", and "dog".

Next, we have to choose how to represent the labels. We have two obvious choices. Perhaps the most natural impulse would be to choose $y \in \{1, 2, 3\}$, where the integers represent {dog, cat, chicken} respectively. This is a great way of *storing* such information on a computer. If the categories had some natural ordering among them, say if we were trying to predict {baby, toddler, adolescent, young adult, adult, geriatric}, then it might even make sense to cast this as an ordinal regression [87] problem and keep the labels in this format. See Moon *et al.* (2010) for an overview of different types of ranking loss functions and Beutel *et al.* (2014) for a Bayesian approach that addresses responses with more than one mode.

In general, classification problems do not come with natural orderings among the classes. Fortunately, statisticians long ago invented a simple way to represent categorical data: the *one-hot encoding*. A one-hot encoding is a vector with as many components as we have categories. The component corresponding to a particular instance's category is set to 1 and all other components are set to 0. In our case, a label y would be a three-dimensional vector, with $(1, 0, 0)$ corresponding to "cat", $(0, 1, 0)$ to "chicken", and $(0, 0, 1)$ to "dog":

$$y \in \{(1, 0, 0), (0, 1, 0), (0, 0, 1)\}. \tag{4.1.1}$$

Linear Model

In order to estimate the conditional probabilities associated with all the possible classes, we need a model with multiple outputs, one per class. To address classification with linear models, we will need as many affine functions as we have outputs. Strictly speaking, we only need one fewer, since the final category has to be the difference between 1 and the sum of the other categories, but for reasons of symmetry we use a slightly redundant parametrization. Each output corresponds to its own affine function. In our case, since we have 4 features and 3 possible output categories, we need 12 scalars to represent the weights (w with subscripts), and 3 scalars to represent the biases (b with subscripts). This yields:

$$o_1 = x_1 w_{11} + x_2 w_{12} + x_3 w_{13} + x_4 w_{14} + b_1,$$
$$o_2 = x_1 w_{21} + x_2 w_{22} + x_3 w_{23} + x_4 w_{24} + b_2, \qquad (4.1.2)$$
$$o_3 = x_1 w_{31} + x_2 w_{32} + x_3 w_{33} + x_4 w_{34} + b_3.$$

The corresponding neural network diagram is shown in Fig. 4.1.1. Just as in linear regression, we use a single-layer neural network. And since the calculation of each output, o_1, o_2, and o_3, depends on every input, x_1, x_2, x_3, and x_4, the output layer can also be described as a *fully connected layer*.

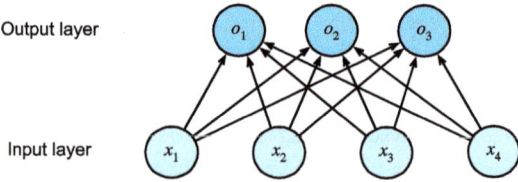

Fig. 4.1.1   Softmax regression is a single-layer neural network.

For a more concise notation we use vectors and matrices: $\mathbf{o} = \mathbf{W}\mathbf{x} + \mathbf{b}$ is much better suited for mathematics and code. Note that we have gathered all of our weights into a 3×4 matrix and all biases $\mathbf{b} \in \mathbb{R}^3$ in a vector.

The Softmax

Assuming a suitable loss function, we could try, directly, to minimize the difference between $\mathbf{o}$ and the labels $\mathbf{y}$. While it turns out that treating classification as a vector-valued regression problem works surprisingly well, it is nonetheless unsatisfactory in the following ways:

- There is no guarantee that the outputs o_i sum up to 1 in the way we expect probabilities to behave.

- There is no guarantee that the outputs o_i are even nonnegative, even if their outputs sum up to 1, or that they do not exceed 1.

Both aspects render the estimation problem difficult to solve and the solution very brittle to outliers. For instance, if we assume that there is a positive linear dependency between the number of bedrooms and the likelihood that someone will buy a house, the probability might exceed 1 when it comes to buying a mansion! As such, we need a mechanism to "squish" the outputs.

There are many ways we might accomplish this goal. For instance, we could assume that the outputs $\mathbf{o}$ are corrupted versions of $\mathbf{y}$, where the corruption occurs by means of adding noise ϵ drawn from a normal distribution. In other words, $\mathbf{y} = \mathbf{o} + \epsilon$, where $\epsilon_i \sim \mathcal{N}(0, \sigma^2)$. This is the so-called probit model[88], first introduced by Fechner (1860). While appealing, it does not work quite as well nor lead to a particularly nice optimization problem, when compared to the softmax.

88

Another way to accomplish this goal (and to ensure nonnegativity) is to use an exponential function $P(y = i) \propto \exp o_i$. This does indeed satisfy the requirement that the conditional class probability increases with increasing o_i, it is monotonic, and all probabilities are nonnegative. We can then transform these values so that they add up to 1 by dividing each by their sum. This process is called *normalization*. Putting these two pieces together gives us the *softmax* function:

$$\hat{\mathbf{y}} = \text{softmax}(\mathbf{o}) \quad \text{where} \quad \hat{y}_i = \frac{\exp(o_i)}{\sum_j \exp(o_j)}. \tag{4.1.3}$$

Note that the largest coordinate of $\mathbf{o}$ corresponds to the most likely class according to $\hat{\mathbf{y}}$. Moreover, because the softmax operation preserves the ordering among its arguments, we do not need to compute the softmax to determine which class has been assigned the highest probability. Thus,

$$\underset{j}{\text{argmax}}\, \hat{y}_j = \underset{j}{\text{argmax}}\, o_j. \tag{4.1.4}$$

The idea of a softmax dates back to Gibbs (1902), who adapted ideas from physics. Dating even further back, Boltzmann, the father of modern statistical physics, used this trick to model a distribution over energy states in gas molecules. In particular, he discovered that the prevalence of a state of energy in a thermodynamic ensemble, such as the molecules in a gas, is proportional to $\exp(-E/kT)$. Here, E is the energy of a state, T is the temperature, and k is the Boltzmann constant. When statisticians talk about increasing or decreasing the "temperature" of a statistical system, they refer to changing T in order to favor lower or higher energy states. Following Gibbs' idea, energy equates to error. Energy-based models (Ranzato *et al.*, 2007) use this point of view when describing problems in deep learning.

Vectorization

To improve computational efficiency, we vectorize calculations in minibatches of data. Assume that we are given a minibatch $\mathbf{X} \in \mathbb{R}^{n \times d}$ of n examples with dimensionality (number of inputs) d. Moreover, assume that we have q categories in the output. Then the weights satisfy $\mathbf{W} \in \mathbb{R}^{d \times q}$ and the bias satisfies $\mathbf{b} \in \mathbb{R}^{1 \times q}$.

$$\begin{aligned} \mathbf{O} &= \mathbf{X}\mathbf{W} + \mathbf{b}, \\ \hat{\mathbf{Y}} &= \text{softmax}(\mathbf{O}). \end{aligned} \tag{4.1.5}$$

This accelerates the dominant operation into a matrix–matrix product $\mathbf{XW}$. Moreover, since each row in $\mathbf{X}$ represents a data example, the softmax operation itself can be computed *rowwise*: for each row of $\mathbf{O}$, exponentiate all entries and then normalize them by the sum. Note, though, that care must be taken to avoid exponentiating and taking logarithms of large numbers, since this can cause numerical overflow or underflow. Deep learning frameworks take care of this automatically.

4.1.2 Loss Function

Now that we have a mapping from features $\mathbf{x}$ to probabilities $\hat{\mathbf{y}}$, we need a way to optimize the accuracy of this mapping. We will rely on maximum likelihood estimation, the very same method that we encountered when providing a probabilistic justification for the mean squared error loss in Section 3.1.3.

Log-Likelihood

The softmax function gives us a vector $\hat{\mathbf{y}}$, which we can interpret as the (estimated) conditional probabilities of each class, given any input $\mathbf{x}$, such as $\hat{y}_1 = P(y = \text{cat} \mid \mathbf{x})$. In the following we assume that for a dataset with features $\mathbf{X}$ the labels $\mathbf{Y}$ are represented using a one-hot encoding label vector. We can compare the estimates with reality by checking how probable the actual classes are according to our model, given the features:

$$P(\mathbf{Y} \mid \mathbf{X}) = \prod_{i=1}^{n} P(\mathbf{y}^{(i)} \mid \mathbf{x}^{(i)}). \qquad (4.1.6)$$

We are allowed to use the factorization since we assume that each label is drawn independently from its respective distribution $P(\mathbf{y} \mid \mathbf{x}^{(i)})$. Since maximizing the product of terms is awkward, we take the negative logarithm to obtain the equivalent problem of minimizing the negative log-likelihood:

$$-\log P(\mathbf{Y} \mid \mathbf{X}) = \sum_{i=1}^{n} -\log P(\mathbf{y}^{(i)} \mid \mathbf{x}^{(i)}) = \sum_{i=1}^{n} l(\mathbf{y}^{(i)}, \hat{\mathbf{y}}^{(i)}), \qquad (4.1.7)$$

where for any pair of label $\mathbf{y}$ and model prediction $\hat{\mathbf{y}}$ over q classes, the loss function l is

$$l(\mathbf{y}, \hat{\mathbf{y}}) = -\sum_{j=1}^{q} y_j \log \hat{y}_j. \qquad (4.1.8)$$

For reasons explained later on, the loss function in (4.1.8) is commonly called the *cross-entropy loss*. Since $\mathbf{y}$ is a one-hot vector of length q, the sum over all its coordinates j vanishes for all but one term. Note that the loss $l(\mathbf{y}, \hat{\mathbf{y}})$ is bounded from below by 0 whenever $\hat{\mathbf{y}}$ is a probability vector: no single entry is larger than 1, hence their negative logarithm cannot be lower than 0; $l(\mathbf{y}, \hat{\mathbf{y}}) = 0$ only if we predict the actual label with *certainty*. This can never happen for any finite setting of the weights because taking a softmax output towards 1 requires taking the corresponding input o_i to infinity (or all other outputs o_j for $j \neq i$ to negative infinity). Even if our model could assign an output probability of 0, any error made when assigning such high confidence would incur infinite loss ($-\log 0 = \infty$).

Softmax and Cross-Entropy Loss

Since the softmax function and the corresponding cross-entropy loss are so common, it is worth understanding a bit better how they are computed. Plugging (4.1.3) into the definition of the loss in (4.1.8) and using the definition of the softmax we obtain

$$
\begin{aligned}
l(\mathbf{y}, \hat{\mathbf{y}}) &= -\sum_{j=1}^{q} y_j \log \frac{\exp(o_j)}{\sum_{k=1}^{q} \exp(o_k)} \\
&= \sum_{j=1}^{q} y_j \log \sum_{k=1}^{q} \exp(o_k) - \sum_{j=1}^{q} y_j o_j \\
&= \log \sum_{k=1}^{q} \exp(o_k) - \sum_{j=1}^{q} y_j o_j.
\end{aligned} \tag{4.1.9}
$$

To understand a bit better what is going on, consider the derivative with respect to any logit o_j. We get

$$
\partial_{o_j} l(\mathbf{y}, \hat{\mathbf{y}}) = \frac{\exp(o_j)}{\sum_{k=1}^{q} \exp(o_k)} - y_j = \mathrm{softmax}(\mathbf{o})_j - y_j. \tag{4.1.10}
$$

In other words, the derivative is the difference between the probability assigned by our model, as expressed by the softmax operation, and what actually happened, as expressed by elements in the one-hot label vector. In this sense, it is very similar to what we saw in regression, where the gradient was the difference between the observation y and estimate $\hat{y}$. This is not a coincidence. In any exponential family model, the gradients of the log-likelihood are given by precisely this term. This fact makes computing gradients easy in practice.

Now consider the case where we observe not just a single outcome but an entire distribution over outcomes. We can use the same representation as before for the label $\mathbf{y}$. The only difference is that rather than a vector containing only binary entries, say $(0, 0, 1)$, we now have a generic probability vector, say $(0.1, 0.2, 0.7)$. The math that we used previously to define the loss l in (4.1.8) still works well, just that the interpretation is slightly more general. It is the expected value of the loss for a distribution over labels. This loss is called the *cross-entropy loss* and it is one of the most commonly used losses for classification problems. We can demystify the name by introducing just the basics of information theory. In a nutshell, it measures the number of bits needed to encode what we see, $\mathbf{y}$, relative to what we predict that should happen, $\hat{\mathbf{y}}$. We provide a very basic explanation in the following. For further details on information theory see Cover and Thomas (1999) or MacKay (2003).

4.1.3 Information Theory Basics

Many deep learning papers use intuition and terms from information theory. To make sense of them, we need some common language. This is a survival guide. *Information theory* deals with the problem of encoding, decoding, transmitting, and manipulating information (also known as data).

Entropy

The central idea in information theory is to quantify the amount of information contained in data. This places a limit on our ability to compress data. For a distribution P its *entropy*, $H[P]$, is defined as:

$$H[P] = \sum_j -P(j) \log P(j).$$
(4.1.11)

One of the fundamental theorems of information theory states that in order to encode data drawn randomly from the distribution P, we need at least $H[P]$ "nats" to encode it (Shannon, 1948). If you wonder what a "nat" is, it is the equivalent of bit but when using a code with base e rather than one with base 2. Thus, one nat is $\frac{1}{\log(2)} \approx 1.44$ bit.

Surprisal

You might be wondering what compression has to do with prediction. Imagine that we have a stream of data that we want to compress. If it is always easy for us to predict the next token, then this data is easy to compress. Take the extreme example where every token in the stream always takes the same value. That is a very boring data stream! And not only it is boring, but it is also easy to predict. Because the tokens are always the same, we do not have to transmit any information to communicate the contents of the stream. Easy to predict, easy to compress.

However if we cannot perfectly predict every event, then we might sometimes be surprised. Our surprise is greater when an event is assigned lower probability. Claude Shannon settled on $\log \frac{1}{P(j)} = -\log P(j)$ to quantify one's *surprisal* at observing an event j having assigned it a (subjective) probability $P(j)$. The entropy defined in (4.1.11) is then the *expected surprisal* when one assigned the correct probabilities that truly match the data-generating process.

Cross-Entropy Revisited

So if entropy is the level of surprise experienced by someone who knows the true probability, then you might be wondering, what is cross-entropy? The cross-entropy *from P to Q*, denoted $H(P, Q)$, is the expected surprisal of an observer with subjective probabilities Q upon seeing data that was actually generated according to probabilities P. This is given by $H(P, Q) \stackrel{\text{def}}{=} \sum_j -P(j) \log Q(j)$. The lowest possible cross-entropy is achieved when $P = Q$. In this case, the cross-entropy from P to Q is $H(P, P) = H(P)$.

In short, we can think of the cross-entropy classification objective in two ways: (i) as maximizing the likelihood of the observed data; and (ii) as minimizing our surprisal (and thus the number of bits) required to communicate the labels.

4.1.4 Summary and Discussion

In this section, we encountered the first nontrivial loss function, allowing us to optimize over *discrete* output spaces. Key in its design was that we took a probabilistic approach, treating discrete categories as instances of draws from a probability distribution. As a side effect, we encountered the softmax, a convenient activation function that transforms outputs of an ordinary neural network layer into valid discrete probability distributions. We saw that the derivative of the cross-entropy loss when combined with softmax behaves very similarly to the derivative of squared error; namely by taking the difference between the expected behavior and its prediction. And, while we were only able to scratch the very surface of it, we encountered exciting connections to statistical physics and information theory.

While this is enough to get you on your way, and hopefully enough to whet your appetite, we hardly dived deep here. Among other things, we skipped over computational considerations. Specifically, for any fully connected layer with d inputs and q outputs, the parametrization and computational cost is $O(dq)$, which can be prohibitively high in practice. Fortunately, this cost of transforming d inputs into q outputs can be reduced through approximation and compression. For instance Deep Fried Convnets (Yang *et al.*, 2015) uses a combination of permutations, Fourier transforms, and scaling to reduce the cost from quadratic to log-linear. Similar techniques work for more advanced structural matrix approximations (Sindhwani *et al.*, 2015). Lastly, we can use quaternion-like decompositions to reduce the cost to $O(\frac{dq}{n})$, again if we are willing to trade off a small amount of accuracy for computational and storage cost (Zhang *et al.*, 2021) based on a compression factor n. This is an active area of research. What makes it challenging is that we do not necessarily strive for the most compact representation or the smallest number of floating point operations but rather for the solution that can be executed most efficiently on modern GPUs.

4.1.5 Exercises

1. We can explore the connection between exponential families and softmax in some more depth.

   1. Compute the second derivative of the cross-entropy loss $l(\mathbf{y}, \hat{\mathbf{y}})$ for softmax.

   2. Compute the variance of the distribution given by softmax($\mathbf{o}$) and show that it matches the second derivative computed above.

2. Assume that we have three classes which occur with equal probability, i.e., the probability vector is $(\frac{1}{3}, \frac{1}{3}, \frac{1}{3})$.

   1. What is the problem if we try to design a binary code for it?

   2. Can you design a better code? Hint: what happens if we try to encode two independent observations? What if we encode n observations jointly?

3. When encoding signals transmitted over a physical wire, engineers do not always use binary codes. For instance, PAM-3[89] uses three signal levels $\{-1, 0, 1\}$ as opposed to two levels $\{0, 1\}$. How many ternary units do you need to transmit an integer in the range $\{0, \ldots, 7\}$? Why might this be a better idea in terms of electronics?

4. The Bradley–Terry model[90] uses a logistic model to capture preferences. For a user to

choose between apples and oranges one assumes scores o_{apple} and o_{orange}. Our requirements are that larger scores should lead to a higher likelihood in choosing the associated item and that the item with the largest score is the most likely one to be chosen (Bradley and Terry, 1952).

1. Prove that softmax satisfies this requirement.

2. What happens if you want to allow for a default option of choosing neither apples nor oranges? Hint: now the user has three choices.

5. Softmax gets its name from the following mapping: $\text{RealSoftMax}(a, b) = \log(\exp(a) + \exp(b))$.

1. Prove that $\text{RealSoftMax}(a, b) > \max(a, b)$.

2. How small can you make the difference between both functions? Hint: without loss of generality you can set $b = 0$ and $a \geq b$.

3. Prove that this holds for $\lambda^{-1}\text{RealSoftMax}(\lambda a, \lambda b)$, provided that $\lambda > 0$.

4. Show that for $\lambda \to \infty$ we have $\lambda^{-1}\text{RealSoftMax}(\lambda a, \lambda b) \to \max(a, b)$.

5. Construct an analogous softmin function.

6. Extend this to more than two numbers.

6. The function $g(\mathbf{x}) \overset{\text{def}}{=} \log \sum_i \exp x_i$ is sometimes also referred to as the log-partition function[91].

1. Prove that the function is convex. Hint: to do so, use the fact that the first derivative amounts to the probabilities from the softmax function and show that the second derivative is the variance.

2. Show that g is translation invariant, i.e., $g(\mathbf{x} + b) = g(\mathbf{x})$.

3. What happens if some of the coordinates x_i are very large? What happens if they're all very small?

4. Show that if we choose $b = \max_i x_i$ we end up with a numerically stable implementation.

7. Assume that we have some probability distribution P. Suppose we pick another distribution Q with $Q(i) \propto P(i)^\alpha$ for $\alpha > 0$.

1. Which choice of α corresponds to doubling the temperature? Which choice corresponds to halving it?

2. What happens if we let the temperature approach 0?

3. What happens if we let the temperature approach ∞?

Discussions[92].

4.2 The Image Classification Dataset

One widely used dataset for image classification is the MNIST dataset[93] (LeCun *et al.*, 1998) of handwritten digits. At the time of its release in the 1990s it posed a formidable challenge to most machine learning algorithms, consisting of 60,000 images of 28×28 pixels resolution (plus a test dataset of 10,000 images). To put things into perspective, back in 1995, a Sun SPARCStation 5 with a whopping 64MB of RAM and a blistering 5 MFLOPs was considered state of the art equipment for machine learning at AT&T Bell Laboratories. Achieving high accuracy on digit recognition was a key component in automating letter sorting for the USPS in the 1990s. Deep networks such as LeNet-5 (LeCun *et al.*, 1995), support vector machines with invariances (Schölkopf *et al.*, 1996), and tangent distance classifiers (Simard *et al.*, 1998) all could reach error rates below 1%.

For over a decade, MNIST served as *the* point of reference for comparing machine learning algorithms. While it had a good run as a benchmark dataset, even simple models by today's standards achieve classification accuracy over 95%, making it unsuitable for distinguishing between strong models and weaker ones. Even more, the dataset allows for *very* high levels of accuracy, not typically seen in many classification problems. This skewed algorithmic development towards specific families of algorithms that can take advantage of clean datasets, such as active set methods and boundary-seeking active set algorithms. Today, MNIST serves as more of a sanity check than as a benchmark. ImageNet (Deng *et al.*, 2009) poses a much more relevant challenge. Unfortunately, ImageNet is too large for many of the examples and illustrations in this book, as it would take too long to train to make the examples interactive. As a substitute we will focus our discussion in the coming sections on the qualitatively similar, but much smaller Fashion-MNIST dataset (Xiao *et al.*, 2017) which was released in 2017. It contains images of 10 categories of clothing at 28×28 pixels resolution.

```
%matplotlib inline
import time
import torch
import torchvision
from torchvision import transforms
from d2l import torch as d2l

d2l.use_svg_display()
```

4.2.1 Loading the Dataset

Since the Fashion-MNIST dataset is so useful, all major frameworks provide preprocessed versions of it. We can download and read it into memory using built-in framework utilities.

```
class FashionMNIST(d2l.DataModule):   #@save
    """The Fashion-MNIST dataset."""
    def __init__(self, batch_size=64, resize=(28, 28)):
        super().__init__()
        self.save_hyperparameters()
        trans = transforms.Compose([transforms.Resize(resize),
                                      transforms.ToTensor()])
        self.train = torchvision.datasets.FashionMNIST(
            root=self.root, train=True, transform=trans, download=True)
        self.val = torchvision.datasets.FashionMNIST(
            root=self.root, train=False, transform=trans, download=True)
```

Fashion-MNIST consists of images from 10 categories, each represented by 6000 images in the training dataset and by 1000 in the test dataset. A *test dataset* is used for evaluating model performance (it must not be used for training). Consequently the training set and the test set contain 60,000 and 10,000 images, respectively.

```
data = FashionMNIST(resize=(32, 32))
len(data.train), len(data.val)
```

```
(60000, 10000)
```

The images are grayscale and upscaled to 32×32 pixels in resolution above. This is similar to the original MNIST dataset which consisted of (binary) black and white images. Note, though, that most modern image data has three channels (red, green, blue) and that hyperspectral images can have in excess of 100 channels (the HyMap sensor has 126 channels). By convention we store an image as a $c \times h \times w$ tensor, where c is the number of color channels, h is the height and w is the width.

```
data.train[0][0].shape
```

```
torch.Size([1, 32, 32])
```

The categories of Fashion-MNIST have human-understandable names. The following convenience method converts between numeric labels and their names.

```
@d2l.add_to_class(FashionMNIST)   #@save
def text_labels(self, indices):
    """Return text labels."""
    labels = ['t-shirt', 'trouser', 'pullover', 'dress', 'coat',
              'sandal', 'shirt', 'sneaker', 'bag', 'ankle boot']
    return [labels[int(i)] for i in indices]
```

4.2.2 Reading a Minibatch

To make our life easier when reading from the training and test sets, we use the built-in data iterator rather than creating one from scratch. Recall that at each iteration, a data iterator reads a minibatch of data with size `batch_size`. We also randomly shuffle the examples for the training data iterator.

```python
@d2l.add_to_class(FashionMNIST)  #@save
def get_dataloader(self, train):
    data = self.train if train else self.val
    return torch.utils.data.DataLoader(data, self.batch_size, shuffle=train,
                                       num_workers=self.num_workers)
```

To see how this works, let's load a minibatch of images by invoking the `train_dataloader` method. It contains 64 images.

```python
X, y = next(iter(data.train_dataloader()))
print(X.shape, X.dtype, y.shape, y.dtype)
```

```
torch.Size([64, 1, 32, 32]) torch.float32 torch.Size([64]) torch.int64
```

Let's look at the time it takes to read the images. Even though it is a built-in loader, it is not blazingly fast. Nonetheless, this is sufficient since processing images with a deep network takes quite a bit longer. Hence it is good enough that training a network will not be I/O constrained.

```python
tic = time.time()
for X, y in data.train_dataloader():
    continue
f'{time.time() - tic:.2f} sec'
```

```
'5.42 sec'
```

4.2.3 Visualization

We will often be using the Fashion-MNIST dataset. A convenience function `show_images` can be used to visualize the images and the associated labels. Skipping implementation details, we just show the interface below: we only need to know how to invoke `d2l.show_images` rather than how it works for such utility functions.

```python
def show_images(imgs, num_rows, num_cols, titles=None, scale=1.5):  #@save
    """Plot a list of images."""
    raise NotImplementedError
```

Let's put it to good use. In general, it is a good idea to visualize and inspect data that you are training on. Humans are very good at spotting oddities and because of that, visualization serves as an additional safeguard against mistakes and errors in the design of experiments. Here are the images and their corresponding labels (in text) for the first few examples in the training dataset.

```
@d2l.add_to_class(FashionMNIST)  #@save
def visualize(self, batch, nrows=1, ncols=8, labels=[]):
    X, y = batch
    if not labels:
        labels = self.text_labels(y)
    d2l.show_images(X.squeeze(1), nrows, ncols, titles=labels)
batch = next(iter(data.val_dataloader()))
data.visualize(batch)
```

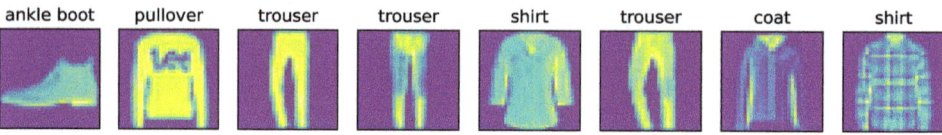

We are now ready to work with the Fashion-MNIST dataset in the sections that follow.

4.2.4 Summary

We now have a slightly more realistic dataset to use for classification. Fashion-MNIST is an apparel classification dataset consisting of images representing 10 categories. We will use this dataset in subsequent sections and chapters to evaluate various network designs, from a simple linear model to advanced residual networks. As we commonly do with images, we read them as a tensor of shape (batch size, number of channels, height, width). For now, we only have one channel as the images are grayscale (the visualization above uses a false color palette for improved visibility).

Lastly, data iterators are a key component for efficient performance. For instance, we might use GPUs for efficient image decompression, video transcoding, or other preprocessing. Whenever possible, you should rely on well-implemented data iterators that exploit high-performance computing to avoid slowing down your training loop.

4.2.5 Exercises

1. Does reducing the batch_size (for instance, to 1) affect the reading performance?

2. The data iterator performance is important. Do you think the current implementation is fast enough? Explore various options to improve it. Use a system profiler to find out where the bottlenecks are.

94

3. Check out the framework's online API documentation. Which other datasets are available?

Discussions[94].

4.3 The Base Classification Model

You may have noticed that the implementations from scratch and the concise implementation using framework functionality were quite similar in the case of regression. The same is true for classification. Since many models in this book deal with classification, it is worth adding functionalities to support this setting specifically. This section provides a base class for classification models to simplify future code.

```
import torch
from d2l import torch as d2l
```

4.3.1 The Classifier Class

We define the Classifier class below. In the validation_step we report both the loss value and the classification accuracy on a validation batch. We draw an update for every num_val_batches batches. This has the benefit of generating the averaged loss and accuracy on the whole validation data. These average numbers are not exactly correct if the final batch contains fewer examples, but we ignore this minor difference to keep the code simple.

```
class Classifier(d2l.Module):  #@save
    """The base class of classification models."""
    def validation_step(self, batch):
        Y_hat = self(*batch[:-1])
        self.plot('loss', self.loss(Y_hat, batch[-1]), train=False)
        self.plot('acc', self.accuracy(Y_hat, batch[-1]), train=False)
```

By default we use a stochastic gradient descent optimizer, operating on minibatches, just as we did in the context of linear regression.

```
@d2l.add_to_class(d2l.Module)  #@save
def configure_optimizers(self):
    return torch.optim.SGD(self.parameters(), lr=self.lr)
```

4.3.2 Accuracy

Given the predicted probability distribution y_hat, we typically choose the class with the highest predicted probability whenever we must output a hard prediction. Indeed, many applications require that we make a choice. For instance, Gmail must categorize an email into "Primary", "Social", "Updates", "Forums", or "Spam". It might estimate probabilities internally, but at the end of the day it has to choose one among the classes.

When predictions are consistent with the label class y, they are correct. The classification accuracy is the fraction of all predictions that are correct. Although it can be difficult to optimize accuracy directly (it is not differentiable), it is often the performance measure that we care about the most. It is often *the* relevant quantity in benchmarks. As such, we will nearly always report it when training classifiers.

Accuracy is computed as follows. First, if y_hat is a matrix, we assume that the second dimension stores prediction scores for each class. We use argmax to obtain the predicted class by the index for the largest entry in each row. Then we compare the predicted class with the ground truth y elementwise. Since the equality operator == is sensitive to data types, we convert y_hat's data type to match that of y. The result is a tensor containing entries of 0 (false) and 1 (true). Taking the sum yields the number of correct predictions.

```
@d21.add_to_class(Classifier)   #@save
def accuracy(self, Y_hat, Y, averaged=True):
    """Compute the number of correct predictions."""
    Y_hat = Y_hat.reshape((-1, Y_hat.shape[-1]))
    preds = Y_hat.argmax(axis=1).type(Y.dtype)
    compare = (preds == Y.reshape(-1)).type(torch.float32)
    return compare.mean() if averaged else compare
```

4.3.3 Summary

Classification is a sufficiently common problem that it warrants its own convenience functions. Of central importance in classification is the *accuracy* of the classifier. Note that while we often care primarily about accuracy, we train classifiers to optimize a variety of other objectives for statistical and computational reasons. However, regardless of which loss function was minimized during training, it is useful to have a convenience method for assessing the accuracy of our classifier empirically.

4.3.4 Exercises

1. Denote by L_v the validation loss, and let L_v^q be its quick and dirty estimate computed by the loss function averaging in this section. Lastly, denote by l_v^b the loss on the last minibatch. Express L_v in terms of L_v^q, l_v^b, and the sample and minibatch sizes.

2. Show that the quick and dirty estimate L_v^q is unbiased. That is, show that $E[L_v] = E[L_v^q]$. Why would you still want to use L_v instead?

3. Given a multiclass classification loss, denoting by $l(y, y')$ the penalty of estimating y' when we see y and given a probabilty $p(y \mid x)$, formulate the rule for an optimal selection of y'. Hint: express the expected loss, using l and $p(y \mid x)$.

Discussions[95].

4.4 Softmax Regression Implementation from Scratch

Because softmax regression is so fundamental, we believe that you ought to know how to implement it yourself. Here, we limit ourselves to defining the softmax-specific aspects of the model and reuse the other components from our linear regression section, including the training loop.

```
import torch
from d2l import torch as d2l
```

4.4.1 The Softmax

Let's begin with the most important part: the mapping from scalars to probabilities. For a refresher, recall the operation of the sum operator along specific dimensions in a tensor, as discussed in Section 2.3.6 and Section 2.3.7. Given a matrix X we can sum over all elements (by default) or only over elements in the same axis. The `axis` variable lets us compute row and column sums:

```
X = torch.tensor([[1.0, 2.0, 3.0], [4.0, 5.0, 6.0]])
X.sum(0, keepdims=True), X.sum(1, keepdims=True)
```

```
(tensor([[5., 7., 9.]]),
 tensor([[ 6.],
         [15.]]))
```

Computing the softmax requires three steps: (i) exponentiation of each term; (ii) a sum over each row to compute the normalization constant for each example; (iii) division of each row by its normalization constant, ensuring that the result sums to 1:

$$\mathrm{softmax}(\mathbf{X})_{ij} = \frac{\exp(\mathbf{X}_{ij})}{\sum_k \exp(\mathbf{X}_{ik})}. \tag{4.4.1}$$

The (logarithm of the) denominator is called the (log) *partition function*. It was introduced in statistical physics[96] to sum over all possible states in a thermodynamic ensemble. The implementation is straightforward:

96

```
def softmax(X):
    X_exp = torch.exp(X)
    partition = X_exp.sum(1, keepdims=True)
    return X_exp / partition  # The broadcasting mechanism is applied here
```

For any input X, we turn each element into a nonnegative number. Each row sums up to 1, as is required for a probability. Caution: the code above is *not* robust against very large or very small arguments. While it is sufficient to illustrate what is happening, you should *not* use this code verbatim for any serious purpose. Deep learning frameworks have such protections built in and we will be using the built-in softmax going forward.

```
X = torch.rand((2, 5))
X_prob = softmax(X)
X_prob, X_prob.sum(1)
```

```
(tensor([[0.1785, 0.1864, 0.1874, 0.2815, 0.1663],
         [0.1830, 0.2041, 0.2380, 0.1292, 0.2457]]),
 tensor([1., 1.]))
```

4.4.2 The Model

We now have everything that we need to implement the softmax regression model. As in our linear regression example, each instance will be represented by a fixed-length vector. Since the raw data here consists of 28×28 pixel images, we flatten each image, treating them as vectors of length 784. In later chapters, we will introduce convolutional neural networks, which exploit the spatial structure in a more satisfying way.

In softmax regression, the number of outputs from our network should be equal to the number of classes. Since our dataset has 10 classes, our network has an output dimension of 10. Consequently, our weights constitute a 784×10 matrix plus a 1×10 row vector for the biases. As with linear regression, we initialize the weights W with Gaussian noise. The biases are initialized as zeros.

```
class SoftmaxRegressionScratch(d2l.Classifier):
    def __init__(self, num_inputs, num_outputs, lr, sigma=0.01):
        super().__init__()
        self.save_hyperparameters()
        self.W = torch.normal(0, sigma, size=(num_inputs, num_outputs),
                              requires_grad=True)
        self.b = torch.zeros(num_outputs, requires_grad=True)

    def parameters(self):
        return [self.W, self.b]
```

The code below defines how the network maps each input to an output. Note that we flatten each 28×28 pixel image in the batch into a vector using `reshape` before passing the data through our model.

```
@d2l.add_to_class(SoftmaxRegressionScratch)
def forward(self, X):
    X = X.reshape((-1, self.W.shape[0]))
    return softmax(torch.matmul(X, self.W) + self.b)
```

4.4.3 The Cross-Entropy Loss

Next we need to implement the cross-entropy loss function (introduced in Section 4.1.2). This may be the most common loss function in all of deep learning. At the moment, applications of deep learning easily cast as classification problems far outnumber those better treated as regression problems.

Recall that cross-entropy takes the negative log-likelihood of the predicted probability assigned to the true label. For efficiency we avoid Python for-loops and use indexing instead. In particular, the one-hot encoding in $\mathbf{y}$ allows us to select the matching terms in $\hat{\mathbf{y}}$.

To see this in action we create sample data y_hat with 2 examples of predicted probabilities over 3 classes and their corresponding labels y. The correct labels are 0 and 2 respectively (i.e., the first and third class). Using y as the indices of the probabilities in y_hat, we can pick out terms efficiently.

```
y = torch.tensor([0, 2])
y_hat = torch.tensor([[0.1, 0.3, 0.6], [0.3, 0.2, 0.5]])
y_hat[[0, 1], y]
```

```
tensor([0.1000, 0.5000])
```

Now we can implement the cross-entropy loss function by averaging over the logarithms of the selected probabilities.

```
def cross_entropy(y_hat, y):
    return -torch.log(y_hat[list(range(len(y_hat))), y]).mean()

cross_entropy(y_hat, y)
```

```
tensor(1.4979)
```

```
@d21.add_to_class(SoftmaxRegressionScratch)
def loss(self, y_hat, y):
    return cross_entropy(y_hat, y)
```

4.4.4 Training

We reuse the fit method defined in Section 3.4 to train the model with 10 epochs. Note that the number of epochs (max_epochs), the minibatch size (batch_size), and learning rate (lr) are adjustable hyperparameters. That means that while these values are not learned during our primary training loop, they still influence the performance of our model, both vis-à-vis training and generalization performance. In practice you will want to choose these values based on the *validation* split of the data and then, ultimately, to evaluate your final model on the *test* split. As discussed in Section 3.6.3, we will regard the test data of Fashion-MNIST as the validation set, thus reporting validation loss and validation accuracy on this split.

```
data = d2l.FashionMNIST(batch_size=256)
model = SoftmaxRegressionScratch(num_inputs=784, num_outputs=10, lr=0.1)
trainer = d2l.Trainer(max_epochs=10)
trainer.fit(model, data)
```

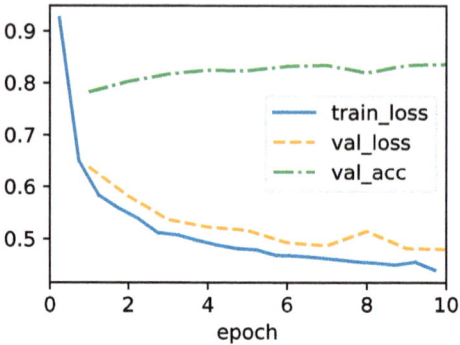

4.4.5 Prediction

Now that training is complete, our model is ready to classify some images.

```
X, y = next(iter(data.val_dataloader()))
preds = model(X).argmax(axis=1)
preds.shape
```

```
torch.Size([256])
```

We are more interested in the images we label *incorrectly*. We visualize them by comparing their actual labels (first line of text output) with the predictions from the model (second line of text output).

```
wrong = preds.type(y.dtype) != y
X, y, preds = X[wrong], y[wrong], preds[wrong]
labels = [a+'\n'+b for a, b in zip(
    data.text_labels(y), data.text_labels(preds))]
data.visualize([X, y], labels=labels)
```

sneaker   coat   pullover   sandal   ankle boot   coat   dress   shirt
sandal   pullover   t-shirt   sneaker   sneaker   pullover   coat   t-shirt

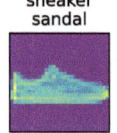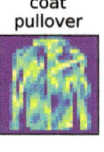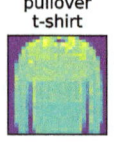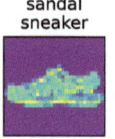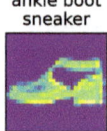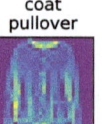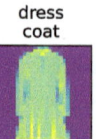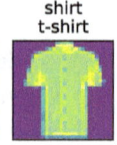

4.4.6 Summary

By now we are starting to get some experience with solving linear regression and classification problems. With it, we have reached what would arguably be the state of the art of 1960–1970s of statistical modeling. In the next section, we will show you how to leverage deep learning frameworks to implement this model much more efficiently.

4.4.7 Exercises

1. In this section, we directly implemented the softmax function based on the mathematical definition of the softmax operation. As discussed in Section 4.1 this can cause numerical instabilities.

    1. Test whether `softmax` still works correctly if an input has a value of 100.

    2. Test whether `softmax` still works correctly if the largest of all inputs is smaller than -100.

    3. Implement a fix by looking at the value relative to the largest entry in the argument.

2. Implement a `cross_entropy` function that follows the definition of the cross-entropy loss function $\sum_i y_i \log \hat{y}_i$.

    1. Try it out in the code example of this section.

    2. Why do you think it runs more slowly?

    3. Should you use it? When would it make sense to?

    4. What do you need to be careful of? Hint: consider the domain of the logarithm.

3. Is it always a good idea to return the most likely label? For example, would you do this for medical diagnosis? How would you try to address this?

4. Assume that we want to use softmax regression to predict the next word based on some features. What are some problems that might arise from a large vocabulary?

5. Experiment with the hyperparameters of the code in this section. In particular:

    1. Plot how the validation loss changes as you change the learning rate.

    2. Do the validation and training loss change as you change the minibatch size? How large or small do you need to go before you see an effect?

Discussions[97].

97

4.5 Concise Implementation of Softmax Regression

Just as high-level deep learning frameworks made it easier to implement linear regression (see Section 3.5), they are similarly convenient here.

```
import torch
from torch import nn
from torch.nn import functional as F
from d2l import torch as d2l
```

4.5.1 Defining the Model

As in Section 3.5, we construct our fully connected layer using the built-in layer. The built-in `__call__` method then invokes `forward` whenever we need to apply the network to some input.

We use a `Flatten` layer to convert the fourth-order tensor X to second order by keeping the dimensionality along the first axis unchanged.

```
class SoftmaxRegression(d2l.Classifier):    #@save
    """The softmax regression model."""
    def __init__(self, num_outputs, lr):
        super().__init__()
        self.save_hyperparameters()
        self.net = nn.Sequential(nn.Flatten(),
                                 nn.LazyLinear(num_outputs))

    def forward(self, X):
        return self.net(X)
```

4.5.2 Softmax Revisited

In Section 4.4 we calculated our model's output and applied the cross-entropy loss. While this is perfectly reasonable mathematically, it is risky computationally, because of numerical underflow and overflow in the exponentiation.

Recall that the softmax function computes probabilities via $\hat{y}_j = \frac{\exp(o_j)}{\sum_k \exp(o_k)}$. If some of the o_k are very large, i.e., very positive, then $\exp(o_k)$ might be larger than the largest number we can have for certain data types. This is called *overflow*. Likewise, if every argument is a very large negative number, we will get *underflow*. For instance, single precision floating point numbers approximately cover the range of 10^{-38} to 10^{38}. As such, if the largest term in o lies outside the interval $[-90, 90]$, the result will not be stable. A way round this problem is to subtract $\bar{o} \overset{\text{def}}{=} \max_k o_k$ from all entries:

$$\hat{y}_j = \frac{\exp o_j}{\sum_k \exp o_k} = \frac{\exp(o_j - \bar{o})\exp \bar{o}}{\sum_k \exp(o_k - \bar{o})\exp \bar{o}} = \frac{\exp(o_j - \bar{o})}{\sum_k \exp(o_k - \bar{o})}. \tag{4.5.1}$$

By construction we know that $o_j - \bar{o} \leq 0$ for all j. As such, for a q-class classification problem, the denominator is contained in the interval $[1, q]$. Moreover, the numerator never exceeds 1, thus preventing numerical overflow. Numerical underflow only occurs when $\exp(o_j - \bar{o})$ numerically evaluates as 0. Nonetheless, a few steps down the road we might find ourselves in trouble when we want to compute $\log \hat{y}_j$ as $\log 0$. In particular, in backpropagation, we might find ourselves faced with a screenful of the dreaded NaN (Not a Number) results.

Fortunately, we are saved by the fact that even though we are computing exponential functions, we ultimately intend to take their log (when calculating the cross-entropy loss). By combining softmax and cross-entropy, we can escape the numerical stability issues altogether. We have:

$$\log \hat{y}_j = \log \frac{\exp(o_j - \bar{o})}{\sum_k \exp(o_k - \bar{o})} = o_j - \bar{o} - \log \sum_k \exp(o_k - \bar{o}). \qquad (4.5.2)$$

This avoids both overflow and underflow. We will want to keep the conventional softmax function handy in case we ever want to evaluate the output probabilities by our model. But instead of passing softmax probabilities into our new loss function, we just pass the logits and compute the softmax and its log all at once inside the cross-entropy loss function, which does smart things like the "LogSumExp trick"[98].

```
@d21.add_to_class(d21.Classifier)  #@save
def loss(self, Y_hat, Y, averaged=True):
    Y_hat = Y_hat.reshape((-1, Y_hat.shape[-1]))
    Y = Y.reshape((-1,))
    return F.cross_entropy(
        Y_hat, Y, reduction='mean' if averaged else 'none')
```

4.5.3 Training

Next we train our model. We use Fashion-MNIST images, flattened to 784-dimensional feature vectors.

```
data = d21.FashionMNIST(batch_size=256)
model = SoftmaxRegression(num_outputs=10, lr=0.1)
trainer = d21.Trainer(max_epochs=10)
trainer.fit(model, data)
```

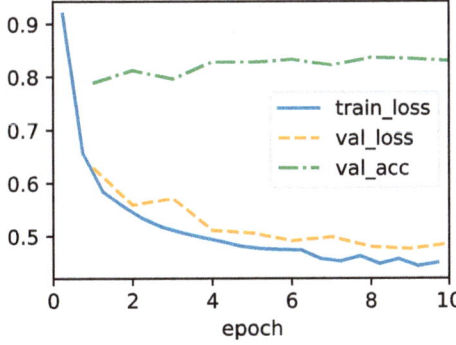

As before, this algorithm converges to a solution that is reasonably accurate, albeit this time with fewer lines of code than before.

4.5.4 Summary

High-level APIs are very convenient at hiding from their user potentially dangerous aspects, such as numerical stability. Moreover, they allow users to design models concisely with very few lines of code. This is both a blessing and a curse. The obvious benefit is that it makes things highly accessible, even to engineers who never took a single class of statistics in their life (in fact, they are part of the target audience of the book). But hiding the sharp edges also comes with a price: a disincentive to add new and different components on your own, since there is little muscle memory for doing it. Moreover, it makes it more difficult to *fix* things whenever the protective padding of a framework fails to cover all the corner cases entirely. Again, this is due to lack of familiarity.

As such, we strongly urge you to review *both* the bare bones and the elegant versions of many of the implementations that follow. While we emphasize ease of understanding, the implementations are nonetheless usually quite performant (convolutions are the big exception here). It is our intention to allow you to build on these when you invent something new that no framework can give you.

4.5.5 Exercises

1. Deep learning uses many different number formats, including FP64 double precision (used extremely rarely), FP32 single precision, BFLOAT16 (good for compressed representations), FP16 (very unstable), TF32 (a new format from NVIDIA), and INT8. Compute the smallest and largest argument of the exponential function for which the result does not lead to numerical underflow or overflow.

2. INT8 is a very limited format consisting of nonzero numbers from 1 to 255. How could you extend its dynamic range without using more bits? Do standard multiplication and addition still work?

3. Increase the number of epochs for training. Why might the validation accuracy decrease after a while? How could we fix this?

4. What happens as you increase the learning rate? Compare the loss curves for several learning rates. Which one works better? When?

Discussions[99].

4.6 Generalization in Classification

So far, we have focused on how to tackle multiclass classification problems by training (linear) neural networks with multiple outputs and softmax functions. Interpreting our model's outputs as probabilistic predictions, we motivated and derived the cross-entropy loss function, which calculates the negative log-likelihood that our model (for a fixed set of parameters) assigns to the actual labels. And finally, we put these tools into practice by fitting our model to the training set. However, as always, our goal is to learn *general patterns*, as assessed empirically on previously unseen data (the test set). High accuracy on the training set means nothing. Whenever each of our inputs is unique (and indeed this is true for most high-dimensional datasets), we can attain perfect accuracy on the training set by just memorizing the dataset on the first training epoch, and subsequently looking up the label whenever we see a new image. And yet, memorizing the exact labels associated with the exact training examples does not tell us how to classify new examples. Absent further guidance, we might have to fall back on random guessing whenever we encounter new examples.

A number of burning questions demand immediate attention:

1. How many test examples do we need to give a good estimate of the accuracy of our classifiers on the underlying population?

2. What happens if we keep evaluating models on the same test repeatedly?

3. Why should we expect that fitting our linear models to the training set should fare any better than our naive memorization scheme?

Whereas Section 3.6 introduced the basics of overfitting and generalization in the context of linear regression, this chapter will go a little deeper, introducing some of the foundational ideas of statistical learning theory. It turns out that we often can guarantee generalization *a priori*: for many models, and for any desired upper bound on the generalization gap ϵ, we can often determine some required number of samples n such that if our training set contains at least n samples, our empirical error will lie within ϵ of the true error, *for any data generating distribution*. Unfortunately, it also turns out that while these sorts of guarantees provide a profound set of intellectual building blocks, they are of limited practical utility to the deep learning practitioner. In short, these guarantees suggest that ensuring generalization of deep neural networks *a priori* requires an absurd number of examples (perhaps trillions or more), even when we find that, on the tasks we care about, deep neural networks typically generalize remarkably well with far fewer examples (thousands). Thus deep learning practitioners often forgo *a priori* guarantees altogether, instead employing methods that have generalized well on similar problems in the past, and certifying generalization *post hoc* through empirical evaluations. When we get to Chapter 5, we will revisit generalization and provide a light introduction to the vast scientific literature that has sprung in attempts to explain why deep neural networks generalize in practice.

4.6.1 The Test Set

Since we have already begun to rely on test sets as the gold standard method for assessing generalization error, let's get started by discussing the properties of such error estimates. Let's focus on a fixed classifier f, without worrying about how it was obtained. Moreover suppose that we possess a *fresh* dataset of examples $\mathcal{D} = (\mathbf{x}^{(i)}, y^{(i)})_{i=1}^{n}$ that were not used to train the classifier f. The *empirical error* of our classifier f on $\mathcal{D}$ is simply the fraction of instances for which the prediction $f(\mathbf{x}^{(i)})$ disagrees with the true label $y^{(i)}$, and is given by the following expression:

$$\epsilon_{\mathcal{D}}(f) = \frac{1}{n} \sum_{i=1}^{n} \mathbf{1}(f(\mathbf{x}^{(i)}) \neq y^{(i)}). \tag{4.6.1}$$

By contrast, the *population error* is the *expected* fraction of examples in the underlying population (some distribution $P(X, Y)$ characterized by probability density function $p(\mathbf{x}, y)$) for which our classifier disagrees with the true label:

$$\epsilon(f) = E_{(\mathbf{x},y)\sim P} \mathbf{1}(f(\mathbf{x}) \neq y) = \int \int \mathbf{1}(f(\mathbf{x}) \neq y) p(\mathbf{x}, y) \, d\mathbf{x} dy. \tag{4.6.2}$$

While $\epsilon(f)$ is the quantity that we actually care about, we cannot observe it directly, just as we cannot directly observe the average height in a large population without measuring every single person. We can only estimate this quantity based on samples. Because our test set $\mathcal{D}$ is statistically representative of the underlying population, we can view $\epsilon_{\mathcal{D}}(f)$ as a statistical estimator of the population error $\epsilon(f)$. Moreover, because our quantity of interest $\epsilon(f)$ is an expectation (of the random variable $\mathbf{1}(f(X) \neq Y)$) and the corresponding estimator $\epsilon_{\mathcal{D}}(f)$ is the sample average, estimating the population error is simply the classic problem of mean estimation, which you may recall from Section 2.6.

An important classical result from probability theory called the *central limit theorem* guarantees that whenever we possess n random samples $a_1, \ldots, a_n$ drawn from any distribution with mean μ and standard deviation σ, then, as the number of samples n approaches infinity, the sample average $\hat{\mu}$ approximately tends towards a normal distribution centered at the true mean and with standard deviation $\sigma/\sqrt{n}$. Already, this tells us something important: as the number of examples grows large, our test error $\epsilon_{\mathcal{D}}(f)$ should approach the true error $\epsilon(f)$ at a rate of $O(1/\sqrt{n})$. Thus, to estimate our test error twice as precisely, we must collect four times as large a test set. To reduce our test error by a factor of one hundred, we must collect ten thousand times as large a test set. In general, such a rate of $O(1/\sqrt{n})$ is often the best we can hope for in statistics.

Now that we know something about the asymptotic rate at which our test error $\epsilon_{\mathcal{D}}(f)$ converges to the true error $\epsilon(f)$, we can zoom in on some important details. Recall that the random variable of interest $\mathbf{1}(f(X) \neq Y)$ can only take values 0 and 1 and thus is a Bernoulli random variable, characterized by a parameter indicating the probability that it takes value 1. Here, 1 means that our classifier made an error, so the parameter of our random variable is actually the true error rate $\epsilon(f)$. The variance σ^2 of a Bernoulli depends on its parameter (here, $\epsilon(f)$) according to the expression $\epsilon(f)(1 - \epsilon(f))$. While $\epsilon(f)$ is initially unknown, we know that it cannot be greater than 1. A little investigation of this function reveals that our variance is highest when the true error rate is close to 0.5 and can be far lower when it is close to 0 or close to 1. This tells us that the asymptotic standard deviation of our estimate $\epsilon_{\mathcal{D}}(f)$ of the error $\epsilon(f)$ (over the choice of the n test samples) cannot be any greater than $\sqrt{0.25/n}$.

If we ignore the fact that this rate characterizes behavior as the test set size approaches infinity rather than when we possess finite samples, this tells us that if we want our test error $\epsilon_{\mathcal{D}}(f)$ to approximate the population error $\epsilon(f)$ such that one standard deviation corresponds to an interval of ± 0.01, then we should collect roughly 2500 samples. If we want to fit two standard deviations in that range and thus be 95% confident that $\epsilon_{\mathcal{D}}(f) \in \epsilon(f) \pm 0.01$, then we will need 10,000 samples!

This turns out to be the size of the test sets for many popular benchmarks in machine learning. You might be surprised to find out that thousands of applied deep learning papers get published every year making a big deal out of error rate improvements of 0.01 or less. Of course, when the error rates are much closer to 0, then an improvement of 0.01 can indeed be a big deal.

One pesky feature of our analysis thus far is that it really only tells us about asymptotics, i.e., how the relationship between $\epsilon_{\mathcal{D}}$ and ϵ evolves as our sample size goes to infinity. Fortunately, because our random variable is bounded, we can obtain valid finite sample bounds by applying an inequality due to Hoeffding (1963):

$$P(\epsilon_{\mathcal{D}}(f) - \epsilon(f) \geq t) < \exp\left(-2nt^2\right). \tag{4.6.3}$$

Solving for the smallest dataset size that would allow us to conclude with 95% confidence that the distance t between our estimate $\epsilon_{\mathcal{D}}(f)$ and the true error rate $\epsilon(f)$ does not exceed 0.01, you will find that roughly 15,000 examples are required as compared to the 10,000 examples suggested by the asymptotic analysis above. If you go deeper into statistics you will find that this trend holds generally. Guarantees that hold even in finite samples are typically slightly more conservative. Note that in the scheme of things, these numbers are not so far apart, reflecting the general usefulness of asymptotic analysis for giving us ballpark figures even if they are not guarantees we can take to court.

4.6.2 Test Set Reuse

In some sense, you are now set up to succeed at conducting empirical machine learning research. Nearly all practical models are developed and validated based on test set performance and you are now a master of the test set. For any fixed classifier f, you know how to evaluate its test error $\epsilon_{\mathcal{D}}(f)$, and know precisely what can (and cannot) be said about its population error $\epsilon(f)$.

So let's say that you take this knowledge and prepare to train your first model f_1. Knowing just how confident you need to be in the performance of your classifier's error rate you apply our analysis above to determine an appropriate number of examples to set aside for the test set. Moreover, let's assume that you took the lessons from Section 3.6 to heart and made sure to preserve the sanctity of the test set by conducting all of your preliminary analysis, hyperparameter tuning, and even selection among multiple competing model architectures on a validation set. Finally you evaluate your model f_1 on the test set and report an unbiased estimate of the population error with an associated confidence interval.

So far everything seems to be going well. However, that night you wake up at 3am with a brilliant idea for a new modeling approach. The next day, you code up your new model, tune its hyperparameters on the validation set and not only are you getting your new model f_2 to work but its error rate appears to be much lower than f_1's. However, the thrill of discovery suddenly fades as you prepare for the final evaluation. You do not have a test set!

Even though the original test set $\mathcal{D}$ is still sitting on your server, you now face two formidable problems. First, when you collected your test set, you determined the required level of precision under the assumption that you were evaluating a single classifier f. However, if you get into the business of evaluating multiple classifiers $f_1, ..., f_k$ on the same test set, you must consider the problem of false discovery. Before, you might have been 95% sure that $\epsilon_{\mathcal{D}}(f) \in \epsilon(f) \pm 0.01$ for a single classifier f and thus the probability of a misleading result was a mere 5%. With k classifiers in the mix, it can be hard to guarantee that there is not even one among them whose test set performance is misleading. With 20 classifiers under consideration, you might have no power at all to rule out the possibility that at least one among them received a misleading score. This problem relates to multiple hypothesis testing, which despite a vast literature in statistics, remains a persistent problem plaguing scientific research.

If that is not enough to worry you, there is a special reason to distrust the results that you get on subsequent evaluations. Recall that our analysis of test set performance rested on the assumption that the classifier was chosen absent any contact with the test set and thus we could view the test set as drawn randomly from the underlying population. Here, not only are you testing multiple functions, the subsequent function f_2 was chosen after you observed the test set performance of f_1. Once information from the test set has leaked to the modeler, it can never be a true test set again in the strictest sense. This problem is called *adaptive overfitting* and has recently emerged as a topic of intense interest to learning theorists and statisticians (Dwork *et al.*, 2015). Fortunately, while it is possible to leak all information out of a holdout set, and the theoretical worst case scenarios are bleak, these analyses may be too conservative. In practice, take care to create real test sets, to consult them as infrequently as possible, to account for multiple hypothesis testing when reporting confidence intervals, and to dial up your vigilance more aggressively when the stakes are high and your dataset size is small. When running a series of benchmark challenges, it is often good practice to maintain several test sets so that after each round, the old test set can be demoted to a validation set.

4.6.3 Statistical Learning Theory

Put simply, *test sets are all that we really have*, and yet this fact seems strangely unsatisfying. First, we seldom possess a *true test set*—unless we are the ones creating the dataset, someone else has probably already evaluated their own classifier on our ostensible "test set". And even when we have first dibs, we soon find ourselves frustrated, wishing we could evaluate our subsequent modeling attempts without the gnawing feeling that we cannot trust our numbers. Moreover, even a true test set can only tell us *post hoc* whether a classifier has in fact generalized to the population, not whether we have any reason to expect *a priori* that it should generalize.

With these misgivings in mind, you might now be sufficiently primed to see the appeal of *statistical learning theory*, the mathematical subfield of machine learning whose practitioners aim to elucidate the fundamental principles that explain why/when models trained on empirical data can/will generalize to unseen data. One of the primary aims of statistical learning researchers has been to bound the generalization gap, relating the properties of the model class to the number of samples in the dataset.

Learning theorists aim to bound the difference between the *empirical error* $\epsilon_S(f_S)$ of a learned classifier f_S, both trained and evaluated on the training set S, and the true error $\epsilon(f_S)$ of that same classifier on the underlying population. This might look similar to the evaluation problem that we just addressed but there is a major difference. Earlier, the classifier f was fixed and we only needed a dataset for evaluative purposes. And indeed, any fixed classifier does generalize: its error on a (previously unseen) dataset is an unbiased estimate of the population error. But what can we say when a classifier is trained and evaluated on the same dataset? Can we ever be confident that the training error will be close to the testing error?

Suppose that our learned classifier f_S must be chosen from some pre-specified set of functions $\mathcal{F}$. Recall from our discussion of test sets that while it is easy to estimate the error of a single classifier, things get hairy when we begin to consider collections of classifiers. Even if the empirical error of any one (fixed) classifier will be close to its true error with high probability, once we consider a collection of classifiers, we need to worry about the possibility that *just one* of them will receive a badly estimated error. The worry is that we might pick such a classifier and thereby grossly underestimate the population error. Moreover, even for linear models, because their parameters are continuously valued, we are typically choosing from an infinite class of functions ($|\mathcal{F}| = \infty$).

One ambitious solution to the problem is to develop analytic tools for proving uniform convergence, i.e., that with high probability, the empirical error rate for every classifier in the class $f \in \mathcal{F}$ will *simultaneously* converge to its true error rate. In other words, we seek a theoretical principle that would allow us to state that with probability at least $1 - \delta$ (for some small δ) no classifier's error rate $\epsilon(f)$ (among all classifiers in the class $\mathcal{F}$) will be misestimated by more than some small amount α. Clearly, we cannot make such statements for all model classes $\mathcal{F}$. Recall the class of memorization machines that always achieve empirical error 0 but never outperform random guessing on the underlying population.

In a sense the class of memorizers is too flexible. No such a uniform convergence result could possibly hold. On the other hand, a fixed classifier is useless—it generalizes perfectly, but fits neither the training data nor the test data. The central question of learning has thus historically been framed as a trade-off between more flexible (higher variance) model classes that better fit the training data but risk overfitting, versus more rigid (higher bias) model classes that generalize well but risk underfitting. A central question in learning theory has been to develop the appropriate mathematical analysis to quantify where a model sits along this spectrum, and to provide the associated guarantees.

In a series of seminal papers, Vapnik and Chervonenkis extended the theory on the convergence of relative frequencies to more general classes of functions (Vapnik and Chervonenkis, 1964, 1968, 1971, 1974, 1981, 1991). One of the key contributions of this line of work is the Vapnik–Chervonenkis (VC) dimension, which measures (one notion of) the complexity (flexibility) of a model class. Moreover, one of their key results bounds the difference between the empirical error and the population error as a function of the VC dimension and the number of samples:

$$P\left(R[p, f] - R_{\mathrm{emp}}[\mathbf{X}, \mathbf{Y}, f] < \alpha\right) \geq 1 - \delta \text{ for } \alpha \geq c\sqrt{(\mathrm{VC} - \log \delta)/n}. \qquad (4.6.4)$$

Here $\delta > 0$ is the probability that the bound is violated, α is the upper bound on the generalization gap, and n is the dataset size. Lastly, $c > 0$ is a constant that depends only on the scale of the loss that can be incurred. One use of the bound might be to plug in desired values of δ and α to determine how many samples to collect. The VC dimension quantifies the largest number of data points for which we can assign any arbitrary (binary) labeling and for each find some model f in the class that agrees with that labeling. For example, linear models on d-dimensional inputs have VC dimension $d + 1$. It is easy to see that a line can assign any possible labeling to three points in two dimensions, but not to four. Unfortunately, the theory tends to be overly pessimistic for more complex models and obtaining this guarantee typically requires far more examples than are actually needed to achieve the desired error rate. Note also that fixing the model class and δ, our error rate again decays with the usual $O(1/\sqrt{n})$ rate. It seems unlikely that we could do better in terms of n. However, as we vary the model class, VC dimension can present a pessimistic picture of the generalization gap.

4.6.4 Summary

The most straightforward way to evaluate a model is to consult a test set comprised of previously unseen data. Test set evaluations provide an unbiased estimate of the true error and converge at the desired $O(1/\sqrt{n})$ rate as the test set grows. We can provide approximate confidence intervals based on exact asymptotic distributions or valid finite sample confidence intervals based on (more conservative) finite sample guarantees. Indeed test set evaluation is the bedrock of modern machine learning research. However, test sets are seldom true test sets (used by multiple researchers again and again). Once the same test set is used to evaluate multiple models, controlling for false discovery can be difficult. This can cause huge problems in theory. In practice, the significance of the problem depends on the size of the holdout sets in question and whether they are merely being used to choose hyperparameters or if they are leaking information more directly. Nevertheless, it is good practice to curate real test sets (or multiple) and to be as conservative as possible about how often they are used.

Hoping to provide a more satisfying solution, statistical learning theorists have developed methods for guaranteeing uniform convergence over a model class. If indeed every model's empirical error simultaneously converges to its true error, then we are free to choose the model that performs best, minimizing the training error, knowing that it too will perform similarly well on the holdout data. Crucially, any one of such results must depend on some property of the model class. Vladimir Vapnik and Alexey Chernovenkis introduced the VC dimension, presenting uniform convergence results that hold for all models in a VC class. The training errors for all models in the class are (simultaneously) guaranteed to be close to their true errors, and guaranteed to grow even closer at $O(1/\sqrt{n})$ rates. Following the revolutionary discovery of VC dimension, numerous alternative complexity measures have been proposed, each facilitating an analogous generalization guarantee. See Boucheron *et al.* (2005) for a detailed discussion of several advanced ways of measuring function complexity. Unfortunately, while these complexity measures have become broadly useful tools in statistical theory, they turn out to be powerless (as straightforwardly applied) for explaining why deep neural networks generalize. Deep neural networks often have millions of parameters (or more), and can easily assign random labels to large collections of points. Nevertheless, they generalize well on practical problems and, surprisingly, they often generalize better, when they are larger and deeper, despite incurring higher VC dimensions. In the next chapter, we will revisit generalization in the context of deep learning.

4.6.5 Exercises

1. If we wish to estimate the error of a fixed model f to within 0.0001 with probability greater than 99.9%, how many samples do we need?

2. Suppose that somebody else possesses a labeled test set $\mathcal{D}$ and only makes available the unlabeled inputs (features). Now suppose that you can only access the test set labels by running a model f (with no restrictions placed on the model class) on each of the unlabeled inputs and receiving the corresponding error $\epsilon_{\mathcal{D}}(f)$. How many models would you need to evaluate before you leak the entire test set and thus could appear to have error 0, regardless of your true error?

3. What is the VC dimension of the class of fifth-order polynomials?

4. What is the VC dimension of axis-aligned rectangles on two-dimensional data?

Discussions[100].

100

4.7 Environment and Distribution Shift

In the previous sections, we worked through a number of hands-on applications of machine learning, fitting models to a variety of datasets. And yet, we never stopped to contemplate either where data came from in the first place or what we ultimately plan to do with the outputs from our models. Too often, machine learning developers in possession of data rush to develop models without pausing to consider these fundamental issues.

Many failed machine learning deployments can be traced back to this failure. Sometimes models appear to perform marvelously as measured by test set accuracy but fail catastrophically in deployment when the distribution of data suddenly shifts. More insidiously, sometimes the very deployment of a model can be the catalyst that perturbs the data distribution. Say, for example, that we trained a model to predict who will repay rather than default on a loan, finding that an applicant's choice of footwear was associated with the risk of default (Oxfords indicate repayment, sneakers indicate default). We might be inclined thereafter to grant a loan to any applicant wearing Oxfords and to deny all applicants wearing sneakers.

In this case, our ill-considered leap from pattern recognition to decision-making and our failure to critically consider the environment might have disastrous consequences. For starters, as soon as we began making decisions based on footwear, customers would catch on and change their behavior. Before long, all applicants would be wearing Oxfords, without any coincident improvement in credit-worthiness. Take a minute to digest this because similar issues abound in many applications of machine learning: by introducing our model-based decisions to the environment, we might break the model.

While we cannot possibly give these topics a complete treatment in one section, we aim here to expose some common concerns, and to stimulate the critical thinking required to detect such situations early, mitigate damage, and use machine learning responsibly. Some of the solutions are simple (ask for the "right" data), some are technically difficult (implement a reinforcement learning system), and others require that we step outside the realm of statistical prediction altogether and grapple with difficult philosophical questions concerning the ethical application of algorithms.

4.7.1 Types of Distribution Shift

To begin, we stick with the passive prediction setting considering the various ways that data distributions might shift and what might be done to salvage model performance. In one classic setup, we assume that our training data was sampled from some distribution $p_S(\mathbf{x}, y)$ but that our test data will consist of unlabeled examples drawn from some different distribution $p_T(\mathbf{x}, y)$. Already, we must confront a sobering reality. Absent any assumptions on how p_S and p_T relate to each other, learning a robust classifier is impossible.

Consider a binary classification problem, where we wish to distinguish between dogs and cats. If the distribution can shift in arbitrary ways, then our setup permits the pathological case in which the distribution over inputs remains constant: $p_S(\mathbf{x}) = p_T(\mathbf{x})$, but the labels are all flipped: $p_S(y \mid \mathbf{x}) = 1 - p_T(y \mid \mathbf{x})$. In other words, if God can suddenly decide that in the future all "cats" are now dogs and what we previously called "dogs" are now cats—without any change in the distribution of inputs $p(\mathbf{x})$, then we cannot possibly distinguish this setting from one in which the distribution did not change at all.

Fortunately, under some restricted assumptions on the ways our data might change in the future, principled algorithms can detect shift and sometimes even adapt on the fly, improving on the accuracy of the original classifier.

Covariate Shift

Among categories of distribution shift, covariate shift may be the most widely studied. Here, we assume that while the distribution of inputs may change over time, the labeling function, i.e., the conditional distribution $P(y \mid \mathbf{x})$ does not change. Statisticians call this *covariate shift* because the problem arises due to a shift in the distribution of the covariates (features). While we can sometimes reason about distribution shift without invoking causality, we note that covariate shift is the natural assumption to invoke in settings where we believe that $\mathbf{x}$ causes y.

Consider the challenge of distinguishing cats and dogs. Our training data might consist of images of the kind in Fig. 4.7.1.

cat   cat   dog   dog

Fig. 4.7.1   Training data for distinguishing cats and dogs (illustrations: Lafeez Hossain / 500px / Getty Images; ilkermetinkursova / iStock / Getty Images Plus; GlobalP / iStock / Getty Images Plus; Musthafa Aboobakuru / 500px / Getty Images).

At test time we are asked to classify the images in Fig. 4.7.2.

cat   cat   dog   dog

Fig. 4.7.2   Test data for distinguishing cats and dogs (illustrations: SIBAS_minich / iStock / Getty Images Plus; Ghrzuzudu / iStock / Getty Images Plus; id-work / DigitalVision Vectors / Getty Images; Yime / iStock / Getty Images Plus).

The training set consists of photos, while the test set contains only cartoons. Training on a dataset with substantially different characteristics from the test set can spell trouble absent a coherent plan for how to adapt to the new domain.

Label Shift

Label shift describes the converse problem. Here, we assume that the label marginal $P(y)$ can change but the class-conditional distribution $P(\mathbf{x} \mid y)$ remains fixed across domains. Label shift is a reasonable assumption to make when we believe that y causes $\mathbf{x}$. For example, we may want to predict diagnoses given their symptoms (or other manifestations), even as the relative prevalence of diagnoses are changing over time. Label shift is the appropriate assumption here because diseases cause symptoms. In some degenerate cases the label shift and covariate shift assumptions can hold simultaneously. For example, when the label is deterministic, the covariate shift assumption will be satisfied, even when y causes

x. Interestingly, in these cases, it is often advantageous to work with methods that flow from the label shift assumption. That is because these methods tend to involve manipulating objects that look like labels (often low-dimensional), as opposed to objects that look like inputs, which tend to be high-dimensional in deep learning.

Concept Shift

We may also encounter the related problem of *concept shift*, which arises when the very definitions of labels can change. This sounds weird—a *cat* is a *cat*, no? However, other categories are subject to changes in usage over time. Diagnostic criteria for mental illness, what passes for fashionable, and job titles, are all subject to considerable amounts of concept shift. It turns out that if we navigate around the United States, shifting the source of our data by geography, we will find considerable concept shift regarding the distribution of names for *soft drinks* as shown in Fig. 4.7.3.

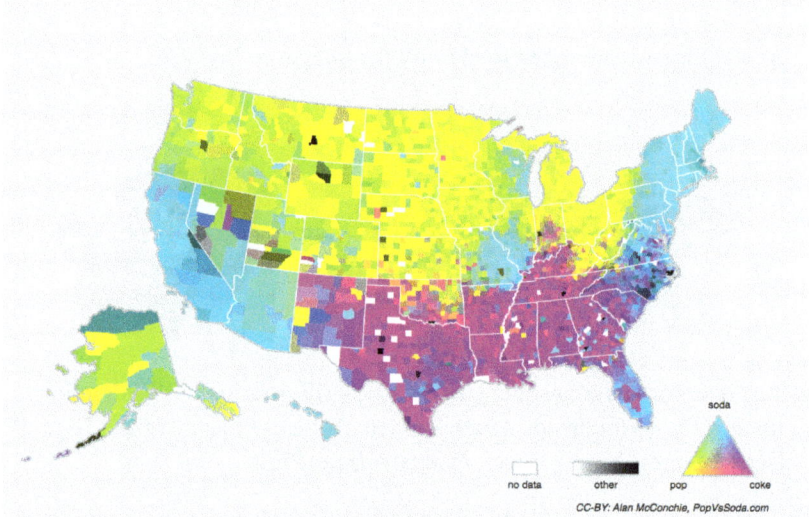

Fig. 4.7.3   Concept shift for soft drink names in the United States (CC-BY: Alan McConchie, PopVsSoda.com).

If we were to build a machine translation system, the distribution $P(y \mid \mathbf{x})$ might be different depending on our location. This problem can be tricky to spot. We might hope to exploit knowledge that shift only takes place gradually either in a temporal or geographic sense.

4.7.2 Examples of Distribution Shift

Before delving into formalism and algorithms, we can discuss some concrete situations where covariate or concept shift might not be obvious.

Medical Diagnostics

Imagine that you want to design an algorithm to detect cancer. You collect data from healthy and sick people and you train your algorithm. It works fine, giving you high accuracy and you conclude that you are ready for a successful career in medical diagnostics. *Not so fast.*

The distributions that gave rise to the training data and those you will encounter in the wild might differ considerably. This happened to an unfortunate startup that some of we authors worked with years ago. They were developing a blood test for a disease that predominantly affects older men and hoped to study it using blood samples that they had collected from patients. However, it is considerably more difficult to obtain blood samples from healthy men than from sick patients already in the system. To compensate, the startup solicited blood donations from students on a university campus to serve as healthy controls in developing their test. Then they asked whether we could help them to build a classifier for detecting the disease.

As we explained to them, it would indeed be easy to distinguish between the healthy and sick cohorts with near-perfect accuracy. However, that is because the test subjects differed in age, hormone levels, physical activity, diet, alcohol consumption, and many more factors unrelated to the disease. This was unlikely to be the case with real patients. Due to their sampling procedure, we could expect to encounter extreme covariate shift. Moreover, this case was unlikely to be correctable via conventional methods. In short, they wasted a significant sum of money.

Self-Driving Cars

Say a company wanted to leverage machine learning for developing self-driving cars. One key component here is a roadside detector. Since real annotated data is expensive to get, they had the (smart and questionable) idea to use synthetic data from a game rendering engine as additional training data. This worked really well on "test data" drawn from the rendering engine. Alas, inside a real car it was a disaster. As it turned out, the roadside had been rendered with a very simplistic texture. More importantly, *all* the roadside had been rendered with the *same* texture and the roadside detector learned about this "feature" very quickly.

A similar thing happened to the US Army when they first tried to detect tanks in the forest. They took aerial photographs of the forest without tanks, then drove the tanks into the forest and took another set of pictures. The classifier appeared to work *perfectly*. Unfortunately, it had merely learned how to distinguish trees with shadows from trees without shadows—the first set of pictures was taken in the early morning, the second set at noon.

Nonstationary Distributions

A much more subtle situation arises when the distribution changes slowly (also known as *nonstationary distribution*) and the model is not updated adequately. Below are some typical cases.

- We train a computational advertising model and then fail to update it frequently (e.g., we forget to incorporate that an obscure new device called an iPad was just launched).

- We build a spam filter. It works well at detecting all spam that we have seen so far. But then the spammers wise up and craft new messages that look unlike anything we have seen before.

- We build a product recommendation system. It works throughout the winter but then continues to recommend Santa hats long after Christmas.

More Anecdotes

- We build a face detector. It works well on all benchmarks. Unfortunately it fails on test data—the offending examples are close-ups where the face fills the entire image (no such data was in the training set).

- We build a web search engine for the US market and want to deploy it in the UK.

- We train an image classifier by compiling a large dataset where each among a large set of classes is equally represented in the dataset, say 1000 categories, represented by 1000 images each. Then we deploy the system in the real world, where the actual label distribution of photographs is decidedly non-uniform.

4.7.3 Correction of Distribution Shift

As we have discussed, there are many cases where training and test distributions $P(\mathbf{x}, y)$ are different. In some cases, we get lucky and the models work despite covariate, label, or concept shift. In other cases, we can do better by employing principled strategies to cope with the shift. The remainder of this section grows considerably more technical. The impatient reader could continue on to the next section as this material is not prerequisite to subsequent concepts.

Empirical Risk and Risk

Let's first reflect on what exactly is happening during model training: we iterate over features and associated labels of training data $\{(\mathbf{x}_1, y_1), \ldots, (\mathbf{x}_n, y_n)\}$ and update the parameters of a model f after every minibatch. For simplicity we do not consider regularization, so we largely minimize the loss on the training:

$$\underset{f}{\text{minimize}} \, \frac{1}{n} \sum_{i=1}^{n} l(f(\mathbf{x}_i), y_i), \tag{4.7.1}$$

where l is the loss function measuring "how bad" the prediction $f(\mathbf{x}_i)$ is given the associated label y_i. Statisticians call the term in (4.7.1) *empirical risk*. The *empirical risk* is an average loss over the training data for approximating the *risk*, which is the expectation of the loss over the entire population of data drawn from their true distribution $p(\mathbf{x}, y)$:

$$E_{p(\mathbf{x}, y)}[l(f(\mathbf{x}), y)] = \int \int l(f(\mathbf{x}), y) p(\mathbf{x}, y) \, d\mathbf{x} dy. \tag{4.7.2}$$

However, in practice we typically cannot obtain the entire population of data. Thus, *empirical risk minimization*, which is minimizing the empirical risk in (4.7.1), is a practical strategy for machine learning, with the hope of approximately minimizing the risk.

Covariate Shift Correction

Assume that we want to estimate some dependency $P(y \mid \mathbf{x})$ for which we have labeled data $(\mathbf{x}_i, y_i)$. Unfortunately, the observations $\mathbf{x}_i$ are drawn from some *source distribution* $q(\mathbf{x})$ rather than the *target distribution* $p(\mathbf{x})$. Fortunately, the dependency assumption means that the conditional distribution does not change: $p(y \mid \mathbf{x}) = q(y \mid \mathbf{x})$. If the source distribution $q(\mathbf{x})$ is "wrong", we can correct for that by using the following simple identity in the risk:

$$\int \int l(f(\mathbf{x}), y) p(y \mid \mathbf{x}) p(\mathbf{x}) \, d\mathbf{x} dy = \int \int l(f(\mathbf{x}), y) q(y \mid \mathbf{x}) q(\mathbf{x}) \frac{p(\mathbf{x})}{q(\mathbf{x})} \, d\mathbf{x} dy.$$
$$(4.7.3)$$

In other words, we need to reweigh each data example by the ratio of the probability that it would have been drawn from the correct distribution to that from the wrong one:

$$\beta_i \overset{\text{def}}{=} \frac{p(\mathbf{x}_i)}{q(\mathbf{x}_i)}.$$
$$(4.7.4)$$

Plugging in the weight β_i for each data example $(\mathbf{x}_i, y_i)$ we can train our model using *weighted empirical risk minimization*:

$$\underset{f}{\text{minimize}} \frac{1}{n} \sum_{i=1}^{n} \beta_i l(f(\mathbf{x}_i), y_i).$$
$$(4.7.5)$$

Alas, we do not know that ratio, so before we can do anything useful we need to estimate it. Many methods are available, including some fancy operator-theoretic approaches that attempt to recalibrate the expectation operator directly using a minimum-norm or a maximum entropy principle. Note that for any such approach, we need samples drawn from both distributions—the "true" p, e.g., by access to test data, and the one used for generating the training set q (the latter is trivially available). Note however, that we only need features $\mathbf{x} \sim p(\mathbf{x})$; we do not need to access labels $y \sim p(y)$.

In this case, there exists a very effective approach that will give almost as good results as the original: namely, logistic regression, which is a special case of softmax regression (see Section 4.1) for binary classification. This is all that is needed to compute estimated probability ratios. We learn a classifier to distinguish between data drawn from $p(\mathbf{x})$ and data drawn from $q(\mathbf{x})$. If it is impossible to distinguish between the two distributions then it means that the associated instances are equally likely to come from either one of those two distributions. On the other hand, any instances that can be well discriminated should be significantly overweighted or underweighted accordingly.

For simplicity's sake assume that we have an equal number of instances from both distributions $p(\mathbf{x})$ and $q(\mathbf{x})$, respectively. Now denote by z labels that are 1 for data drawn from p

and -1 for data drawn from q. Then the probability in a mixed dataset is given by

$$P(z = 1 \mid \mathbf{x}) = \frac{p(\mathbf{x})}{p(\mathbf{x}) + q(\mathbf{x})} \text{ and hence } \frac{P(z = 1 \mid \mathbf{x})}{P(z = -1 \mid \mathbf{x})} = \frac{p(\mathbf{x})}{q(\mathbf{x})}. \qquad (4.7.6)$$

Thus, if we use a logistic regression approach, where $P(z = 1 \mid \mathbf{x}) = \frac{1}{1+\exp(-h(\mathbf{x}))}$ (h is a parametrized function), it follows that

$$\beta_i = \frac{1/(1 + \exp(-h(\mathbf{x}_i)))}{\exp(-h(\mathbf{x}_i))/(1 + \exp(-h(\mathbf{x}_i)))} = \exp(h(\mathbf{x}_i)). \qquad (4.7.7)$$

As a result, we need to solve two problems: the first, to distinguish between data drawn from both distributions, and then a weighted empirical risk minimization problem in (4.7.5) where we weigh terms by β_i.

Now we are ready to describe a correction algorithm. Suppose that we have a training set $\{(\mathbf{x}_1, y_1), \ldots, (\mathbf{x}_n, y_n)\}$ and an unlabeled test set $\{\mathbf{u}_1, \ldots, \mathbf{u}_m\}$. For covariate shift, we assume that $\mathbf{x}_i$ for all $1 \leq i \leq n$ are drawn from some source distribution and $\mathbf{u}_i$ for all $1 \leq i \leq m$ are drawn from the target distribution. Here is a prototypical algorithm for correcting covariate shift:

1. Create a binary-classification training set: $\{(\mathbf{x}_1, -1), \ldots, (\mathbf{x}_n, -1), (\mathbf{u}_1, 1), \ldots, (\mathbf{u}_m, 1)\}$.

2. Train a binary classifier using logistic regression to get the function h.

3. Weigh training data using $\beta_i = \exp(h(\mathbf{x}_i))$ or better $\beta_i = \min(\exp(h(\mathbf{x}_i)), c)$ for some constant c.

4. Use weights β_i for training on $\{(\mathbf{x}_1, y_1), \ldots, (\mathbf{x}_n, y_n)\}$ in (4.7.5).

Note that the above algorithm relies on a crucial assumption. For this scheme to work, we need that each data example in the target (e.g., test time) distribution had nonzero probability of occurring at training time. If we find a point where $p(\mathbf{x}) > 0$ but $q(\mathbf{x}) = 0$, then the corresponding importance weight should be infinity.

Label Shift Correction

Assume that we are dealing with a classification task with k categories. Using the same notation in Section 4.7.3, q and p are the source distribution (e.g., training time) and target distribution (e.g., test time), respectively. Assume that the distribution of labels shifts over time: $q(y) \neq p(y)$, but the class-conditional distribution stays the same: $q(\mathbf{x} \mid y) = p(\mathbf{x} \mid y)$. If the source distribution $q(y)$ is "wrong", we can correct for that according to the following identity in the risk as defined in (4.7.2):

$$\int \int l(f(\mathbf{x}), y) p(\mathbf{x} \mid y) p(y) \, d\mathbf{x} dy = \int \int l(f(\mathbf{x}), y) q(\mathbf{x} \mid y) q(y) \frac{p(y)}{q(y)} \, d\mathbf{x} dy. \qquad (4.7.8)$$

Here, our importance weights will correspond to the label likelihood ratios:

$$\beta_i \overset{\text{def}}{=} \frac{p(y_i)}{q(y_i)}. \qquad (4.7.9)$$

One nice thing about label shift is that if we have a reasonably good model on the source distribution, then we can get consistent estimates of these weights without ever having to deal with the ambient dimension. In deep learning, the inputs tend to be high-dimensional objects like images, while the labels are often simpler objects like categories.

To estimate the target label distribution, we first take our reasonably good off-the-shelf classifier (typically trained on the training data) and compute its "confusion" matrix using the validation set (also from the training distribution). The *confusion matrix*, $\mathbf{C}$, is simply a $k \times k$ matrix, where each column corresponds to the label category (ground truth) and each row corresponds to our model's predicted category. Each cell's value c_{ij} is the fraction of total predictions on the validation set where the true label was j and our model predicted i.

Now, we cannot calculate the confusion matrix on the target data directly because we do not get to see the labels for the examples that we see in the wild, unless we invest in a complex real-time annotation pipeline. What we can do, however, is average all of our model's predictions at test time together, yielding the mean model outputs $\mu(\hat{\mathbf{y}}) \in \mathbb{R}^k$, where the ith element $\mu(\hat{y}_i)$ is the fraction of the total predictions on the test set where our model predicted i.

It turns out that under some mild conditions—if our classifier was reasonably accurate in the first place, and if the target data contains only categories that we have seen before, and if the label shift assumption holds in the first place (the strongest assumption here)—we can estimate the test set label distribution by solving a simple linear system

$$\mathbf{C}p(\mathbf{y}) = \mu(\hat{\mathbf{y}}), \tag{4.7.10}$$

because as an estimate $\sum_{j=1}^{k} c_{ij} p(y_j) = \mu(\hat{y}_i)$ holds for all $1 \leq i \leq k$, where $p(y_j)$ is the jth element of the k-dimensional label distribution vector $p(\mathbf{y})$. If our classifier is sufficiently accurate to begin with, then the confusion matrix $\mathbf{C}$ will be invertible, and we get a solution $p(\mathbf{y}) = \mathbf{C}^{-1}\mu(\hat{\mathbf{y}})$.

Because we observe the labels on the source data, it is easy to estimate the distribution $q(y)$. Then, for any training example i with label y_i, we can take the ratio of our estimated $p(y_i)/q(y_i)$ to calculate the weight β_i, and plug this into weighted empirical risk minimization in (4.7.5).

Concept Shift Correction

Concept shift is much harder to fix in a principled manner. For instance, in a situation where suddenly the problem changes from distinguishing cats from dogs to one of distinguishing white from black animals, it will be unreasonable to assume that we can do much better than just collecting new labels and training from scratch. Fortunately, in practice, such extreme shifts are rare. Instead, what usually happens is that the task keeps on changing slowly. To make things more concrete, here are some examples:

• In computational advertising, new products are launched, old products become less popular. This means that the distribution over ads and their popularity changes gradually and any click-through rate predictor needs to change gradually with it.

- Traffic camera lenses degrade gradually due to environmental wear, affecting image quality progressively.

- News content changes gradually (i.e., most of the news remains unchanged but new stories appear).

In such cases, we can use the same approach that we used for training networks to make them adapt to the change in the data. In other words, we use the existing network weights and simply perform a few update steps with the new data rather than training from scratch.

4.7.4 A Taxonomy of Learning Problems

Armed with knowledge about how to deal with changes in distributions, we can now consider some other aspects of machine learning problem formulation.

Batch Learning

In *batch learning*, we have access to training features and labels $\{(\mathbf{x}_1, y_1), \ldots, (\mathbf{x}_n, y_n)\}$, which we use to train a model $f(\mathbf{x})$. Later on, we deploy this model to score new data $(\mathbf{x}, y)$ drawn from the same distribution. This is the default assumption for any of the problems that we discuss here. For instance, we might train a cat detector based on lots of pictures of cats and dogs. Once we have trained it, we ship it as part of a smart catdoor computer vision system that lets only cats in. This is then installed in a customer's home and is never updated again (barring extreme circumstances).

Online Learning

Now imagine that the data $(\mathbf{x}_i, y_i)$ arrives one sample at a time. More specifically, assume that we first observe $\mathbf{x}_i$, then we need to come up with an estimate $f(\mathbf{x}_i)$. Only once we have done this do we observe y_i and so receive a reward or incur a loss, given our decision. Many real problems fall into this category. For example, we need to predict tomorrow's stock price, which allows us to trade based on that estimate and at the end of the day we find out whether our estimate made us a profit. In other words, in *online learning*, we have the following cycle where we are continuously improving our model given new observations:

$$\text{model } f_t \longrightarrow \text{data } \mathbf{x}_t \longrightarrow \text{estimate } f_t(\mathbf{x}_t) \longrightarrow$$
$$\text{observation } y_t \longrightarrow \text{loss } l(y_t, f_t(\mathbf{x}_t)) \longrightarrow \text{model } f_{t+1} \tag{4.7.11}$$

Bandits

Bandits are a special case of the problem above. While in most learning problems we have a continuously parametrized function f where we want to learn its parameters (e.g., a deep network), in a *bandit* problem we only have a finite number of arms that we can pull, i.e., a finite number of actions that we can take. It is not very surprising that for this simpler problem stronger theoretical guarantees in terms of optimality can be obtained. We list it mainly since this problem is often (confusingly) treated as if it were a distinct learning setting.

Control

In many cases the environment remembers what we did. Not necessarily in an adversarial manner but it will just remember and the response will depend on what happened before. For instance, a coffee boiler controller will observe different temperatures depending on whether it was heating the boiler previously. PID (proportional-integral-derivative) controller algorithms are a popular choice there. Likewise, a user's behavior on a news site will depend on what we showed them previously (e.g., they will read most news only once). Many such algorithms form a model of the environment in which they act so as to make their decisions appear less random. Recently, control theory (e.g., PID variants) has also been used to automatically tune hyperparameters to achieve better disentangling and reconstruction quality, and improve the diversity of generated text and the reconstruction quality of generated images (Shao *et al.*, 2020).

Reinforcement Learning

In the more general case of an environment with memory, we may encounter situations where the environment is trying to cooperate with us (cooperative games, in particular for non-zero-sum games), or others where the environment will try to win. Chess, Go, Backgammon, or StarCraft are some of the cases in *reinforcement learning*. Likewise, we might want to build a good controller for autonomous cars. Other cars are likely to respond to the autonomous car's driving style in nontrivial ways, e.g., trying to avoid it, trying to cause an accident, or trying to cooperate with it.

Considering the Environment

One key distinction between the different situations above is that a strategy that might have worked throughout in the case of a stationary environment, might not work throughout in an environment that can adapt. For instance, an arbitrage opportunity discovered by a trader is likely to disappear once it is exploited. The speed and manner at which the environment changes determines to a large extent the type of algorithms that we can bring to bear. For instance, if we know that things may only change slowly, we can force any estimate to change only slowly, too. If we know that the environment might change instantaneously, but only very infrequently, we can make allowances for that. These types of knowledge are crucial for the aspiring data scientist in dealing with concept shift, i.e., when the problem that is being solved can change over time.

4.7.5 Fairness, Accountability, and Transparency in Machine Learning

Finally, it is important to remember that when you deploy machine learning systems you are not merely optimizing a predictive model—you are typically providing a tool that will be used to (partially or fully) automate decisions. These technical systems can impact the lives of individuals who are subject to the resulting decisions. The leap from considering predictions to making decisions raises not only new technical questions, but also a slew of ethical questions that must be carefully considered. If we are deploying a medical diagnostic system, we need to know for which populations it may work and for which it may not. Overlooking foreseeable risks to the welfare of a subpopulation could cause us to administer inferior care. Moreover, once we contemplate decision-making systems, we must step back and reconsider how we evaluate our technology. Among other consequences of this change of scope, we will find that *accuracy* is seldom the right measure. For instance, when translating predictions into actions, we will often want to take into account the potential cost sensitivity of erring in various ways. If one way of misclassifying an image could be perceived as a racial sleight of hand, while misclassification to a different category would be harmless, then we might want to adjust our thresholds accordingly, accounting for societal values in designing the decision-making protocol. We also want to be careful about how prediction systems can lead to feedback loops. For example, consider predictive policing systems, which allocate patrol officers to areas with high forecasted crime. It is easy to see how a worrying pattern can emerge:

1. Neighborhoods with more crime get more patrols.

2. Consequently, more crimes are discovered in these neighborhoods, entering the training data available for future iterations.

3. Exposed to more positives, the model predicts yet more crime in these neighborhoods.

4. In the next iteration, the updated model targets the same neighborhood even more heavily leading to yet more crimes discovered, etc.

Often, the various mechanisms by which a model's predictions become coupled to its training data are unaccounted for in the modeling process. This can lead to what researchers call *runaway feedback loops*. Additionally, we want to be careful about whether we are addressing the right problem in the first place. Predictive algorithms now play an outsize role in mediating the dissemination of information. Should the news that an individual encounters be determined by the set of Facebook pages they have *Liked*? These are just a few among the many pressing ethical dilemmas that you might encounter in a career in machine learning.

4.7.6 Summary

In many cases training and test sets do not come from the same distribution. This is called distribution shift. The risk is the expectation of the loss over the entire population of data drawn from their true distribution. However, this entire population is usually unavailable.

Empirical risk is an average loss over the training data to approximate the risk. In practice, we perform empirical risk minimization.

Under the corresponding assumptions, covariate and label shift can be detected and corrected for at test time. Failure to account for this bias can become problematic at test time. In some cases, the environment may remember automated actions and respond in surprising ways. We must account for this possibility when building models and continue to monitor live systems, open to the possibility that our models and the environment will become entangled in unanticipated ways.

4.7.7 Exercises

1. What could happen when we change the behavior of a search engine? What might the users do? What about the advertisers?

2. Implement a covariate shift detector. Hint: build a classifier.

3. Implement a covariate shift corrector.

4. Besides distribution shift, what else could affect how the empirical risk approximates the risk?

Discussions [101] .

101

## 5

Multilayer Perceptrons

In this chapter, we will introduce your first truly *deep* network. The simplest deep networks are called *multilayer perceptrons*, and they consist of multiple layers of neurons each fully connected to those in the layer below (from which they receive input) and those above (which they, in turn, influence). Although automatic differentiation significantly simplifies the implementation of deep learning algorithms, we will dive deep into how these gradients are calculated in deep networks. Then we will be ready to discuss issues relating to numerical stability and parameter initialization that are key to successfully training deep networks. When we train such high-capacity models we run the risk of overfitting. Thus, we will revisit regularization and generalization for deep networks. Throughout, we aim to give you a firm grasp not just of the concepts but also of the practice of using deep networks. At the end of this chapter, we apply what we have introduced so far to a real case: house price prediction. We punt matters relating to the computational performance, scalability, and efficiency of our models to subsequent chapters.

5.1 Multilayer Perceptrons

In Section 4.1, we introduced softmax regression, implementing the algorithm from scratch (Section 4.4) and using high-level APIs (Section 4.5). This allowed us to train classifiers capable of recognizing 10 categories of clothing from low-resolution images. Along the way, we learned how to wrangle data, coerce our outputs into a valid probability distribution, apply an appropriate loss function, and minimize it with respect to our model's parameters. Now that we have mastered these mechanics in the context of simple linear models, we can launch our exploration of deep neural networks, the comparatively rich class of models with which this book is primarily concerned.

```
%matplotlib inline
import torch
from d2l import torch as d2l
```

5.1.1 Hidden Layers

We described affine transformations in Section 3.1.1 as linear transformations with added bias. To begin, recall the model architecture corresponding to our softmax regression example, illustrated in Fig. 4.1.1. This model maps inputs directly to outputs via a single affine transformation, followed by a softmax operation. If our labels truly were related to the input data by a simple affine transformation, then this approach would be sufficient. However, linearity (in affine transformations) is a *strong* assumption.

Limitations of Linear Models

For example, linearity implies the *weaker* assumption of *monotonicity*, i.e., that any increase in our feature must either always cause an increase in our model's output (if the corresponding weight is positive), or always cause a decrease in our model's output (if the corresponding weight is negative). Sometimes that makes sense. For example, if we were trying to predict whether an individual will repay a loan, we might reasonably assume that all other things being equal, an applicant with a higher income would always be more likely to repay than one with a lower income. While monotonic, this relationship likely is not linearly associated with the probability of repayment. An increase in income from $0 to $50,000 likely corresponds to a bigger increase in likelihood of repayment than an increase from $1 million to $1.05 million. One way to handle this might be to postprocess our outcome such that linearity becomes more plausible, by using the logistic map (and thus the logarithm of the probability of outcome).

Note that we can easily come up with examples that violate monotonicity. Say for example that we want to predict health as a function of body temperature. For individuals with a normal body temperature above 37°C (98.6°F), higher temperatures indicate greater risk. However, if the body temperatures drops below 37°C, lower temperatures indicate greater risk! Again, we might resolve the problem with some clever preprocessing, such as using the distance from 37°C as a feature.

But what about classifying images of cats and dogs? Should increasing the intensity of the pixel at location (13, 17) always increase (or always decrease) the likelihood that the image depicts a dog? Reliance on a linear model corresponds to the implicit assumption that the only requirement for differentiating cats and dogs is to assess the brightness of individual pixels. This approach is doomed to fail in a world where inverting an image preserves the category.

And yet despite the apparent absurdity of linearity here, as compared with our previous examples, it is less obvious that we could address the problem with a simple preprocessing fix. That is, because the significance of any pixel depends in complex ways on its context (the values of the surrounding pixels). While there might exist a representation of our data that would take into account the relevant interactions among our features, on top of which a linear model would be suitable, we simply do not know how to calculate it by hand. With deep neural networks, we used observational data to jointly learn both a representation via hidden layers and a linear predictor that acts upon that representation.

This problem of nonlinearity has been studied for at least a century (Fisher, 1925). For instance, decision trees in their most basic form use a sequence of binary decisions to decide upon class membership (Quinlan, 1993). Likewise, kernel methods have been used for many decades to model nonlinear dependencies (Aronszajn, 1950). This has found its way into nonparametric spline models (Wahba, 1990) and kernel methods (Schölkopf and Smola, 2002). It is also something that the brain solves quite naturally. After all, neurons feed into other neurons which, in turn, feed into other neurons again (Ramón y Cajal and Azoulay, 1894). Consequently we have a sequence of relatively simple transformations.

Incorporating Hidden Layers

We can overcome the limitations of linear models by incorporating one or more hidden layers. The easiest way to do this is to stack many fully connected layers on top of one another. Each layer feeds into the layer above it, until we generate outputs. We can think of the first $L - 1$ layers as our representation and the final layer as our linear predictor. This architecture is commonly called a *multilayer perceptron*, often abbreviated as *MLP* (Fig. 5.1.1).

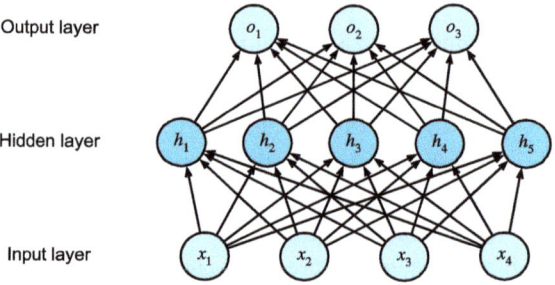

Fig. 5.1.1   An MLP with a hidden layer of five hidden units.

This MLP has four inputs, three outputs, and its hidden layer contains five hidden units. Since the input layer does not involve any calculations, producing outputs with this network requires implementing the computations for both the hidden and output layers; thus, the number of layers in this MLP is two. Note that both layers are fully connected. Every input influences every neuron in the hidden layer, and each of these in turn influences every neuron in the output layer. Alas, we are not quite done yet.

From Linear to Nonlinear

As before, we denote by the matrix $\mathbf{X} \in \mathbb{R}^{n \times d}$ a minibatch of n examples where each example has d inputs (features). For a one-hidden-layer MLP whose hidden layer has h hidden units, we denote by $\mathbf{H} \in \mathbb{R}^{n \times h}$ the outputs of the hidden layer, which are *hidden representations*. Since the hidden and output layers are both fully connected, we have hidden-layer weights $\mathbf{W}^{(1)} \in \mathbb{R}^{d \times h}$ and biases $\mathbf{b}^{(1)} \in \mathbb{R}^{1 \times h}$ and output-layer weights $\mathbf{W}^{(2)} \in \mathbb{R}^{h \times q}$ and biases $\mathbf{b}^{(2)} \in \mathbb{R}^{1 \times q}$. This allows us to calculate the outputs $\mathbf{O} \in \mathbb{R}^{n \times q}$ of the one-hidden-layer MLP as follows:

$$\mathbf{H} = \mathbf{X}\mathbf{W}^{(1)} + \mathbf{b}^{(1)},$$
$$\mathbf{O} = \mathbf{H}\mathbf{W}^{(2)} + \mathbf{b}^{(2)}. \tag{5.1.1}$$

Note that after adding the hidden layer, our model now requires us to track and update additional sets of parameters. So what have we gained in exchange? You might be surprised to find out that—in the model defined above—*we gain nothing for our troubles*! The reason is plain. The hidden units above are given by an affine function of the inputs, and the outputs (pre-softmax) are just an affine function of the hidden units. An affine function of an affine function is itself an affine function. Moreover, our linear model was already capable of representing any affine function.

To see this formally we can just collapse out the hidden layer in the above definition, yielding an equivalent single-layer model with parameters $\mathbf{W} = \mathbf{W}^{(1)}\mathbf{W}^{(2)}$ and $\mathbf{b} = \mathbf{b}^{(1)}\mathbf{W}^{(2)} + \mathbf{b}^{(2)}$:

$$\mathbf{O} = (\mathbf{X}\mathbf{W}^{(1)} + \mathbf{b}^{(1)})\mathbf{W}^{(2)} + \mathbf{b}^{(2)} = \mathbf{X}\mathbf{W}^{(1)}\mathbf{W}^{(2)} + \mathbf{b}^{(1)}\mathbf{W}^{(2)} + \mathbf{b}^{(2)} = \mathbf{X}\mathbf{W} + \mathbf{b}.$$

$$(5.1.2)$$

In order to realize the potential of multilayer architectures, we need one more key ingredient: a nonlinear *activation function* σ to be applied to each hidden unit following the affine transformation. For instance, a popular choice is the ReLU (rectified linear unit) activation function (Nair and Hinton, 2010) $\sigma(x) = \max(0, x)$ operating on its arguments elementwise. The outputs of activation functions $\sigma(\cdot)$ are called *activations*. In general, with activation functions in place, it is no longer possible to collapse our MLP into a linear model:

$$\mathbf{H} = \sigma(\mathbf{X}\mathbf{W}^{(1)} + \mathbf{b}^{(1)}),$$
$$\mathbf{O} = \mathbf{H}\mathbf{W}^{(2)} + \mathbf{b}^{(2)}.$$

$$(5.1.3)$$

Since each row in $\mathbf{X}$ corresponds to an example in the minibatch, with some abuse of notation, we define the nonlinearity σ to apply to its inputs in a rowwise fashion, i.e., one example at a time. Note that we used the same notation for softmax when we denoted a rowwise operation in Section 4.1.1. Quite frequently the activation functions we use apply not merely rowwise but elementwise. That means that after computing the linear portion of the layer, we can calculate each activation without looking at the values taken by the other hidden units.

To build more general MLPs, we can continue stacking such hidden layers, e.g., $\mathbf{H}^{(1)} = \sigma_1(\mathbf{X}\mathbf{W}^{(1)} + \mathbf{b}^{(1)})$ and $\mathbf{H}^{(2)} = \sigma_2(\mathbf{H}^{(1)}\mathbf{W}^{(2)} + \mathbf{b}^{(2)})$, one atop another, yielding ever more expressive models.

Universal Approximators

We know that the brain is capable of very sophisticated statistical analysis. As such, it is worth asking, just *how powerful* a deep network could be. This question has been answered multiple times, e.g., in Cybenko (1989) in the context of MLPs, and in Micchelli (1984) in the context of reproducing kernel Hilbert spaces in a way that could be seen as radial basis function (RBF) networks with a single hidden layer. These (and related results) suggest that even with a single-hidden-layer network, given enough nodes (possibly absurdly many), and the right set of weights, we can model any function. Actually learning that function is the hard part, though. You might think of your neural network as being a bit like the

C programming language. The language, like any other modern language, is capable of expressing any computable program. But actually coming up with a program that meets your specifications is the hard part.

Moreover, just because a single-hidden-layer network *can* learn any function does not mean that you should try to solve all of your problems with one. In fact, in this case kernel methods are way more effective, since they are capable of solving the problem *exactly* even in infinite-dimensional spaces (Kimeldorf and Wahba, 1971, Schölkopf *et al.*, 2001). In fact, we can approximate many functions much more compactly by using deeper (rather than wider) networks (Simonyan and Zisserman, 2014). We will touch upon more rigorous arguments in subsequent chapters.

5.1.2 Activation Functions

Activation functions decide whether a neuron should be activated or not by calculating the weighted sum and further adding bias to it. They are differentiable operators for transforming input signals to outputs, while most of them add nonlinearity. Because activation functions are fundamental to deep learning, let's briefly survey some common ones.

ReLU Function

The most popular choice, due to both simplicity of implementation and its good performance on a variety of predictive tasks, is the *rectified linear unit* (*ReLU*) (Nair and Hinton, 2010). ReLU provides a very simple nonlinear transformation. Given an element x, the function is defined as the maximum of that element and 0:

$$\text{ReLU}(x) = \max(x, 0). \tag{5.1.4}$$

Informally, the ReLU function retains only positive elements and discards all negative elements by setting the corresponding activations to 0. To gain some intuition, we can plot the function. As you can see, the activation function is piecewise linear.

```
x = torch.arange(-8.0, 8.0, 0.1, requires_grad=True)
y = torch.relu(x)
d2l.plot(x.detach(), y.detach(), 'x', 'relu(x)', figsize=(5, 2.5))
```

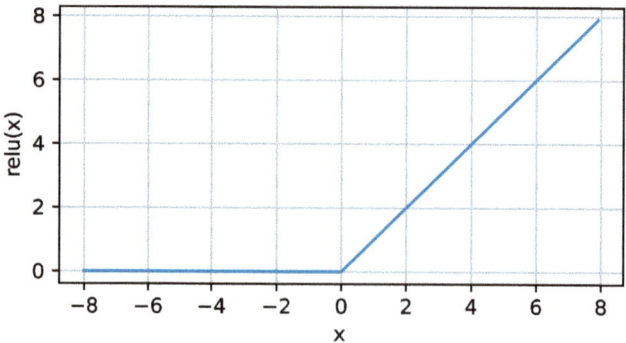

When the input is negative, the derivative of the ReLU function is 0, and when the input is positive, the derivative of the ReLU function is 1. Note that the ReLU function is not differentiable when the input takes value precisely equal to 0. In these cases, we default to the left-hand-side derivative and say that the derivative is 0 when the input is 0. We can get away with this because the input may never actually be zero (mathematicians would say that it is nondifferentiable on a set of measure zero). There is an old adage that if subtle boundary conditions matter, we are probably doing (*real*) mathematics, not engineering. That conventional wisdom may apply here, or at least, the fact that we are not performing constrained optimization (Mangasarian, 1965, Rockafellar, 1970). We plot the derivative of the ReLU function below.

```
y.backward(torch.ones_like(x), retain_graph=True)
d2l.plot(x.detach(), x.grad, 'x', 'grad of relu', figsize=(5, 2.5))
```

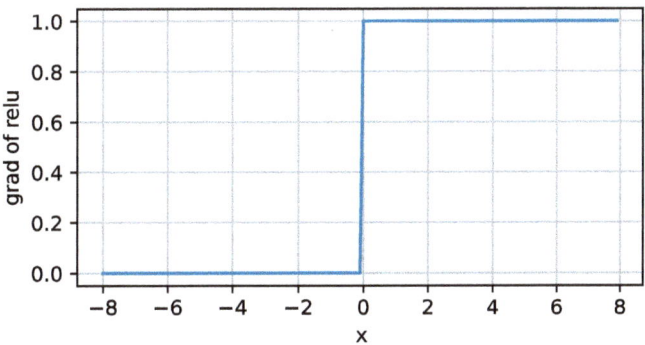

The reason for using ReLU is that its derivatives are particularly well behaved: either they vanish or they just let the argument through. This makes optimization better behaved and it mitigated the well-documented problem of vanishing gradients that plagued previous versions of neural networks (more on this later).

Note that there are many variants to the ReLU function, including the *parametrized ReLU* (*pReLU*) function (He *et al.*, 2015). This variation adds a linear term to ReLU, so some information still gets through, even when the argument is negative:

$$\text{pReLU}(x) = \max(0, x) + \alpha \min(0, x). \tag{5.1.5}$$

Sigmoid Function

The *sigmoid function* transforms those inputs whose values lie in the domain $\mathbb{R}$, to outputs that lie on the interval $(0, 1)$. For that reason, the sigmoid is often called a *squashing function*: it squashes any input in the range (-inf, inf) to some value in the range (0, 1):

$$\text{sigmoid}(x) = \frac{1}{1 + \exp(-x)}. \tag{5.1.6}$$

In the earliest neural networks, scientists were interested in modeling biological neurons that either *fire* or *do not fire*. Thus the pioneers of this field, going all the way back to

McCulloch and Pitts, the inventors of the artificial neuron, focused on thresholding units (McCulloch and Pitts, 1943). A thresholding activation takes value 0 when its input is below some threshold and value 1 when the input exceeds the threshold.

When attention shifted to gradient-based learning, the sigmoid function was a natural choice because it is a smooth, differentiable approximation to a thresholding unit. Sigmoids are still widely used as activation functions on the output units when we want to interpret the outputs as probabilities for binary classification problems: you can think of the sigmoid as a special case of the softmax. However, the sigmoid has largely been replaced by the simpler and more easily trainable ReLU for most use in hidden layers. Much of this has to do with the fact that the sigmoid poses challenges for optimization (LeCun *et al.*, 1998) since its gradient vanishes for large positive *and* negative arguments. This can lead to plateaus that are difficult to escape from. Nonetheless sigmoids are important. In later chapters (e.g., Section 10.1) on recurrent neural networks, we will describe architectures that leverage sigmoid units to control the flow of information across time.

Below, we plot the sigmoid function. Note that when the input is close to 0, the sigmoid function approaches a linear transformation.

```
y = torch.sigmoid(x)
d2l.plot(x.detach(), y.detach(), 'x', 'sigmoid(x)', figsize=(5, 2.5))
```

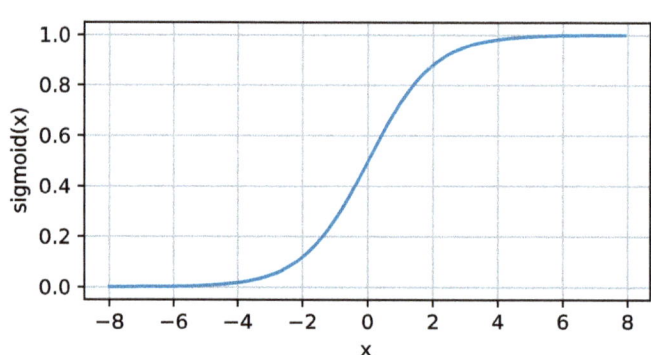

The derivative of the sigmoid function is given by the following equation:

$$\frac{d}{dx}\,\text{sigmoid}(x) = \frac{\exp(-x)}{(1+\exp(-x))^2} = \text{sigmoid}(x)\,(1-\text{sigmoid}(x)). \qquad (5.1.7)$$

The derivative of the sigmoid function is plotted below. Note that when the input is 0, the derivative of the sigmoid function reaches a maximum of 0.25. As the input diverges from 0 in either direction, the derivative approaches 0.

```
# Clear out previous gradients
x.grad.data.zero_()
y.backward(torch.ones_like(x),retain_graph=True)
d2l.plot(x.detach(), x.grad, 'x', 'grad of sigmoid', figsize=(5, 2.5))
```

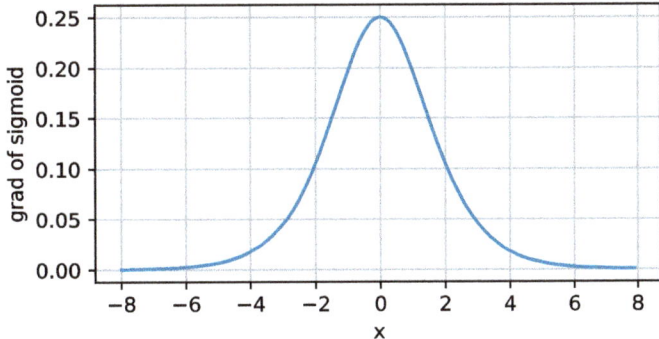

Tanh Function

Like the sigmoid function, the tanh (hyperbolic tangent) function also squashes its inputs, transforming them into elements on the interval between -1 and 1:

$$\tanh(x) = \frac{1 - \exp(-2x)}{1 + \exp(-2x)}. \tag{5.1.8}$$

We plot the tanh function below. Note that as input nears 0, the tanh function approaches a linear transformation. Although the shape of the function is similar to that of the sigmoid function, the tanh function exhibits point symmetry about the origin of the coordinate system (Kalman and Kwasny, 1992).

```
y = torch.tanh(x)
d2l.plot(x.detach(), y.detach(), 'x', 'tanh(x)', figsize=(5, 2.5))
```

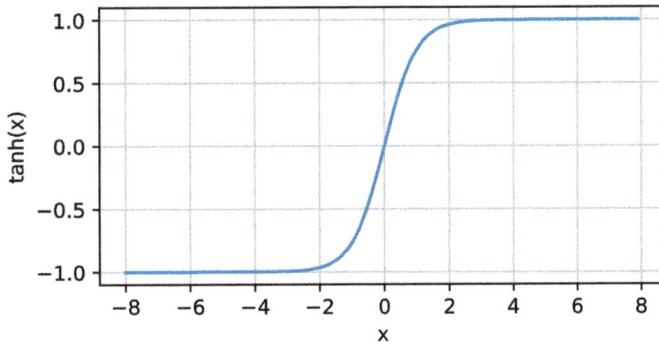

The derivative of the tanh function is:

$$\frac{d}{dx}\tanh(x) = 1 - \tanh^2(x). \tag{5.1.9}$$

It is plotted below. As the input nears 0, the derivative of the tanh function approaches a maximum of 1. And as we saw with the sigmoid function, as input moves away from 0 in either direction, the derivative of the tanh function approaches 0.

```
# Clear out previous gradients
x.grad.data.zero_()
y.backward(torch.ones_like(x),retain_graph=True)
d2l.plot(x.detach(), x.grad, 'x', 'grad of tanh', figsize=(5, 2.5))
```

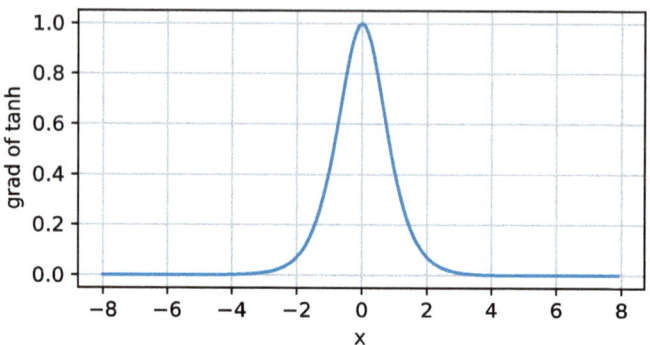

5.1.3 Summary and Discussion

We now know how to incorporate nonlinearities to build expressive multilayer neural network architectures. As a side note, your knowledge already puts you in command of a similar toolkit to a practitioner circa 1990. In some ways, you have an advantage over anyone working back then, because you can leverage powerful open-source deep learning frameworks to build models rapidly, using only a few lines of code. Previously, training these networks required researchers to code up layers and derivatives explicitly in C, Fortran, or even Lisp (in the case of LeNet).

A secondary benefit is that ReLU is significantly more amenable to optimization than the sigmoid or the tanh function. One could argue that this was one of the key innovations that helped the resurgence of deep learning over the past decade. Note, though, that research in activation functions has not stopped. For instance, the GELU (Gaussian error linear unit) activation function $x\Phi(x)$ by Hendrycks and Gimpel (2016) ($\Phi(x)$ is the standard Gaussian cumulative distribution function) and the Swish activation function $\sigma(x) = x\,\mathrm{sigmoid}(\beta x)$ as proposed in Ramachandran *et al.* (2017) can yield better accuracy in many cases.

5.1.4 Exercises

1. Show that adding layers to a *linear* deep network, i.e., a network without nonlinearity σ can never increase the expressive power of the network. Give an example where it actively reduces it.

2. Compute the derivative of the pReLU activation function.

3. Compute the derivative of the Swish activation function $x\,\mathrm{sigmoid}(\beta x)$.

4. Show that an MLP using only ReLU (or pReLU) constructs a continuous piecewise linear function.

5. Sigmoid and tanh are very similar.

   1. Show that $\tanh(x) + 1 = 2\,\text{sigmoid}(2x)$.

   2. Prove that the function classes parametrized by both nonlinearities are identical. Hint: affine layers have bias terms, too.

6. Assume that we have a nonlinearity that applies to one minibatch at a time, such as the batch normalization (Ioffe and Szegedy, 2015). What kinds of problems do you expect this to cause?

7. Provide an example where the gradients vanish for the sigmoid activation function.

 Discussions[102].

5.2 Implementation of Multilayer Perceptrons

Multilayer perceptrons (MLPs) are not much more complex to implement than simple linear models. The key conceptual difference is that we now concatenate multiple layers.

```
import torch
from torch import nn
from d2l import torch as d2l
```

5.2.1 Implementation from Scratch

Let's begin again by implementing such a network from scratch.

Initializing Model Parameters

Recall that Fashion-MNIST contains 10 classes, and that each image consists of a $28 \times 28 = 784$ grid of grayscale pixel values. As before we will disregard the spatial structure among the pixels for now, so we can think of this as a classification dataset with 784 input features and 10 classes. To begin, we will implement an MLP with one hidden layer and 256 hidden units. Both the number of layers and their width are adjustable (they are considered hyperparameters). Typically, we choose the layer widths to be divisible by larger powers of 2. This is computationally efficient due to the way memory is allocated and addressed in hardware.

Again, we will represent our parameters with several tensors. Note that *for every layer*, we must keep track of one weight matrix and one bias vector. As always, we allocate memory for the gradients of the loss with respect to these parameters.

In the code below we use `nn.Parameter` to automatically register a class attribute as a parameter to be tracked by `autograd` (Section 2.5).

```
class MLPScratch(d2l.Classifier):
    def __init__(self, num_inputs, num_outputs, num_hiddens, lr, sigma=0.01):
        super().__init__()
        self.save_hyperparameters()
        self.W1 = nn.Parameter(torch.randn(num_inputs, num_hiddens) * sigma)
        self.b1 = nn.Parameter(torch.zeros(num_hiddens))
        self.W2 = nn.Parameter(torch.randn(num_hiddens, num_outputs) * sigma)
        self.b2 = nn.Parameter(torch.zeros(num_outputs))
```

Model

To make sure we know how everything works, we will implement the ReLU activation ourselves rather than invoking the built-in `relu` function directly.

```
def relu(X):
    a = torch.zeros_like(X)
    return torch.max(X, a)
```

Since we are disregarding spatial structure, we `reshape` each two-dimensional image into a flat vector of length `num_inputs`. Finally, we implement our model with just a few lines of code. Since we use the framework built-in autograd this is all that it takes.

```
@d2l.add_to_class(MLPScratch)
def forward(self, X):
    X = X.reshape((-1, self.num_inputs))
    H = relu(torch.matmul(X, self.W1) + self.b1)
    return torch.matmul(H, self.W2) + self.b2
```

Training

Fortunately, the training loop for MLPs is exactly the same as for softmax regression. We define the model, data, and trainer, then finally invoke the `fit` method on model and data.

```
model = MLPScratch(num_inputs=784, num_outputs=10, num_hiddens=256, lr=0.1)
data = d2l.FashionMNIST(batch_size=256)
trainer = d2l.Trainer(max_epochs=10)
trainer.fit(model, data)
```

5.2.2 Concise Implementation

As you might expect, by relying on the high-level APIs, we can implement MLPs even more concisely.

Model

Compared with our concise implementation of softmax regression implementation (Section 4.5), the only difference is that we add *two* fully connected layers where we previously added only *one*. The first is the hidden layer, the second is the output layer.

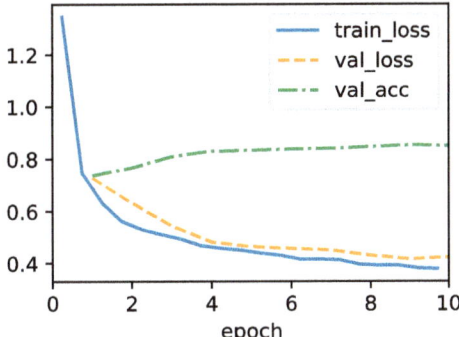

```
class MLP(d2l.Classifier):
    def __init__(self, num_outputs, num_hiddens, lr):
        super().__init__()
        self.save_hyperparameters()
        self.net = nn.Sequential(nn.Flatten(), nn.LazyLinear(num_hiddens),
                                 nn.ReLU(), nn.LazyLinear(num_outputs))
```

Previously, we defined forward methods for models to transform input using the model parameters. These operations are essentially a pipeline: you take an input and apply a transformation (e.g., matrix multiplication with weights followed by bias addition), then repetitively use the output of the current transformation as input to the next transformation. However, you may have noticed that no forward method is defined here. In fact, MLP inherits the forward method from the Module class (Section 3.2.2) to simply invoke self.net(X) (X is input), which is now defined as a sequence of transformations via the Sequential class. The Sequential class abstracts the forward process enabling us to focus on the transformations. We will further discuss how the Sequential class works in Section 6.1.2.

Training

The training loop is exactly the same as when we implemented softmax regression. This modularity enables us to separate matters concerning the model architecture from orthogonal considerations.

```
model = MLP(num_outputs=10, num_hiddens=256, lr=0.1)
trainer.fit(model, data)
```

5.2.3 Summary

Now that we have more practice in designing deep networks, the step from a single to multiple layers of deep networks does not pose such a significant challenge any longer. In particular, we can reuse the training algorithm and data loader. Note, though, that implementing MLPs from scratch is nonetheless messy: naming and keeping track of the model parameters makes it difficult to extend models. For instance, imagine wanting to insert another layer between layers 42 and 43. This might now be layer 42b, unless we are willing

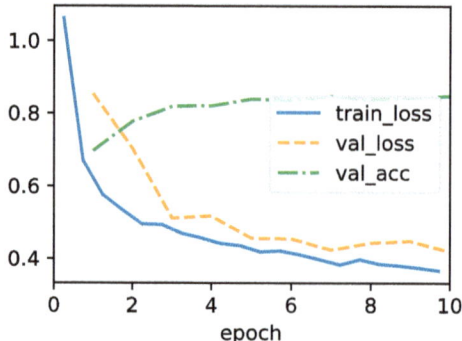

to perform sequential renaming. Moreover, if we implement the network from scratch, it is much more difficult for the framework to perform meaningful performance optimizations.

Nonetheless, you have now reached the state of the art of the late 1980s when fully connected deep networks were the method of choice for neural network modeling. Our next conceptual step will be to consider images. Before we do so, we need to review a number of statistical basics and details on how to compute models efficiently.

5.2.4 Exercises

1. Change the number of hidden units num_hiddens and plot how its number affects the accuracy of the model. What is the best value of this hyperparameter?

2. Try adding a hidden layer to see how it affects the results.

3. Why is it a bad idea to insert a hidden layer with a single neuron? What could go wrong?

4. How does changing the learning rate alter your results? With all other parameters fixed, which learning rate gives you the best results? How does this relate to the number of epochs?

5. Let's optimize over all hyperparameters jointly, i.e., learning rate, number of epochs, number of hidden layers, and number of hidden units per layer.

   1. What is the best result you can get by optimizing over all of them?

   2. Why it is much more challenging to deal with multiple hyperparameters?

   3. Describe an efficient strategy for optimizing over multiple parameters jointly.

6. Compare the speed of the framework and the from-scratch implementation for a challenging problem. How does it change with the complexity of the network?

7. Measure the speed of tensor–matrix multiplications for well-aligned and misaligned matrices. For instance, test for matrices with dimension 1024, 1025, 1026, 1028, and 1032.

    1. How does this change between GPUs and CPUs?

    2. Determine the memory bus width of your CPU and GPU.

8. Try out different activation functions. Which one works best?

9. Is there a difference between weight initializations of the network? Does it matter?

 Discussions [103].

5.3 Forward Propagation, Backward Propagation, and Computational Graphs

So far, we have trained our models with minibatch stochastic gradient descent. However, when we implemented the algorithm, we only worried about the calculations involved in *forward propagation* through the model. When it came time to calculate the gradients, we just invoked the backpropagation function provided by the deep learning framework.

The automatic calculation of gradients profoundly simplifies the implementation of deep learning algorithms. Before automatic differentiation, even small changes to complicated models required recalculating complicated derivatives by hand. Surprisingly often, academic papers had to allocate numerous pages to deriving update rules. While we must continue to rely on automatic differentiation so we can focus on the interesting parts, you ought to know how these gradients are calculated under the hood if you want to go beyond a shallow understanding of deep learning.

In this section, we take a deep dive into the details of *backward propagation* (more commonly called *backpropagation*). To convey some insight for both the techniques and their implementations, we rely on some basic mathematics and computational graphs. To start, we focus our exposition on a one-hidden-layer MLP with weight decay (ℓ_2 regularization, to be described in subsequent chapters).

5.3.1 Forward Propagation

Forward propagation (or *forward pass*) refers to the calculation and storage of intermediate variables (including outputs) for a neural network in order from the input layer to the output layer. We now work step-by-step through the mechanics of a neural network with one hidden layer. This may seem tedious but in the eternal words of funk virtuoso James Brown, you must "pay the cost to be the boss".

For the sake of simplicity, let's assume that the input example is $\mathbf{x} \in \mathbb{R}^d$ and that our hidden layer does not include a bias term. Here the intermediate variable is:

$$\mathbf{z} = \mathbf{W}^{(1)}\mathbf{x}, \tag{5.3.1}$$

where $\mathbf{W}^{(1)} \in \mathbb{R}^{h \times d}$ is the weight parameter of the hidden layer. After running the intermediate variable $\mathbf{z} \in \mathbb{R}^h$ through the activation function ϕ we obtain our hidden activation vector of length h:

$$\mathbf{h} = \phi(\mathbf{z}).\tag{5.3.2}$$

The hidden layer output $\mathbf{h}$ is also an intermediate variable. Assuming that the parameters of the output layer possess only a weight of $\mathbf{W}^{(2)} \in \mathbb{R}^{q \times h}$, we can obtain an output layer variable with a vector of length q:

$$\mathbf{o} = \mathbf{W}^{(2)}\mathbf{h}.\tag{5.3.3}$$

Assuming that the loss function is l and the example label is y, we can then calculate the loss term for a single data example,

$$L = l(\mathbf{o}, y).\tag{5.3.4}$$

As we will see the definition of ℓ_2 regularization to be introduced later, given the hyperparameter λ, the regularization term is

$$s = \frac{\lambda}{2}\left(\|\mathbf{W}^{(1)}\|_F^2 + \|\mathbf{W}^{(2)}\|_F^2\right),\tag{5.3.5}$$

where the Frobenius norm of the matrix is simply the ℓ_2 norm applied after flattening the matrix into a vector. Finally, the model's regularized loss on a given data example is:

$$J = L + s.\tag{5.3.6}$$

We refer to J as the *objective function* in the following discussion.

5.3.2 Computational Graph of Forward Propagation

Plotting *computational graphs* helps us visualize the dependencies of operators and variables within the calculation. Fig. 5.3.1 contains the graph associated with the simple network described above, where squares denote variables and circles denote operators. The lower-left corner signifies the input and the upper-right corner is the output. Notice that the directions of the arrows (which illustrate data flow) are primarily rightward and upward.

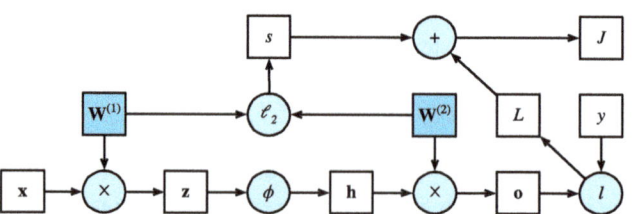

Fig. 5.3.1   Computational graph of forward propagation.

5.3.3 Backpropagation

Backpropagation refers to the method of calculating the gradient of neural network parameters. In short, the method traverses the network in reverse order, from the output to the input layer, according to the *chain rule* from calculus. The algorithm stores any intermediate variables (partial derivatives) required while calculating the gradient with respect to some parameters. Assume that we have functions $Y = f(X)$ and $Z = g(Y)$, in which the input and the output X, Y, Z are tensors of arbitrary shapes. By using the chain rule, we can compute the derivative of Z with respect to X via

$$\frac{\partial Z}{\partial X} = \text{prod}\left(\frac{\partial Z}{\partial Y}, \frac{\partial Y}{\partial X}\right). \tag{5.3.7}$$

Here we use the prod operator to multiply its arguments after the necessary operations, such as transposition and swapping input positions, have been carried out. For vectors, this is straightforward: it is simply matrix–matrix multiplication. For higher-dimensional tensors, we use the appropriate counterpart. The operator prod hides all the notational overhead.

Recall that the parameters of the simple network with one hidden layer, whose computational graph is in Fig. 5.3.1, are $\mathbf{W}^{(1)}$ and $\mathbf{W}^{(2)}$. The objective of backpropagation is to calculate the gradients $\partial J/\partial \mathbf{W}^{(1)}$ and $\partial J/\partial \mathbf{W}^{(2)}$. To accomplish this, we apply the chain rule and calculate, in turn, the gradient of each intermediate variable and parameter. The order of calculations are reversed relative to those performed in forward propagation, since we need to start with the outcome of the computational graph and work our way towards the parameters. The first step is to calculate the gradients of the objective function $J = L + s$ with respect to the loss term L and the regularization term s:

$$\frac{\partial J}{\partial L} = 1 \text{ and } \frac{\partial J}{\partial s} = 1. \tag{5.3.8}$$

Next, we compute the gradient of the objective function with respect to variable of the output layer $\mathbf{o}$ according to the chain rule:

$$\frac{\partial J}{\partial \mathbf{o}} = \text{prod}\left(\frac{\partial J}{\partial L}, \frac{\partial L}{\partial \mathbf{o}}\right) = \frac{\partial L}{\partial \mathbf{o}} \in \mathbb{R}^q. \tag{5.3.9}$$

Next, we calculate the gradients of the regularization term with respect to both parameters:

$$\frac{\partial s}{\partial \mathbf{W}^{(1)}} = \lambda \mathbf{W}^{(1)} \text{ and } \frac{\partial s}{\partial \mathbf{W}^{(2)}} = \lambda \mathbf{W}^{(2)}. \tag{5.3.10}$$

Now we are able to calculate the gradient $\partial J/\partial \mathbf{W}^{(2)} \in \mathbb{R}^{q \times h}$ of the model parameters closest to the output layer. Using the chain rule yields:

$$\frac{\partial J}{\partial \mathbf{W}^{(2)}} = \text{prod}\left(\frac{\partial J}{\partial \mathbf{o}}, \frac{\partial \mathbf{o}}{\partial \mathbf{W}^{(2)}}\right) + \text{prod}\left(\frac{\partial J}{\partial s}, \frac{\partial s}{\partial \mathbf{W}^{(2)}}\right) = \frac{\partial J}{\partial \mathbf{o}}\mathbf{h}^\top + \lambda \mathbf{W}^{(2)}. \tag{5.3.11}$$

To obtain the gradient with respect to $\mathbf{W}^{(1)}$ we need to continue backpropagation along the output layer to the hidden layer. The gradient with respect to the hidden layer output

$\partial J / \partial \mathbf{h} \in \mathbb{R}^h$ is given by

$$
\frac{\partial J}{\partial \mathbf{h}} = \text{prod}\left(\frac{\partial J}{\partial \mathbf{o}}, \frac{\partial \mathbf{o}}{\partial \mathbf{h}}\right) = \mathbf{W}^{(2)\top} \frac{\partial J}{\partial \mathbf{o}}.
\tag{5.3.12}
$$

Since the activation function ϕ applies elementwise, calculating the gradient $\partial J / \partial \mathbf{z} \in \mathbb{R}^h$ of the intermediate variable $\mathbf{z}$ requires that we use the elementwise multiplication operator, which we denote by $\odot$:

$$
\frac{\partial J}{\partial \mathbf{z}} = \text{prod}\left(\frac{\partial J}{\partial \mathbf{h}}, \frac{\partial \mathbf{h}}{\partial \mathbf{z}}\right) = \frac{\partial J}{\partial \mathbf{h}} \odot \phi'\left(\mathbf{z}\right).
\tag{5.3.13}
$$

Finally, we can obtain the gradient $\partial J / \partial \mathbf{W}^{(1)} \in \mathbb{R}^{h \times d}$ of the model parameters closest to the input layer. According to the chain rule, we get

$$
\frac{\partial J}{\partial \mathbf{W}^{(1)}} = \text{prod}\left(\frac{\partial J}{\partial \mathbf{z}}, \frac{\partial \mathbf{z}}{\partial \mathbf{W}^{(1)}}\right) + \text{prod}\left(\frac{\partial J}{\partial s}, \frac{\partial s}{\partial \mathbf{W}^{(1)}}\right) = \frac{\partial J}{\partial \mathbf{z}} \mathbf{x}^\top + \lambda \mathbf{W}^{(1)}.
\tag{5.3.14}
$$

5.3.4 Training Neural Networks

When training neural networks, forward and backward propagation depend on each other. In particular, for forward propagation, we traverse the computational graph in the direction of dependencies and compute all the variables on its path. These are then used for backpropagation where the compute order on the graph is reversed.

Take the aforementioned simple network as an illustrative example. On the one hand, computing the regularization term (5.3.5) during forward propagation depends on the current values of model parameters $\mathbf{W}^{(1)}$ and $\mathbf{W}^{(2)}$. They are given by the optimization algorithm according to backpropagation in the most recent iteration. On the other hand, the gradient calculation for the parameter (5.3.11) during backpropagation depends on the current value of the hidden layer output $\mathbf{h}$, which is given by forward propagation.

Therefore when training neural networks, once model parameters are initialized, we alternate forward propagation with backpropagation, updating model parameters using gradients given by backpropagation. Note that backpropagation reuses the stored intermediate values from forward propagation to avoid duplicate calculations. One of the consequences is that we need to retain the intermediate values until backpropagation is complete. This is also one of the reasons why training requires significantly more memory than plain prediction. Besides, the size of such intermediate values is roughly proportional to the number of network layers and the batch size. Thus, training deeper networks using larger batch sizes more easily leads to *out-of-memory* errors.

5.3.5 Summary

Forward propagation sequentially calculates and stores intermediate variables within the computational graph defined by the neural network. It proceeds from the input to the output layer. Backpropagation sequentially calculates and stores the gradients of intermediate variables and parameters within the neural network in the reversed order. When training deep learning models, forward propagation and backpropagation are interdependent, and training requires significantly more memory than prediction.

5.3.6 Exercises

1. Assume that the inputs $\mathbf{X}$ to some scalar function f are $n \times m$ matrices. What is the dimensionality of the gradient of f with respect to $\mathbf{X}$?

2. Add a bias to the hidden layer of the model described in this section (you do not need to include bias in the regularization term).

   1. Draw the corresponding computational graph.

   2. Derive the forward and backward propagation equations.

3. Compute the memory footprint for training and prediction in the model described in this section.

4. Assume that you want to compute second derivatives. What happens to the computational graph? How long do you expect the calculation to take?

5. Assume that the computational graph is too large for your GPU.

   1. Can you partition it over more than one GPU?

   2. What are the advantages and disadvantages over training on a smaller minibatch?

Discussions [104].

104

5.4 Numerical Stability and Initialization

Thus far, every model that we have implemented required that we initialize its parameters according to some pre-specified distribution. Until now, we took the initialization scheme for granted, glossing over the details of how these choices are made. You might have even gotten the impression that these choices are not especially important. On the contrary, the choice of initialization scheme plays a significant role in neural network learning, and it can be crucial for maintaining numerical stability. Moreover, these choices can be tied up in interesting ways with the choice of the nonlinear activation function. Which function we choose and how we initialize parameters can determine how quickly our optimization algorithm converges. Poor choices here can cause us to encounter exploding or vanishing gradients while training. In this section, we delve into these topics in greater detail and discuss some useful heuristics that you will find useful throughout your career in deep learning.

```
%matplotlib inline
import torch
from d2l import torch as d2l
```

5.4.1 Vanishing and Exploding Gradients

Consider a deep network with L layers, input $\mathbf{x}$ and output $\mathbf{o}$. With each layer l defined by a transformation f_l parametrized by weights $\mathbf{W}^{(l)}$, whose hidden layer output is $\mathbf{h}^{(l)}$ (let $\mathbf{h}^{(0)} = \mathbf{x}$), our network can be expressed as:

$$\mathbf{h}^{(l)} = f_l(\mathbf{h}^{(l-1)}) \text{ and thus } \mathbf{o} = f_L \circ \cdots \circ f_1(\mathbf{x}). \tag{5.4.1}$$

If all the hidden layer output and the input are vectors, we can write the gradient of $\mathbf{o}$ with respect to any set of parameters $\mathbf{W}^{(l)}$ as follows:

$$\partial_{\mathbf{W}^{(l)}} \mathbf{o} = \underbrace{\partial_{\mathbf{h}^{(L-1)}} \mathbf{h}^{(L)}}_{\mathbf{M}^{(L)} \stackrel{\text{def}}{=}} \cdots \underbrace{\partial_{\mathbf{h}^{(l)}} \mathbf{h}^{(l+1)}}_{\mathbf{M}^{(l+1)} \stackrel{\text{def}}{=}} \underbrace{\partial_{\mathbf{W}^{(l)}} \mathbf{h}^{(l)}}_{\mathbf{v}^{(l)} \stackrel{\text{def}}{=}} . \tag{5.4.2}$$

In other words, this gradient is the product of $L - l$ matrices $\mathbf{M}^{(L)} \cdots \mathbf{M}^{(l+1)}$ and the gradient vector $\mathbf{v}^{(l)}$. Thus we are susceptible to the same problems of numerical underflow that often crop up when multiplying together too many probabilities. When dealing with probabilities, a common trick is to switch into log-space, i.e., shifting pressure from the mantissa to the exponent of the numerical representation. Unfortunately, our problem above is more serious: initially the matrices $\mathbf{M}^{(l)}$ may have a wide variety of eigenvalues. They might be small or large, and their product might be *very large* or *very small*.

The risks posed by unstable gradients go beyond numerical representation. Gradients of unpredictable magnitude also threaten the stability of our optimization algorithms. We may be facing parameter updates that are either (i) excessively large, destroying our model (the *exploding gradient* problem); or (ii) excessively small (the *vanishing gradient* problem), rendering learning impossible as parameters hardly move on each update.

Vanishing Gradients

One frequent culprit causing the vanishing gradient problem is the choice of the activation function σ that is appended following each layer's linear operations. Historically, the sigmoid function $1/(1+\exp(-x))$ (introduced in Section 5.1) was popular because it resembles a thresholding function. Since early artificial neural networks were inspired by biological neural networks, the idea of neurons that fire either *fully* or *not at all* (like biological neurons) seemed appealing. Let's take a closer look at the sigmoid to see why it can cause vanishing gradients.

```
x = torch.arange(-8.0, 8.0, 0.1, requires_grad=True)
y = torch.sigmoid(x)
y.backward(torch.ones_like(x))

d2l.plot(x.detach().numpy(), [y.detach().numpy(), x.grad.numpy()],
         legend=['sigmoid', 'gradient'], figsize=(4.5, 2.5))
```

As you can see, the sigmoid's gradient vanishes both when its inputs are large and when they are small. Moreover, when backpropagating through many layers, unless we are in the Goldilocks zone, where the inputs to many of the sigmoids are close to zero, the gradients

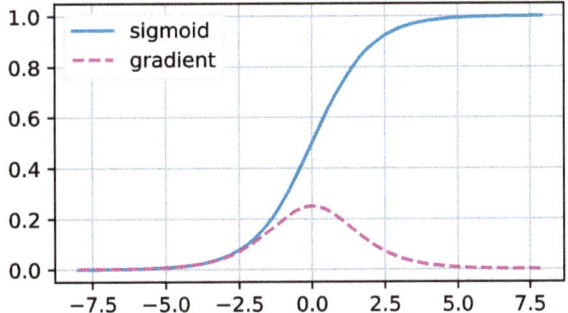

of the overall product may vanish. When our network boasts many layers, unless we are careful, the gradient will likely be cut off at some layer. Indeed, this problem used to plague deep network training. Consequently, ReLUs, which are more stable (but less neurally plausible), have emerged as the default choice for practitioners.

Exploding Gradients

The opposite problem, when gradients explode, can be similarly vexing. To illustrate this a bit better, we draw 100 Gaussian random matrices and multiply them with some initial matrix. For the scale that we picked (the choice of the variance $\sigma^2 = 1$), the matrix product explodes. When this happens because of the initialization of a deep network, we have no chance of getting a gradient descent optimizer to converge.

```
M = torch.normal(0, 1, size=(4, 4))
print('a single matrix \n',M)
for i in range(100):
    M = M @ torch.normal(0, 1, size=(4, 4))
print('after multiplying 100 matrices\n', M)
```

```
a single matrix
 tensor([[-0.0835,  2.0290, -1.2583, -0.7280],
         [ 0.4690,  0.4682,  1.0634,  0.1073],
         [-0.0918,  0.8248, -1.6865, -1.3911],
         [-0.4725,  0.4498, -0.1694,  0.2922]])
after multiplying 100 matrices
 tensor([[-1.9035e+23, -2.5684e+23,  1.9399e+22, -3.3161e+22],
         [ 5.8544e+22,  7.8992e+22, -5.9660e+21,  1.0199e+22],
         [-1.7436e+23, -2.3526e+23,  1.7769e+22, -3.0375e+22],
         [-5.4481e+22, -7.3511e+22,  5.5520e+21, -9.4910e+21]])
```

Breaking the Symmetry

Another problem in neural network design is the symmetry inherent in their parametrization. Assume that we have a simple MLP with one hidden layer and two units. In this case, we could permute the weights $\mathbf{W}^{(1)}$ of the first layer and likewise permute the weights of the output layer to obtain the same function. There is nothing special differentiating the first and second hidden units. In other words, we have permutation symmetry among the hidden units of each layer.

This is more than just a theoretical nuisance. Consider the aforementioned one-hidden-layer MLP with two hidden units. For illustration, suppose that the output layer transforms the two hidden units into only one output unit. Imagine what would happen if we initialized all the parameters of the hidden layer as $\mathbf{W}^{(1)} = c$ for some constant c. In this case, during forward propagation either hidden unit takes the same inputs and parameters producing the same activation which is fed to the output unit. During backpropagation, differentiating the output unit with respect to parameters $\mathbf{W}^{(1)}$ gives a gradient all of whose elements take the same value. Thus, after gradient-based iteration (e.g., minibatch stochastic gradient descent), all the elements of $\mathbf{W}^{(1)}$ still take the same value. Such iterations would never *break the symmetry* on their own and we might never be able to realize the network's expressive power. The hidden layer would behave as if it had only a single unit. Note that while minibatch stochastic gradient descent would not break this symmetry, dropout regularization (to be introduced later) would!

5.4.2 Parameter Initialization

One way of addressing—or at least mitigating—the issues raised above is through careful initialization. As we will see later, additional care during optimization and suitable regularization can further enhance stability.

Default Initialization

In the previous sections, e.g., in Section 3.5, we used a normal distribution to initialize the values of our weights. If we do not specify the initialization method, the framework will use a default random initialization method, which often works well in practice for moderate problem sizes.

Xavier Initialization

Let's look at the scale distribution of an output o_i for some fully connected layer *without nonlinearities*. With n_{in} inputs x_j and their associated weights w_{ij} for this layer, an output is given by

$$o_i = \sum_{j=1}^{n_{\text{in}}} w_{ij} x_j. \tag{5.4.3}$$

The weights w_{ij} are all drawn independently from the same distribution. Furthermore, let's assume that this distribution has zero mean and variance σ^2. Note that this does not mean that the distribution has to be Gaussian, just that the mean and variance need to exist. For now, let's assume that the inputs to the layer x_j also have zero mean and variance γ^2 and that they are independent of w_{ij} and independent of each other. In this case, we can compute the mean of o_i:

$$E[o_i] = \sum_{j=1}^{n_{in}} E[w_{ij}x_j] = \sum_{j=1}^{n_{in}} E[w_{ij}]E[x_j] = 0, \qquad (5.4.4)$$

and the variance:

$$\text{Var}[o_i] = E[o_i^2] - (E[o_i])^2 = \sum_{j=1}^{n_{in}} E[w_{ij}^2 x_j^2] - 0 = \sum_{j=1}^{n_{in}} E[w_{ij}^2]E[x_j^2] = n_{in}\sigma^2\gamma^2. \quad (5.4.5)$$

One way to keep the variance fixed is to set $n_{in}\sigma^2 = 1$. Now consider backpropagation. There we face a similar problem, albeit with gradients being propagated from the layers closer to the output. Using the same reasoning as for forward propagation, we see that the gradients' variance can blow up unless $n_{out}\sigma^2 = 1$, where n_{out} is the number of outputs of this layer. This leaves us in a dilemma: we cannot possibly satisfy both conditions simultaneously. Instead, we simply try to satisfy:

$$\frac{1}{2}(n_{in} + n_{out})\sigma^2 = 1 \text{ or equivalently } \sigma = \sqrt{\frac{2}{n_{in} + n_{out}}}. \qquad (5.4.6)$$

This is the reasoning underlying the now-standard and practically beneficial *Xavier initialization*, named after the first author of its creators (Glorot and Bengio, 2010). Typically, the Xavier initialization samples weights from a Gaussian distribution with zero mean and variance $\sigma^2 = \frac{2}{n_{in}+n_{out}}$. We can also adapt this to choose the variance when sampling weights from a uniform distribution. Note that the uniform distribution $U(-a, a)$ has variance $\frac{a^2}{3}$. Plugging $\frac{a^2}{3}$ into our condition on σ^2 prompts us to initialize according to

$$U\left(-\sqrt{\frac{6}{n_{in} + n_{out}}}, \sqrt{\frac{6}{n_{in} + n_{out}}}\right). \qquad (5.4.7)$$

Though the assumption for nonexistence of nonlinearities in the above mathematical reasoning can be easily violated in neural networks, the Xavier initialization method turns out to work well in practice.

Beyond

The reasoning above barely scratches the surface of modern approaches to parameter initialization. A deep learning framework often implements over a dozen different heuristics. Moreover, parameter initialization continues to be a hot area of fundamental research in deep learning. Among these are heuristics specialized for tied (shared) parameters, super-resolution, sequence models, and other situations. For instance, Xiao *et al.* (2018) demonstrated the possibility of training 10,000-layer neural networks without architectural tricks by using a carefully-designed initialization method.

If the topic interests you we suggest a deep dive into this module's offerings, reading the papers that proposed and analyzed each heuristic, and then exploring the latest publications on the topic. Perhaps you will stumble across or even invent a clever idea and contribute an implementation to deep learning frameworks.

5.4.3 Summary

Vanishing and exploding gradients are common issues in deep networks. Great care in parameter initialization is required to ensure that gradients and parameters remain well controlled. Initialization heuristics are needed to ensure that the initial gradients are neither too large nor too small. Random initialization is key to ensuring that symmetry is broken before optimization. Xavier initialization suggests that, for each layer, variance of any output is not affected by the number of inputs, and variance of any gradient is not affected by the number of outputs. ReLU activation functions mitigate the vanishing gradient problem. This can accelerate convergence.

5.4.4 Exercises

1. Can you design other cases where a neural network might exhibit symmetry that needs breaking, besides the permutation symmetry in an MLP's layers?

2. Can we initialize all weight parameters in linear regression or in softmax regression to the same value?

3. Look up analytic bounds on the eigenvalues of the product of two matrices. What does this tell you about ensuring that gradients are well conditioned?

4. If we know that some terms diverge, can we fix this after the fact? Look at the paper on layerwise adaptive rate scaling for inspiration (You *et al.*, 2017).

 Discussions[105].

105

5.5 Generalization in Deep Learning

In Chapter 3 and Chapter 4, we tackled regression and classification problems by fitting linear models to training data. In both cases, we provided practical algorithms for finding the parameters that maximized the likelihood of the observed training labels. And then, towards the end of each chapter, we recalled that fitting the training data was only an intermediate goal. Our real quest all along was to discover *general patterns* on the basis of which we can make accurate predictions even on new examples drawn from the same underlying population. Machine learning researchers are *consumers* of optimization algorithms. Sometimes, we must even develop new optimization algorithms. But at the end of the day, optimization is merely a means to an end. At its core, machine learning is a statistical discipline and we wish to optimize training loss only insofar as some statistical principle (known or unknown) leads the resulting models to generalize beyond the training set.

On the bright side, it turns out that deep neural networks trained by stochastic gradient descent generalize remarkably well across myriad prediction problems, spanning computer vision; natural language processing; time series data; recommender systems; electronic health records; protein folding; value function approximation in video games and board games; and numerous other domains. On the downside, if you were looking for a straightforward account of either the optimization story (why we can fit them to training data) or the generalization story (why the resulting models generalize to unseen examples), then you might want to pour yourself a drink. While our procedures for optimizing linear models and the statistical properties of the solutions are both described well by a comprehensive body of theory, our understanding of deep learning still resembles the wild west on both fronts.

Both the theory and practice of deep learning are rapidly evolving, with theorists adopting new strategies to explain what's going on, even as practitioners continue to innovate at a blistering pace, building arsenals of heuristics for training deep networks and a body of intuitions and folk knowledge that provide guidance for deciding which techniques to apply in which situations.

The summary of the present moment is that the theory of deep learning has produced promising lines of attack and scattered fascinating results, but still appears far from a comprehensive account of both (i) why we are able to optimize neural networks and (ii) how models learned by gradient descent manage to generalize so well, even on high-dimensional tasks. However, in practice, (i) is seldom a problem (we can always find parameters that will fit all of our training data) and thus understanding generalization is far the bigger problem. On the other hand, even absent the comfort of a coherent scientific theory, practitioners have developed a large collection of techniques that may help you to produce models that generalize well in practice. While no pithy summary can possibly do justice to the vast topic of generalization in deep learning, and while the overall state of research is far from resolved, we hope, in this section, to present a broad overview of the state of research and practice.

5.5.1 Revisiting Overfitting and Regularization

According to the "no free lunch" theorem of Wolpert and Macready (1995), any learning algorithm generalizes better on data with certain distributions, and worse with other distributions. Thus, given a finite training set, a model relies on certain assumptions: to achieve human-level performance it may be useful to identify *inductive biases* that reflect how humans think about the world. Such inductive biases show preferences for solutions with certain properties. For example, a deep MLP has an inductive bias towards building up a complicated function by the composition of simpler functions.

With machine learning models encoding inductive biases, our approach to training them typically consists of two phases: (i) fit the training data; and (ii) estimate the *generalization error* (the true error on the underlying population) by evaluating the model on holdout data. The difference between our fit on the training data and our fit on the test data is called the *generalization gap* and when this is large, we say that our models *overfit* to the training data. In extreme cases of overfitting, we might exactly fit the training data, even when the test error remains significant. And in the classical view, the interpretation is that our models are too complex, requiring that we either shrink the number of features, the number of nonzero parameters learned, or the size of the parameters as quantified. Recall the plot of model complexity compared with loss (Fig. 3.6.1) from Section 3.6.

However deep learning complicates this picture in counterintuitive ways. First, for classification problems, our models are typically expressive enough to perfectly fit every training example, even in datasets consisting of millions (Zhang *et al.*, 2021). In the classical picture, we might think that this setting lies on the far right extreme of the model complexity axis, and that any improvements in generalization error must come by way of regularization, either by reducing the complexity of the model class, or by applying a penalty, severely constraining the set of values that our parameters might take. But that is where things start to get weird.

Strangely, for many deep learning tasks (e.g., image recognition and text classification) we are typically choosing among model architectures, all of which can achieve arbitrarily low training loss (and zero training error). Because all models under consideration achieve zero training error, *the only avenue for further gains is to reduce overfitting*. Even stranger, it is often the case that despite fitting the training data perfectly, we can actually *reduce the generalization error* further by making the model *even more expressive*, e.g., adding layers, nodes, or training for a larger number of epochs. Stranger yet, the pattern relating the generalization gap to the *complexity* of the model (as captured, for example, in the depth or width of the networks) can be non-monotonic, with greater complexity hurting at first but subsequently helping in a so-called "double-descent" pattern (Nakkiran *et al.*, 2021). Thus the deep learning practitioner possesses a bag of tricks, some of which seemingly restrict the model in some fashion and others that seemingly make it even more expressive, and all of which, in some sense, are applied to mitigate overfitting.

Complicating things even further, while the guarantees provided by classical learning theory can be conservative even for classical models, they appear powerless to explain why it is that deep neural networks generalize in the first place. Because deep neural networks are capable of fitting arbitrary labels even for large datasets, and despite the use of familiar methods such as ℓ_2 regularization, traditional complexity-based generalization bounds, e.g., those based on the VC dimension or Rademacher complexity of a hypothesis class cannot explain why neural networks generalize.

5.5.2 Inspiration from Nonparametrics

Approaching deep learning for the first time, it is tempting to think of them as parametric models. After all, the models *do* have millions of parameters. When we update the models, we update their parameters. When we save the models, we write their parameters to disk. However, mathematics and computer science are riddled with counterintuitive changes of perspective, and surprising isomorphisms between seemingly different problems. While neural networks clearly *have* parameters, in some ways it can be more fruitful to think of them as behaving like nonparametric models. So what precisely makes a model nonparametric? While the name covers a diverse set of approaches, one common theme is that nonparametric methods tend to have a level of complexity that grows as the amount of available data grows.

Perhaps the simplest example of a nonparametric model is the k-nearest neighbor algorithm (we will cover more nonparametric models later, for example in Section 11.2). Here, at training time, the learner simply memorizes the dataset. Then, at prediction time, when confronted with a new point $\mathbf{x}$, the learner looks up the k nearest neighbors (the k points $\mathbf{x}'_i$ that minimize some distance $d(\mathbf{x}, \mathbf{x}'_i)$). When $k = 1$, this algorithm is called 1-nearest neighbors, and the algorithm will always achieve a training error of zero. That however, does not mean that the algorithm will not generalize. In fact, it turns out that under some mild conditions, the 1-nearest neighbor algorithm is consistent (eventually converging to the optimal predictor).

Note that 1-nearest neighbor requires that we specify some distance function d, or equivalently, that we specify some vector-valued basis function $\phi(\mathbf{x})$ for featurizing our data. For any choice of the distance metric, we will achieve zero training error and eventually reach an optimal predictor, but different distance metrics d encode different inductive biases and with a finite amount of available data will yield different predictors. Different choices of the distance metric d represent different assumptions about the underlying patterns and the performance of the different predictors will depend on how compatible the assumptions are with the observed data.

In a sense, because neural networks are over-parametrized, possessing many more parameters than are needed to fit the training data, they tend to *interpolate* the training data (fitting it perfectly) and thus behave, in some ways, more like nonparametric models. More recent theoretical research has established deep connection between large neural networks and nonparametric methods, notably kernel methods. In particular, Jacot *et al.* (2018) demonstrated that in the limit, as multilayer perceptrons with randomly initialized weights grow infinitely wide, they become equivalent to (nonparametric) kernel methods for a specific choice of the kernel function (essentially, a distance function), which they call the neural tangent kernel. While current neural tangent kernel models may not fully explain the behavior of modern deep networks, their success as an analytical tool underscores the usefulness of nonparametric modeling for understanding the behavior of over-parametrized deep networks.

5.5.3 Early Stopping

While deep neural networks are capable of fitting arbitrary labels, even when labels are assigned incorrectly or randomly (Zhang *et al.*, 2021), this capability only emerges over many iterations of training. A new line of work (Rolnick *et al.*, 2017) has revealed that in the setting of label noise, neural networks tend to fit cleanly labeled data first and only subsequently to interpolate the mislabeled data. Moreover, it has been established that this phenomenon translates directly into a guarantee on generalization: whenever a model has fitted the cleanly labeled data but not randomly labeled examples included in the training set, it has in fact generalized (Garg *et al.*, 2021).

Together these findings help to motivate *early stopping*, a classic technique for regularizing deep neural networks. Here, rather than directly constraining the values of the weights, one constrains the number of epochs of training. The most common way to determine the stopping criterion is to monitor validation error throughout training (typically by checking once after each epoch) and to cut off training when the validation error has not decreased by more than some small amount ϵ for some number of epochs. This is sometimes called a *patience criterion*. As well as the potential to lead to better generalization in the setting of noisy labels, another benefit of early stopping is the time saved. Once the patience criterion is met, one can terminate training. For large models that might require days of training simultaneously across eight or more GPUs, well-tuned early stopping can save researchers days of time and can save their employers many thousands of dollars.

Notably, when there is no label noise and datasets are *realizable* (the classes are truly separable, e.g., distinguishing cats from dogs), early stopping tends not to lead to significant improvements in generalization. On the other hand, when there is label noise, or intrinsic variability in the label (e.g., predicting mortality among patients), early stopping is crucial. Training models until they interpolate noisy data is typically a bad idea.

5.5.4 Classical Regularization Methods for Deep Networks

In Chapter 3, we described several classical regularization techniques for constraining the complexity of our models. In particular, Section 3.7 introduced a method called weight decay, which consists of adding a regularization term to the loss function in order to penalize large values of the weights. Depending on which weight norm is penalized this technique is known either as ridge regularization (for ℓ_2 penalty) or lasso regularization (for an ℓ_1 penalty). In the classical analysis of these regularizers, they are considered as sufficiently restrictive on the values that the weights can take to prevent the model from fitting arbitrary labels.

In deep learning implementations, weight decay remains a popular tool. However, researchers have noted that typical strengths of ℓ_2 regularization are insufficient to prevent the networks from interpolating the data (Zhang *et al.*, 2021) and thus the benefits if interpreted as regularization might only make sense in combination with the early stopping criterion. Absent early stopping, it is possible that just like the number of layers or number of nodes (in deep learning) or the distance metric (in 1-nearest neighbor), these methods may lead to better generalization not because they meaningfully constrain the power of the neural network but rather because they somehow encode inductive biases that are better compatible with the patterns found in datasets of interests. Thus, classical regularizers remain popular in deep learning implementations, even if the theoretical rationale for their efficacy may be radically different.

Notably, deep learning researchers have also built on techniques first popularized in classical regularization contexts, such as adding noise to model inputs. In the next section we will introduce the famous dropout technique (invented by Srivastava *et al.* (2014)), which has become a mainstay of deep learning, even as the theoretical basis for its efficacy remains similarly mysterious.

5.5.5 Summary

Unlike classical linear models, which tend to have fewer parameters than examples, deep networks tend to be over-parametrized, and for most tasks are capable of perfectly fitting the training set. This *interpolation regime* challenges many hard fast-held intuitions. Functionally, neural networks look like parametric models. But thinking of them as nonparametric models can sometimes be a more reliable source of intuition. Because it is often the case that all deep networks under consideration are capable of fitting all of the training labels, nearly all gains must come by mitigating overfitting (closing the *generalization gap*). Paradoxically, the interventions that reduce the generalization gap sometimes appear to increase model complexity and at other times appear to decrease complexity. However, these methods seldom decrease complexity sufficiently for classical theory to explain the generalization of deep networks, and *why certain choices lead to improved generalization* remains for the most part a massive open question despite the concerted efforts of many brilliant researchers.

5.5.6 Exercises

1. In what sense do traditional complexity-based measures fail to account for generalization of deep neural networks?

2. Why might *early stopping* be considered a regularization technique?

3. How do researchers typically determine the stopping criterion?

4. What important factor seems to differentiate cases when early stopping leads to big improvements in generalization?

5. Beyond generalization, describe another benefit of early stopping.

Discussions[106].

106

5.6 Dropout

Let's think briefly about what we expect from a good predictive model. We want it to peform well on unseen data. Classical generalization theory suggests that to close the gap between train and test performance, we should aim for a simple model. Simplicity can come in the form of a small number of dimensions. We explored this when discussing the monomial basis functions of linear models in Section 3.6. Additionally, as we saw when discussing weight decay (ℓ_2 regularization) in Section 3.7, the (inverse) norm of the parameters also represents a useful measure of simplicity. Another useful notion of simplicity is smoothness, i.e., that the function should not be sensitive to small changes to its inputs. For instance, when we classify images, we would expect that adding some random noise to the pixels should be mostly harmless.

Bishop (1995) formalized this idea when he proved that training with input noise is equivalent to Tikhonov regularization. This work drew a clear mathematical connection between the requirement that a function be smooth (and thus simple), and the requirement that it be resilient to perturbations in the input.

Then, Srivastava *et al.* (2014) developed a clever idea for how to apply Bishop's idea to the internal layers of a network, too. Their idea, called *dropout*, involves injecting noise while computing each internal layer during forward propagation, and it has become a standard technique for training neural networks. The method is called *dropout* because we literally *drop out* some neurons during training. Throughout training, on each iteration, standard dropout consists of zeroing out some fraction of the nodes in each layer before calculating the subsequent layer.

To be clear, we are imposing our own narrative with the link to Bishop. The original paper on dropout offers intuition through a surprising analogy to sexual reproduction. The authors argue that neural network overfitting is characterized by a state in which each layer relies on a specific pattern of activations in the previous layer, calling this condition *co-adaptation*. Dropout, they claim, breaks up co-adaptation just as sexual reproduction is argued to break up co-adapted genes. While such an justification of this theory is certainly up for debate, the dropout technique itself has proved enduring, and various forms of dropout are implemented in most deep learning libraries.

The key challenge is how to inject this noise. One idea is to inject it in an *unbiased* manner so that the expected value of each layer—while fixing the others—equals the value it would have taken absent noise. In Bishop's work, he added Gaussian noise to the inputs to a linear model. At each training iteration, he added noise sampled from a distribution with mean zero $\epsilon \sim \mathcal{N}(0, \sigma^2)$ to the input $\mathbf{x}$, yielding a perturbed point $\mathbf{x}' = \mathbf{x} + \epsilon$. In expectation, $E[\mathbf{x}'] = \mathbf{x}$.

In standard dropout regularization, one zeros out some fraction of the nodes in each layer and then *debiases* each layer by normalizing by the fraction of nodes that were retained (not dropped out). In other words, with *dropout probability* p, each intermediate activation h is replaced by a random variable h' as follows:

$$h' = \begin{cases} 0 & \text{with probability } p \\ \frac{h}{1-p} & \text{otherwise} \end{cases} \tag{5.6.1}$$

By design, the expectation remains unchanged, i.e., $E[h'] = h$.

```
import torch
from torch import nn
from d2l import torch as d2l
```

5.6.1 Dropout in Practice

Recall the MLP with a hidden layer and five hidden units from Fig. 5.1.1. When we apply dropout to a hidden layer, zeroing out each hidden unit with probability p, the result can be viewed as a network containing only a subset of the original neurons. In Fig. 5.6.1, h_2 and h_5 are removed. Consequently, the calculation of the outputs no longer depends on h_2 or h_5 and their respective gradient also vanishes when performing backpropagation. In this way, the calculation of the output layer cannot be overly dependent on any one element of $h_1, \ldots, h_5$.

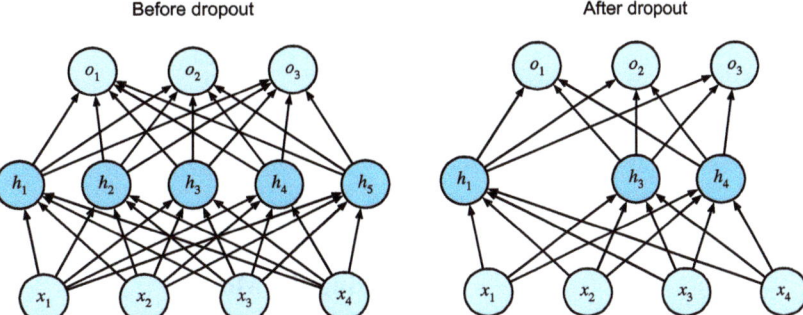

Fig. 5.6.1 MLP before and after dropout.

Typically, we disable dropout at test time. Given a trained model and a new example, we do not drop out any nodes and thus do not need to normalize. However, there are some exceptions: some researchers use dropout at test time as a heuristic for estimating the *uncertainty* of neural network predictions: if the predictions agree across many different dropout outputs, then we might say that the network is more confident.

5.6.2 Implementation from Scratch

To implement the dropout function for a single layer, we must draw as many samples from a Bernoulli (binary) random variable as our layer has dimensions, where the random variable takes value 1 (keep) with probability $1 - p$ and 0 (drop) with probability p. One easy way to implement this is to first draw samples from the uniform distribution $U[0, 1]$. Then we can keep those nodes for which the corresponding sample is greater than p, dropping the rest.

In the following code, we implement a `dropout_layer` function that drops out the elements in the tensor input X with probability `dropout`, rescaling the remainder as described above: dividing the survivors by `1.0-dropout`.

```
def dropout_layer(X, dropout):
    assert 0 <= dropout <= 1
    if dropout == 1: return torch.zeros_like(X)
    mask = (torch.rand(X.shape) > dropout).float()
    return mask * X / (1.0 - dropout)
```

We can test out the `dropout_layer` function on a few examples. In the following lines of code, we pass our input X through the dropout operation, with probabilities 0, 0.5, and 1, respectively.

```
X = torch.arange(16, dtype = torch.float32).reshape((2, 8))
print('dropout_p = 0:', dropout_layer(X, 0))
print('dropout_p = 0.5:', dropout_layer(X, 0.5))
print('dropout_p = 1:', dropout_layer(X, 1))
```

```
dropout_p = 0: tensor([[ 0.,  1.,  2.,  3.,  4.,  5.,  6.,  7.],
            [ 8.,  9., 10., 11., 12., 13., 14., 15.]])
dropout_p = 0.5: tensor([[ 0.,  2.,  4.,  6.,  8., 10., 12.,  0.],
            [16., 18.,  0., 22.,  0., 26., 28., 30.]])
dropout_p = 1: tensor([[0., 0., 0., 0., 0., 0., 0., 0.],
            [0., 0., 0., 0., 0., 0., 0., 0.]])
```

Defining the Model

The model below applies dropout to the output of each hidden layer (following the activation function). We can set dropout probabilities for each layer separately. A common choice is to set a lower dropout probability closer to the input layer. We ensure that dropout is only active during training.

```python
class DropoutMLPScratch(d2l.Classifier):
    def __init__(self, num_outputs, num_hiddens_1, num_hiddens_2,
                 dropout_1, dropout_2, lr):
        super().__init__()
        self.save_hyperparameters()
        self.lin1 = nn.LazyLinear(num_hiddens_1)
        self.lin2 = nn.LazyLinear(num_hiddens_2)
        self.lin3 = nn.LazyLinear(num_outputs)
        self.relu = nn.ReLU()

    def forward(self, X):
        H1 = self.relu(self.lin1(X.reshape((X.shape[0], -1))))
        if self.training:
            H1 = dropout_layer(H1, self.dropout_1)
        H2 = self.relu(self.lin2(H1))
        if self.training:
            H2 = dropout_layer(H2, self.dropout_2)
        return self.lin3(H2)
```

Training

The following is similar to the training of MLPs described previously.

```python
hparams = {'num_outputs':10, 'num_hiddens_1':256, 'num_hiddens_2':256,
           'dropout_1':0.5, 'dropout_2':0.5, 'lr':0.1}
model = DropoutMLPScratch(**hparams)
data = d2l.FashionMNIST(batch_size=256)
trainer = d2l.Trainer(max_epochs=10)
trainer.fit(model, data)
```

5.6.3 Concise Implementation

With high-level APIs, all we need to do is add a Dropout layer after each fully connected layer, passing in the dropout probability as the only argument to its constructor. During

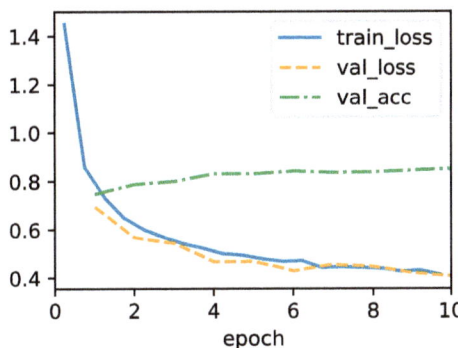

training, the Dropout layer will randomly drop out outputs of the previous layer (or equivalently, the inputs to the subsequent layer) according to the specified dropout probability.
When not in training mode, the Dropout layer simply passes the data through during testing.

```
class DropoutMLP(d2l.Classifier):
    def __init__(self, num_outputs, num_hiddens_1, num_hiddens_2,
                 dropout_1, dropout_2, lr):
        super().__init__()
        self.save_hyperparameters()
        self.net = nn.Sequential(
            nn.Flatten(), nn.LazyLinear(num_hiddens_1), nn.ReLU(),
            nn.Dropout(dropout_1), nn.LazyLinear(num_hiddens_2), nn.ReLU(),
            nn.Dropout(dropout_2), nn.LazyLinear(num_outputs))
```

Next, we train the model.

```
model = DropoutMLP(**hparams)
trainer.fit(model, data)
```

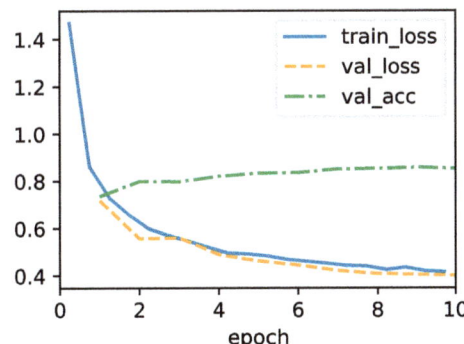

5.6.4 Summary

Beyond controlling the number of dimensions and the size of the weight vector, dropout is yet another tool for avoiding overfitting. Often tools are used jointly. Note that dropout is used only during training: it replaces an activation h with a random variable with expected value h.

5.6.5 Exercises

1. What happens if you change the dropout probabilities for the first and second layers? In particular, what happens if you switch the ones for both layers? Design an experiment to answer these questions, describe your results quantitatively, and summarize the qualitative takeaways.

2. Increase the number of epochs and compare the results obtained when using dropout with those when not using it.

3. What is the variance of the activations in each hidden layer when dropout is and is not applied? Draw a plot to show how this quantity evolves over time for both models.

4. Why is dropout not typically used at test time?

5. Using the model in this section as an example, compare the effects of using dropout and weight decay. What happens when dropout and weight decay are used at the same time? Are the results additive? Are there diminished returns (or worse)? Do they cancel each other out?

6. What happens if we apply dropout to the individual weights of the weight matrix rather than the activations?

7. Invent another technique for injecting random noise at each layer that is different from the standard dropout technique. Can you develop a method that outperforms dropout on the Fashion-MNIST dataset (for a fixed architecture)?

Discussions[107].

5.7 Predicting House Prices on Kaggle

Now that we have introduced some basic tools for building and training deep networks and regularizing them with techniques including weight decay and dropout, we are ready to put all this knowledge into practice by participating in a Kaggle competition. The house price prediction competition is a great place to start. The data is fairly generic and do not exhibit exotic structure that might require specialized models (as audio or video might). This dataset, collected by De Cock (2011), covers house prices in Ames, Iowa from the period 2006–2010. It is considerably larger than the famous Boston housing dataset[108] of Harrison and Rubinfeld (1978), boasting both more examples and more features.

In this section, we will walk you through details of data preprocessing, model design, and hyperparameter selection. We hope that through a hands-on approach, you will gain some intuitions that will guide you in your career as a data scientist.

```python
%matplotlib inline
import pandas as pd
import torch
from torch import nn
from d2l import torch as d2l
```

5.7.1 Downloading Data

Throughout the book, we will train and test models on various downloaded datasets. Here, we implement two utility functions for downloading and extracting zip or tar files. Again, we skip implementation details of such utility functions.

```python
def download(url, folder, sha1_hash=None):
    """Download a file to folder and return the local filepath."""

def extract(filename, folder):
    """Extract a zip/tar file into folder."""
```

5.7.2 Kaggle

Kaggle[109] is a popular platform that hosts machine learning competitions. Each competition centers on a dataset and many are sponsored by stakeholders who offer prizes to the winning solutions. The platform helps users to interact via forums and shared code, fostering both collaboration and competition. While leaderboard chasing often spirals out of control, with researchers focusing myopically on preprocessing steps rather than asking fundamental questions, there is also tremendous value in the objectivity of a platform that facilitates direct quantitative comparisons among competing approaches as well as code sharing so that everyone can learn what did and did not work. If you want to participate in a Kaggle competition, you will first need to register for an account (see Fig. 5.7.1).

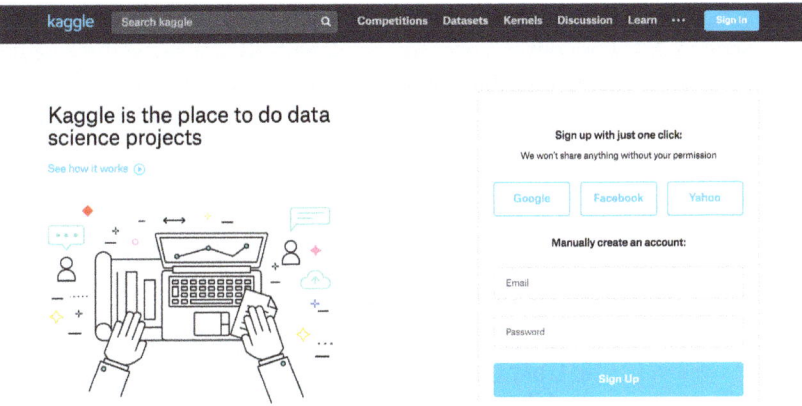

Fig. 5.7.1   The Kaggle website.

On the house price prediction competition page, as illustrated in Fig. 5.7.2, you can find the dataset (under the "Data" tab), submit predictions, and see your ranking, The URL is right here:

https://www.kaggle.com/c/house-prices-advanced-regression-techniques

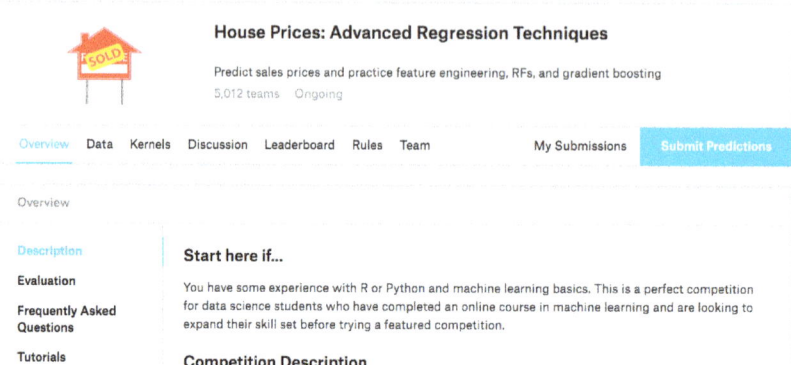

Fig. 5.7.2 The house price prediction competition page.

5.7.3 Accessing and Reading the Dataset

Note that the competition data is separated into training and test sets. Each record includes the property value of the house and attributes such as street type, year of construction, roof type, basement condition, etc. The features consist of various data types. For example, the year of construction is represented by an integer, the roof type by discrete categorical assignments, and other features by floating point numbers. And here is where reality complicates things: for some examples, some data is altogether missing with the missing value marked simply as "na". The price of each house is included for the training set only (it is a competition after all). We will want to partition the training set to create a validation set, but we only get to evaluate our models on the official test set after uploading predictions to Kaggle. The "Data" tab on the competition tab in Fig. 5.7.2 has links for downloading the data.

To get started, we will read in and process the data using pandas, which we introduced in Section 2.2. For convenience, we can download and cache the Kaggle housing dataset. If a file corresponding to this dataset already exists in the cache directory and its SHA-1 matches sha1_hash, our code will use the cached file to avoid clogging up your Internet with redundant downloads.

```
class KaggleHouse(d2l.DataModule):
    def __init__(self, batch_size, train=None, val=None):
        super().__init__()
        self.save_hyperparameters()
        if self.train is None:
            self.raw_train = pd.read_csv(d2l.download(
                d2l.DATA_URL + 'kaggle_house_pred_train.csv', self.root,
                sha1_hash='585e9cc93e70b39160e7921475f9bcd7d31219ce'))
```

(continues on next page)

(continued from previous page)

```
        self.raw_val = pd.read_csv(d2l.download(
            d2l.DATA_URL + 'kaggle_house_pred_test.csv', self.root,
            sha1_hash='fa19780a7b011d9b009e8bff8e99922a8ee2eb90'))
```

The training dataset includes 1460 examples, 80 features, and one label, while the validation data contains 1459 examples and 80 features.

```
data = KaggleHouse(batch_size=64)
print(data.raw_train.shape)
print(data.raw_val.shape)
```

```
(1460, 81)
(1459, 80)
```

5.7.4 Data Preprocessing

Let's take a look at the first four and final two features as well as the label (SalePrice) from the first four examples.

```
print(data.raw_train.iloc[:4, [0, 1, 2, 3, -3, -2, -1]])
```

	Id	MSSubClass	MSZoning	LotFrontage	SaleType	SaleCondition	SalePrice
0	1	60	RL	65.0	WD	Normal	208500
1	2	20	RL	80.0	WD	Normal	181500
2	3	60	RL	68.0	WD	Normal	223500
3	4	70	RL	60.0	WD	Abnorml	140000

We can see that in each example, the first feature is the identifier. This helps the model determine each training example. While this is convenient, it does not carry any information for prediction purposes. Hence, we will remove it from the dataset before feeding the data into the model. Furthermore, given a wide variety of data types, we will need to preprocess the data before we can start modeling.

Let's start with the numerical features. First, we apply a heuristic, replacing all missing values by the corresponding feature's mean. Then, to put all features on a common scale, we *standardize* the data by rescaling features to zero mean and unit variance:

$$x \leftarrow \frac{x - \mu}{\sigma}, \tag{5.7.1}$$

where μ and σ denote mean and standard deviation, respectively. To verify that this indeed transforms our feature (variable) such that it has zero mean and unit variance, note that $E[\frac{x-\mu}{\sigma}] = \frac{\mu-\mu}{\sigma} = 0$ and that $E[(x - \mu)^2] = (\sigma^2 + \mu^2) - 2\mu^2 + \mu^2 = \sigma^2$. Intuitively, we standardize the data for two reasons. First, it proves convenient for optimization. Second, because we do not know *a priori* which features will be relevant, we do not want to penalize coefficients assigned to one feature more than any other.

Next we deal with discrete values. These include features such as "MSZoning". We replace them by a one-hot encoding in the same way that we earlier transformed multiclass labels into vectors (see Section 4.1.1). For instance, "MSZoning" assumes the values "RL" and "RM". Dropping the "MSZoning" feature, two new indicator features "MSZoning_RL" and "MSZoning_RM" are created with values being either 0 or 1. According to one-hot encoding, if the original value of "MSZoning" is "RL", then "MSZoning_RL" is 1 and "MSZoning_RM" is 0. The pandas package does this automatically for us.

```python
@d2l.add_to_class(KaggleHouse)
def preprocess(self):
    # Remove the ID and label columns
    label = 'SalePrice'
    features = pd.concat(
        (self.raw_train.drop(columns=['Id', label]),
         self.raw_val.drop(columns=['Id'])))
    # Standardize numerical columns
    numeric_features = features.dtypes[features.dtypes!='object'].index
    features[numeric_features] = features[numeric_features].apply(
        lambda x: (x - x.mean()) / (x.std()))
    # Replace NAN numerical features by 0
    features[numeric_features] = features[numeric_features].fillna(0)
    # Replace discrete features by one-hot encoding
    features = pd.get_dummies(features, dummy_na=True)
    # Save preprocessed features
    self.train = features[:self.raw_train.shape[0]].copy()
    self.train[label] = self.raw_train[label]
    self.val = features[self.raw_train.shape[0]:].copy()
```

You can see that this conversion increases the number of features from 79 to 331 (excluding ID and label columns).

```python
data.preprocess()
data.train.shape
```

```
(1460, 331)
```

5.7.5 Error Measure

To get started we will train a linear model with squared loss. Not surprisingly, our linear model will not lead to a competition-winning submission but it does provide a sanity check to see whether there is meaningful information in the data. If we cannot do better than random guessing here, then there might be a good chance that we have a data processing bug. And if things work, the linear model will serve as a baseline giving us some intuition about how close the simple model gets to the best reported models, giving us a sense of how much gain we should expect from fancier models.

With house prices, as with stock prices, we care about relative quantities more than absolute quantities. Thus we tend to care more about the relative error $\frac{y-\hat{y}}{y}$ than about the absolute error $y - \hat{y}$. For instance, if our prediction is off by $100,000 when estimating the price of a house in rural Ohio, where the value of a typical house is $125,000, then we are probably doing a horrible job. On the other hand, if we err by this amount in Los Altos Hills, California, this might represent a stunningly accurate prediction (there, the median house price exceeds $4 million).

One way to address this problem is to measure the discrepancy in the logarithm of the price estimates. In fact, this is also the official error measure used by the competition to evaluate the quality of submissions. After all, a small value δ for $|\log y - \log \hat{y}| \leq \delta$ translates into $e^{-\delta} \leq \frac{\hat{y}}{y} \leq e^{\delta}$. This leads to the following root-mean-squared-error between the logarithm of the predicted price and the logarithm of the label price:

$$\sqrt{\frac{1}{n} \sum_{i=1}^{n} (\log y_i - \log \hat{y}_i)^2}. \tag{5.7.2}$$

```
@d2l.add_to_class(KaggleHouse)
def get_dataloader(self, train):
    label = 'SalePrice'
    data = self.train if train else self.val
    if label not in data: return
    get_tensor = lambda x: torch.tensor(x.values.astype(float),
                                        dtype=torch.float32)
    # Logarithm of prices
    tensors = (get_tensor(data.drop(columns=[label])),  # X
               torch.log(get_tensor(data[label])).reshape((-1, 1)))  # Y
    return self.get_tensorloader(tensors, train)
```

5.7.6 K-Fold Cross-Validation

You might recall that we introduced cross-validation in Section 3.6.3, where we discussed how to deal with model selection. We will put this to good use to select the model design and to adjust the hyperparameters. We first need a function that returns the ith fold of the data in a K-fold cross-validation procedure. It proceeds by slicing out the ith segment as validation data and returning the rest as training data. Note that this is not the most efficient way of handling data and we would definitely do something much smarter if our dataset was considerably larger. But this added complexity might obfuscate our code unnecessarily so we can safely omit it here owing to the simplicity of our problem.

```
def k_fold_data(data, k):
    rets = []
    fold_size = data.train.shape[0] // k
    for j in range(k):
        idx = range(j * fold_size, (j+1) * fold_size)
        rets.append(KaggleHouse(data.batch_size, data.train.drop(index=idx),
                                data.train.loc[idx]))
    return rets
```

The average validation error is returned when we train K times in the K-fold cross-validation.

```
def k_fold(trainer, data, k, lr):
    val_loss, models = [], []
    for i, data_fold in enumerate(k_fold_data(data, k)):
        model = d2l.LinearRegression(lr)
        model.board.yscale='log'
        if i != 0: model.board.display = False
        trainer.fit(model, data_fold)
        val_loss.append(float(model.board.data['val_loss'][-1].y))
        models.append(model)
    print(f'average validation log mse = {sum(val_loss)/len(val_loss)}')
    return models
```

5.7.7 Model Selection

In this example, we pick an untuned set of hyperparameters and leave it up to the reader to improve the model. Finding a good choice can take time, depending on how many variables one optimizes over. With a large enough dataset, and the normal sorts of hyperparameters, K-fold cross-validation tends to be reasonably resilient against multiple testing. However, if we try an unreasonably large number of options we might find that our validation performance is no longer representative of the true error.

```
trainer = d2l.Trainer(max_epochs=10)
models = k_fold(trainer, data, k=5, lr=0.01)
```

```
average validation log mse = 0.18012988299131394
```

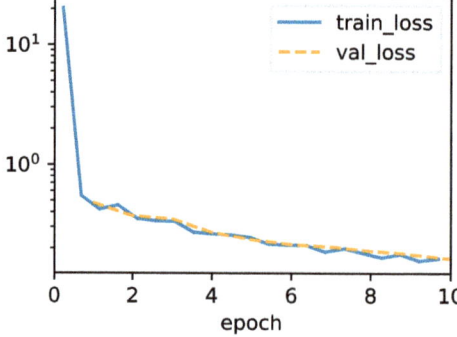

Notice that sometimes the number of training errors for a set of hyperparameters can be very low, even as the number of errors on K-fold cross-validation grows considerably higher. This indicates that we are overfitting. Throughout training you will want to monitor both numbers. Less overfitting might indicate that our data can support a more powerful model. Massive overfitting might suggest that we can gain by incorporating regularization techniques.

5.7.8 Submitting Predictions on Kaggle

Now that we know what a good choice of hyperparameters should be, we might calculate the average predictions on the test set by all the K models. Saving the predictions in a csv file will simplify uploading the results to Kaggle. The following code will generate a file called `submission.csv`.

```
preds = [model(torch.tensor(data.val.values.astype(float), dtype=torch.
↪float32))
        for model in models]
# Taking exponentiation of predictions in the logarithm scale
ensemble_preds = torch.exp(torch.cat(preds, 1)).mean(1)
submission = pd.DataFrame({'Id':data.raw_val.Id,
                            'SalePrice':ensemble_preds.detach().numpy()})
submission.to_csv('submission.csv', index=False)
```

Next, as demonstrated in Fig. 5.7.3, we can submit our predictions on Kaggle and see how they compare with the actual house prices (labels) on the test set. The steps are quite simple:

• Log in to the Kaggle website and visit the house price prediction competition page.

• Click the "Submit Predictions" or "Late Submission" button.

• Click the "Upload Submission File" button in the dashed box at the bottom of the page and select the prediction file you wish to upload.

• Click the "Make Submission" button at the bottom of the page to view your results.

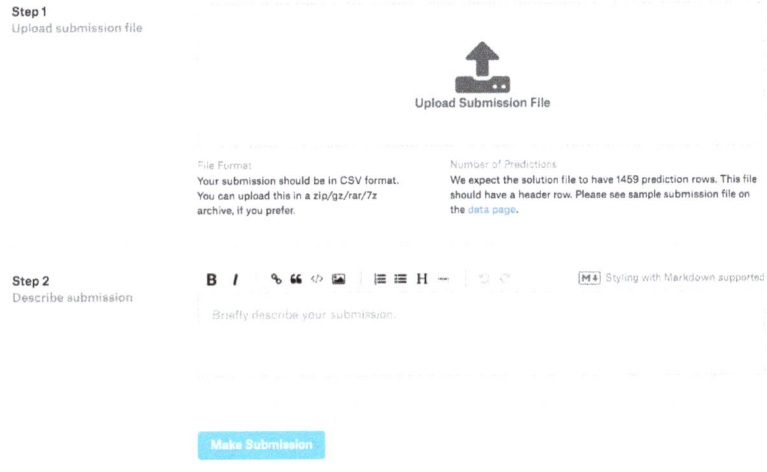

Fig. 5.7.3 Submitting data to Kaggle.

5.7.9 Summary and Discussion

Real data often contains a mix of different data types and needs to be preprocessed. Rescaling real-valued data to zero mean and unit variance is a good default. So is replacing missing values with their mean. Furthermore, transforming categorical features into indicator features allows us to treat them like one-hot vectors. When we tend to care more about the relative error than about the absolute error, we can measure the discrepancy in the logarithm of the prediction. To select the model and adjust the hyperparameters, we can use K-fold cross-validation .

5.7.10 Exercises

1. Submit your predictions for this section to Kaggle. How good are they?

2. Is it always a good idea to replace missing values by a mean? Hint: can you construct a situation where the values are not missing at random?

3. Improve the score by tuning the hyperparameters through K-fold cross-validation.

4. Improve the score by improving the model (e.g., layers, weight decay, and dropout).

5. What happens if we do not standardize the continuous numerical features as we have done in this section?

Discussions[110].

110

## 6

Builders' Guide

Alongside giant datasets and powerful hardware, great software tools have played an indispensable role in the rapid progress of deep learning. Starting with the pathbreaking Theano library released in 2007, flexible open-source tools have enabled researchers to rapidly prototype models, avoiding repetitive work when recycling standard components while still maintaining the ability to make low-level modifications. Over time, deep learning's libraries have evolved to offer increasingly coarse abstractions. Just as semiconductor designers went from specifying transistors to logical circuits to writing code, neural networks researchers have moved from thinking about the behavior of individual artificial neurons to conceiving of networks in terms of whole layers, and now often design architectures with far coarser *blocks* in mind.

So far, we have introduced some basic machine learning concepts, ramping up to fully-functional deep learning models. In the last chapter, we implemented each component of an MLP from scratch and even showed how to leverage high-level APIs to roll out the same models effortlessly. To get you that far that fast, we *called upon* the libraries, but skipped over more advanced details about *how they work*. In this chapter, we will peel back the curtain, digging deeper into the key components of deep learning computation, namely model construction, parameter access and initialization, designing custom layers and blocks, reading and writing models to disk, and leveraging GPUs to achieve dramatic speedups. These insights will move you from *end user* to *power user*, giving you the tools needed to reap the benefits of a mature deep learning library while retaining the flexibility to implement more complex models, including those you invent yourself! While this chapter does not introduce any new models or datasets, the advanced modeling chapters that follow rely heavily on these techniques.

6.1 Layers and Modules

When we first introduced neural networks, we focused on linear models with a single output. Here, the entire model consists of just a single neuron. Note that a single neuron (i) takes some set of inputs; (ii) generates a corresponding scalar output; and (iii) has a set of associated parameters that can be updated to optimize some objective function of interest. Then, once we started thinking about networks with multiple outputs, we leveraged vectorized arithmetic to characterize an entire layer of neurons. Just like individual neurons, layers (i) take a set of inputs, (ii) generate corresponding outputs, and (iii) are described by a set of tunable parameters. When we worked through softmax regression, a single layer was itself the model. However, even when we subsequently introduced MLPs, we could still think of the model as retaining this same basic structure.

Interestingly, for MLPs, both the entire model and its constituent layers share this structure. The entire model takes in raw inputs (the features), generates outputs (the predictions), and possesses parameters (the combined parameters from all constituent layers). Likewise, each individual layer ingests inputs (supplied by the previous layer) generates outputs (the inputs to the subsequent layer), and possesses a set of tunable parameters that are updated according to the signal that flows backwards from the subsequent layer.

While you might think that neurons, layers, and models give us enough abstractions to go about our business, it turns out that we often find it convenient to speak about components that are larger than an individual layer but smaller than the entire model. For example, the ResNet-152 architecture, which is wildly popular in computer vision, possesses hundreds of layers. These layers consist of repeating patterns of *groups of layers*. Implementing such a network one layer at a time can grow tedious. This concern is not just hypothetical—such design patterns are common in practice. The ResNet architecture mentioned above won the 2015 ImageNet and COCO computer vision competitions for both recognition and detection (He *et al.*, 2016) and remains a go-to architecture for many vision tasks. Similar architectures in which layers are arranged in various repeating patterns are now ubiquitous in other domains, including natural language processing and speech.

To implement these complex networks, we introduce the concept of a neural network *module*. A module could describe a single layer, a component consisting of multiple layers, or the entire model itself! One benefit of working with the module abstraction is that they can be combined into larger artifacts, often recursively. This is illustrated in Fig. 6.1.1. By defining code to generate modules of arbitrary complexity on demand, we can write surprisingly compact code and still implement complex neural networks.

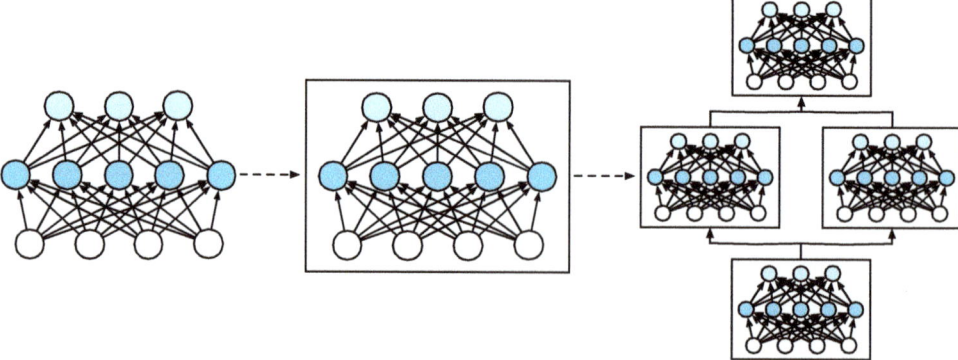

Fig. 6.1.1   Multiple layers are combined into modules, forming repeating patterns of larger models.

From a programming standpoint, a module is represented by a *class*. Any subclass of it must define a forward propagation method that transforms its input into output and must store any necessary parameters. Note that some modules do not require any parameters at all. Finally a module must possess a backpropagation method, for purposes of calculating gradients. Fortunately, due to some behind-the-scenes magic supplied by the auto differentiation (introduced in Section 2.5) when defining our own module, we only need to worry about parameters and the forward propagation method.

```
import torch
from torch import nn
from torch.nn import functional as F
```

To begin, we revisit the code that we used to implement MLPs (Section 5.1). The following code generates a network with one fully connected hidden layer with 256 units and ReLU activation, followed by a fully connected output layer with ten units (no activation function).

```
net = nn.Sequential(nn.LazyLinear(256), nn.ReLU(), nn.LazyLinear(10))

X = torch.rand(2, 20)
net(X).shape
```

```
torch.Size([2, 10])
```

In this example, we constructed our model by instantiating an nn.Sequential, with layers in the order that they should be executed passed as arguments. In short, nn.Sequential defines a special kind of Module, the class that presents a module in PyTorch. It maintains an ordered list of constituent Modules. Note that each of the two fully connected layers is an instance of the Linear class which is itself a subclass of Module. The forward propagation (forward) method is also remarkably simple: it chains each module in the list together, passing the output of each as input to the next. Note that until now, we have been invoking our models via the construction net(X) to obtain their outputs. This is actually just shorthand for net.__call__(X).

6.1.1 A Custom Module

Perhaps the easiest way to develop intuition about how a module works is to implement one ourselves. Before we do that, we briefly summarize the basic functionality that each module must provide:

1. Ingest input data as arguments to its forward propagation method.

2. Generate an output by having the forward propagation method return a value. Note that the output may have a different shape from the input. For example, the first fully connected layer in our model above ingests an input of arbitrary dimension but returns an output of dimension 256.

3. Calculate the gradient of its output with respect to its input, which can be accessed via its backpropagation method. Typically this happens automatically.

4. Store and provide access to those parameters necessary for executing the forward propagation computation.

5. Initialize model parameters as needed.

In the following snippet, we code up a module from scratch corresponding to an MLP with one hidden layer with 256 hidden units, and a 10-dimensional output layer. Note that the MLP class below inherits the class that represents a module. We will heavily rely on the parent class's methods, supplying only our own constructor (the __init__ method in Python) and the forward propagation method.

```python
class MLP(nn.Module):
    def __init__(self):
        # Call the constructor of the parent class nn.Module to perform
        # the necessary initialization
        super().__init__()
        self.hidden = nn.LazyLinear(256)
        self.out = nn.LazyLinear(10)

    # Define the forward propagation of the model, that is, how to return the
    # required model output based on the input X
    def forward(self, X):
        return self.out(F.relu(self.hidden(X)))
```

Let's first focus on the forward propagation method. Note that it takes X as input, calculates the hidden representation with the activation function applied, and outputs its logits. In this MLP implementation, both layers are instance variables. To see why this is reasonable, imagine instantiating two MLPs, net1 and net2, and training them on different data. Naturally, we would expect them to represent two different learned models.

We instantiate the MLP's layers in the constructor and subsequently invoke these layers on each call to the forward propagation method. Note a few key details. First, our customized __init__ method invokes the parent class's __init__ method via super().__init__() sparing us the pain of restating boilerplate code applicable to most modules. We then instantiate our two fully connected layers, assigning them to self.hidden and self.out. Note that unless we implement a new layer, we need not worry about the backpropagation method or parameter initialization. The system will generate these methods automatically. Let's try this out.

```python
net = MLP()
net(X).shape
```

```
torch.Size([2, 10])
```

A key virtue of the module abstraction is its versatility. We can subclass a module to create layers (such as the fully connected layer class), entire models (such as the MLP class above), or various components of intermediate complexity. We exploit this versatility throughout the coming chapters, such as when addressing convolutional neural networks.

6.1.2 The Sequential Module

We can now take a closer look at how the Sequential class works. Recall that Sequential was designed to daisy-chain other modules together. To build our own simplified MySequential, we just need to define two key methods:

1. A method for appending modules one by one to a list.

2. A forward propagation method for passing an input through the chain of modules, in the same order as they were appended.

The following MySequential class delivers the same functionality of the default Sequential class.

```python
class MySequential(nn.Module):
    def __init__(self, *args):
        super().__init__()
        for idx, module in enumerate(args):
            self.add_module(str(idx), module)

    def forward(self, X):
        for module in self.children():
            X = module(X)
        return X
```

In the __init__ method, we add every module by calling the add_modules method. These modules can be accessed by the children method at a later date. In this way the system knows the added modules, and it will properly initialize each module's parameters.

When our MySequential's forward propagation method is invoked, each added module is executed in the order in which they were added. We can now reimplement an MLP using our MySequential class.

```python
net = MySequential(nn.LazyLinear(256), nn.ReLU(), nn.LazyLinear(10))
net(X).shape
```

```
torch.Size([2, 10])
```

Note that this use of MySequential is identical to the code we previously wrote for the Sequential class (as described in Section 5.1).

6.1.3 Executing Code in the Forward Propagation Method

The Sequential class makes model construction easy, allowing us to assemble new architectures without having to define our own class. However, not all architectures are simple daisy chains. When greater flexibility is required, we will want to define our own blocks. For example, we might want to execute Python's control flow within the forward propagation method. Moreover, we might want to perform arbitrary mathematical operations, not simply relying on predefined neural network layers.

You may have noticed that until now, all of the operations in our networks have acted upon our network's activations and its parameters. Sometimes, however, we might want to incorporate terms that are neither the result of previous layers nor updatable parameters. We call these *constant parameters*. Say for example that we want a layer that calculates the function $f(\mathbf{x}, \mathbf{w}) = c \cdot \mathbf{w}^\top \mathbf{x}$, where $\mathbf{x}$ is the input, $\mathbf{w}$ is our parameter, and c is some specified constant that is not updated during optimization. So we implement a FixedHiddenMLP class as follows.

```python
class FixedHiddenMLP(nn.Module):
    def __init__(self):
        super().__init__()
        # Random weight parameters that will not compute gradients and
        # therefore keep constant during training
        self.rand_weight = torch.rand((20, 20))
        self.linear = nn.LazyLinear(20)

    def forward(self, X):
        X = self.linear(X)
        X = F.relu(X @ self.rand_weight + 1)
        # Reuse the fully connected layer. This is equivalent to sharing
        # parameters with two fully connected layers
        X = self.linear(X)
        # Control flow
        while X.abs().sum() > 1:
            X /= 2
        return X.sum()
```

In this model, we implement a hidden layer whose weights (self.rand_weight) are initialized randomly at instantiation and are thereafter constant. This weight is not a model parameter and thus it is never updated by backpropagation. The network then passes the output of this "fixed" layer through a fully connected layer.

Note that before returning the output, our model did something unusual. We ran a while-loop, testing on the condition its ℓ_1 norm is larger than 1, and dividing our output vector by 2 until it satisfied the condition. Finally, we returned the sum of the entries in X. To our knowledge, no standard neural network performs this operation. Note that this particular operation may not be useful in any real-world task. Our point is only to show you how to integrate arbitrary code into the flow of your neural network computations.

```python
net = FixedHiddenMLP()
net(X)
```

```
tensor(-0.1759, grad_fn=<SumBackward0>)
```

We can mix and match various ways of assembling modules together. In the following example, we nest modules in some creative ways.

```python
class NestMLP(nn.Module):
```

(continues on next page)

(continued from previous page)

```
    def __init__(self):
        super().__init__()
        self.net = nn.Sequential(nn.LazyLinear(64), nn.ReLU(),
                                 nn.LazyLinear(32), nn.ReLU())
        self.linear = nn.LazyLinear(16)

    def forward(self, X):
        return self.linear(self.net(X))

chimera = nn.Sequential(NestMLP(), nn.LazyLinear(20), FixedHiddenMLP())
chimera(X)
```

```
tensor(0.0856, grad_fn=<SumBackward0>)
```

6.1.4 Summary

Individual layers can be modules. Many layers can comprise a module. Many modules can comprise a module.

A module can contain code. Modules take care of lots of housekeeping, including parameter initialization and backpropagation. Sequential concatenations of layers and modules are handled by the Sequential module.

6.1.5 Exercises

1. What kinds of problems will occur if you change MySequential to store modules in a Python list?

2. Implement a module that takes two modules as an argument, say net1 and net2 and returns the concatenated output of both networks in the forward propagation. This is also called a *parallel module*.

3. Assume that you want to concatenate multiple instances of the same network. Implement a factory function that generates multiple instances of the same module and build a larger network from it.

 Discussions[111].

6.2 Parameter Management

Once we have chosen an architecture and set our hyperparameters, we proceed to the training loop, where our goal is to find parameter values that minimize our loss function. After training, we will need these parameters in order to make future predictions. Additionally, we will sometimes wish to extract the parameters perhaps to reuse them in some other context, to save our model to disk so that it may be executed in other software, or for examination in the hope of gaining scientific understanding.

Most of the time, we will be able to ignore the nitty-gritty details of how parameters are declared and manipulated, relying on deep learning frameworks to do the heavy lifting. However, when we move away from stacked architectures with standard layers, we will sometimes need to get into the weeds of declaring and manipulating parameters. In this section, we cover the following:

- Accessing parameters for debugging, diagnostics, and visualizations.

- Sharing parameters across different model components.

```python
import torch
from torch import nn
```

We start by focusing on an MLP with one hidden layer.

```python
net = nn.Sequential(nn.LazyLinear(8),
                    nn.ReLU(),
                    nn.LazyLinear(1))

X = torch.rand(size=(2, 4))
net(X).shape
```

```
torch.Size([2, 1])
```

6.2.1 Parameter Access

Let's start with how to access parameters from the models that you already know.

When a model is defined via the Sequential class, we can first access any layer by indexing into the model as though it were a list. Each layer's parameters are conveniently located in its attribute.

We can inspect the parameters of the second fully connected layer as follows.

```python
net[2].state_dict()
```

```
OrderedDict([('weight',
              tensor([[ 0.0496,  0.0269, -0.2629,  0.0424, -0.0263, -0.3302, ↵
→0.2941, -0.2763]])),
             ('bias', tensor([-0.1453]))])
```

We can see that this fully connected layer contains two parameters, corresponding to that layer's weights and biases, respectively.

Targeted Parameters

Note that each parameter is represented as an instance of the parameter class. To do anything useful with the parameters, we first need to access the underlying numerical values. There are several ways to do this. Some are simpler while others are more general. The following code extracts the bias from the second neural network layer, which returns a parameter class instance, and further accesses that parameter's value.

```
type(net[2].bias), net[2].bias.data
```

```
(torch.nn.parameter.Parameter, tensor([-0.1453]))
```

Parameters are complex objects, containing values, gradients, and additional information. That is why we need to request the value explicitly.

In addition to the value, each parameter also allows us to access the gradient. Because we have not invoked backpropagation for this network yet, it is in its initial state.

```
net[2].weight.grad == None
```

```
True
```

All Parameters at Once

When we need to perform operations on all parameters, accessing them one-by-one can grow tedious. The situation can grow especially unwieldy when we work with more complex, e.g., nested, modules, since we would need to recurse through the entire tree to extract each sub-module's parameters. Below we demonstrate accessing the parameters of all layers.

```
[(name, param.shape) for name, param in net.named_parameters()]
```

```
[('0.weight', torch.Size([8, 4])),
 ('0.bias', torch.Size([8])),
 ('2.weight', torch.Size([1, 8])),
 ('2.bias', torch.Size([1]))]
```

6.2.2 Tied Parameters

Often, we want to share parameters across multiple layers. Let's see how to do this elegantly. In the following we allocate a fully connected layer and then use its parameters specifically to set those of another layer. Here we need to run the forward propagation net(X) before accessing the parameters.

```
# We need to give the shared layer a name so that we can refer to its
# parameters
shared = nn.LazyLinear(8)
net = nn.Sequential(nn.LazyLinear(8), nn.ReLU(),
                    shared, nn.ReLU(),
                    shared, nn.ReLU(),
                    nn.LazyLinear(1))

net(X)
# Check whether the parameters are the same
print(net[2].weight.data[0] == net[4].weight.data[0])
net[2].weight.data[0, 0] = 100
# Make sure that they are actually the same object rather than just having the
# same value
print(net[2].weight.data[0] == net[4].weight.data[0])
```

```
tensor([True, True, True, True, True, True, True, True])
tensor([True, True, True, True, True, True, True, True])
```

This example shows that the parameters of the second and third layer are tied. They are not just equal, they are represented by the same exact tensor. Thus, if we change one of the parameters, the other one changes, too.

You might wonder, when parameters are tied what happens to the gradients? Since the model parameters contain gradients, the gradients of the second hidden layer and the third hidden layer are added together during backpropagation.

6.2.3 Summary

We have several ways of accessing and tying model parameters.

6.2.4 Exercises

1. Use the `NestMLP` model defined in Section 6.1 and access the parameters of the various layers.

2. Construct an MLP containing a shared parameter layer and train it. During the training process, observe the model parameters and gradients of each layer.

3. Why is sharing parameters a good idea?

Discussions[112].

112

6.3 Parameter Initialization

Now that we know how to access the parameters, let's look at how to initialize them properly. We discussed the need for proper initialization in Section 5.4. The deep learning framework provides default random initializations to its layers. However, we often want to initialize our weights according to various other protocols. The framework provides most commonly used protocols, and also allows to create a custom initializer.

```python
import torch
from torch import nn
```

By default, PyTorch initializes weight and bias matrices uniformly by drawing from a range that is computed according to the input and output dimension. PyTorch's nn.init module provides a variety of preset initialization methods.

```python
net = nn.Sequential(nn.LazyLinear(8), nn.ReLU(), nn.LazyLinear(1))
X = torch.rand(size=(2, 4))
net(X).shape
```

```
torch.Size([2, 1])
```

6.3.1 Built-in Initialization

Let's begin by calling on built-in initializers. The code below initializes all weight parameters as Gaussian random variables with standard deviation 0.01, while bias parameters are cleared to zero.

```python
def init_normal(module):
    if type(module) == nn.Linear:
        nn.init.normal_(module.weight, mean=0, std=0.01)
        nn.init.zeros_(module.bias)

net.apply(init_normal)
net[0].weight.data[0], net[0].bias.data[0]
```

```
(tensor([ 0.0098,  0.0059, -0.0102, -0.0040]), tensor(0.))
```

We can also initialize all the parameters to a given constant value (say, 1).

```python
def init_constant(module):
    if type(module) == nn.Linear:
        nn.init.constant_(module.weight, 1)
        nn.init.zeros_(module.bias)

net.apply(init_constant)
net[0].weight.data[0], net[0].bias.data[0]
```

```
(tensor([1., 1., 1., 1.]), tensor(0.))
```

We can also apply different initializers for certain blocks. For example, below we initialize the first layer with the Xavier initializer and initialize the second layer to a constant value of 42.

```
def init_xavier(module):
    if type(module) == nn.Linear:
        nn.init.xavier_uniform_(module.weight)

def init_42(module):
    if type(module) == nn.Linear:
        nn.init.constant_(module.weight, 42)

net[0].apply(init_xavier)
net[2].apply(init_42)
print(net[0].weight.data[0])
print(net[2].weight.data)
```

```
tensor([-0.1618, -0.3734, -0.5363, -0.2550])
tensor([[42., 42., 42., 42., 42., 42., 42., 42.]])
```

Custom Initialization

Sometimes, the initialization methods we need are not provided by the deep learning framework. In the example below, we define an initializer for any weight parameter w using the following strange distribution:

$$w \sim \begin{cases} U(5, 10) & \text{with probability } \frac{1}{4} \\ 0 & \text{with probability } \frac{1}{2} \\ U(-10, -5) & \text{with probability } \frac{1}{4} \end{cases} \tag{6.3.1}$$

Again, we implement a `my_init` function to apply to `net`.

```
def my_init(module):
    if type(module) == nn.Linear:
        print("Init", *[(name, param.shape)
                        for name, param in module.named_parameters()][0])
        nn.init.uniform_(module.weight, -10, 10)
        module.weight.data *= module.weight.data.abs() >= 5

net.apply(my_init)
net[0].weight[:2]
```

```
Init weight torch.Size([8, 4])
Init weight torch.Size([1, 8])
```

```
tensor([[ 0.0000, -0.0000, -5.7099, -7.7277],
        [ 0.0000,  0.0000,  0.0000, -0.0000]], grad_fn=<SliceBackward0>)
```

Note that we always have the option of setting parameters directly.

```
net[0].weight.data[:] += 1
net[0].weight.data[0, 0] = 42
net[0].weight.data[0]
```

```
tensor([42.0000,  1.0000, -4.7099, -6.7277])
```

6.3.2 Summary

We can initialize parameters using built-in and custom initializers.

6.3.3 Exercises

Look up the online documentation for more built-in initializers.

Discussions[113].

113

6.4 Lazy Initialization

So far, it might seem that we got away with being sloppy in setting up our networks. Specifically, we did the following unintuitive things, which might not seem like they should work:

- We defined the network architectures without specifying the input dimensionality.

- We added layers without specifying the output dimension of the previous layer.

- We even "initialized" these parameters before providing enough information to determine how many parameters our models should contain.

You might be surprised that our code runs at all. After all, there is no way the deep learning framework could tell what the input dimensionality of a network would be. The trick here is that the framework *defers initialization*, waiting until the first time we pass data through the model, to infer the sizes of each layer on the fly.

Later on, when working with convolutional neural networks, this technique will become even more convenient since the input dimensionality (e.g., the resolution of an image) will affect the dimensionality of each subsequent layer. Hence the ability to set parameters without the need to know, at the time of writing the code, the value of the dimension can greatly simplify the task of specifying and subsequently modifying our models. Next, we go deeper into the mechanics of initialization.

```
import torch
from torch import nn
from d2l import torch as d2l
```

To begin, let's instantiate an MLP.

```
net = nn.Sequential(nn.LazyLinear(256), nn.ReLU(), nn.LazyLinear(10))
```

At this point, the network cannot possibly know the dimensions of the input layer's weights because the input dimension remains unknown.

Consequently the framework has not yet initialized any parameters. We confirm by attempting to access the parameters below.

```
net[0].weight
```

```
<UninitializedParameter>
```

Next let's pass data through the network to make the framework finally initialize parameters.

```
X = torch.rand(2, 20)
net(X)
```

```
net[0].weight.shape
```

```
torch.Size([256, 20])
```

As soon as we know the input dimensionality, 20, the framework can identify the shape of the first layer's weight matrix by plugging in the value of 20. Having recognized the first layer's shape, the framework proceeds to the second layer, and so on through the computational graph until all shapes are known. Note that in this case, only the first layer requires lazy initialization, but the framework initializes sequentially. Once all parameter shapes are known, the framework can finally initialize the parameters.

The following method passes in dummy inputs through the network for a dry run to infer all parameter shapes and subsequently initializes the parameters. It will be used later when default random initializations are not desired.

```
@d2l.add_to_class(d2l.Module)  #@save
def apply_init(self, inputs, init=None):
    self.forward(*inputs)
    if init is not None:
        self.net.apply(init)
```

6.4.1 Summary

Lazy initialization can be convenient, allowing the framework to infer parameter shapes automatically, making it easy to modify architectures and eliminating one common source of errors. We can pass data through the model to make the framework finally initialize parameters.

6.4.2 Exercises

1. What happens if you specify the input dimensions to the first layer but not to subsequent layers? Do you get immediate initialization?

2. What happens if you specify mismatching dimensions?

3. What would you need to do if you have input of varying dimensionality? Hint: look at the parameter tying.

 Discussions[114].

6.5 Custom Layers

One factor behind deep learning's success is the availability of a wide range of layers that can be composed in creative ways to design architectures suitable for a wide variety of tasks. For instance, researchers have invented layers specifically for handling images, text, looping over sequential data, and performing dynamic programming. Sooner or later, you will need a layer that does not exist yet in the deep learning framework. In these cases, you must build a custom layer. In this section, we show you how.

```python
import torch
from torch import nn
from torch.nn import functional as F
from d2l import torch as d2l
```

6.5.1 Layers without Parameters

To start, we construct a custom layer that does not have any parameters of its own. This should look familiar if you recall our introduction to modules in Section 6.1. The following CenteredLayer class simply subtracts the mean from its input. To build it, we simply need to inherit from the base layer class and implement the forward propagation function.

```python
class CenteredLayer(nn.Module):
    def __init__(self):
        super().__init__()

    def forward(self, X):
        return X - X.mean()
```

Let's verify that our layer works as intended by feeding some data through it.

```python
layer = CenteredLayer()
layer(torch.tensor([1.0, 2, 3, 4, 5]))
```

```
tensor([-2., -1.,  0.,  1.,  2.])
```

We can now incorporate our layer as a component in constructing more complex models.

```
net = nn.Sequential(nn.LazyLinear(128), CenteredLayer())
```

As an extra sanity check, we can send random data through the network and check that the mean is in fact 0. Because we are dealing with floating point numbers, we may still see a very small nonzero number due to quantization.

```
Y = net(torch.rand(4, 8))
Y.mean()
```

```
tensor(-1.8626e-09, grad_fn=<MeanBackward0>)
```

6.5.2 Layers with Parameters

Now that we know how to define simple layers, let's move on to defining layers with parameters that can be adjusted through training. We can use built-in functions to create parameters, which provide some basic housekeeping functionality. In particular, they govern access, initialization, sharing, saving, and loading model parameters. This way, among other benefits, we will not need to write custom serialization routines for every custom layer.

Now let's implement our own version of the fully connected layer. Recall that this layer requires two parameters, one to represent the weight and the other for the bias. In this implementation, we bake in the ReLU activation as a default. This layer requires two input arguments: in_units and units, which denote the number of inputs and outputs, respectively.

```
class MyLinear(nn.Module):
    def __init__(self, in_units, units):
        super().__init__()
        self.weight = nn.Parameter(torch.randn(in_units, units))
        self.bias = nn.Parameter(torch.randn(units,))

    def forward(self, X):
        linear = torch.matmul(X, self.weight.data) + self.bias.data
        return F.relu(linear)
```

Next, we instantiate the MyLinear class and access its model parameters.

```
linear = MyLinear(5, 3)
linear.weight
```

```
Parameter containing:
tensor([[ 0.0375, -1.5261,  0.0223],
        [ 0.3962, -0.8338,  1.4341],
        [-0.8831,  0.4161,  0.0845],
        [ 0.0527,  1.7500,  0.9908],
        [ 1.1151,  0.4590,  0.6598]], requires_grad=True)
```

We can directly carry out forward propagation calculations using custom layers.

```
linear(torch.rand(2, 5))
```

```
tensor([[0.0000, 0.0203, 1.1094],
        [0.0000, 0.8137, 1.6988]])
```

We can also construct models using custom layers. Once we have that we can use it just like the built-in fully connected layer.

```
net = nn.Sequential(MyLinear(64, 8), MyLinear(8, 1))
net(torch.rand(2, 64))
```

```
tensor([[5.4649],
        [0.0000]])
```

6.5.3 Summary

We can design custom layers via the basic layer class. This allows us to define flexible new layers that behave differently from any existing layers in the library. Once defined, custom layers can be invoked in arbitrary contexts and architectures. Layers can have local parameters, which can be created through built-in functions.

6.5.4 Exercises

1. Design a layer that takes an input and computes a tensor reduction, i.e., it returns $y_k = \sum_{i,j} W_{ijk} x_i x_j$.

2. Design a layer that returns the leading half of the Fourier coefficients of the data.

115

Discussions[115].

6.6 File I/O

So far we have discussed how to process data and how to build, train, and test deep learning models. However, at some point we will hopefully be happy enough with the learned models that we will want to save the results for later use in various contexts (perhaps even to make predictions in deployment). Additionally, when running a long training process, the best practice is to periodically save intermediate results (checkpointing) to ensure that we do not lose several days' worth of computation if we trip over the power cord of our server. Thus it is time to learn how to load and store both individual weight vectors and entire models. This section addresses both issues.

```python
import torch
from torch import nn
from torch.nn import functional as F
```

6.6.1 Loading and Saving Tensors

For individual tensors, we can directly invoke the load and save functions to read and write them respectively. Both functions require that we supply a name, and save requires as input the variable to be saved.

```python
x = torch.arange(4)
torch.save(x, 'x-file')
```

We can now read the data from the stored file back into memory.

```python
x2 = torch.load('x-file')
x2
```

```
tensor([0, 1, 2, 3])
```

We can store a list of tensors and read them back into memory.

```python
y = torch.zeros(4)
torch.save([x, y],'x-files')
x2, y2 = torch.load('x-files')
(x2, y2)
```

```
(tensor([0, 1, 2, 3]), tensor([0., 0., 0., 0.]))
```

We can even write and read a dictionary that maps from strings to tensors. This is convenient when we want to read or write all the weights in a model.

```python
mydict = {'x': x, 'y': y}
torch.save(mydict, 'mydict')
mydict2 = torch.load('mydict')
mydict2
```

```
{'x': tensor([0, 1, 2, 3]), 'y': tensor([0., 0., 0., 0.])}
```

6.6.2 Loading and Saving Model Parameters

Saving individual weight vectors (or other tensors) is useful, but it gets very tedious if we
want to save (and later load) an entire model. After all, we might have hundreds of parameter groups sprinkled throughout. For this reason the deep learning framework provides
built-in functionalities to load and save entire networks. An important detail to note is that
this saves model *parameters* and not the entire model. For example, if we have a 3-layer
MLP, we need to specify the architecture separately. The reason for this is that the models
themselves can contain arbitrary code, hence they cannot be serialized as naturally. Thus,
in order to reinstate a model, we need to generate the architecture in code and then load the
parameters from disk. Let's start with our familiar MLP.

```python
class MLP(nn.Module):
    def __init__(self):
        super().__init__()
        self.hidden = nn.LazyLinear(256)
        self.output = nn.LazyLinear(10)

    def forward(self, x):
        return self.output(F.relu(self.hidden(x)))

net = MLP()
X = torch.randn(size=(2, 20))
Y = net(X)
```

Next, we store the parameters of the model as a file with the name "mlp.params".

```python
torch.save(net.state_dict(), 'mlp.params')
```

To recover the model, we instantiate a clone of the original MLP model. Instead of randomly initializing the model parameters, we read the parameters stored in the file directly.

```python
clone = MLP()
clone.load_state_dict(torch.load('mlp.params'))
clone.eval()
```

```
MLP(
  (hidden): LazyLinear(in_features=0, out_features=256, bias=True)
  (output): LazyLinear(in_features=0, out_features=10, bias=True)
)
```

Since both instances have the same model parameters, the computational result of the same
input X should be the same. Let's verify this.

```
Y_clone = clone(X)
Y_clone == Y
```

```
tensor([[True, True, True, True, True, True, True, True, True, True],
        [True, True, True, True, True, True, True, True, True, True]])
```

6.6.3 Summary

The save and load functions can be used to perform file I/O for tensor objects. We can save and load the entire sets of parameters for a network via a parameter dictionary. Saving the architecture has to be done in code rather than in parameters.

6.6.4 Exercises

1. Even if there is no need to deploy trained models to a different device, what are the practical benefits of storing model parameters?

2. Assume that we want to reuse only parts of a network to be incorporated into a network having a different architecture. How would you go about using, say the first two layers from a previous network in a new network?

3. How would you go about saving the network architecture and parameters? What restrictions would you impose on the architecture?

Discussions[116].

6.7 GPUs

In Table 1.5.1 , we illustrated the rapid growth of computation over the past two decades. In a nutshell, GPU performance has increased by a factor of 1000 every decade since 2000. This offers great opportunities but it also suggests that there was significant demand for such performance.

In this section, we begin to discuss how to harness this computational performance for your research. First by using a single GPU and at a later point, how to use multiple GPUs and multiple servers (with multiple GPUs).

Specifically, we will discuss how to use a single NVIDIA GPU for calculations. First, make sure you have at least one NVIDIA GPU installed. Then, download the NVIDIA driver and CUDA[117] and follow the prompts to set the appropriate path. Once these preparations are complete, the nvidia-smi command can be used to view the graphics card information.

In PyTorch, every array has a device; we often refer it as a *context*. So far, by default, all variables and associated computation have been assigned to the CPU. Typically, other contexts might be various GPUs. Things can get even hairier when we deploy jobs across multiple servers. By assigning arrays to contexts intelligently, we can minimize the time spent transferring data between devices. For example, when training neural networks on a server with a GPU, we typically prefer for the model's parameters to live on the GPU.

To run the programs in this section, you need at least two GPUs. Note that this might be extravagant for most desktop computers but it is easily available in the cloud, e.g., by using the AWS EC2 multi-GPU instances. Almost all other sections do *not* require multiple GPUs, but here we simply wish to illustrate data flow between different devices.

```
import torch
from torch import nn
from d2l import torch as d2l
```

6.7.1 Computing Devices

We can specify devices, such as CPUs and GPUs, for storage and calculation. By default, tensors are created in the main memory and then the CPU is used for calculations.

In PyTorch, the CPU and GPU can be indicated by `torch.device('cpu')` and `torch.device('cuda')`. It should be noted that the cpu device means all physical CPUs and memory. This means that PyTorch's calculations will try to use all CPU cores. However, a gpu device only represents one card and the corresponding memory. If there are multiple GPUs, we use `torch.device(f'cuda:{i}')` to represent the ith GPU (i starts at 0). Also, gpu:0 and gpu are equivalent.

```
def cpu():  #@save
    """Get the CPU device."""
    return torch.device('cpu')

def gpu(i=0):  #@save
    """Get a GPU device."""
    return torch.device(f'cuda:{i}')

cpu(), gpu(), gpu(1)
```

```
(device(type='cpu'),
 device(type='cuda', index=0),
 device(type='cuda', index=1))
```

We can query the number of available GPUs.

```
def num_gpus():  #@save
    """Get the number of available GPUs."""
    return torch.cuda.device_count()

num_gpus()
```

```
2
```

Now we define two convenient functions that allow us to run code even if the requested GPUs do not exist.

```
def try_gpu(i=0):  #@save
    """Return gpu(i) if exists, otherwise return cpu()."""
    if num_gpus() >= i + 1:
        return gpu(i)
    return cpu()

def try_all_gpus():  #@save
    """Return all available GPUs, or [cpu(),] if no GPU exists."""
    return [gpu(i) for i in range(num_gpus())]

try_gpu(), try_gpu(10), try_all_gpus()
```

```
(device(type='cuda', index=0),
 device(type='cpu'),
 [device(type='cuda', index=0), device(type='cuda', index=1)])
```

6.7.2 Tensors and GPUs

By default, tensors are created on the CPU. We can query the device where the tensor is located.

```
x = torch.tensor([1, 2, 3])
x.device
```

```
device(type='cpu')
```

It is important to note that whenever we want to operate on multiple terms, they need to be on the same device. For instance, if we sum two tensors, we need to make sure that both arguments live on the same device—otherwise the framework would not know where to store the result or even how to decide where to perform the computation.

Storage on the GPU

There are several ways to store a tensor on the GPU. For example, we can specify a storage device when creating a tensor. Next, we create the tensor variable X on the first gpu. The tensor created on a GPU only consumes the memory of this GPU. We can use the nvidia-smi command to view GPU memory usage. In general, we need to make sure that we do not create data that exceeds the GPU memory limit.

```
X = torch.ones(2, 3, device=try_gpu())
X
```

```
tensor([[1., 1., 1.],
        [1., 1., 1.]], device='cuda:0')
```

Assuming that you have at least two GPUs, the following code will create a random tensor, Y, on the second GPU.

```
Y = torch.rand(2, 3, device=try_gpu(1))
Y
```

```
tensor([[0.1005, 0.0277, 0.0528],
        [0.0024, 0.0898, 0.9373]], device='cuda:1')
```

Copying

If we want to compute X + Y, we need to decide where to perform this operation. For instance, as shown in Fig. 6.7.1, we can transfer X to the second GPU and perform the operation there. *Do not* simply add X and Y, since this will result in an exception. The runtime engine would not know what to do: it cannot find data on the same device and it fails. Since Y lives on the second GPU, we need to move X there before we can add the two.

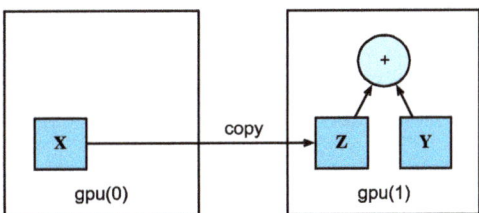

Fig. 6.7.1   Copy data to perform an operation on the same device.

```
Z = X.cuda(1)
print(X)
print(Z)
```

```
tensor([[1., 1., 1.],
        [1., 1., 1.]], device='cuda:0')
tensor([[1., 1., 1.],
        [1., 1., 1.]], device='cuda:1')
```

Now that the data (both Z and Y) are on the same GPU), we can add them up.

```
Y + Z
```

```
tensor([[1.1005, 1.0277, 1.0528],
        [1.0024, 1.0898, 1.9373]], device='cuda:1')
```

But what if your variable Z already lived on your second GPU? What happens if we still call Z.cuda(1)? It will return Z instead of making a copy and allocating new memory.

```
Z.cuda(1) is Z
```

```
True
```

Side Notes

People use GPUs to do machine learning because they expect them to be fast. But transferring variables between devices is slow: much slower than computation. So we want you to be 100% certain that you want to do something slow before we let you do it. If the deep learning framework just did the copy automatically without crashing then you might not realize that you had written some slow code.

Transferring data is not only slow, it also makes parallelization a lot more difficult, since we have to wait for data to be sent (or rather to be received) before we can proceed with more operations. This is why copy operations should be taken with great care. As a rule of thumb, many small operations are much worse than one big operation. Moreover, several operations at a time are much better than many single operations interspersed in the code unless you know what you are doing. This is the case since such operations can block if one device has to wait for the other before it can do something else. It is a bit like ordering your coffee in a queue rather than pre-ordering it by phone and finding out that it is ready when you are.

Last, when we print tensors or convert tensors to the NumPy format, if the data is not in the main memory, the framework will copy it to the main memory first, resulting in additional transmission overhead. Even worse, it is now subject to the dreaded global interpreter lock that makes everything wait for Python to complete.

6.7.3 Neural Networks and GPUs

Similarly, a neural network model can specify devices. The following code puts the model parameters on the GPU.

```
net = nn.Sequential(nn.LazyLinear(1))
net = net.to(device=try_gpu())
```

We will see many more examples of how to run models on GPUs in the following chapters, simply because the models will become somewhat more computationally intensive.

For example, when the input is a tensor on the GPU, the model will calculate the result on the same GPU.

```
net(X)
```

```
tensor([[0.0087],
        [0.0087]], device='cuda:0', grad_fn=<AddmmBackward0>)
```

Let's confirm that the model parameters are stored on the same GPU.

```
net[0].weight.data.device
```

```
device(type='cuda', index=0)
```

Let the trainer support GPU.

```
@d2l.add_to_class(d2l.Trainer)  #@save
def __init__(self, max_epochs, num_gpus=0, gradient_clip_val=0):
    self.save_hyperparameters()
    self.gpus = [d2l.gpu(i) for i in range(min(num_gpus, d2l.num_gpus()))]

@d2l.add_to_class(d2l.Trainer)  #@save
def prepare_batch(self, batch):
    if self.gpus:
        batch = [a.to(self.gpus[0]) for a in batch]
    return batch

@d2l.add_to_class(d2l.Trainer)  #@save
def prepare_model(self, model):
    model.trainer = self
    model.board.xlim = [0, self.max_epochs]
    if self.gpus:
        model.to(self.gpus[0])
    self.model = model
```

In short, as long as all data and parameters are on the same device, we can learn models efficiently. In the following chapters we will see several such examples.

6.7.4 Summary

We can specify devices for storage and calculation, such as the CPU or GPU. By default, data is created in the main memory and then uses the CPU for calculations. The deep learning framework requires all input data for calculation to be on the same device, be it CPU or the same GPU. You can lose significant performance by moving data without care. A typical mistake is as follows: computing the loss for every minibatch on the GPU and reporting it back to the user on the command line (or logging it in a NumPy ndarray) will trigger a global interpreter lock which stalls all GPUs. It is much better to allocate memory for logging inside the GPU and only move larger logs.

6.7.5 Exercises

1. Try a larger computation task, such as the multiplication of large matrices, and see the difference in speed between the CPU and GPU. What about a task with a small number of calculations?

2. How should we read and write model parameters on the GPU?

3. Measure the time it takes to compute 1000 matrix–matrix multiplications of 100×100 matrices and log the Frobenius norm of the output matrix one result at a time. Compare it with keeping a log on the GPU and transferring only the final result.

4. Measure how much time it takes to perform two matrix–matrix multiplications on two GPUs at the same time. Compare it with computing in in sequence on one GPU. Hint: you should see almost linear scaling.

Discussions[118].

118

## 7

Convolutional Neural Networks

Image data is represented as a two-dimensional grid of pixels, be the image monochromatic or in color. Accordingly each pixel corresponds to one or multiple numerical values respectively. So far we have ignored this rich structure and treated images as vectors of numbers by *flattening* them, irrespective of the spatial relation between pixels. This deeply unsatisfying approach was necessary in order to feed the resulting one-dimensional vectors through a fully connected MLP.

Because these networks are invariant to the order of the features, we could get similar results regardless of whether we preserve an order corresponding to the spatial structure of the pixels or if we permute the columns of our design matrix before fitting the MLP's parameters. Ideally, we would leverage our prior knowledge that nearby pixels are typically related to each other, to build efficient models for learning from image data.

This chapter introduces *convolutional neural networks* (CNNs) (LeCun *et al.*, 1995), a powerful family of neural networks that are designed for precisely this purpose. CNN-based architectures are now ubiquitous in the field of computer vision. For instance, on the Imagnet collection (Deng *et al.*, 2009) it was only the use of convolutional neural networks, in short Convnets, that provided significant performance improvements (Krizhevsky *et al.*, 2012).

Modern CNNs, as they are called colloquially, owe their design to inspirations from biology, group theory, and a healthy dose of experimental tinkering. In addition to their sample efficiency in achieving accurate models, CNNs tend to be computationally efficient, both because they require fewer parameters than fully connected architectures and because convolutions are easy to parallelize across GPU cores (Chetlur *et al.*, 2014). Consequently, practitioners often apply CNNs whenever possible, and increasingly they have emerged as credible competitors even on tasks with a one-dimensional sequence structure, such as audio (Abdel-Hamid *et al.*, 2014), text (Kalchbrenner *et al.*, 2014), and time series analysis (LeCun *et al.*, 1995), where recurrent neural networks are conventionally used. Some clever adaptations of CNNs have also brought them to bear on graph-structured data (Kipf and Welling, 2016) and in recommender systems.

First, we will dive more deeply into the motivation for convolutional neural networks. This is followed by a walk through the basic operations that comprise the backbone of all convolutional networks. These include the convolutional layers themselves, nitty-gritty details including padding and stride, the pooling layers used to aggregate information across adjacent spatial regions, the use of multiple channels at each layer, and a careful discussion of the structure of modern architectures. We will conclude the chapter with a full working example of LeNet, the first convolutional network successfully deployed, long before the rise of modern deep learning. In the next chapter, we will dive into full implementations of some popular and comparatively recent CNN architectures whose designs represent most of the techniques commonly used by modern practitioners.

7.1 From Fully Connected Layers to Convolutions

To this day, the models that we have discussed so far remain appropriate options when we are dealing with tabular data. By tabular, we mean that the data consist of rows corresponding to examples and columns corresponding to features. With tabular data, we might anticipate that the patterns we seek could involve interactions among the features, but we do not assume any structure *a priori* concerning how the features interact.

Sometimes, we truly lack the knowledge to be able to guide the construction of fancier architectures. In these cases, an MLP may be the best that we can do. However, for high-dimensional perceptual data, such structureless networks can grow unwieldy.

For instance, let's return to our running example of distinguishing cats from dogs. Say that we do a thorough job in data collection, collecting an annotated dataset of one-megapixel photographs. This means that each input to the network has one million dimensions. Even an aggressive reduction to one thousand hidden dimensions would require a fully connected layer characterized by $10^6 \times 10^3 = 10^9$ parameters. Unless we have lots of GPUs, a talent for distributed optimization, and an extraordinary amount of patience, learning the parameters of this network may turn out to be infeasible.

A careful reader might object to this argument on the basis that one megapixel resolution may not be necessary. However, while we might be able to get away with one hundred thousand pixels, our hidden layer of size 1000 grossly underestimates the number of hidden units that it takes to learn good representations of images, so a practical system will still require billions of parameters. Moreover, learning a classifier by fitting so many parameters might require collecting an enormous dataset. And yet today both humans and computers are able to distinguish cats from dogs quite well, seemingly contradicting these intuitions. That is because images exhibit rich structure that can be exploited by humans and machine learning models alike. Convolutional neural networks (CNNs) are one creative way that machine learning has embraced for exploiting some of the known structure in natural images.

7.1.1 Invariance

Imagine that we want to detect an object in an image. It seems reasonable that whatever method we use to recognize objects should not be overly concerned with the precise location of the object in the image. Ideally, our system should exploit this knowledge. Pigs usually do not fly and planes usually do not swim. Nonetheless, we should still recognize a pig were one to appear at the top of the image. We can draw some inspiration here from the children's game "Where's Waldo" (which itself has inspired many real-life imitations, such as that depicted in Fig. 7.1.1). The game consists of a number of chaotic scenes bursting with activities. Waldo shows up somewhere in each, typically lurking in some unlikely location. The reader's goal is to locate him. Despite his characteristic outfit, this can be surprisingly difficult, due to the large number of distractions. However, *what Waldo looks like* does not depend upon *where Waldo is located*. We could sweep the image with a Waldo detector that could assign a score to each patch, indicating the likelihood that the patch contains Waldo. In fact, many object detection and segmentation algorithms are based on this approach (Long *et al.*, 2015). CNNs systematize this idea of *spatial invariance*, exploiting it to learn useful representations with fewer parameters.

Fig. 7.1.1   Can you find Waldo? (Image courtesy of William Murphy (Infomatique).)

We can now make these intuitions more concrete by enumerating a few desiderata to guide our design of a neural network architecture suitable for computer vision:

1. In the earliest layers, our network should respond similarly to the same patch, regardless of where it appears in the image. This principle is called *translation invariance* (or *translation equivariance*).

2. The earliest layers of the network should focus on local regions, without regard for the contents of the image in distant regions. This is the *locality* principle. Eventually, these local representations can be aggregated to make predictions at the whole image level.

3. As we proceed, deeper layers should be able to capture longer-range features of the image, in a way similar to higher-level vision in nature.

Let's see how this translates into mathematics.

7.1.2 Constraining the MLP

To start off, we can consider an MLP with two-dimensional images $\mathbf{X}$ as inputs and their immediate hidden representations $\mathbf{H}$ similarly represented as matrices (they are two-dimensional tensors in code), where both $\mathbf{X}$ and $\mathbf{H}$ have the same shape. Let that sink in. We now imagine that not only the inputs but also the hidden representations possess spatial structure.

Let $[\mathbf{X}]_{i,j}$ and $[\mathbf{H}]_{i,j}$ denote the pixel at location (i, j) in the input image and hidden representation, respectively. Consequently, to have each of the hidden units receive input from each of the input pixels, we would switch from using weight matrices (as we did previously in MLPs) to representing our parameters as fourth-order weight tensors $\mathbf{W}$. Suppose that $\mathbf{U}$ contains biases, we could formally express the fully connected layer as

$$
\begin{aligned}
[\mathbf{H}]_{i,j} &= [\mathbf{U}]_{i,j} + \sum_k \sum_l [\mathbf{W}]_{i,j,k,l}[\mathbf{X}]_{k,l} \\
&= [\mathbf{U}]_{i,j} + \sum_a \sum_b [\mathsf{V}]_{i,j,a,b}[\mathbf{X}]_{i+a,j+b}.
\end{aligned}
\tag{7.1.1}
$$

The switch from $\mathbf{W}$ to V is entirely cosmetic for now since there is a one-to-one correspondence between coefficients in both fourth-order tensors. We simply re-index the subscripts (k, l) such that $k = i + a$ and $l = j + b$. In other words, we set $[\mathsf{V}]_{i,j,a,b} = [\mathbf{W}]_{i,j,i+a,j+b}$. The indices a and b run over both positive and negative offsets, covering the entire image. For any given location (i, j) in the hidden representation $[\mathbf{H}]_{i,j}$, we compute its value by summing over pixels in x, centered around (i, j) and weighted by $[\mathsf{V}]_{i,j,a,b}$. Before we carry on, let's consider the total number of parameters required for a *single* layer in this parametrization: a 1000×1000 image (1 megapixel) is mapped to a 1000×1000 hidden representation. This requires 10^{12} parameters, far beyond what computers currently can handle.

Translation Invariance

Now let's invoke the first principle established above: translation invariance (Zhang *et al.*, 1988). This implies that a shift in the input $\mathbf{X}$ should simply lead to a shift in the hidden representation $\mathbf{H}$. This is only possible if V and $\mathbf{U}$ do not actually depend on (i, j). As such, we have $[\mathsf{V}]_{i,j,a,b} = [\mathbf{V}]_{a,b}$ and $\mathbf{U}$ is a constant, say u. As a result, we can simplify the definition for $\mathbf{H}$:

$$
[\mathbf{H}]_{i,j} = u + \sum_a \sum_b [\mathbf{V}]_{a,b}[\mathbf{X}]_{i+a,j+b}.
\tag{7.1.2}
$$

This is a *convolution*! We are effectively weighting pixels at $(i + a, j + b)$ in the vicinity of location (i, j) with coefficients $[\mathbf{V}]_{a,b}$ to obtain the value $[\mathbf{H}]_{i,j}$. Note that $[\mathbf{V}]_{a,b}$ needs many fewer coefficients than $[\mathsf{V}]_{i,j,a,b}$ since it no longer depends on the location within the image. Consequently, the number of parameters required is no longer 10^{12} but a much more reasonable 4×10^6: we still have the dependency on $a, b \in (-1000, 1000)$. In short, we have made significant progress. Time-delay neural networks (TDNNs) are some of the first examples to exploit this idea (Waibel *et al.*, 1989).

Locality

Now let's invoke the second principle: locality. As motivated above, we believe that we should not have to look very far away from location (i, j) in order to glean relevant information to assess what is going on at $[\mathbf{H}]_{i,j}$. This means that outside some range $|a| > \Delta$ or $|b| > \Delta$, we should set $[\mathbf{V}]_{a,b} = 0$. Equivalently, we can rewrite $[\mathbf{H}]_{i,j}$ as

$$[\mathbf{H}]_{i,j} = u + \sum_{a=-\Delta}^{\Delta} \sum_{b=-\Delta}^{\Delta} [\mathbf{V}]_{a,b} [\mathbf{X}]_{i+a,j+b}. \tag{7.1.3}$$

This reduces the number of parameters from 4×10^6 to $4\Delta^2$, where Δ is typically smaller than 10. As such, we reduced the number of parameters by another four orders of magnitude. Note that (7.1.3), is what is called, in a nutshell, a *convolutional layer*. *Convolutional neural networks* (CNNs) are a special family of neural networks that contain convolutional layers. In the deep learning research community, $\mathbf{V}$ is referred to as a *convolution kernel*, a *filter*, or simply the layer's *weights* that are learnable parameters.

While previously, we might have required billions of parameters to represent just a single layer in an image-processing network, we now typically need just a few hundred, without altering the dimensionality of either the inputs or the hidden representations. The price paid for this drastic reduction in parameters is that our features are now translation invariant and that our layer can only incorporate local information, when determining the value of each hidden activation. All learning depends on imposing inductive bias. When that bias agrees with reality, we get sample-efficient models that generalize well to unseen data. But of course, if those biases do not agree with reality, e.g., if images turned out not to be translation invariant, our models might struggle even to fit our training data.

This dramatic reduction in parameters brings us to our last desideratum, namely that deeper layers should represent larger and more complex aspects of an image. This can be achieved by interleaving nonlinearities and convolutional layers repeatedly.

7.1.3 Convolutions

Let's briefly review why (7.1.3) is called a convolution. In mathematics, the *convolution* between two functions (Rudin, 1973), say $f, g : \mathbb{R}^d \to \mathbb{R}$ is defined as

$$(f * g)(\mathbf{x}) = \int f(\mathbf{z})g(\mathbf{x} - \mathbf{z})d\mathbf{z}. \tag{7.1.4}$$

That is, we measure the overlap between f and g when one function is "flipped" and shifted by $\mathbf{x}$. Whenever we have discrete objects, the integral turns into a sum. For instance, for vectors from the set of square-summable infinite-dimensional vectors with index running over $\mathbb{Z}$ we obtain the following definition:

$$(f * g)(i) = \sum_a f(a)g(i - a). \tag{7.1.5}$$

For two-dimensional tensors, we have a corresponding sum with indices (a, b) for f and $(i - a, j - b)$ for g, respectively:

$$(f * g)(i, j) = \sum_a \sum_b f(a, b)g(i - a, j - b). \tag{7.1.6}$$

This looks similar to (7.1.3), with one major difference. Rather than using $(i + a, j + b)$, we are using the difference instead. Note, though, that this distinction is mostly cosmetic since we can always match the notation between (7.1.3) and (7.1.6). Our original definition in (7.1.3) more properly describes a *cross-correlation*. We will come back to this in the following section.

7.1.4 Channels

Returning to our Waldo detector, let's see what this looks like. The convolutional layer picks windows of a given size and weighs intensities according to the filter V, as demonstrated in Fig. 7.1.2. We might aim to learn a model so that wherever the "waldoness" is highest, we should find a peak in the hidden layer representations.

Fig. 7.1.2   Detect Waldo. (Image courtesy of William Murphy (Infomatique).)

There is just one problem with this approach. So far, we blissfully ignored that images consist of three channels: red, green, and blue. In sum, images are not two-dimensional objects but rather third-order tensors, characterized by a height, width, and channel, e.g., with shape $1024 \times 1024 \times 3$ pixels. While the first two of these axes concern spatial relationships, the third can be regarded as assigning a multidimensional representation to each pixel location. We thus index X as $[X]_{i,j,k}$. The convolutional filter has to adapt accordingly. Instead of $[\mathbf{V}]_{a,b}$, we now have $[V]_{a,b,c}$.

Moreover, just as our input consists of a third-order tensor, it turns out to be a good idea to similarly formulate our hidden representations as third-order tensors H. In other words, rather than just having a single hidden representation corresponding to each spatial location, we want an entire vector of hidden representations corresponding to each spatial location. We could think of the hidden representations as comprising a number of two-dimensional grids stacked on top of each other. As in the inputs, these are sometimes called *channels*. They are also sometimes called *feature maps*, as each provides a spatialized set of learned features for the subsequent layer. Intuitively, you might imagine that at lower layers that are closer to inputs, some channels could become specialized to recognize edges while others could recognize textures.

To support multiple channels in both inputs (X) and hidden representations (H), we can add a fourth coordinate to V: $[V]_{a,b,c,d}$. Putting everything together we have:

$$[H]_{i,j,d} = \sum_{a=-\Delta}^{\Delta} \sum_{b=-\Delta}^{\Delta} \sum_{c} [V]_{a,b,c,d}[X]_{i+a,j+b,c}, \qquad (7.1.7)$$

where d indexes the output channels in the hidden representations H. The subsequent convolutional layer will go on to take a third-order tensor, H, as input. We take (7.1.7), because of its generality, as the definition of a convolutional layer for multiple channels, where V is a kernel or filter of the layer.

There are still many operations that we need to address. For instance, we need to figure out how to combine all the hidden representations to a single output, e.g., whether there is a Waldo *anywhere* in the image. We also need to decide how to compute things efficiently, how to combine multiple layers, appropriate activation functions, and how to make reasonable design choices to yield networks that are effective in practice. We turn to these issues in the remainder of the chapter.

7.1.5 Summary and Discussion

In this section we derived the structure of convolutional neural networks from first principles. While it is unclear whether this was the route taken to the invention of CNNs, it is satisfying to know that they are the *right* choice when applying reasonable principles to how image processing and computer vision algorithms should operate, at least at lower levels. In particular, translation invariance in images implies that all patches of an image will be treated in the same manner. Locality means that only a small neighborhood of pixels will be used to compute the corresponding hidden representations. Some of the earliest references to CNNs are in the form of the Neocognitron (Fukushima, 1982).

A second principle that we encountered in our reasoning is how to reduce the number of parameters in a function class without limiting its expressive power, at least, whenever certain assumptions on the model hold. We saw a dramatic reduction of complexity as a result of this restriction, turning computationally and statistically infeasible problems into tractable models.

Adding channels allowed us to bring back some of the complexity that was lost due to the restrictions imposed on the convolutional kernel by locality and translation invariance. Note that it is quite natural to add channels other than just red, green, and blue. Many satellite images, in particular for agriculture and meteorology, have tens to hundreds of channels, generating hyperspectral images instead. They report data on many different wavelengths. In the following we will see how to use convolutions effectively to manipulate the dimensionality of the images they operate on, how to move from location-based to channel-based representations, and how to deal with large numbers of categories efficiently.

7.1.6 Exercises

1. Assume that the size of the convolution kernel is $\Delta = 0$. Show that in this case the convolution kernel implements an MLP independently for each set of channels. This leads to the Network in Network architectures (Lin *et al.*, 2013).

2. Audio data is often represented as a one-dimensional sequence.

    1. When might you want to impose locality and translation invariance for audio?

    2. Derive the convolution operations for audio.

    3. Can you treat audio using the same tools as computer vision? Hint: use the spectrogram.

3. Why might translation invariance not be a good idea after all? Give an example.

4. Do you think that convolutional layers might also be applicable for text data? Which problems might you encounter with language?

5. What happens with convolutions when an object is at the boundary of an image?

6. Prove that the convolution is symmetric, i.e., $f * g = g * f$.

 Discussions[119].

7.2 Convolutions for Images

Now that we understand how convolutional layers work in theory, we are ready to see how they work in practice. Building on our motivation of convolutional neural networks as efficient architectures for exploring structure in image data, we stick with images as our running example.

```
import torch
from torch import nn
from d2l import torch as d2l
```

7.2.1 The Cross-Correlation Operation

Recall that strictly speaking, convolutional layers are a misnomer, since the operations they express are more accurately described as cross-correlations. Based on our descriptions of convolutional layers in Section 7.1, in such a layer, an input tensor and a kernel tensor are combined to produce an output tensor through a cross-correlation operation.

Let's ignore channels for now and see how this works with two-dimensional data and hidden representations. In Fig. 7.2.1, the input is a two-dimensional tensor with a height of 3 and width of 3. We mark the shape of the tensor as 3×3 or $(3, 3)$. The height and width of the kernel are both 2. The shape of the *kernel window* (or *convolution window*) is given by the height and width of the kernel (here it is 2×2).

In the two-dimensional cross-correlation operation, we begin with the convolution window positioned at the upper-left corner of the input tensor and slide it across the input tensor, both from left to right and top to bottom. When the convolution window slides to a certain

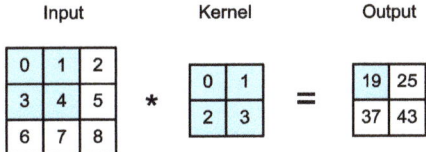

Fig. 7.2.1 Two-dimensional cross-correlation operation. The shaded portions are the first output element as well as the input and kernel tensor elements used for the output computation: $0 \times 0 + 1 \times 1 + 3 \times 2 + 4 \times 3 = 19$.

position, the input subtensor contained in that window and the kernel tensor are multiplied elementwise and the resulting tensor is summed up yielding a single scalar value. This result gives the value of the output tensor at the corresponding location. Here, the output tensor has a height of 2 and width of 2 and the four elements are derived from the two-dimensional cross-correlation operation:

$$
\begin{aligned}
0 \times 0 + 1 \times 1 + 3 \times 2 + 4 \times 3 &= 19, \\
1 \times 0 + 2 \times 1 + 4 \times 2 + 5 \times 3 &= 25, \\
3 \times 0 + 4 \times 1 + 6 \times 2 + 7 \times 3 &= 37, \\
4 \times 0 + 5 \times 1 + 7 \times 2 + 8 \times 3 &= 43.
\end{aligned}
\tag{7.2.1}
$$

Note that along each axis, the output size is slightly smaller than the input size. Because the kernel has width and height greater than 1, we can only properly compute the cross-correlation for locations where the kernel fits wholly within the image, the output size is given by the input size $n_h \times n_w$ minus the size of the convolution kernel $k_h \times k_w$ via

$$
(n_h - k_h + 1) \times (n_w - k_w + 1). \tag{7.2.2}
$$

This is the case since we need enough space to "shift" the convolution kernel across the image. Later we will see how to keep the size unchanged by padding the image with zeros around its boundary so that there is enough space to shift the kernel. Next, we implement this process in the corr2d function, which accepts an input tensor X and a kernel tensor K and returns an output tensor Y.

```
def corr2d(X, K):   #@save
    """Compute 2D cross-correlation."""
    h, w = K.shape
    Y = torch.zeros((X.shape[0] - h + 1, X.shape[1] - w + 1))
    for i in range(Y.shape[0]):
        for j in range(Y.shape[1]):
            Y[i, j] = (X[i:i + h, j:j + w] * K).sum()
    return Y
```

We can construct the input tensor X and the kernel tensor K from Fig. 7.2.1 to validate the output of the above implementation of the two-dimensional cross-correlation operation.

```
X = torch.tensor([[0.0, 1.0, 2.0], [3.0, 4.0, 5.0], [6.0, 7.0, 8.0]])
K = torch.tensor([[0.0, 1.0], [2.0, 3.0]])
corr2d(X, K)
```

```
tensor([[19., 25.],
        [37., 43.]])
```

7.2.2 Convolutional Layers

A convolutional layer cross-correlates the input and kernel and adds a scalar bias to produce an output. The two parameters of a convolutional layer are the kernel and the scalar bias. When training models based on convolutional layers, we typically initialize the kernels randomly, just as we would with a fully connected layer.

We are now ready to implement a two-dimensional convolutional layer based on the `corr2d` function defined above. In the `__init__` constructor method, we declare `weight` and `bias` as the two model parameters. The forward propagation method calls the `corr2d` function and adds the bias.

```
class Conv2D(nn.Module):
    def __init__(self, kernel_size):
        super().__init__()
        self.weight = nn.Parameter(torch.rand(kernel_size))
        self.bias = nn.Parameter(torch.zeros(1))

    def forward(self, x):
        return corr2d(x, self.weight) + self.bias
```

In $h \times w$ convolution or an $h \times w$ convolution kernel, the height and width of the convolution kernel are h and w, respectively. We also refer to a convolutional layer with an $h \times w$ convolution kernel simply as an $h \times w$ convolutional layer.

7.2.3 Object Edge Detection in Images

Let's take a moment to parse a simple application of a convolutional layer: detecting the edge of an object in an image by finding the location of the pixel change. First, we construct an "image" of 6×8 pixels. The middle four columns are black (0) and the rest are white (1).

```
X = torch.ones((6, 8))
X[:, 2:6] = 0
X
```

```
tensor([[1., 1., 0., 0., 0., 0., 1., 1.],
        [1., 1., 0., 0., 0., 0., 1., 1.],
        [1., 1., 0., 0., 0., 0., 1., 1.],
```

(continues on next page)

(continued from previous page)

```
        [1., 1., 0., 0., 0., 0., 1., 1.],
        [1., 1., 0., 0., 0., 0., 1., 1.],
        [1., 1., 0., 0., 0., 0., 1., 1.]])
```

Next, we construct a kernel K with a height of 1 and a width of 2. When we perform the cross-correlation operation with the input, if the horizontally adjacent elements are the same, the output is 0. Otherwise, the output is nonzero. Note that this kernel is a special case of a finite difference operator. At location (i, j) it computes $x_{i,j} - x_{(i+1),j}$, i.e., it computes the difference between the values of horizontally adjacent pixels. This is a discrete approximation of the first derivative in the horizontal direction. After all, for a function $f(i, j)$ its derivative $-\partial_i f(i, j) = \lim_{\epsilon \to 0} \frac{f(i,j) - f(i+\epsilon,j)}{\epsilon}$. Let's see how this works in practice.

```
K = torch.tensor([[1.0, -1.0]])
```

We are ready to perform the cross-correlation operation with arguments X (our input) and K (our kernel). As you can see, we detect 1 for the edge from white to black and -1 for the edge from black to white. All other outputs take value 0.

```
Y = corr2d(X, K)
Y
```

```
tensor([[ 0.,  1.,  0.,  0.,  0., -1.,  0.],
        [ 0.,  1.,  0.,  0.,  0., -1.,  0.],
        [ 0.,  1.,  0.,  0.,  0., -1.,  0.],
        [ 0.,  1.,  0.,  0.,  0., -1.,  0.],
        [ 0.,  1.,  0.,  0.,  0., -1.,  0.],
        [ 0.,  1.,  0.,  0.,  0., -1.,  0.]])
```

We can now apply the kernel to the transposed image. As expected, it vanishes. The kernel K only detects vertical edges.

```
corr2d(X.t(), K)
```

```
tensor([[0., 0., 0., 0., 0.],
        [0., 0., 0., 0., 0.],
        [0., 0., 0., 0., 0.],
        [0., 0., 0., 0., 0.],
        [0., 0., 0., 0., 0.],
        [0., 0., 0., 0., 0.],
        [0., 0., 0., 0., 0.],
        [0., 0., 0., 0., 0.]])
```

7.2.4 Learning a Kernel

Designing an edge detector by finite differences [1, -1] is neat if we know this is precisely what we are looking for. However, as we look at larger kernels, and consider successive layers of convolutions, it might be impossible to specify precisely what each filter should be doing manually.

Now let's see whether we can learn the kernel that generated Y from X by looking at the input–output pairs only. We first construct a convolutional layer and initialize its kernel as a random tensor. Next, in each iteration, we will use the squared error to compare Y with the output of the convolutional layer. We can then calculate the gradient to update the kernel. For the sake of simplicity, in the following we use the built-in class for two-dimensional convolutional layers and ignore the bias.

```python
# Construct a two-dimensional convolutional layer with 1 output channel and a
# kernel of shape (1, 2). For the sake of simplicity, we ignore the bias here
conv2d = nn.LazyConv2d(1, kernel_size=(1, 2), bias=False)

# The two-dimensional convolutional layer uses four-dimensional input and
# output in the format of (example, channel, height, width), where the batch
# size (number of examples in the batch) and the number of channels are both 1
X = X.reshape((1, 1, 6, 8))
Y = Y.reshape((1, 1, 6, 7))
lr = 3e-2  # Learning rate

for i in range(10):
    Y_hat = conv2d(X)
    l = (Y_hat - Y) ** 2
    conv2d.zero_grad()
    l.sum().backward()
    # Update the kernel
    conv2d.weight.data[:] -= lr * conv2d.weight.grad
    if (i + 1) % 2 == 0:
        print(f'epoch {i + 1}, loss {l.sum():.3f}')
```

```
epoch 2, loss 3.067
epoch 4, loss 0.611
epoch 6, loss 0.142
epoch 8, loss 0.040
epoch 10, loss 0.013
```

Note that the error has dropped to a small value after 10 iterations. Now we will take a look at the kernel tensor we learned.

```python
conv2d.weight.data.reshape((1, 2))
```

```
tensor([[ 0.9806, -1.0024]])
```

Indeed, the learned kernel tensor is remarkably close to the kernel tensor K we defined earlier.

7.2.5 Cross-Correlation and Convolution

Recall our observation from Section 7.1 of the correspondence between the cross-correlation and convolution operations. Here let's continue to consider two-dimensional convolutional layers. What if such layers perform strict convolution operations as defined in (7.1.6) instead of cross-correlations? In order to obtain the output of the strict *convolution* operation, we only need to flip the two-dimensional kernel tensor both horizontally and vertically, and then perform the *cross-correlation* operation with the input tensor.

It is noteworthy that since kernels are learned from data in deep learning, the outputs of convolutional layers remain unaffected no matter such layers perform either the strict convolution operations or the cross-correlation operations.

To illustrate this, suppose that a convolutional layer performs *cross-correlation* and learns the kernel in Fig. 7.2.1, which is here denoted as the matrix $\mathbf{K}$. Assuming that other conditions remain unchanged, when this layer instead performs strict *convolution*, the learned kernel $\mathbf{K}'$ will be the same as $\mathbf{K}$ after $\mathbf{K}'$ is flipped both horizontally and vertically. That is to say, when the convolutional layer performs strict *convolution* for the input in Fig. 7.2.1 and $\mathbf{K}'$, the same output in Fig. 7.2.1 (cross-correlation of the input and $\mathbf{K}$) will be obtained.

In keeping with standard terminology in deep learning literature, we will continue to refer to the cross-correlation operation as a convolution even though, strictly-speaking, it is slightly different. Furthermore, we use the term *element* to refer to an entry (or component) of any tensor representing a layer representation or a convolution kernel.

7.2.6 Feature Map and Receptive Field

As described in Section 7.1.4, the convolutional layer output in Fig. 7.2.1 is sometimes called a *feature map*, as it can be regarded as the learned representations (features) in the spatial dimensions (e.g., width and height) to the subsequent layer. In CNNs, for any element x of some layer, its *receptive field* refers to all the elements (from all the previous layers) that may affect the calculation of x during the forward propagation. Note that the receptive field may be larger than the actual size of the input.

Let's continue to use Fig. 7.2.1 to explain the receptive field. Given the 2×2 convolution kernel, the receptive field of the shaded output element (of value 19) is the four elements in the shaded portion of the input. Now let's denote the 2×2 output as $\mathbf{Y}$ and consider a deeper CNN with an additional 2×2 convolutional layer that takes $\mathbf{Y}$ as its input, outputting a single element z. In this case, the receptive field of z on $\mathbf{Y}$ includes all the four elements of $\mathbf{Y}$, while the receptive field on the input includes all the nine input elements. Thus, when any element in a feature map needs a larger receptive field to detect input features over a broader area, we can build a deeper network.

Receptive fields derive their name from neurophysiology. A series of experiments on a range of animals using different stimuli (Hubel and Wiesel, 1959, Hubel and Wiesel, 1962, Hubel and Wiesel, 1968) explored the response of what is called the visual cortex on said stimuli. By and large they found that lower levels respond to edges and related shapes.

Later on, Field (1987) illustrated this effect on natural images with, what can only be called, convolutional kernels. We reprint a key figure in Fig. 7.2.2 to illustrate the striking similarities.

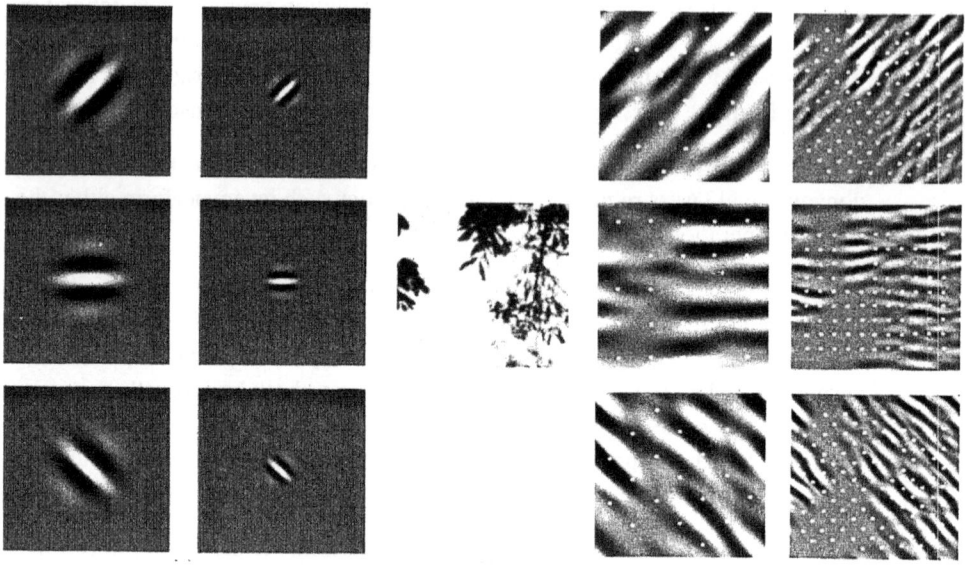

Fig. 7.2.2 Figure and caption taken from Field (1987): An example of coding with six different channels. (Left) Examples of the six types of sensor associated with each channel. (Right) Convolution of the image in (Middle) with the six sensors shown in (Left). The response of the individual sensors is determined by sampling these filtered images at a distance proportional to the size of the sensor (shown with dots). This diagram shows the response of only the even symmetric sensors.

As it turns out, this relation even holds for the features computed by deeper layers of networks trained on image classification tasks, as demonstrated in, for example, Kuzovkin *et al.* (2018). Suffice it to say, convolutions have proven to be an incredibly powerful tool for computer vision, both in biology and in code. As such, it is not surprising (in hindsight) that they heralded the recent success in deep learning.

7.2.7 Summary

The core computation required for a convolutional layer is a cross-correlation operation. We saw that a simple nested for-loop is all that is required to compute its value. If we have multiple input and multiple output channels, we are performing a matrix–matrix operation between channels. As can be seen, the computation is straightforward and, most importantly, highly *local*. This affords significant hardware optimization and many recent results in computer vision are only possible because of that. After all, it means that chip designers can invest in fast computation rather than memory when it comes to optimizing for convolutions. While this may not lead to optimal designs for other applications, it does open the door to ubiquitous and affordable computer vision.

In terms of convolutions themselves, they can be used for many purposes, for example detecting edges and lines, blurring images, or sharpening them. Most importantly, it is not necessary that the statistician (or engineer) invents suitable filters. Instead, we can simply *learn* them from data. This replaces feature engineering heuristics by evidence-based statistics. Lastly, and quite delightfully, these filters are not just advantageous for building deep networks but they also correspond to receptive fields and feature maps in the brain. This gives us confidence that we are on the right track.

7.2.8 Exercises

1. Construct an image X with diagonal edges.

    1. What happens if you apply the kernel K in this section to it?

    2. What happens if you transpose X?

    3. What happens if you transpose K?

2. Design some kernels manually.

    1. Given a directional vector $\mathbf{v} = (v_1, v_2)$, derive an edge-detection kernel that detects edges orthogonal to $\mathbf{v}$, i.e., edges in the direction $(v_2, -v_1)$.

    2. Derive a finite difference operator for the second derivative. What is the minimum size of the convolutional kernel associated with it? Which structures in images respond most strongly to it?

    3. How would you design a blur kernel? Why might you want to use such a kernel?

    4. What is the minimum size of a kernel to obtain a derivative of order d?

3. When you try to automatically find the gradient for the Conv2D class we created, what kind of error message do you see?

4. How do you represent a cross-correlation operation as a matrix multiplication by changing the input and kernel tensors?

 Discussions[120].

7.3 Padding and Stride

Recall the example of a convolution in Fig. 7.2.1. The input had both a height and width of 3 and the convolution kernel had both a height and width of 2, yielding an output representation with dimension 2×2. Assuming that the input shape is $n_h \times n_w$ and the convolution kernel shape is $k_h \times k_w$, the output shape will be $(n_h - k_h + 1) \times (n_w - k_w + 1)$: we can only shift the convolution kernel so far until it runs out of pixels to apply the convolution to.

In the following we will explore a number of techniques, including padding and strided convolutions, that offer more control over the size of the output. As motivation, note that since kernels generally have width and height greater than 1, after applying many successive convolutions, we tend to wind up with outputs that are considerably smaller than our input. If we start with a 240×240 pixel image, ten layers of 5×5 convolutions reduce the image to 200×200 pixels, slicing off 30% of the image and with it obliterating any interesting information on the boundaries of the original image. *Padding* is the most popular tool for handling this issue. In other cases, we may want to reduce the dimensionality drastically, e.g., if we find the original input resolution to be unwieldy. *Strided convolutions* are a popular technique that can help in these instances.

```
import torch
from torch import nn
```

7.3.1 Padding

As described above, one tricky issue when applying convolutional layers is that we tend to lose pixels on the perimeter of our image. Consider Fig. 7.3.1 that depicts the pixel utilization as a function of the convolution kernel size and the position within the image. The pixels in the corners are hardly used at all.

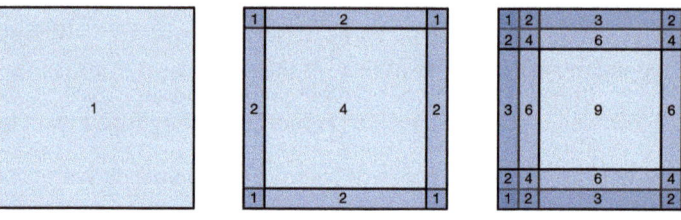

Fig. 7.3.1 Pixel utilization for convolutions of size 1×1, 2×2, and 3×3 respectively.

Since we typically use small kernels, for any given convolution we might only lose a few pixels but this can add up as we apply many successive convolutional layers. One straightforward solution to this problem is to add extra pixels of filler around the boundary of our input image, thus increasing the effective size of the image. Typically, we set the values of the extra pixels to zero. In Fig. 7.3.2, we pad a 3×3 input, increasing its size to 5×5. The corresponding output then increases to a 4×4 matrix. The shaded portions are the first output element as well as the input and kernel tensor elements used for the output computation: $0 \times 0 + 0 \times 1 + 0 \times 2 + 0 \times 3 = 0$.

In general, if we add a total of p_h rows of padding (roughly half on top and half on bottom) and a total of p_w columns of padding (roughly half on the left and half on the right), the output shape will be

$$(n_h - k_h + p_h + 1) \times (n_w - k_w + p_w + 1).$$ (7.3.1)

This means that the height and width of the output will increase by p_h and p_w, respectively.

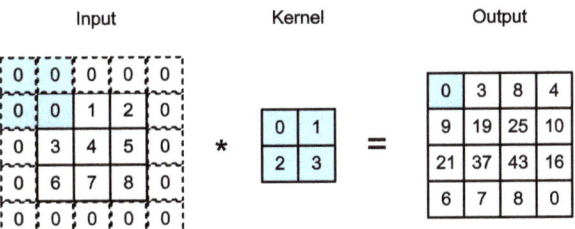

Fig. 7.3.2 Two-dimensional cross-correlation with padding.

In many cases, we will want to set $p_h = k_h - 1$ and $p_w = k_w - 1$ to give the input and output the same height and width. This will make it easier to predict the output shape of each layer when constructing the network. Assuming that k_h is odd here, we will pad $p_h/2$ rows on both sides of the height. If k_h is even, one possibility is to pad $\lceil p_h/2 \rceil$ rows on the top of the input and $\lfloor p_h/2 \rfloor$ rows on the bottom. We will pad both sides of the width in the same way.

CNNs commonly use convolution kernels with odd height and width values, such as 1, 3, 5, or 7. Choosing odd kernel sizes has the benefit that we can preserve the dimensionality while padding with the same number of rows on top and bottom, and the same number of columns on left and right.

Moreover, this practice of using odd kernels and padding to precisely preserve dimensionality offers a clerical benefit. For any two-dimensional tensor X, when the kernel's size is odd and the number of padding rows and columns on all sides are the same, thereby producing an output with the same height and width as the input, we know that the output Y[i, j] is calculated by cross-correlation of the input and convolution kernel with the window centered on X[i, j].

In the following example, we create a two-dimensional convolutional layer with a height and width of 3 and apply 1 pixel of padding on all sides. Given an input with a height and width of 8, we find that the height and width of the output is also 8.

```
# We define a helper function to calculate convolutions. It initializes the
# convolutional layer weights and performs corresponding dimensionality
# elevations and reductions on the input and output
def comp_conv2d(conv2d, X):
    # (1, 1) indicates that batch size and the number of channels are both 1
    X = X.reshape((1, 1) + X.shape)
    Y = conv2d(X)
    # Strip the first two dimensions: examples and channels
    return Y.reshape(Y.shape[2:])

# 1 row and column is padded on either side, so a total of 2 rows or columns
# are added
conv2d = nn.LazyConv2d(1, kernel_size=3, padding=1)
X = torch.rand(size=(8, 8))
comp_conv2d(conv2d, X).shape
```

```
torch.Size([8, 8])
```

When the height and width of the convolution kernel are different, we can make the output and input have the same height and width by setting different padding numbers for height and width.

```
# We use a convolution kernel with height 5 and width 3. The padding on either
# side of the height and width are 2 and 1, respectively
conv2d = nn.LazyConv2d(1, kernel_size=(5, 3), padding=(2, 1))
comp_conv2d(conv2d, X).shape
```

```
torch.Size([8, 8])
```

7.3.2 Stride

When computing the cross-correlation, we start with the convolution window at the upper-left corner of the input tensor, and then slide it over all locations both down and to the right. In the previous examples, we defaulted to sliding one element at a time. However, sometimes, either for computational efficiency or because we wish to downsample, we move our window more than one element at a time, skipping the intermediate locations. This is particularly useful if the convolution kernel is large since it captures a large area of the underlying image.

We refer to the number of rows and columns traversed per slide as *stride*. So far, we have used strides of 1, both for height and width. Sometimes, we may want to use a larger stride. Fig. 7.3.3 shows a two-dimensional cross-correlation operation with a stride of 3 vertically and 2 horizontally. The shaded portions are the output elements as well as the input and kernel tensor elements used for the output computation: $0 \times 0 + 0 \times 1 + 1 \times 2 + 2 \times 3 = 8$, $0 \times 0 + 6 \times 1 + 0 \times 2 + 0 \times 3 = 6$. We can see that when the second element of the first column is generated, the convolution window slides down three rows. The convolution window slides two columns to the right when the second element of the first row is generated. When the convolution window continues to slide two columns to the right on the input, there is no output because the input element cannot fill the window (unless we add another column of padding).

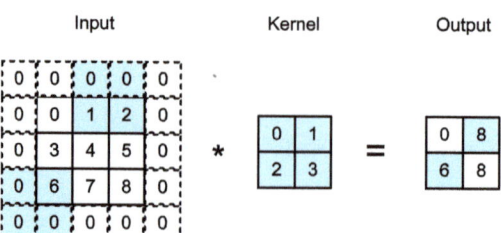

Fig. 7.3.3   Cross-correlation with strides of 3 and 2 for height and width, respectively.

In general, when the stride for the height is s_h and the stride for the width is s_w, the output shape is

$$\lfloor (n_h - k_h + p_h + s_h)/s_h \rfloor \times \lfloor (n_w - k_w + p_w + s_w)/s_w \rfloor. \qquad (7.3.2)$$

If we set $p_h = k_h - 1$ and $p_w = k_w - 1$, then the output shape can be simplified to $\lfloor (n_h + s_h - 1)/s_h \rfloor \times \lfloor (n_w + s_w - 1)/s_w \rfloor$. Going a step further, if the input height and width are divisible by the strides on the height and width, then the output shape will be $(n_h/s_h) \times (n_w/s_w)$.

Below, we set the strides on both the height and width to 2, thus halving the input height and width.

```
conv2d = nn.LazyConv2d(1, kernel_size=3, padding=1, stride=2)
comp_conv2d(conv2d, X).shape
```

```
torch.Size([4, 4])
```

Let's look at a slightly more complicated example.

```
conv2d = nn.LazyConv2d(1, kernel_size=(3, 5), padding=(0, 1), stride=(3, 4))
comp_conv2d(conv2d, X).shape
```

```
torch.Size([2, 2])
```

7.3.3 Summary and Discussion

Padding can increase the height and width of the output. This is often used to give the output the same height and width as the input to avoid undesirable shrinkage of the output. Moreover, it ensures that all pixels are used equally frequently. Typically we pick symmetric padding on both sides of the input height and width. In this case we refer to (p_h, p_w) padding. Most commonly we set $p_h = p_w$, in which case we simply state that we choose padding p.

A similar convention applies to strides. When horizontal stride s_h and vertical stride s_w match, we simply talk about stride s. The stride can reduce the resolution of the output, for example reducing the height and width of the output to only $1/n$ of the height and width of the input for $n > 1$. By default, the padding is 0 and the stride is 1.

So far all padding that we discussed simply extended images with zeros. This has significant computational benefit since it is trivial to accomplish. Moreover, operators can be engineered to take advantage of this padding implicitly without the need to allocate additional memory. At the same time, it allows CNNs to encode implicit position information within an image, simply by learning where the "whitespace" is. There are many alternatives to zero-padding. Alsallakh *et al.* (2020) provided an extensive overview of those (albeit without a clear case for when to use nonzero paddings unless artifacts occur).

7.3.4 Exercises

1. Given the final code example in this section with kernel size $(3, 5)$, padding $(0, 1)$, and stride $(3, 4)$, calculate the output shape to check if it is consistent with the experimental result.

2. For audio signals, what does a stride of 2 correspond to?

3. Implement mirror padding, i.e., padding where the border values are simply mirrored to extend tensors.

4. What are the computational benefits of a stride larger than 1?

5. What might be statistical benefits of a stride larger than 1?

6. How would you implement a stride of $\frac{1}{2}$? What does it correspond to? When would this be useful?

 Discussions[121].

[121]

7.4 Multiple Input and Multiple Output Channels

While we described the multiple channels that comprise each image (e.g., color images have the standard RGB channels to indicate the amount of red, green and blue) and convolutional layers for multiple channels in Section 7.1.4, until now, we simplified all of our numerical examples by working with just a single input and a single output channel. This allowed us to think of our inputs, convolution kernels, and outputs each as two-dimensional tensors.

When we add channels into the mix, our inputs and hidden representations both become three-dimensional tensors. For example, each RGB input image has shape $3 \times h \times w$. We refer to this axis, with a size of 3, as the *channel* dimension. The notion of channels is as old as CNNs themselves: for instance LeNet-5 (LeCun *et al.*, 1995) uses them. In this section, we will take a deeper look at convolution kernels with multiple input and multiple output channels.

```
import torch
from d2l import torch as d2l
```

7.4.1 Multiple Input Channels

When the input data contains multiple channels, we need to construct a convolution kernel with the same number of input channels as the input data, so that it can perform cross-correlation with the input data. Assuming that the number of channels for the input data is c_i, the number of input channels of the convolution kernel also needs to be c_i. If our convolution kernel's window shape is $k_h \times k_w$, then, when $c_i = 1$, we can think of our convolution kernel as just a two-dimensional tensor of shape $k_h \times k_w$.

However, when $c_i > 1$, we need a kernel that contains a tensor of shape $k_h \times k_w$ for *every* input channel. Concatenating these c_i tensors together yields a convolution kernel of shape $c_i \times k_h \times k_w$. Since the input and convolution kernel each have c_i channels, we can perform a cross-correlation operation on the two-dimensional tensor of the input and the two-dimensional tensor of the convolution kernel for each channel, adding the c_i results together (summing over the channels) to yield a two-dimensional tensor. This is the result of a two-dimensional cross-correlation between a multi-channel input and a multi-input-channel convolution kernel.

Fig. 7.4.1 provides an example of a two-dimensional cross-correlation with two input channels. The shaded portions are the first output element as well as the input and kernel tensor elements used for the output computation: $(1 \times 1 + 2 \times 2 + 4 \times 3 + 5 \times 4) + (0 \times 0 + 1 \times 1 + 3 \times 2 + 4 \times 3) = 56$.

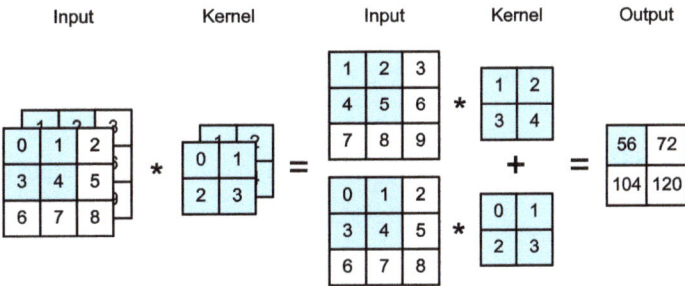

Fig. 7.4.1   Cross-correlation computation with two input channels.

To make sure we really understand what is going on here, we can implement cross-correlation operations with multiple input channels ourselves. Notice that all we are doing is performing a cross-correlation operation per channel and then adding up the results.

```
def corr2d_multi_in(X, K):
    # Iterate through the 0th dimension (channel) of K first, then add them up
    return sum(d2l.corr2d(x, k) for x, k in zip(X, K))
```

We can construct the input tensor X and the kernel tensor K corresponding to the values in Fig. 7.4.1 to validate the output of the cross-correlation operation.

```
X = torch.tensor([[[0.0, 1.0, 2.0], [3.0, 4.0, 5.0], [6.0, 7.0, 8.0]],
               [[1.0, 2.0, 3.0], [4.0, 5.0, 6.0], [7.0, 8.0, 9.0]]])
K = torch.tensor([[[0.0, 1.0], [2.0, 3.0]], [[1.0, 2.0], [3.0, 4.0]]])

corr2d_multi_in(X, K)
```

```
tensor([[ 56.,   72.],
        [104., 120.]])
```

7.4.2 Multiple Output Channels

Regardless of the number of input channels, so far we always ended up with one output channel. However, as we discussed in Section 7.1.4, it turns out to be essential to have multiple channels at each layer. In the most popular neural network architectures, we actually increase the channel dimension as we go deeper in the neural network, typically downsampling to trade off spatial resolution for greater *channel depth*. Intuitively, you could think of each channel as responding to a different set of features. The reality is a bit more complicated than this. A naive interpretation would suggest that representations are learned independently per pixel or per channel. Instead, channels are optimized to be jointly useful. This means that rather than mapping a single channel to an edge detector, it may simply mean that some direction in channel space corresponds to detecting edges.

Denote by c_i and c_o the number of input and output channels, respectively, and by k_h and k_w the height and width of the kernel. To get an output with multiple channels, we can create a kernel tensor of shape $c_i \times k_h \times k_w$ for *every* output channel. We concatenate them on the output channel dimension, so that the shape of the convolution kernel is $c_o \times c_i \times k_h \times k_w$. In cross-correlation operations, the result on each output channel is calculated from the convolution kernel corresponding to that output channel and takes input from all channels in the input tensor.

We implement a cross-correlation function to calculate the output of multiple channels as shown below.

```
def corr2d_multi_in_out(X, K):
    # Iterate through the 0th dimension of K, and each time, perform
    # cross-correlation operations with input X. All of the results are
    # stacked together
    return torch.stack([corr2d_multi_in(X, k) for k in K], 0)
```

We construct a trivial convolution kernel with three output channels by concatenating the kernel tensor for K with K+1 and K+2.

```
K = torch.stack((K, K + 1, K + 2), 0)
K.shape
```

```
torch.Size([3, 2, 2, 2])
```

Below, we perform cross-correlation operations on the input tensor X with the kernel tensor K. Now the output contains three channels. The result of the first channel is consistent with the result of the previous input tensor X and the multi-input channel, single-output channel kernel.

```
corr2d_multi_in_out(X, K)
```

```
tensor([[[ 56.,   72.],
         [104., 120.]],

        [[ 76., 100.],
         [148., 172.]],

        [[ 96., 128.],
         [192., 224.]]])
```

7.4.3 1×1 Convolutional Layer

At first, a 1×1 convolution, i.e., $k_h = k_w = 1$, does not seem to make much sense. After all, a convolution correlates adjacent pixels. A 1×1 convolution obviously does not. Nonetheless, they are popular operations that are sometimes included in the designs of complex deep networks (Lin *et al.*, 2013, Szegedy *et al.*, 2017). Let's see in some detail what it actually does.

Because the minimum window is used, the 1×1 convolution loses the ability of larger convolutional layers to recognize patterns consisting of interactions among adjacent elements in the height and width dimensions. The only computation of the 1×1 convolution occurs on the channel dimension.

Fig. 7.4.2 shows the cross-correlation computation using the 1×1 convolution kernel with 3 input channels and 2 output channels. Note that the inputs and outputs have the same height and width. Each element in the output is derived from a linear combination of elements *at the same position* in the input image. You could think of the 1×1 convolutional layer as constituting a fully connected layer applied at every single pixel location to transform the c_i corresponding input values into c_o output values. Because this is still a convolutional layer, the weights are tied across pixel location. Thus the 1×1 convolutional layer requires $c_o \times c_i$ weights (plus the bias). Also note that convolutional layers are typically followed by nonlinearities. This ensures that 1×1 convolutions cannot simply be folded into other convolutions.

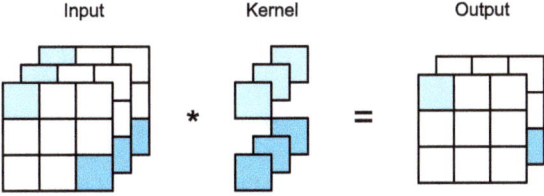

Fig. 7.4.2   The cross-correlation computation uses the 1×1 convolution kernel with three input channels and two output channels. The input and output have the same height and width.

Let's check whether this works in practice: we implement a 1×1 convolution using a fully connected layer. The only thing is that we need to make some adjustments to the data shape before and after the matrix multiplication.

```
def corr2d_multi_in_out_1x1(X, K):
    c_i, h, w = X.shape
    c_o = K.shape[0]
    X = X.reshape((c_i, h * w))
    K = K.reshape((c_o, c_i))
    # Matrix multiplication in the fully connected layer
    Y = torch.matmul(K, X)
    return Y.reshape((c_o, h, w))
```

When performing 1×1 convolutions, the above function is equivalent to the previously implemented cross-correlation function corr2d_multi_in_out. Let's check this with some sample data.

```
X = torch.normal(0, 1, (3, 3, 3))
K = torch.normal(0, 1, (2, 3, 1, 1))
Y1 = corr2d_multi_in_out_1x1(X, K)
Y2 = corr2d_multi_in_out(X, K)
assert float(torch.abs(Y1 - Y2).sum()) < 1e-6
```

7.4.4 Discussion

Channels allow us to combine the best of both worlds: MLPs that allow for significant nonlinearities and convolutions that allow for *localized* analysis of features. In particular, channels allow the CNN to reason with multiple features, such as edge and shape detectors at the same time. They also offer a practical trade-off between the drastic parameter reduction arising from translation invariance and locality, and the need for expressive and diverse models in computer vision.

Note, though, that this flexibility comes at a price. Given an image of size $(h \times w)$, the cost for computing a $k \times k$ convolution is $O(h \cdot w \cdot k^2)$. For c_i and c_o input and output channels respectively this increases to $O(h \cdot w \cdot k^2 \cdot c_i \cdot c_o)$. For a 256×256 pixel image with a 5×5 kernel and 128 input and output channels respectively this amounts to over 53 billion operations (we count multiplications and additions separately). Later on we will encounter effective strategies to cut down on the cost, e.g., by requiring the channel-wise operations to be block-diagonal, leading to architectures such as ResNeXt (Xie *et al.*, 2017).

7.4.5 Exercises

1. Assume that we have two convolution kernels of size k_1 and k_2, respectively (with no nonlinearity in between).

    1. Prove that the result of the operation can be expressed by a single convolution.

    2. What is the dimensionality of the equivalent single convolution?

    3. Is the converse true, i.e., can you always decompose a convolution into two smaller ones?

2. Assume an input of shape $c_i \times h \times w$ and a convolution kernel of shape $c_o \times c_i \times k_h \times k_w$, padding of (p_h, p_w), and stride of (s_h, s_w).

  1. What is the computational cost (multiplications and additions) for the forward propagation?

  2. What is the memory footprint?

  3. What is the memory footprint for the backward computation?

  4. What is the computational cost for the backpropagation?

3. By what factor does the number of calculations increase if we double both the number of input channels c_i and the number of output channels c_o? What happens if we double the padding?

4. Are the variables Y1 and Y2 in the final example of this section exactly the same? Why?

5. Express convolutions as a matrix multiplication, even when the convolution window is not 1×1.

6. Your task is to implement fast convolutions with a $k \times k$ kernel. One of the algorithm candidates is to scan horizontally across the source, reading a k-wide strip and computing the 1-wide output strip one value at a time. The alternative is to read a $k + \Delta$ wide strip and compute a Δ-wide output strip. Why is the latter preferable? Is there a limit to how large you should choose Δ?

7. Assume that we have a $c \times c$ matrix.

  1. How much faster is it to multiply with a block-diagonal matrix if the matrix is broken up into b blocks?

  2. What is the downside of having b blocks? How could you fix it, at least partly?

 Discussions[122].

7.5 Pooling

In many cases our ultimate task asks some global question about the image, e.g., *does it contain a cat?* Consequently, the units of our final layer should be sensitive to the entire input. By gradually aggregating information, yielding coarser and coarser maps, we accomplish this goal of ultimately learning a global representation, while keeping all of the advantages of convolutional layers at the intermediate layers of processing. The deeper we go in the network, the larger the receptive field (relative to the input) to which each hidden node is sensitive. Reducing spatial resolution accelerates this process, since the convolution kernels cover a larger effective area.

Moreover, when detecting lower-level features, such as edges (as discussed in Section 7.2), we often want our representations to be somewhat invariant to translation. For instance, if we take the image X with a sharp delineation between black and white and shift the whole image by one pixel to the right, i.e., Z[i, j] = X[i, j + 1], then the output for the new image Z might be vastly different. The edge will have shifted by one pixel. In reality, objects hardly ever occur exactly at the same place. In fact, even with a tripod and a stationary object, vibration of the camera due to the movement of the shutter might shift everything by a pixel or so (high-end cameras are loaded with special features to address this problem).

This section introduces *pooling layers*, which serve the dual purposes of mitigating the sensitivity of convolutional layers to location and of spatially downsampling representations.

```
import torch
from torch import nn
from d2l import torch as d2l
```

7.5.1 Maximum Pooling and Average Pooling

Like convolutional layers, *pooling* operators consist of a fixed-shape window that is slid over all regions in the input according to its stride, computing a single output for each location traversed by the fixed-shape window (sometimes known as the *pooling window*). However, unlike the cross-correlation computation of the inputs and kernels in the convolutional layer, the pooling layer contains no parameters (there is no *kernel*). Instead, pooling operators are deterministic, typically calculating either the maximum or the average value of the elements in the pooling window. These operations are called *maximum pooling* (*max-pooling* for short) and *average pooling*, respectively.

Average pooling is essentially as old as CNNs. The idea is akin to downsampling an image. Rather than just taking the value of every second (or third) pixel for the lower resolution image, we can average over adjacent pixels to obtain an image with better signal-to-noise ratio since we are combining the information from multiple adjacent pixels. *Max-pooling* was introduced in Riesenhuber and Poggio (1999) in the context of cognitive neuroscience to describe how information aggregation might be aggregated hierarchically for the purpose of object recognition; there already was an earlier version in speech recognition (Yamaguchi *et al.*, 1990). In almost all cases, max-pooling, as it is also referred to, is preferable to average pooling.

In both cases, as with the cross-correlation operator, we can think of the pooling window as starting from the upper-left of the input tensor and sliding across it from left to right and top to bottom. At each location that the pooling window hits, it computes the maximum or average value of the input subtensor in the window, depending on whether max or average pooling is employed.

The output tensor in Fig. 7.5.1 has a height of 2 and a width of 2. The four elements are

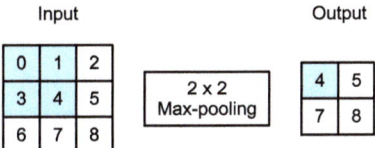

Fig. 7.5.1 Max-pooling with a pooling window shape of 2×2. The shaded portions are the first output element as well as the input tensor elements used for the output computation: $\max(0, 1, 3, 4) = 4$.

derived from the maximum value in each pooling window:

$$
\begin{aligned}
\max(0, 1, 3, 4) &= 4, \\
\max(1, 2, 4, 5) &= 5, \\
\max(3, 4, 6, 7) &= 7, \\
\max(4, 5, 7, 8) &= 8.
\end{aligned}
\tag{7.5.1}
$$

More generally, we can define a $p \times q$ pooling layer by aggregating over a region of said size. Returning to the problem of edge detection, we use the output of the convolutional layer as input for 2×2 max-pooling. Denote by X the input of the convolutional layer input and Y the pooling layer output. Regardless of whether or not the values of X[i, j], X[i, j + 1], X[i+1, j] and X[i+1, j + 1] are different, the pooling layer always outputs Y[i, j] = 1. That is to say, using the 2×2 max-pooling layer, we can still detect if the pattern recognized by the convolutional layer moves no more than one element in height or width.

In the code below, we implement the forward propagation of the pooling layer in the pool2d function. This function is similar to the corr2d function in Section 7.2. However, no kernel is needed, computing the output as either the maximum or the average of each region in the input.

```python
def pool2d(X, pool_size, mode='max'):
    p_h, p_w = pool_size
    Y = torch.zeros((X.shape[0] - p_h + 1, X.shape[1] - p_w + 1))
    for i in range(Y.shape[0]):
        for j in range(Y.shape[1]):
            if mode == 'max':
                Y[i, j] = X[i: i + p_h, j: j + p_w].max()
            elif mode == 'avg':
                Y[i, j] = X[i: i + p_h, j: j + p_w].mean()
    return Y
```

We can construct the input tensor X in Fig. 7.5.1 to validate the output of the two-dimensional max-pooling layer.

```python
X = torch.tensor([[0.0, 1.0, 2.0], [3.0, 4.0, 5.0], [6.0, 7.0, 8.0]])
pool2d(X, (2, 2))
```

```
tensor([[4., 5.],
        [7., 8.]])
```

Also, we can experiment with the average pooling layer.

```
pool2d(X, (2, 2), 'avg')
```

```
tensor([[2., 3.],
        [5., 6.]])
```

7.5.2 Padding and Stride

As with convolutional layers, pooling layers change the output shape. And as before, we can adjust the operation to achieve a desired output shape by padding the input and adjusting the stride. We can demonstrate the use of padding and strides in pooling layers via the built-in two-dimensional max-pooling layer from the deep learning framework. We first construct an input tensor X whose shape has four dimensions, where the number of examples (batch size) and number of channels are both 1.

```
X = torch.arange(16, dtype=torch.float32).reshape((1, 1, 4, 4))
X
```

```
tensor([[[[ 0.,  1.,  2.,  3.],
          [ 4.,  5.,  6.,  7.],
          [ 8.,  9., 10., 11.],
          [12., 13., 14., 15.]]]])
```

Since pooling aggregates information from an area, deep learning frameworks default to matching pooling window sizes and stride. For instance, if we use a pooling window of shape (3, 3) we get a stride shape of (3, 3) by default.

```
pool2d = nn.MaxPool2d(3)
# Pooling has no model parameters, hence it needs no initialization
pool2d(X)
```

```
tensor([[[[10.]]]])
```

Needless to say, the stride and padding can be manually specified to override framework defaults if required.

```
pool2d = nn.MaxPool2d(3, padding=1, stride=2)
pool2d(X)
```

```
tensor([[[[ 5.,   7.],
          [13., 15.]]]])
```

Of course, we can specify an arbitrary rectangular pooling window with arbitrary height and width respectively, as the example below shows.

```
pool2d = nn.MaxPool2d((2, 3), stride=(2, 3), padding=(0, 1))
pool2d(X)
```

```
tensor([[[[ 5.,   7.],
          [13., 15.]]]])
```

7.5.3 Multiple Channels

When processing multi-channel input data, the pooling layer pools each input channel separately, rather than summing the inputs up over channels as in a convolutional layer. This means that the number of output channels for the pooling layer is the same as the number of input channels. Below, we will concatenate tensors X and X + 1 on the channel dimension to construct an input with two channels.

```
X = torch.cat((X, X + 1), 1)
X
```

```
tensor([[[[ 0.,   1.,   2.,   3.],
          [ 4.,   5.,   6.,   7.],
          [ 8.,   9., 10., 11.],
          [12., 13., 14., 15.]],

         [[ 1.,   2.,   3.,   4.],
          [ 5.,   6.,   7.,   8.],
          [ 9., 10., 11., 12.],
          [13., 14., 15., 16.]]]])
```

As we can see, the number of output channels is still two after pooling.

```
pool2d = nn.MaxPool2d(3, padding=1, stride=2)
pool2d(X)
```

```
tensor([[[[ 5.,   7.],
          [13., 15.]],

         [[ 6.,   8.],
          [14., 16.]]]])
```

7.5.4 Summary

Pooling is an exceedingly simple operation. It does exactly what its name indicates, aggregate results over a window of values. All convolution semantics, such as strides and padding apply in the same way as they did previously. Note that pooling is indifferent to channels, i.e., it leaves the number of channels unchanged and it applies to each channel separately. Lastly, of the two popular pooling choices, max-pooling is preferable to average pooling, as it confers some degree of invariance to output. A popular choice is to pick a pooling window size of 2×2 to quarter the spatial resolution of output.

Note that there are many more ways of reducing resolution beyond pooling. For instance, in stochastic pooling (Zeiler and Fergus, 2013) and fractional max-pooling (Graham, 2014) aggregation is combined with randomization. This can slightly improve the accuracy in some cases. Lastly, as we will see later with the attention mechanism, there are more refined ways of aggregating over outputs, e.g., by using the alignment between a query and representation vectors.

7.5.5 Exercises

1. Implement average pooling through a convolution.

2. Prove that max-pooling cannot be implemented through a convolution alone.

3. Max-pooling can be accomplished using ReLU operations, i.e., $\text{ReLU}(x) = \max(0, x)$.

   1. Express $\max(a, b)$ by using only ReLU operations.

   2. Use this to implement max-pooling by means of convolutions and ReLU layers.

   3. How many channels and layers do you need for a 2×2 convolution? How many for a 3×3 convolution?

4. What is the computational cost of the pooling layer? Assume that the input to the pooling layer is of size $c \times h \times w$, the pooling window has a shape of $p_h \times p_w$ with a padding of (p_h, p_w) and a stride of (s_h, s_w).

5. Why do you expect max-pooling and average pooling to work differently?

6. Do we need a separate minimum pooling layer? Can you replace it with another operation?

7. We could use the softmax operation for pooling. Why might it not be so popular?

 Discussions [123].

7.6 Convolutional Neural Networks (LeNet)

We now have all the ingredients required to assemble a fully-functional CNN. In our earlier encounter with image data, we applied a linear model with softmax regression (Section 4.4) and an MLP (Section 5.2) to pictures of clothing in the Fashion-MNIST dataset. To make such data amenable we first flattened each image from a 28×28 matrix into a fixed-length 784-dimensional vector, and thereafter processed them in fully connected layers. Now that we have a handle on convolutional layers, we can retain the spatial structure in our images. As an additional benefit of replacing fully connected layers with convolutional layers, we will enjoy more parsimonious models that require far fewer parameters.

In this section, we will introduce *LeNet*, among the first published CNNs to capture wide attention for its performance on computer vision tasks. The model was introduced by (and named for) Yann LeCun, then a researcher at AT&T Bell Labs, for the purpose of recognizing handwritten digits in images (LeCun *et al.*, 1998). This work represented the culmination of a decade of research developing the technology; LeCun's team published the first study to successfully train CNNs via backpropagation (LeCun *et al.*, 1989).

At the time LeNet achieved outstanding results matching the performance of support vector machines, then a dominant approach in supervised learning, achieving an error rate of less than 1% per digit. LeNet was eventually adapted to recognize digits for processing deposits in ATM machines. To this day, some ATMs still run the code that Yann LeCun and his colleague Leon Bottou wrote in the 1990s!

```python
import torch
from torch import nn
from d2l import torch as d2l
```

7.6.1 LeNet

At a high level, LeNet (LeNet-5) consists of two parts: (i) a convolutional encoder consisting of two convolutional layers; and (ii) a dense block consisting of three fully connected layers. The architecture is summarized in Fig. 7.6.1.

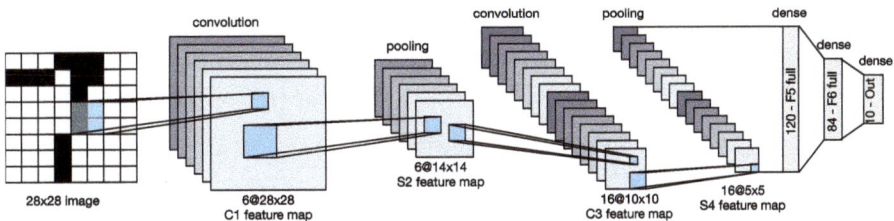

Fig. 7.6.1   Data flow in LeNet. The input is a handwritten digit, the output is a probability over 10 possible outcomes.

The basic units in each convolutional block are a convolutional layer, a sigmoid activation function, and a subsequent average pooling operation. Note that while ReLUs and max-pooling work better, they had not yet been discovered. Each convolutional layer uses a 5×5

kernel and a sigmoid activation function. These layers map spatially arranged inputs to a number of two-dimensional feature maps, typically increasing the number of channels. The first convolutional layer has 6 output channels, while the second has 16. Each 2×2 pooling operation (stride 2) reduces dimensionality by a factor of 4 via spatial downsampling. The convolutional block emits an output with shape given by (batch size, number of channel, height, width).

In order to pass output from the convolutional block to the dense block, we must flatten each example in the minibatch. In other words, we take this four-dimensional input and transform it into the two-dimensional input expected by fully connected layers: as a reminder, the two-dimensional representation that we desire uses the first dimension to index examples in the minibatch and the second to give the flat vector representation of each example. LeNet's dense block has three fully connected layers, with 120, 84, and 10 outputs, respectively. Because we are still performing classification, the 10-dimensional output layer corresponds to the number of possible output classes.

While getting to the point where you truly understand what is going on inside LeNet may have taken a bit of work, we hope that the following code snippet will convince you that implementing such models with modern deep learning frameworks is remarkably simple. We need only to instantiate a Sequential block and chain together the appropriate layers, using Xavier initialization as introduced in Section 5.4.2.

```python
def init_cnn(module):  #@save
    """Initialize weights for CNNs."""
    if type(module) == nn.Linear or type(module) == nn.Conv2d:
        nn.init.xavier_uniform_(module.weight)
```

```python
class LeNet(d2l.Classifier):  #@save
    """The LeNet-5 model."""
    def __init__(self, lr=0.1, num_classes=10):
        super().__init__()
        self.save_hyperparameters()
        self.net = nn.Sequential(
            nn.LazyConv2d(6, kernel_size=5, padding=2), nn.Sigmoid(),
            nn.AvgPool2d(kernel_size=2, stride=2),
            nn.LazyConv2d(16, kernel_size=5), nn.Sigmoid(),
            nn.AvgPool2d(kernel_size=2, stride=2),
            nn.Flatten(),
            nn.LazyLinear(120), nn.Sigmoid(),
            nn.LazyLinear(84), nn.Sigmoid(),
            nn.LazyLinear(num_classes))
```

We have taken some liberty in the reproduction of LeNet insofar as we have replaced the Gaussian activation layer by a softmax layer. This greatly simplifies the implementation, not least due to the fact that the Gaussian decoder is rarely used nowadays. Other than that, this network matches the original LeNet-5 architecture.

Let's see what happens inside the network. By passing a single-channel (black and white) 28×28 image through the network and printing the output shape at each layer, we can inspect the model to ensure that its operations line up with what we expect from Fig. 7.6.2.

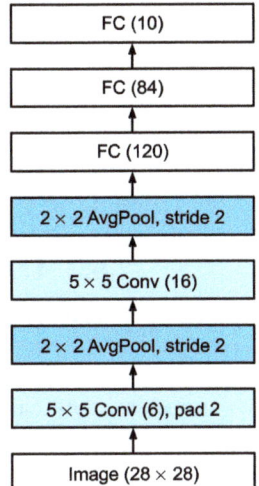

Fig. 7.6.2   Compressed notation for LeNet-5.

```
@d2l.add_to_class(d2l.Classifier)    #@save
def layer_summary(self, X_shape):
    X = torch.randn(*X_shape)
    for layer in self.net:
        X = layer(X)
        print(layer.__class__.__name__, 'output shape:\t', X.shape)

model = LeNet()
model.layer_summary((1, 1, 28, 28))
```

```
Conv2d output shape:          torch.Size([1, 6, 28, 28])
Sigmoid output shape:         torch.Size([1, 6, 28, 28])
AvgPool2d output shape:       torch.Size([1, 6, 14, 14])
Conv2d output shape:          torch.Size([1, 16, 10, 10])
Sigmoid output shape:         torch.Size([1, 16, 10, 10])
AvgPool2d output shape:       torch.Size([1, 16, 5, 5])
Flatten output shape:         torch.Size([1, 400])
Linear output shape:          torch.Size([1, 120])
Sigmoid output shape:         torch.Size([1, 120])
Linear output shape:          torch.Size([1, 84])
Sigmoid output shape:         torch.Size([1, 84])
Linear output shape:          torch.Size([1, 10])
```

Note that the height and width of the representation at each layer throughout the convolutional block is reduced (compared with the previous layer). The first convolutional layer uses two pixels of padding to compensate for the reduction in height and width that would otherwise result from using a 5×5 kernel. As an aside, the image size of 28×28 pixels in the original MNIST OCR dataset is a result of *trimming* two pixel rows (and columns) from the original scans that measured 32×32 pixels. This was done primarily to save space (a 30% reduction) at a time when megabytes mattered.

In contrast, the second convolutional layer forgoes padding, and thus the height and width are both reduced by four pixels. As we go up the stack of layers, the number of channels increases layer-over-layer from 1 in the input to 6 after the first convolutional layer and 16 after the second convolutional layer. However, each pooling layer halves the height and width. Finally, each fully connected layer reduces dimensionality, finally emitting an output whose dimension matches the number of classes.

7.6.2 Training

Now that we have implemented the model, let's run an experiment to see how the LeNet-5 model fares on Fashion-MNIST.

While CNNs have fewer parameters, they can still be more expensive to compute than similarly deep MLPs because each parameter participates in many more multiplications. If you have access to a GPU, this might be a good time to put it into action to speed up training. Note that the d2l.Trainer class takes care of all details. By default, it initializes the model parameters on the available devices. Just as with MLPs, our loss function is cross-entropy, and we minimize it via minibatch stochastic gradient descent.

```
trainer = d2l.Trainer(max_epochs=10, num_gpus=1)
data = d2l.FashionMNIST(batch_size=128)
model = LeNet(lr=0.1)
model.apply_init([next(iter(data.get_dataloader(True)))[0]], init_cnn)
trainer.fit(model, data)
```

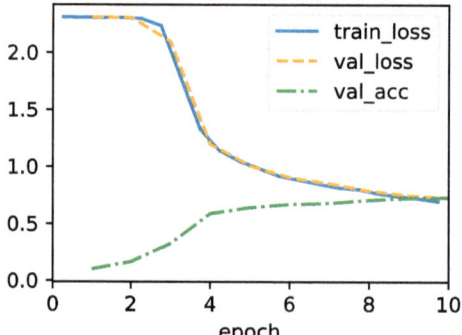

7.6.3 Summary

We have made significant progress in this chapter. We moved from the MLPs of the 1980s to the CNNs of the 1990s and early 2000s. The architectures proposed, e.g., in the form of LeNet-5 remain meaningful, even to this day. It is worth comparing the error rates on Fashion-MNIST achievable with LeNet-5 both to the very best possible with MLPs (Section 5.2) and those with significantly more advanced architectures such as ResNet (Section 8.6). LeNet is much more similar to the latter than to the former. One of the primary differences, as we shall see, is that greater amounts of computation enabled significantly more complex architectures.

A second difference is the relative ease with which we were able to implement LeNet. What used to be an engineering challenge worth months of C++ and assembly code, engineering to improve SN, an early Lisp-based deep learning tool (Bottou and Le Cun, 1988), and finally experimentation with models can now be accomplished in minutes. It is this incredible productivity boost that has democratized deep learning model development tremendously. In the next chapter we will journey down this rabbit to hole to see where it takes us.

7.6.4 Exercises

1. Let's modernize LeNet. Implement and test the following changes:

    1. Replace average pooling with max-pooling.

    2. Replace the softmax layer with ReLU.

2. Try to change the size of the LeNet style network to improve its accuracy in addition to max-pooling and ReLU.

    1. Adjust the convolution window size.

    2. Adjust the number of output channels.

    3. Adjust the number of convolution layers.

    4. Adjust the number of fully connected layers.

    5. Adjust the learning rates and other training details (e.g., initialization and number of epochs).

3. Try out the improved network on the original MNIST dataset.

4. Display the activations of the first and second layer of LeNet for different inputs (e.g., sweaters and coats).

5. What happens to the activations when you feed significantly different images into the network (e.g., cats, cars, or even random noise)?

Discussions [124].

124

8 Modern Convolutional Neural Networks

Now that we understand the basics of wiring together CNNs, let's take a tour of modern CNN architectures. This tour is, by necessity, incomplete, thanks to the plethora of exciting new designs being added. Their importance derives from the fact that not only can they be used directly for vision tasks, but they also serve as basic feature generators for more advanced tasks such as tracking (Zhang *et al.*, 2021), segmentation (Long *et al.*, 2015), object detection (Redmon and Farhadi, 2018), or style transformation (Gatys *et al.*, 2016). In this chapter, most sections correspond to a significant CNN architecture that was at some point (or currently) the base model upon which many research projects and deployed systems were built. Each of these networks was briefly a dominant architecture and many were winners or runners-up in the ImageNet competition [125] which has served as a barometer of progress on supervised learning in computer vision since 2010. It is only recently that Transformers have begun to displace CNNs, starting with Dosovitskiy *et al.* (2021) and followed by the Swin Transformer (Liu *et al.*, 2021). We will cover this development later in Chapter 11.

While the idea of *deep* neural networks is quite simple (stack together a bunch of layers), performance can vary wildly across architectures and hyperparameter choices. The neural networks described in this chapter are the product of intuition, a few mathematical insights, and a lot of trial and error. We present these models in chronological order, partly to convey a sense of the history so that you can form your own intuitions about where the field is heading and perhaps develop your own architectures. For instance, batch normalization and residual connections described in this chapter have offered two popular ideas for training and designing deep models, both of which have since also been applied to architectures beyond computer vision.

We begin our tour of modern CNNs with AlexNet (Krizhevsky *et al.*, 2012), the first large-scale network deployed to beat conventional computer vision methods on a large-scale vision challenge; the VGG network (Simonyan and Zisserman, 2014), which makes use of a number of repeating blocks of elements; the network in network (NiN) that convolves whole neural networks patch-wise over inputs (Lin *et al.*, 2013); GoogLeNet that uses networks with multi-branch convolutions (Szegedy *et al.*, 2015); the residual network (ResNet) (He *et al.*, 2016), which remains one of the most popular off-the-shelf architectures in computer vision; ResNeXt blocks (Xie *et al.*, 2017) for sparser connections; and DenseNet (Huang *et al.*, 2017) for a generalization of the residual architecture. Over time many special optimizations for efficient networks have been developed, such as coordinate shifts (ShiftNet) (Wu *et al.*, 2018). This culminated in the automatic search for efficient architectures such as MobileNet v3 (Howard *et al.*, 2019). It also includes the semi-automatic design exploration of Radosavovic *et al.* (2020) that led to the RegNetX/Y which we will discuss later in this chapter. The work is instructive insofar as it offers a path for marrying brute force computation with the ingenuity of an experimenter in the search for efficient design spaces. Of note is also the work of Liu *et al.* (2022) as it shows that training techniques (e.g., optimizers, data augmentation, and regularization) play a pivotal role in improving accuracy. It also shows that long-held assumptions, such as the size of a convolution window, may need to be revisited, given the increase in computation and data. We will cover this and many more questions in due course throughout this chapter.

8.1 Deep Convolutional Neural Networks (AlexNet)

Although CNNs were well known in the computer vision and machine learning communities following the introduction of LeNet (LeCun *et al.*, 1995), they did not immediately dominate the field. Although LeNet achieved good results on early small datasets, the performance and feasibility of training CNNs on larger, more realistic datasets had yet to be established. In fact, for much of the intervening time between the early 1990s and the watershed results of 2012 (Krizhevsky *et al.*, 2012), neural networks were often surpassed by other machine learning methods, such as kernel methods (Schölkopf and Smola, 2002), ensemble methods (Freund and Schapire, 1996), and structured estimation (Taskar *et al.*, 2004).

For computer vision, this comparison is perhaps not entirely accurate. That is, although the inputs to convolutional networks consist of raw or lightly-processed (e.g., by centering) pixel values, practitioners would never feed raw pixels into traditional models. Instead, typical computer vision pipelines consisted of manually engineering feature extraction pipelines, such as SIFT (Lowe, 2004), SURF (Bay *et al.*, 2006), and bags of visual words (Sivic and Zisserman, 2003). Rather than *learning* the features, the features were *crafted*. Most of the progress came from having more clever ideas for feature extraction on the one hand and deep insight into geometry (Hartley and Zisserman, 2000) on the other. The learning algorithm was often considered an afterthought.

Although some neural network accelerators were available in the 1990s, they were not yet sufficiently powerful to make deep multichannel, multilayer CNNs with a large number of parameters. For instance, NVIDIA's GeForce 256 from 1999 was able to process at most 480 million floating-point operations, such as additions and multiplications, per second (MFLOPS), without any meaningful programming framework for operations beyond games. Today's accelerators are able to perform in excess of 1000 TFLOPs per device. Moreover, datasets were still relatively small: OCR on 60,000 low-resolution 28×28 pixel images was considered a highly challenging task. Added to these obstacles, key tricks for training neural networks including parameter initialization heuristics (Glorot and Bengio, 2010), clever variants of stochastic gradient descent (Kingma and Ba, 2014), non-squashing activation functions (Nair and Hinton, 2010), and effective regularization techniques (Srivastava *et al.*, 2014) were still missing.

Thus, rather than training *end-to-end* (pixel to classification) systems, classical pipelines looked more like this:

1. Obtain an interesting dataset. In the early days, these datasets required expensive sensors. For instance, the Apple QuickTake 100 [126] of 1994 sported a whopping 0.3 megapixel (VGA) resolution, capable of storing up to 8 images, all for the price of $1000.

2. Preprocess the dataset with hand-crafted features based on some knowledge of optics, geometry, other analytic tools, and occasionally on the serendipitous discoveries by lucky graduate students.

3. Feed the data through a standard set of feature extractors such as the SIFT (scale-invariant feature transform) (Lowe, 2004), the SURF (speeded up robust features) (Bay *et al.*, 2006), or any number of other hand-tuned pipelines. OpenCV still provides SIFT extractors to this day!

4. Dump the resulting representations into your favorite classifier, likely a linear model or kernel method, to train a classifier.

If you spoke to machine learning researchers, they would reply that machine learning was both important and beautiful. Elegant theories proved the properties of various classifiers (Boucheron *et al.*, 2005) and convex optimization (Boyd and Vandenberghe, 2004) had become the mainstay for obtaining them. The field of machine learning was thriving, rigorous, and eminently useful. However, if you spoke to a computer vision researcher, you would hear a very different story. The dirty truth of image recognition, they would tell you, is that features, geometry (Hartley and Zisserman, 2000, Hartley and Kahl, 2009), and engineering, rather than novel learning algorithms, drove progress. Computer vision researchers justifiably believed that a slightly bigger or cleaner dataset or a slightly improved feature-extraction pipeline mattered far more to the final accuracy than any learning algorithm.

```
import torch
from torch import nn
from d2l import torch as d2l
```

8.1.1 Representation Learning

Another way to cast the state of affairs is that the most important part of the pipeline was the representation. And up until 2012 the representation was calculated mostly mechanically. In fact, engineering a new set of feature functions, improving results, and writing up the method all featured prominently in papers. SIFT (Lowe, 2004), SURF (Bay *et al.*, 2006), HOG (histograms of oriented gradient) (Dalal and Triggs, 2005), bags of visual words (Sivic and Zisserman, 2003), and similar feature extractors ruled the roost.

Another group of researchers, including Yann LeCun, Geoff Hinton, Yoshua Bengio, Andrew Ng, Shun-ichi Amari, and Juergen Schmidhuber, had different plans. They believed that features themselves ought to be learned. Moreover, they believed that to be reasonably complex, the features ought to be hierarchically composed with multiple jointly learned layers, each with learnable parameters. In the case of an image, the lowest layers might come to detect edges, colors, and textures, by analogy with how the visual system in animals processes its input. In particular, the automatic design of visual features such as those obtained by sparse coding (Olshausen and Field, 1996) remained an open challenge until the advent of modern CNNs. It was not until Dean *et al.* (2012), Le (2013) that the idea of generating features from image data automatically gained significant traction.

The first modern CNN (Krizhevsky *et al.*, 2012), named *AlexNet* after one of its inventors, Alex Krizhevsky, is largely an evolutionary improvement over LeNet. It achieved excellent performance in the 2012 ImageNet challenge.

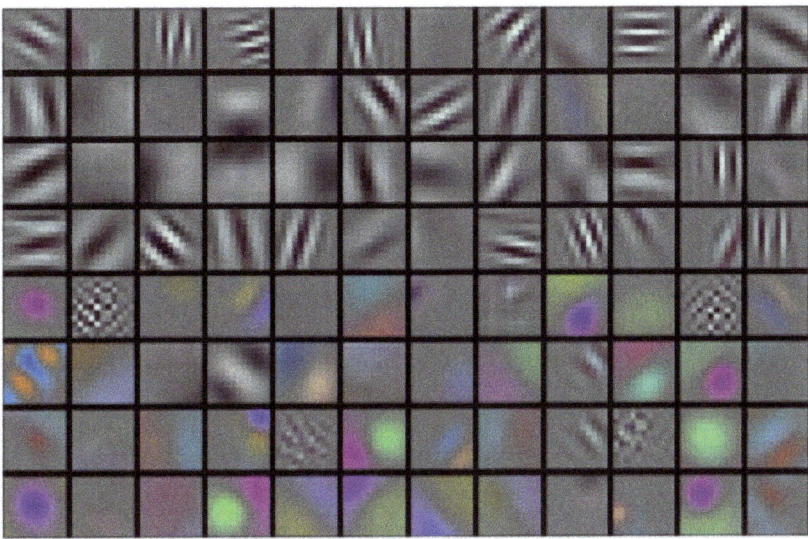

Fig. 8.1.1   Image filters learned by the first layer of AlexNet. Reproduction courtesy of Krizhevsky et al. (2012).

Interestingly, in the lowest layers of the network, the model learned feature extractors that resembled some traditional filters. Fig. 8.1.1 shows lower-level image descriptors. Higher layers in the network might build upon these representations to represent larger structures, like eyes, noses, blades of grass, and so on. Even higher layers might represent whole objects like people, airplanes, dogs, or frisbees. Ultimately, the final hidden state learns a compact representation of the image that summarizes its contents such that data belonging to different categories can be easily separated.

AlexNet (2012) and its precursor LeNet (1995) share many architectural elements. This begs the question: why did it take so long? A key difference was that, over the previous two decades, the amount of data and the computing power available had increased significantly. As such AlexNet was much larger: it was trained on much more data, and on much faster GPUs compared to the CPUs available in 1995.

Missing Ingredient: Data

Deep models with many layers require large amounts of data in order to enter the regime where they significantly outperform traditional methods based on convex optimizations (e.g., linear and kernel methods). However, given the limited storage capacity of computers, the relative expense of (imaging) sensors, and the comparatively tighter research budgets in the 1990s, most research relied on tiny datasets. Numerous papers relied on the UCI collection of datasets, many of which contained only hundreds or (a few) thousands of images captured in low resolution and often with an artificially clean background.

In 2009, the ImageNet dataset was released (Deng *et al.*, 2009), challenging researchers to learn models from 1 million examples, 1000 each from 1000 distinct categories of objects. The categories themselves were based on the most popular noun nodes in WordNet (Miller, 1995). The ImageNet team used Google Image Search to prefilter large candidate sets for each category and employed the Amazon Mechanical Turk crowdsourcing pipeline to confirm for each image whether it belonged to the associated category. This scale was unprecedented, exceeding others by over an order of magnitude (e.g., CIFAR-100 has 60,000 images). Another aspect was that the images were at relatively high resolution of 224×224 pixels, unlike the 80 million-sized TinyImages dataset (Torralba *et al.*, 2008), consisting of 32×32 pixel thumbnails. This allowed for the formation of higher-level features. The associated competition, dubbed the ImageNet Large Scale Visual Recognition Challenge (Russakovsky *et al.*, 2015), pushed computer vision and machine learning research forward, challenging researchers to identify which models performed best at a greater scale than academics had previously considered. The largest vision datasets, such as LAION -5B (Schuhmann *et al.*, 2022) contain billions of images with additional metadata.

Missing Ingredient: Hardware

Deep learning models are voracious consumers of compute cycles. Training can take hundreds of epochs, and each iteration requires passing data through many layers of computationally expensive linear algebra operations. This is one of the main reasons why in the 1990s and early 2000s, simple algorithms based on the more-efficiently optimized convex objectives were preferred.

Graphical processing units (GPUs) proved to be a game changer in making deep learning feasible. These chips had earlier been developed for accelerating graphics processing to benefit computer games. In particular, they were optimized for high throughput 4×4 matrix–vector products, which are needed for many computer graphics tasks. Fortunately, the math is strikingly similar to that required for calculating convolutional layers. Around that time, NVIDIA and ATI had begun optimizing GPUs for general computing operations (Fernando, 2004), going as far as to market them as *general-purpose GPUs* (GPG-PUs).

To provide some intuition, consider the cores of a modern microprocessor (CPU). Each of the cores is fairly powerful running at a high clock frequency and sporting large caches (up to several megabytes of L3). Each core is well-suited to executing a wide range of instructions, with branch predictors, a deep pipeline, specialized execution units, speculative execution, and many other bells and whistles that enable it to run a large variety of programs with sophisticated control flow. This apparent strength, however, is also its Achilles heel: general-purpose cores are very expensive to build. They excel at general-purpose code with lots of control flow. This requires lots of chip area, not just for the actual ALU (arithmetic logical unit) where computation happens, but also for all the aforementioned bells and whistles, plus memory interfaces, caching logic between cores, high-speed interconnects, and so on. CPUs are comparatively bad at any single task when compared with dedicated hardware. Modern laptops have 4–8 cores, and even high-end servers rarely exceed 64 cores per socket, simply because it is not cost-effective.

By comparison, GPUs can consist of thousands of small processing elements (NIVIDA's latest Ampere chips have up to 6912 CUDA cores), often grouped into larger groups (NVIDIA calls them warps). The details differ somewhat between NVIDIA, AMD, ARM and other chip vendors. While each core is relatively weak, running at about 1GHz clock frequency, it is the total number of such cores that makes GPUs orders of magnitude faster than CPUs. For instance, NVIDIA's recent Ampere A100 GPU offers over 300 TFLOPs per chip for specialized 16-bit precision (BFLOAT16) matrix-matrix multiplications, and up to 20 TFLOPs for more general-purpose floating point operations (FP32). At the same time, floating point performance of CPUs rarely exceeds 1 TFLOPs. For instance, Amazon's Graviton 3 reaches 2 TFLOPs peak performance for 16-bit precision operations, a number similar to the GPU performance of Apple's M1 processor.

There are many reasons why GPUs are much faster than CPUs in terms of FLOPs. First, power consumption tends to grow *quadratically* with clock frequency. Hence, for the power budget of a CPU core that runs four times faster (a typical number), you can use 16 GPU cores at $\frac{1}{4}$ the speed, which yields $16 \times \frac{1}{4} = 4$ times the performance. Second, GPU cores are much simpler (in fact, for a long time they were not even *able* to execute general-purpose code), which makes them more energy efficient. For instance, (i) they tend not to support speculative evaluation, (ii) it typically is not possible to program each processing element individually, and (iii) the caches per core tend to be much smaller. Last, many operations in deep learning require high memory bandwidth. Again, GPUs shine here with buses that are at least 10 times as wide as many CPUs.

Back to 2012. A major breakthrough came when Alex Krizhevsky and Ilya Sutskever implemented a deep CNN that could run on GPUs. They realized that the computational bottlenecks in CNNs, convolutions and matrix multiplications, are all operations that could be parallelized in hardware. Using two NVIDIA GTX 580s with 3GB of memory, either of which was capable of 1.5 TFLOPs (still a challenge for most CPUs a decade later), they implemented fast convolutions. The cuda-convnet[127] code was good enough that for several years it was the industry standard and powered the first couple of years of the deep learning boom.

127

8.1.2 AlexNet

AlexNet, which employed an 8-layer CNN, won the ImageNet Large Scale Visual Recognition Challenge 2012 by a large margin (Russakovsky *et al.*, 2013). This network showed, for the first time, that the features obtained by learning can transcend manually-designed features, breaking the previous paradigm in computer vision.

The architectures of AlexNet and LeNet are strikingly similar, as Fig. 8.1.2 illustrates. Note that we provide a slightly streamlined version of AlexNet removing some of the design quirks that were needed in 2012 to make the model fit on two small GPUs.

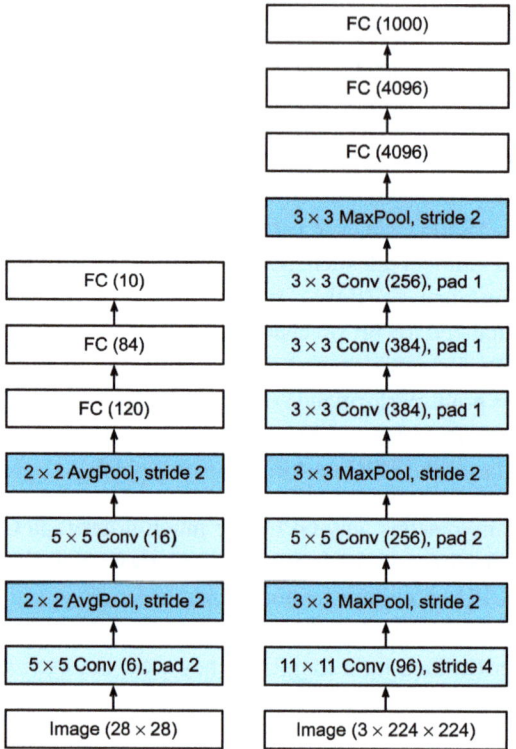

Fig. 8.1.2   From LeNet (left) to AlexNet (right).

There are also significant differences between AlexNet and LeNet. First, AlexNet is much deeper than the comparatively small LeNet-5. AlexNet consists of eight layers: five convolutional layers, two fully connected hidden layers, and one fully connected output layer. Second, AlexNet used the ReLU instead of the sigmoid as its activation function. Let's delve into the details below.

Architecture

In AlexNet's first layer, the convolution window shape is 11×11. Since the images in ImageNet are eight times taller and wider than the MNIST images, objects in ImageNet data tend to occupy more pixels with more visual detail. Consequently, a larger convolution window is needed to capture the object. The convolution window shape in the second layer is reduced to 5×5, followed by 3×3. In addition, after the first, second, and fifth convolutional layers, the network adds max-pooling layers with a window shape of $3 \times$

3 and a stride of 2. Moreover, AlexNet has ten times more convolution channels than LeNet.

After the final convolutional layer, there are two huge fully connected layers with 4096 outputs. These layers require nearly 1GB model parameters. Because of the limited memory in early GPUs, the original AlexNet used a dual data stream design, so that each of their two GPUs could be responsible for storing and computing only its half of the model. Fortunately, GPU memory is comparatively abundant now, so we rarely need to break up models across GPUs these days (our version of the AlexNet model deviates from the original paper in this aspect).

Activation Functions

Furthermore, AlexNet changed the sigmoid activation function to a simpler ReLU activation function. On the one hand, the computation of the ReLU activation function is simpler. For example, it does not have the exponentiation operation found in the sigmoid activation function. On the other hand, the ReLU activation function makes model training easier when using different parameter initialization methods. This is because, when the output of the sigmoid activation function is very close to 0 or 1, the gradient of these regions is almost 0, so that backpropagation cannot continue to update some of the model parameters. By contrast, the gradient of the ReLU activation function in the positive interval is always 1 (Section 5.1.2). Therefore, if the model parameters are not properly initialized, the sigmoid function may obtain a gradient of almost 0 in the positive interval, meaning that the model cannot be effectively trained.

Capacity Control and Preprocessing

AlexNet controls the model complexity of the fully connected layer by dropout (Section 5.6), while LeNet only uses weight decay. To augment the data even further, the training loop of AlexNet added a great deal of image augmentation, such as flipping, clipping, and color changes. This makes the model more robust and the larger sample size effectively reduces overfitting. See Buslaev *et al.* (2020) for an in-depth review of such preprocessing steps.

```
class AlexNet(d2l.Classifier):
    def __init__(self, lr=0.1, num_classes=10):
        super().__init__()
        self.save_hyperparameters()
        self.net = nn.Sequential(
            nn.LazyConv2d(96, kernel_size=11, stride=4, padding=1),
            nn.ReLU(), nn.MaxPool2d(kernel_size=3, stride=2),
            nn.LazyConv2d(256, kernel_size=5, padding=2), nn.ReLU(),
            nn.MaxPool2d(kernel_size=3, stride=2),
            nn.LazyConv2d(384, kernel_size=3, padding=1), nn.ReLU(),
            nn.LazyConv2d(384, kernel_size=3, padding=1), nn.ReLU(),
            nn.LazyConv2d(256, kernel_size=3, padding=1), nn.ReLU(),
            nn.MaxPool2d(kernel_size=3, stride=2), nn.Flatten(),
            nn.LazyLinear(4096), nn.ReLU(), nn.Dropout(p=0.5),
```

(continues on next page)

(continued from previous page)

```
            nn.LazyLinear(4096), nn.ReLU(),nn.Dropout(p=0.5),
            nn.LazyLinear(num_classes))
        self.net.apply(d2l.init_cnn)
```

We construct a single-channel data example with both height and width of 224 to observe the output shape of each layer. It matches the AlexNet architecture in Fig. 8.1.2.

```
AlexNet().layer_summary((1, 1, 224, 224))
```

```
Conv2d output shape:            torch.Size([1, 96, 54, 54])
ReLU output shape:      torch.Size([1, 96, 54, 54])
MaxPool2d output shape:         torch.Size([1, 96, 26, 26])
Conv2d output shape:            torch.Size([1, 256, 26, 26])
ReLU output shape:      torch.Size([1, 256, 26, 26])
MaxPool2d output shape:         torch.Size([1, 256, 12, 12])
Conv2d output shape:            torch.Size([1, 384, 12, 12])
ReLU output shape:      torch.Size([1, 384, 12, 12])
Conv2d output shape:            torch.Size([1, 384, 12, 12])
ReLU output shape:      torch.Size([1, 384, 12, 12])
Conv2d output shape:            torch.Size([1, 256, 12, 12])
ReLU output shape:      torch.Size([1, 256, 12, 12])
MaxPool2d output shape:         torch.Size([1, 256, 5, 5])
Flatten output shape:           torch.Size([1, 6400])
Linear output shape:            torch.Size([1, 4096])
ReLU output shape:      torch.Size([1, 4096])
Dropout output shape:           torch.Size([1, 4096])
Linear output shape:            torch.Size([1, 4096])
ReLU output shape:      torch.Size([1, 4096])
Dropout output shape:           torch.Size([1, 4096])
Linear output shape:            torch.Size([1, 10])
```

8.1.3 Training

Although AlexNet was trained on ImageNet in Krizhevsky *et al.* (2012), we use Fashion-MNIST here since training an ImageNet model to convergence could take hours or days even on a modern GPU. One of the problems with applying AlexNet directly on Fashion-MNIST is that its images have lower resolution (28×28 pixels) than ImageNet images. To make things work, we upsample them to 224×224. This is generally not a smart practice, as it simply increases the computational complexity without adding information. Nonetheless, we do it here to be faithful to the AlexNet architecture. We perform this resizing with the resize argument in the d2l.FashionMNIST constructor.

Now, we can start training AlexNet. Compared to LeNet in Section 7.6, the main change here is the use of a smaller learning rate and much slower training due to the deeper and wider network, the higher image resolution, and the more costly convolutions.

```
model = AlexNet(lr=0.01)
data = d2l.FashionMNIST(batch_size=128, resize=(224, 224))
```

(continues on next page)

(continued from previous page)

```
trainer = d2l.Trainer(max_epochs=10, num_gpus=1)
trainer.fit(model, data)
```

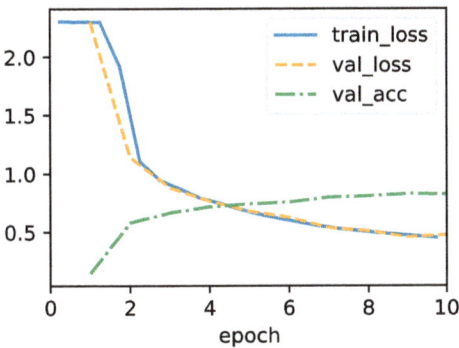

8.1.4 Discussion

AlexNet's structure bears a striking resemblance to LeNet, with a number of critical improvements, both for accuracy (dropout) and for ease of training (ReLU). What is equally striking is the amount of progress that has been made in terms of deep learning tooling. What was several months of work in 2012 can now be accomplished in a dozen lines of code using any modern framework.

Reviewing the architecture, we see that AlexNet has an Achilles heel when it comes to efficiency: the last two hidden layers require matrices of size 6400×4096 and 4096×4096, respectively. This corresponds to 164 MB of memory and 81 MFLOPs of computation, both of which are a nontrivial outlay, especially on smaller devices, such as mobile phones. This is one of the reasons why AlexNet has been surpassed by much more effective architectures that we will cover in the following sections. Nonetheless, it is a key step from shallow to deep networks that are used nowadays. Note that even though the number of parameters exceeds by far the amount of training data in our experiments (the last two layers have more than 40 million parameters, trained on a datasets of 60 thousand images), there is hardly any overfitting: training and validation loss are virtually identical throughout training. This is due to the improved regularization, such as dropout, inherent in modern deep network designs.

Although it seems that there are only a few more lines in AlexNet's implementation than in LeNet's, it took the academic community many years to embrace this conceptual change and take advantage of its excellent experimental results. This was also due to the lack of efficient computational tools. At the time neither DistBelief (Dean *et al.*, 2012) nor Caffe (Jia *et al.*, 2014) existed, and Theano (Bergstra *et al.*, 2010) still lacked many distinguishing features. It was the availability of TensorFlow (Abadi *et al.*, 2016) that dramatically changed the situation.

8.1.5 Exercises

1. Following up on the discussion above, analyze the computational properties of AlexNet.

   1. Compute the memory footprint for convolutions and fully connected layers, respectively. Which one dominates?

   2. Calculate the computational cost for the convolutions and the fully connected layers.

   3. How does the memory (read and write bandwidth, latency, size) affect computation? Is there any difference in its effects for training and inference?

2. You are a chip designer and need to trade off computation and memory bandwidth. For example, a faster chip requires more power and possibly a larger chip area. More memory bandwidth requires more pins and control logic, thus also more area. How do you optimize?

3. Why do engineers no longer report performance benchmarks on AlexNet?

4. Try increasing the number of epochs when training AlexNet. Compared with LeNet, how do the results differ? Why?

5. AlexNet may be too complex for the Fashion-MNIST dataset, in particular due to the low resolution of the initial images.

   1. Try simplifying the model to make the training faster, while ensuring that the accuracy does not drop significantly.

   2. Design a better model that works directly on 28×28 images.

6. Modify the batch size, and observe the changes in throughput (images/s), accuracy, and GPU memory.

7. Apply dropout and ReLU to LeNet-5. Does it improve? Can you improve things further by preprocessing to take advantage of the invariances inherent in the images?

8. Can you make AlexNet overfit? Which feature do you need to remove or change to break training?

Discussions[128].

128

8.2 Networks Using Blocks (VGG)

While AlexNet offered empirical evidence that deep CNNs can achieve good results, it did not provide a general template to guide subsequent researchers in designing new networks. In the following sections, we will introduce several heuristic concepts commonly used to design deep networks.

Progress in this field mirrors that of VLSI (very large scale integration) in chip design where engineers moved from placing transistors to logical elements to logic blocks (Mead, 1980). Similarly, the design of neural network architectures has grown progressively more abstract, with researchers moving from thinking in terms of individual neurons to whole layers, and now to blocks, repeating patterns of layers. A decade later, this has now progressed to researchers using entire trained models to repurpose them for different, albeit related, tasks. Such large pretrained models are typically called *foundation models* (Bommasani *et al.*, 2021).

Back to network design. The idea of using blocks first emerged from the Visual Geometry Group (VGG) at Oxford University, in their eponymously-named *VGG* network (Simonyan and Zisserman, 2014). It is easy to implement these repeated structures in code with any modern deep learning framework by using loops and subroutines.

```
import torch
from torch import nn
from d2l import torch as d2l
```

8.2.1 VGG Blocks

The basic building block of CNNs is a sequence of the following: (i) a convolutional layer with padding to maintain the resolution, (ii) a nonlinearity such as a ReLU, (iii) a pooling layer such as max-pooling to reduce the resolution. One of the problems with this approach is that the spatial resolution decreases quite rapidly. In particular, this imposes a hard limit of $\log_2 d$ convolutional layers on the network before all dimensions (d) are used up. For instance, in the case of ImageNet, it would be impossible to have more than 8 convolutional layers in this way.

The key idea of Simonyan and Zisserman (2014) was to use *multiple* convolutions in between downsampling via max-pooling in the form of a block. They were primarily interested in whether deep or wide networks perform better. For instance, the successive application of two 3×3 convolutions touches the same pixels as a single 5×5 convolution does. At the same time, the latter uses approximately as many parameters ($25 \cdot c^2$) as three 3×3 convolutions do ($3 \cdot 9 \cdot c^2$). In a rather detailed analysis they showed that deep and narrow networks significantly outperform their shallow counterparts. This set deep learning on a quest for ever deeper networks with over 100 layers for typical applications. Stacking 3×3 convolutions has become a gold standard in later deep networks (a design decision only to be revisited recently by Liu *et al.* (2022)). Consequently, fast implementations for small convolutions have become a staple on GPUs (Lavin and Gray, 2016).

Back to VGG: a VGG block consists of a *sequence* of convolutions with 3×3 kernels with padding of 1 (keeping height and width) followed by a 2×2 max-pooling layer with stride of 2 (halving height and width after each block). In the code below, we define a function called vgg_block to implement one VGG block.

The function below takes two arguments, corresponding to the number of convolutional layers num_convs and the number of output channels num_channels.

```
def vgg_block(num_convs, out_channels):
    layers = []
```

(continues on next page)

(continued from previous page)

```
for _ in range(num_convs):
    layers.append(nn.LazyConv2d(out_channels, kernel_size=3, padding=1))
    layers.append(nn.ReLU())
layers.append(nn.MaxPool2d(kernel_size=2,stride=2))
return nn.Sequential(*layers)
```

8.2.2 VGG Network

Like AlexNet and LeNet, the VGG Network can be partitioned into two parts: the first consisting mostly of convolutional and pooling layers and the second consisting of fully connected layers that are identical to those in AlexNet. The key difference is that the convolutional layers are grouped in nonlinear transformations that leave the dimensonality unchanged, followed by a resolution-reduction step, as depicted in Fig. 8.2.1.

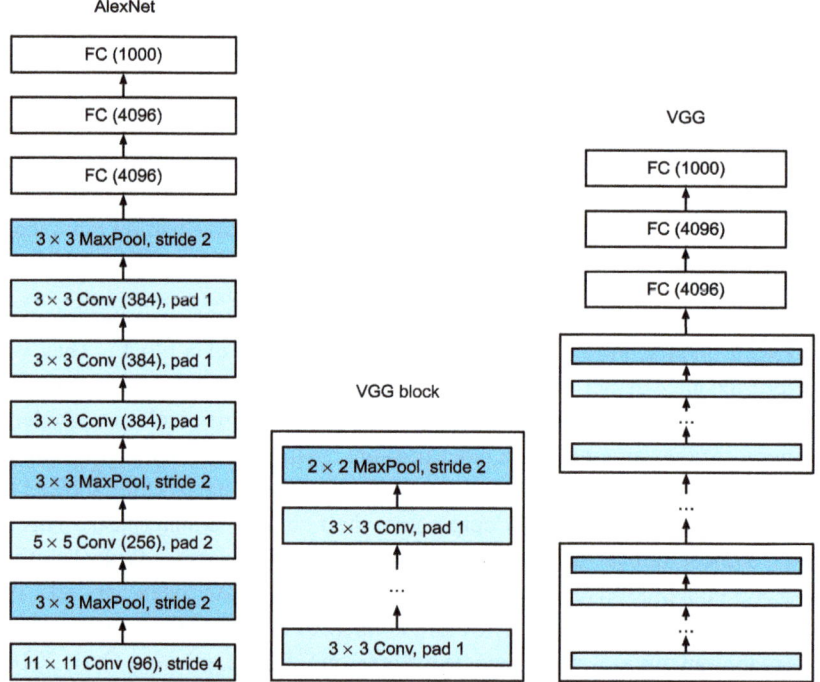

Fig. 8.2.1 From AlexNet to VGG. The key difference is that VGG consists of blocks of layers, whereas AlexNet's layers are all designed individually.

The convolutional part of the network connects several VGG blocks from Fig. 8.2.1 (also defined in the vgg_block function) in succession. This grouping of convolutions is a pattern that has remained almost unchanged over the past decade, although the specific choice of operations has undergone considerable modifications. The variable arch consists of a list of tuples (one per block), where each contains two values: the number of convolutional layers and the number of output channels, which are precisely the arguments required to call the vgg_block function. As such, VGG defines a *family* of networks rather than just a specific manifestation. To build a specific network we simply iterate over arch to compose the blocks.

```
class VGG(d2l.Classifier):
    def __init__(self, arch, lr=0.1, num_classes=10):
        super().__init__()
        self.save_hyperparameters()
        conv_blks = []
        for (num_convs, out_channels) in arch:
            conv_blks.append(vgg_block(num_convs, out_channels))
        self.net = nn.Sequential(
            *conv_blks, nn.Flatten(),
            nn.LazyLinear(4096), nn.ReLU(), nn.Dropout(0.5),
            nn.LazyLinear(4096), nn.ReLU(), nn.Dropout(0.5),
            nn.LazyLinear(num_classes))
        self.net.apply(d2l.init_cnn)
```

The original VGG network had five convolutional blocks, among which the first two have one convolutional layer each and the latter three contain two convolutional layers each. The first block has 64 output channels and each subsequent block doubles the number of output channels, until that number reaches 512. Since this network uses eight convolutional layers and three fully connected layers, it is often called VGG-11.

```
VGG(arch=((1, 64), (1, 128), (2, 256), (2, 512), (2, 512))).layer_summary(
    (1, 1, 224, 224))
```

```
Sequential output shape:       torch.Size([1, 64, 112, 112])
Sequential output shape:       torch.Size([1, 128, 56, 56])
Sequential output shape:       torch.Size([1, 256, 28, 28])
Sequential output shape:       torch.Size([1, 512, 14, 14])
Sequential output shape:       torch.Size([1, 512, 7, 7])
Flatten output shape:          torch.Size([1, 25088])
Linear output shape:           torch.Size([1, 4096])
ReLU output shape:   torch.Size([1, 4096])
Dropout output shape:          torch.Size([1, 4096])
Linear output shape:           torch.Size([1, 4096])
ReLU output shape:   torch.Size([1, 4096])
Dropout output shape:          torch.Size([1, 4096])
Linear output shape:           torch.Size([1, 10])
```

As you can see, we halve height and width at each block, finally reaching a height and width of 7 before flattening the representations for processing by the fully connected part of the network. Simonyan and Zisserman (2014) described several other variants of VGG. In fact, it has become the norm to propose *families* of networks with different speed–accuracy trade-off when introducing a new architecture.

8.2.3 Training

Since VGG-11 is computationally more demanding than AlexNet we construct a network with a smaller number of channels. This is more than sufficient for training on Fashion-

MNIST. The model training process is similar to that of AlexNet in Section 8.1. Again observe the close match between validation and training loss, suggesting only a small amount of overfitting.

```
model = VGG(arch=((1, 16), (1, 32), (2, 64), (2, 128), (2, 128)), lr=0.01)
trainer = d2l.Trainer(max_epochs=10, num_gpus=1)
data = d2l.FashionMNIST(batch_size=128, resize=(224, 224))
model.apply_init([next(iter(data.get_dataloader(True)))[0]], d2l.init_cnn)
trainer.fit(model, data)
```

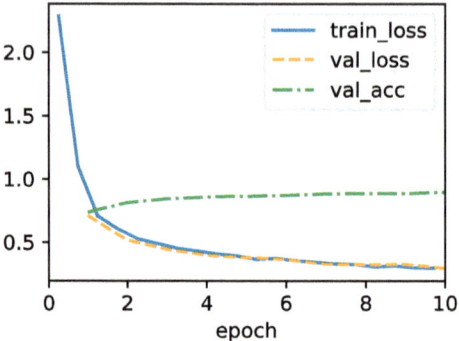

8.2.4 Summary

One might argue that VGG is the first truly modern convolutional neural network. While AlexNet introduced many of the components of what make deep learning effective at scale, it is VGG that arguably introduced key properties such as blocks of multiple convolutions and a preference for deep and narrow networks. It is also the first network that is actually an entire family of similarly parametrized models, giving the practitioner ample trade-off between complexity and speed. This is also the place where modern deep learning frameworks shine. It is no longer necessary to generate XML configuration files to specify a network but rather, to assemble said networks through simple Python code.

More recently ParNet (Goyal *et al.*, 2021) demonstrated that it is possible to achieve competitive performance using a much more shallow architecture through a large number of parallel computations. This is an exciting development and there is hope that it will influence architecture designs in the future. For the remainder of the chapter, though, we will follow the path of scientific progress over the past decade.

8.2.5 Exercises

1. Compared with AlexNet, VGG is much slower in terms of computation, and it also needs more GPU memory.

    1. Compare the number of parameters needed for AlexNet and VGG.

    2. Compare the number of floating point operations used in the convolutional layers and in the fully connected layers.

3. How could you reduce the computational cost created by the fully connected layers?

2. When displaying the dimensions associated with the various layers of the network, we only see the information associated with eight blocks (plus some auxiliary transforms), even though the network has 11 layers. Where did the remaining three layers go?

3. Use Table 1 in the VGG paper (Simonyan and Zisserman, 2014) to construct other common models, such as VGG-16 or VGG-19.

4. Upsampling the resolution in Fashion-MNIST eight-fold from 28×28 to 224×224 dimensions is very wasteful. Try modifying the network architecture and resolution conversion, e.g., to 56 or to 84 dimensions for its input instead. Can you do so without reducing the accuracy of the network? Consult the VGG paper (Simonyan and Zisserman, 2014) for ideas on adding more nonlinearities prior to downsampling.

 Discussions[129].

8.3 Network in Network (NiN)

LeNet, AlexNet, and VGG all share a common design pattern: extract features exploiting *spatial* structure via a sequence of convolutions and pooling layers and post-process the representations via fully connected layers. The improvements upon LeNet by AlexNet and VGG mainly lie in how these later networks widen and deepen these two modules.

This design poses two major challenges. First, the fully connected layers at the end of the architecture consume tremendous numbers of parameters. For instance, even a simple model such as VGG-11 requires a monstrous matrix, occupying almost 400MB of RAM in single precision (FP32). This is a significant impediment to computation, in particular on mobile and embedded devices. After all, even high-end mobile phones sport no more than 8GB of RAM. At the time VGG was invented, this was an order of magnitude less (the iPhone 4S had 512MB). As such, it would have been difficult to justify spending the majority of memory on an image classifier.

Second, it is equally impossible to add fully connected layers earlier in the network to increase the degree of nonlinearity: doing so would destroy the spatial structure and require potentially even more memory.

The *network in network* (*NiN*) blocks (Lin *et al.*, 2013) offer an alternative, capable of solving both problems in one simple strategy. They were proposed based on a very simple insight: (i) use 1×1 convolutions to add local nonlinearities across the channel activations and (ii) use global average pooling to integrate across all locations in the last representation layer. Note that global average pooling would not be effective, were it not for the added nonlinearities. Let's dive into this in detail.

```
import torch
from torch import nn
from d2l import torch as d2l
```

8.3.1 NiN Blocks

Recall Section 7.4.3. In it we said that the inputs and outputs of convolutional layers consist of four-dimensional tensors with axes corresponding to the example, channel, height, and width. Also recall that the inputs and outputs of fully connected layers are typically two-dimensional tensors corresponding to the example and feature. The idea behind NiN is to apply a fully connected layer at each pixel location (for each height and width). The resulting 1×1 convolution can be thought of as a fully connected layer acting independently on each pixel location.

Fig. 8.3.1 illustrates the main structural differences between VGG and NiN, and their blocks. Note both the difference in the NiN blocks (the initial convolution is followed by 1 × 1 convolutions, whereas VGG retains 3 × 3 convolutions) and at the end where we no longer require a giant fully connected layer.

```
def nin_block(out_channels, kernel_size, strides, padding):
    return nn.Sequential(
        nn.LazyConv2d(out_channels, kernel_size, strides, padding), nn.ReLU(),
        nn.LazyConv2d(out_channels, kernel_size=1), nn.ReLU(),
        nn.LazyConv2d(out_channels, kernel_size=1), nn.ReLU())
```

8.3.2 NiN Model

NiN uses the same initial convolution sizes as AlexNet (it was proposed shortly thereafter). The kernel sizes are 11 × 11, 5 × 5, and 3 × 3, respectively, and the numbers of output channels match those of AlexNet. Each NiN block is followed by a max-pooling layer with a stride of 2 and a window shape of 3 × 3.

The second significant difference between NiN and both AlexNet and VGG is that NiN avoids fully connected layers altogether. Instead, NiN uses a NiN block with a number of output channels equal to the number of label classes, followed by a *global* average pooling layer, yielding a vector of logits. This design significantly reduces the number of required model parameters, albeit at the expense of a potential increase in training time.

```
class NiN(d2l.Classifier):
    def __init__(self, lr=0.1, num_classes=10):
        super().__init__()
        self.save_hyperparameters()
        self.net = nn.Sequential(
            nin_block(96, kernel_size=11, strides=4, padding=0),
            nn.MaxPool2d(3, stride=2),
            nin_block(256, kernel_size=5, strides=1, padding=2),
            nn.MaxPool2d(3, stride=2),
```

(continues on next page)

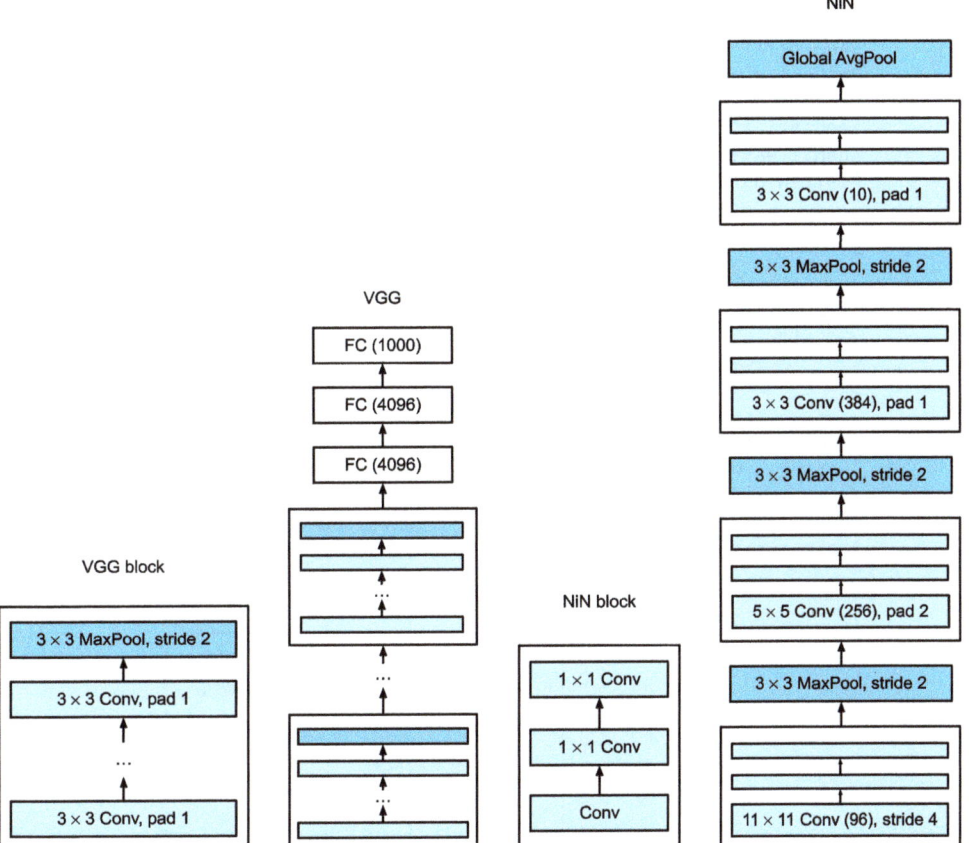

Fig. 8.3.1 Comparing the architectures of VGG and NiN, and of their blocks.

(continued from previous page)

```
        nin_block(384, kernel_size=3, strides=1, padding=1),
        nn.MaxPool2d(3, stride=2),
        nn.Dropout(0.5),
        nin_block(num_classes, kernel_size=3, strides=1, padding=1),
        nn.AdaptiveAvgPool2d((1, 1)),
        nn.Flatten())
    self.net.apply(d2l.init_cnn)
```

We create a data example to see the output shape of each block.

```
NiN().layer_summary((1, 1, 224, 224))
```

```
Sequential output shape:      torch.Size([1, 96, 54, 54])
MaxPool2d output shape:       torch.Size([1, 96, 26, 26])
Sequential output shape:      torch.Size([1, 256, 26, 26])
MaxPool2d output shape:       torch.Size([1, 256, 12, 12])
```

(continues on next page)

(continued from previous page)

```
Sequential output shape:        torch.Size([1, 384, 12, 12])
MaxPool2d output shape:         torch.Size([1, 384, 5, 5])
Dropout output shape:           torch.Size([1, 384, 5, 5])
Sequential output shape:        torch.Size([1, 10, 5, 5])
AdaptiveAvgPool2d output shape:      torch.Size([1, 10, 1, 1])
Flatten output shape:           torch.Size([1, 10])
```

8.3.3 Training

As before we use Fashion-MNIST to train the model using the same optimizer that we used for AlexNet and VGG.

```
model = NiN(lr=0.05)
trainer = d2l.Trainer(max_epochs=10, num_gpus=1)
data = d2l.FashionMNIST(batch_size=128, resize=(224, 224))
model.apply_init([next(iter(data.get_dataloader(True)))[0]], d2l.init_cnn)
trainer.fit(model, data)
```

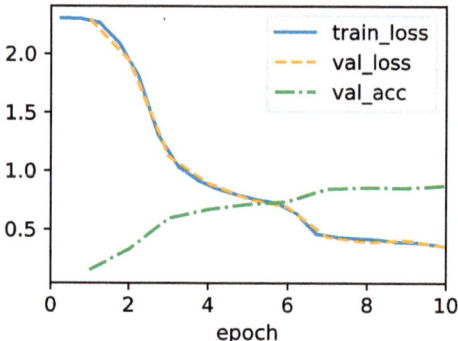

8.3.4 Summary

NiN has dramatically fewer parameters than AlexNet and VGG. This stems primarily from the fact that it needs no giant fully connected layers. Instead, it uses global average pooling to aggregate across all image locations after the last stage of the network body. This obviates the need for expensive (learned) reduction operations and replaces them by a simple average. What surprised researchers at the time was the fact that this averaging operation did not harm accuracy. Note that averaging across a low-resolution representation (with many channels) also adds to the amount of translation invariance that the network can handle.

Choosing fewer convolutions with wide kernels and replacing them by 1×1 convolutions aids the quest for fewer parameters further. It can cater for a significant amount of non-linearity across channels within any given location. Both 1×1 convolutions and global average pooling significantly influenced subsequent CNN designs.

8.3.5 Exercises

1. Why are there two 1×1 convolutional layers per NiN block? Increase their number to three. Reduce their number to one. What changes?

2. What changes if you replace the 1×1 convolutions by 3×3 convolutions?

3. What happens if you replace the global average pooling by a fully connected layer (speed, accuracy, number of parameters)?

4. Calculate the resource usage for NiN.

    1. What is the number of parameters?

    2. What is the amount of computation?

    3. What is the amount of memory needed during training?

    4. What is the amount of memory needed during prediction?

5. What are possible problems with reducing the $384 \times 5 \times 5$ representation to a $10 \times 5 \times 5$ representation in one step?

6. Use the structural design decisions in VGG that led to VGG-11, VGG-16, and VGG-19 to design a family of NiN-like networks.

 Discussions[130].

8.4 Multi-Branch Networks (GoogLeNet)

In 2014, *GoogLeNet* won the ImageNet Challenge (Szegedy *et al.*, 2015), using a structure that combined the strengths of NiN (Lin *et al.*, 2013), repeated blocks (Simonyan and Zisserman, 2014), and a cocktail of convolution kernels. It was arguably also the first network that exhibited a clear distinction among the stem (data ingest), body (data processing), and head (prediction) in a CNN. This design pattern has persisted ever since in the design of deep networks: the *stem* is given by the first two or three convolutions that operate on the image. They extract low-level features from the underlying images. This is followed by a *body* of convolutional blocks. Finally, the *head* maps the features obtained so far to the required classification, segmentation, detection, or tracking problem at hand.

The key contribution in GoogLeNet was the design of the network body. It solved the problem of selecting convolution kernels in an ingenious way. While other works tried to identify which convolution, ranging from 1×1 to 11×11 would be best, it simply *concatenated* multi-branch convolutions. In what follows we introduce a slightly simplified version of GoogLeNet: the original design included a number of tricks for stabilizing training through intermediate loss functions, applied to multiple layers of the network. They are no longer necessary due to the availability of improved training algorithms.

```
import torch
from torch import nn
from torch.nn import functional as F
from d2l import torch as d2l
```

8.4.1 Inception Blocks

The basic convolutional block in GoogLeNet is called an *Inception block*, stemming from
the meme "we need to go deeper" from the movie *Inception*.

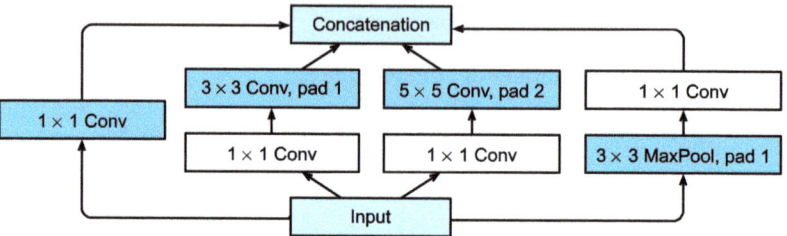

Fig. 8.4.1   Structure of the Inception block.

As depicted in Fig. 8.4.1, the inception block consists of four parallel branches. The first
three branches use convolutional layers with window sizes of 1×1, 3×3, and 5×5 to
extract information from different spatial sizes. The middle two branches also add a 1×1
convolution of the input to reduce the number of channels, reducing the model's complexity. The fourth branch uses a 3×3 max-pooling layer, followed by a 1×1 convolutional
layer to change the number of channels. The four branches all use appropriate padding
to give the input and output the same height and width. Finally, the outputs along each
branch are concatenated along the channel dimension and comprise the block's output. The
commonly-tuned hyperparameters of the Inception block are the number of output channels
per layer, i.e., how to allocate capacity among convolutions of different size.

```
class Inception(nn.Module):
    # c1--c4 are the number of output channels for each branch
    def __init__(self, c1, c2, c3, c4, **kwargs):
        super(Inception, self).__init__(**kwargs)
        # Branch 1
        self.b1_1 = nn.LazyConv2d(c1, kernel_size=1)
        # Branch 2
        self.b2_1 = nn.LazyConv2d(c2[0], kernel_size=1)
        self.b2_2 = nn.LazyConv2d(c2[1], kernel_size=3, padding=1)
        # Branch 3
        self.b3_1 = nn.LazyConv2d(c3[0], kernel_size=1)
        self.b3_2 = nn.LazyConv2d(c3[1], kernel_size=5, padding=2)
        # Branch 4
        self.b4_1 = nn.MaxPool2d(kernel_size=3, stride=1, padding=1)
        self.b4_2 = nn.LazyConv2d(c4, kernel_size=1)

    def forward(self, x):
        b1 = F.relu(self.b1_1(x))
```

(continues on next page)

(continued from previous page)

```
        b2 = F.relu(self.b2_2(F.relu(self.b2_1(x))))
        b3 = F.relu(self.b3_2(F.relu(self.b3_1(x))))
        b4 = F.relu(self.b4_2(self.b4_1(x)))
        return torch.cat((b1, b2, b3, b4), dim=1)
```

To gain some intuition for why this network works so well, consider the combination of the filters. They explore the image in a variety of filter sizes. This means that details at different extents can be recognized efficiently by filters of different sizes. At the same time, we can allocate different amounts of parameters for different filters.

8.4.2 GoogLeNet Model

As shown in Fig. 8.4.2, GoogLeNet uses a stack of a total of 9 inception blocks, arranged into three groups with max-pooling in between, and global average pooling in its head to generate its estimates. Max-pooling between inception blocks reduces the dimensionality. At its stem, the first module is similar to AlexNet and LeNet.

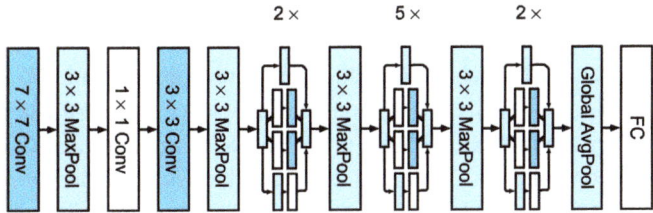

Fig. 8.4.2   The GoogLeNet architecture.

We can now implement GoogLeNet piece by piece. Let's begin with the stem. The first module uses a 64-channel 7×7 convolutional layer.

```
class GoogleNet(d2l.Classifier):
    def b1(self):
        return nn.Sequential(
            nn.LazyConv2d(64, kernel_size=7, stride=2, padding=3),
            nn.ReLU(), nn.MaxPool2d(kernel_size=3, stride=2, padding=1))
```

The second module uses two convolutional layers: first, a 64-channel 1×1 convolutional layer, followed by a 3×3 convolutional layer that triples the number of channels. This corresponds to the second branch in the Inception block and concludes the design of the body. At this point we have 192 channels.

```
@d2l.add_to_class(GoogleNet)
def b2(self):
    return nn.Sequential(
        nn.LazyConv2d(64, kernel_size=1), nn.ReLU(),
        nn.LazyConv2d(192, kernel_size=3, padding=1), nn.ReLU(),
        nn.MaxPool2d(kernel_size=3, stride=2, padding=1))
```

The third module connects two complete Inception blocks in series. The number of output channels of the first Inception block is $64 + 128 + 32 + 32 = 256$. This amounts to a ratio of the number of output channels among the four branches of $2 : 4 : 1 : 1$. To achieve this, we first reduce the input dimensions by $\frac{1}{2}$ and by $\frac{1}{12}$ in the second and third branch respectively to arrive at $96 = 192/2$ and $16 = 192/12$ channels respectively.

The number of output channels of the second Inception block is increased to $128 + 192 + 96 + 64 = 480$, yielding a ratio of $128 : 192 : 96 : 64 = 4 : 6 : 3 : 2$. As before, we need to reduce the number of intermediate dimensions in the second and third channel. A scale of $\frac{1}{2}$ and $\frac{1}{8}$ respectively suffices, yielding 128 and 32 channels respectively. This is captured by the arguments of the following Inception block constructors.

```
@d2l.add_to_class(GoogleNet)
def b3(self):
    return nn.Sequential(Inception(64, (96, 128), (16, 32), 32),
                         Inception(128, (128, 192), (32, 96), 64),
                         nn.MaxPool2d(kernel_size=3, stride=2, padding=1))
```

The fourth module is more complicated. It connects five Inception blocks in series, and they have $192 + 208 + 48 + 64 = 512$, $160 + 224 + 64 + 64 = 512$, $128 + 256 + 64 + 64 = 512$, $112 + 288 + 64 + 64 = 528$, and $256 + 320 + 128 + 128 = 832$ output channels, respectively. The number of channels assigned to these branches is similar to that in the third module: the second branch with the 3×3 convolutional layer outputs the largest number of channels, followed by the first branch with only the 1×1 convolutional layer, the third branch with the 5×5 convolutional layer, and the fourth branch with the 3×3 max-pooling layer. The second and third branches will first reduce the number of channels according to the ratio. These ratios are slightly different in different Inception blocks.

```
@d2l.add_to_class(GoogleNet)
def b4(self):
    return nn.Sequential(Inception(192, (96, 208), (16, 48), 64),
                         Inception(160, (112, 224), (24, 64), 64),
                         Inception(128, (128, 256), (24, 64), 64),
                         Inception(112, (144, 288), (32, 64), 64),
                         Inception(256, (160, 320), (32, 128), 128),
                         nn.MaxPool2d(kernel_size=3, stride=2, padding=1))
```

The fifth module has two Inception blocks with $256 + 320 + 128 + 128 = 832$ and $384 + 384 + 128 + 128 = 1024$ output channels. The number of channels assigned to each branch is the same as that in the third and fourth modules, but differs in specific values. It should be noted that the fifth block is followed by the output layer. This block uses the global average pooling layer to change the height and width of each channel to 1, just as in NiN. Finally, we turn the output into a two-dimensional array followed by a fully connected layer whose number of outputs is the number of label classes.

```
@d2l.add_to_class(GoogleNet)
def b5(self):
    return nn.Sequential(Inception(256, (160, 320), (32, 128), 128),
```

(continues on next page)

(continued from previous page)

```
           Inception(384, (192, 384), (48, 128), 128),
           nn.AdaptiveAvgPool2d((1,1)), nn.Flatten())
```

Now that we defined all blocks b1 through b5, it is just a matter of assembling them all into a full network.

```
@d2l.add_to_class(GoogleNet)
def __init__(self, lr=0.1, num_classes=10):
    super(GoogleNet, self).__init__()
    self.save_hyperparameters()
    self.net = nn.Sequential(self.b1(), self.b2(), self.b3(), self.b4(),
                             self.b5(), nn.LazyLinear(num_classes))
    self.net.apply(d2l.init_cnn)
```

The GoogLeNet model is computationally complex. Note the large number of relatively arbitrary hyperparameters in terms of the number of channels chosen, the number of blocks prior to dimensionality reduction, the relative partitioning of capacity across channels, etc. Much of it is due to the fact that at the time when GoogLeNet was introduced, automatic tools for network definition or design exploration were not yet available. For instance, by now we take it for granted that a competent deep learning framework is capable of inferring dimensionalities of input tensors automatically. At the time, many such configurations had to be specified explicitly by the experimenter, thus often slowing down active experimentation. Moreover, the tools needed for automatic exploration were still in flux and initial experiments largely amounted to costly brute-force exploration, genetic algorithms, and similar strategies.

For now the only modification we will carry out is to reduce the input height and width from 224 to 96 to have a reasonable training time on Fashion-MNIST. This simplifies the computation. Let's have a look at the changes in the shape of the output between the various modules.

```
model = GoogleNet().layer_summary((1, 1, 96, 96))
```

```
Sequential output shape:     torch.Size([1, 64, 24, 24])
Sequential output shape:     torch.Size([1, 192, 12, 12])
Sequential output shape:     torch.Size([1, 480, 6, 6])
Sequential output shape:     torch.Size([1, 832, 3, 3])
Sequential output shape:     torch.Size([1, 1024])
Linear output shape:         torch.Size([1, 10])
```

8.4.3 Training

As before, we train our model using the Fashion-MNIST dataset. We transform it to 96×96 pixel resolution before invoking the training procedure.

```
model = GoogleNet(lr=0.01)
trainer = d2l.Trainer(max_epochs=10, num_gpus=1)
data = d2l.FashionMNIST(batch_size=128, resize=(96, 96))
model.apply_init([next(iter(data.get_dataloader(True)))[0]], d2l.init_cnn)
trainer.fit(model, data)
```

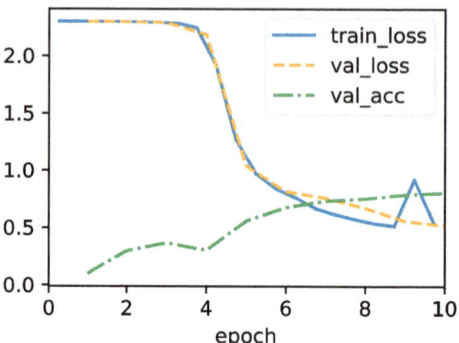

8.4.4 Discussion

A key feature of GoogLeNet is that it is actually *cheaper* to compute than its predecessors while simultaneously providing improved accuracy. This marks the beginning of a much more deliberate network design that trades off the cost of evaluating a network with a reduction in errors. It also marks the beginning of experimentation at a block level with network design hyperparameters, even though it was entirely manual at the time. We will revisit this topic in Section 8.8 when discussing strategies for network structure exploration.

Over the following sections we will encounter a number of design choices (e.g., batch normalization, residual connections, and channel grouping) that allow us to improve networks significantly. For now, you can be proud to have implemented what is arguably the first truly modern CNN.

8.4.5 Exercises

1. GoogLeNet was so successful that it went through a number of iterations, progressively improving speed and accuracy. Try to implement and run some of them. They include the following:

   1. Add a batch normalization layer (Ioffe and Szegedy, 2015), as described later in Section 8.5.

   2. Make adjustments to the Inception block (width, choice and order of convolutions), as described in Szegedy *et al.* (2016).

   3. Use label smoothing for model regularization, as described in Szegedy *et al.* (2016).

   4. Make further adjustments to the Inception block by adding residual connection (Szegedy *et al.*, 2017), as described later in Section 8.6.

2. What is the minimum image size needed for GoogLeNet to work?

3. Can you design a variant of GoogLeNet that works on Fashion-MNIST's native resolution of 28×28 pixels? How would you need to change the stem, the body, and the head of the network, if anything at all?

4. Compare the model parameter sizes of AlexNet, VGG, NiN, and GoogLeNet. How do the latter two network architectures significantly reduce the model parameter size?

5. Compare the amount of computation needed in GoogLeNet and AlexNet. How does this affect the design of an accelerator chip, e.g., in terms of memory size, memory bandwidth, cache size, the amount of computation, and the benefit of specialized operations?

Discussions [131].

131

8.5 Batch Normalization

Training deep neural networks is difficult. Getting them to converge in a reasonable amount of time can be tricky. In this section, we describe *batch normalization*, a popular and effective technique that consistently accelerates the convergence of deep networks (Ioffe and Szegedy, 2015). Together with residual blocks—covered later in Section 8.6—batch normalization has made it possible for practitioners to routinely train networks with over 100 layers. A secondary (serendipitous) benefit of batch normalization lies in its inherent regularization.

```
import torch
from torch import nn
from d2l import torch as d2l
```

8.5.1 Training Deep Networks

When working with data, we often preprocess before training. Choices regarding data preprocessing often make an enormous difference in the final results. Recall our application of MLPs to predicting house prices (Section 5.7). Our first step when working with real data was to standardize our input features to have zero mean $\mu = 0$ and unit variance $\Sigma = 1$ across multiple observations (Friedman, 1987), frequently rescaling the latter so that the diagonal is unity, i.e., $\Sigma_{ii} = 1$. Yet another strategy is to rescale vectors to unit length, possibly zero mean *per observation*. This can work well, e.g., for spatial sensor data. These preprocessing techniques and many others, are beneficial for keeping the estimation problem well controlled. For a review of feature selection and extraction see the article of Guyon *et al.* (2008), for example. Standardizing vectors also has the nice side-effect of constraining the function complexity of functions that act upon it. For instance, the celebrated radius-margin bound (Vapnik, 1995) in support vector machines and the Perceptron Convergence Theorem (Novikoff, 1962) rely on inputs of bounded norm.

Intuitively, this standardization plays nicely with our optimizers since it puts the parameters *a priori* on a similar scale. As such, it is only natural to ask whether a corresponding normalization step *inside* a deep network might not be beneficial. While this is not quite the reasoning that led to the invention of batch normalization (Ioffe and Szegedy, 2015), it is a useful way of understanding it and its cousin, layer normalization (Ba *et al.*, 2016), within a unified framework.

Second, for a typical MLP or CNN, as we train, the variables in intermediate layers (e.g., affine transformation outputs in MLP) may take values with widely varying magnitudes: whether along the layers from input to output, across units in the same layer, and over time due to our updates to the model parameters. The inventors of batch normalization postulated informally that this drift in the distribution of such variables could hamper the convergence of the network. Intuitively, we might conjecture that if one layer has variable activations that are 100 times that of another layer, this might necessitate compensatory adjustments in the learning rates. Adaptive solvers such as AdaGrad (Duchi *et al.*, 2011), Adam (Kingma and Ba, 2014), Yogi (Zaheer *et al.*, 2018), or Distributed Shampoo (Anil *et al.*, 2020) aim to address this from the viewpoint of optimization, e.g., by adding aspects of second-order methods. The alternative is to prevent the problem from occurring, simply by adaptive normalization.

Third, deeper networks are complex and tend to be more liable to overfitting. This means that regularization becomes more critical. A common technique for regularization is noise injection. This has been known for a long time, e.g., with regard to noise injection for the inputs (Bishop, 1995). It also forms the basis of dropout in Section 5.6. As it turns out, quite serendipitously, batch normalization conveys all three benefits: preprocessing, numerical stability, and regularization.

Batch normalization is applied to individual layers, or optionally, to all of them: In each training iteration, we first normalize the inputs (of batch normalization) by subtracting their mean and dividing by their standard deviation, where both are estimated based on the statistics of the current minibatch. Next, we apply a scale coefficient and an offset to recover the lost degrees of freedom. It is precisely due to this *normalization* based on *batch* statistics that *batch normalization* derives its name.

Note that if we tried to apply batch normalization with minibatches of size 1, we would not be able to learn anything. That is because after subtracting the means, each hidden unit would take value 0. As you might guess, since we are devoting a whole section to batch normalization, with large enough minibatches the approach proves effective and stable. One takeaway here is that when applying batch normalization, the choice of batch size is even more significant than without batch normalization, or at least, suitable calibration is needed as we might adjust batch size.

Denote by $\mathcal{B}$ a minibatch and let $\mathbf{x} \in \mathcal{B}$ be an input to batch normalization (BN). In this case the batch normalization is defined as follows:

$$\mathrm{BN}(\mathbf{x}) = \boldsymbol{\gamma} \odot \frac{\mathbf{x} - \hat{\boldsymbol{\mu}}_{\mathcal{B}}}{\hat{\boldsymbol{\sigma}}_{\mathcal{B}}} + \boldsymbol{\beta}. \tag{8.5.1}$$

In (8.5.1), $\hat{\boldsymbol{\mu}}_{\mathcal{B}}$ is the sample mean and $\hat{\boldsymbol{\sigma}}_{\mathcal{B}}$ is the sample standard deviation of the minibatch

$\mathcal{B}$. After applying standardization, the resulting minibatch has zero mean and unit variance. The choice of unit variance (rather than some other magic number) is arbitrary. We recover this degree of freedom by including an elementwise *scale parameter* γ and *shift parameter* β that have the same shape as $\mathbf{x}$. Both are parameters that need to be learned as part of model training.

The variable magnitudes for intermediate layers cannot diverge during training since batch normalization actively centers and rescales them back to a given mean and size (via $\hat{\mu}_{\mathcal{B}}$ and $\hat{\sigma}_{\mathcal{B}}$). Practical experience confirms that, as alluded to when discussing feature rescaling, batch normalization seems to allow for more aggressive learning rates. We calculate $\hat{\mu}_{\mathcal{B}}$ and $\hat{\sigma}_{\mathcal{B}}$ in (8.5.1) as follows:

$$\hat{\mu}_{\mathcal{B}} = \frac{1}{|\mathcal{B}|} \sum_{\mathbf{x} \in \mathcal{B}} \mathbf{x} \text{ and } \hat{\sigma}_{\mathcal{B}}^2 = \frac{1}{|\mathcal{B}|} \sum_{\mathbf{x} \in \mathcal{B}} (\mathbf{x} - \hat{\mu}_{\mathcal{B}})^2 + \epsilon. \tag{8.5.2}$$

Note that we add a small constant $\epsilon > 0$ to the variance estimate to ensure that we never attempt division by zero, even in cases where the empirical variance estimate might be very small or vanish. The estimates $\hat{\mu}_{\mathcal{B}}$ and $\hat{\sigma}_{\mathcal{B}}$ counteract the scaling issue by using noisy estimates of mean and variance. You might think that this noisiness should be a problem. On the contrary, it is actually beneficial.

This turns out to be a recurring theme in deep learning. For reasons that are not yet well-characterized theoretically, various sources of noise in optimization often lead to faster training and less overfitting: this variation appears to act as a form of regularization. Teye *et al.* (2018) and Luo *et al.* (2018) related the properties of batch normalization to Bayesian priors and penalties, respectively. In particular, this sheds some light on the puzzle of why batch normalization works best for moderate minibatch sizes in the 50–100 range. This particular size of minibatch seems to inject just the "right amount" of noise per layer, both in terms of scale via $\hat{\sigma}$, and in terms of offset via $\hat{\mu}$: a larger minibatch regularizes less due to the more stable estimates, whereas tiny minibatches destroy useful signal due to high variance. Exploring this direction further, considering alternative types of preprocessing and filtering may yet lead to other effective types of regularization.

Fixing a trained model, you might think that we would prefer using the entire dataset to estimate the mean and variance. Once training is complete, why would we want the same image to be classified differently, depending on the batch in which it happens to reside? During training, such exact calculation is infeasible because the intermediate variables for all data examples change every time we update our model. However, once the model is trained, we can calculate the means and variances of each layer's variables based on the entire dataset. Indeed this is standard practice for models employing batch normalization; thus batch normalization layers function differently in *training mode* (normalizing by mini-batch statistics) than in *prediction mode* (normalizing by dataset statistics). In this form they closely resemble the behavior of dropout regularization of Section 5.6, where noise is only injected during training.

8.5.2 Batch Normalization Layers

Batch normalization implementations for fully connected layers and convolutional layers are slightly different. One key difference between batch normalization and other layers is that because the former operates on a full minibatch at a time, we cannot just ignore the batch dimension as we did before when introducing other layers.

Fully Connected Layers

When applying batch normalization to fully connected layers, Ioffe and Szegedy (2015), in their original paper inserted batch normalization after the affine transformation and *before* the nonlinear activation function. Later applications experimented with inserting batch normalization right *after* activation functions. Denoting the input to the fully connected layer by $\mathbf{x}$, the affine transformation by $\mathbf{Wx} + \mathbf{b}$ (with the weight parameter $\mathbf{W}$ and the bias parameter $\mathbf{b}$), and the activation function by ϕ, we can express the computation of a batch-normalization-enabled, fully connected layer output $\mathbf{h}$ as follows:

$$\mathbf{h} = \phi(\mathrm{BN}(\mathbf{Wx} + \mathbf{b})). \tag{8.5.3}$$

Recall that mean and variance are computed on the *same* minibatch on which the transformation is applied.

Convolutional Layers

Similarly, with convolutional layers, we can apply batch normalization after the convolution but before the nonlinear activation function. The key difference from batch normalization in fully connected layers is that we apply the operation on a per-channel basis *across all locations*. This is compatible with our assumption of translation invariance that led to convolutions: we assumed that the specific location of a pattern within an image was not critical for the purpose of understanding.

Assume that our minibatches contain m examples and that for each channel, the output of the convolution has height p and width q. For convolutional layers, we carry out each batch normalization over the $m \cdot p \cdot q$ elements per output channel simultaneously. Thus, we collect the values over all spatial locations when computing the mean and variance and consequently apply the same mean and variance within a given channel to normalize the value at each spatial location. Each channel has its own scale and shift parameters, both of which are scalars.

Layer Normalization

Note that in the context of convolutions the batch normalization is well defined even for minibatches of size 1: after all, we have all the locations across an image to average. Consequently, mean and variance are well defined, even if it is just within a single observation. This consideration led Ba *et al.* (2016) to introduce the notion of *layer normalization*. It works just like a batch norm, only that it is applied to one observation at a time. Consequently both the offset and the scaling factor are scalars. For an n-dimensional vector $\mathbf{x}$, layer norms are given by

$$\mathbf{x} \to \mathrm{LN}(\mathbf{x}) = \frac{\mathbf{x} - \hat{\mu}}{\hat{\sigma}}, \tag{8.5.4}$$

where scaling and offset are applied coefficient-wise and given by

$$\hat{\mu} \overset{\text{def}}{=} \frac{1}{n} \sum_{i=1}^{n} x_i \text{ and } \hat{\sigma}^2 \overset{\text{def}}{=} \frac{1}{n} \sum_{i=1}^{n} (x_i - \hat{\mu})^2 + \epsilon. \tag{8.5.5}$$

As before we add a small offset $\epsilon > 0$ to prevent division by zero. One of the major benefits of using layer normalization is that it prevents divergence. After all, ignoring ϵ, the output of the layer normalization is scale independent. That is, we have $\mathrm{LN}(\mathbf{x}) \approx \mathrm{LN}(\alpha\mathbf{x})$ for any choice of $\alpha \neq 0$. This becomes an equality for $|\alpha| \to \infty$ (the approximate equality is due to the offset ϵ for the variance).

Another advantage of the layer normalization is that it does not depend on the minibatch size. It is also independent of whether we are in training or test regime. In other words, it is simply a deterministic transformation that standardizes the activations to a given scale. This can be very beneficial in preventing divergence in optimization. We skip further details and recommend that interested readers consult the original paper.

Batch Normalization During Prediction

As we mentioned earlier, batch normalization typically behaves differently in training mode than in prediction mode. First, the noise in the sample mean and the sample variance arising from estimating each on minibatches is no longer desirable once we have trained the model. Second, we might not have the luxury of computing per-batch normalization statistics. For example, we might need to apply our model to make one prediction at a time.

Typically, after training, we use the entire dataset to compute stable estimates of the variable statistics and then fix them at prediction time. Hence, batch normalization behaves differently during training than at test time. Recall that dropout also exhibits this characteristic.

8.5.3 Implementation from Scratch

To see how batch normalization works in practice, we implement one from scratch below.

```
def batch_norm(X, gamma, beta, moving_mean, moving_var, eps, momentum):
    # Use is_grad_enabled to determine whether we are in training mode
    if not torch.is_grad_enabled():
        # In prediction mode, use mean and variance obtained by moving average
        X_hat = (X - moving_mean) / torch.sqrt(moving_var + eps)
    else:
        assert len(X.shape) in (2, 4)
        if len(X.shape) == 2:
            # When using a fully connected layer, calculate the mean and
            # variance on the feature dimension
```

(continues on next page)

(continued from previous page)

```
        mean = X.mean(dim=0)
        var = ((X - mean) ** 2).mean(dim=0)
    else:
        # When using a two-dimensional convolutional layer, calculate the
        # mean and variance on the channel dimension (axis=1). Here we
        # need to maintain the shape of X, so that the broadcasting
        # operation can be carried out later
        mean = X.mean(dim=(0, 2, 3), keepdim=True)
        var = ((X - mean) ** 2).mean(dim=(0, 2, 3), keepdim=True)
    # In training mode, the current mean and variance are used
    X_hat = (X - mean) / torch.sqrt(var + eps)
    # Update the mean and variance using moving average
    moving_mean = (1.0 - momentum) * moving_mean + momentum * mean
    moving_var = (1.0 - momentum) * moving_var + momentum * var
Y = gamma * X_hat + beta  # Scale and shift
return Y, moving_mean.data, moving_var.data
```

We can now create a proper `BatchNorm` layer. Our layer will maintain proper parameters for scale gamma and shift `beta`, both of which will be updated in the course of training. Additionally, our layer will maintain moving averages of the means and variances for subsequent use during model prediction.

Putting aside the algorithmic details, note the design pattern underlying our implementation of the layer. Typically, we define the mathematics in a separate function, say `batch_norm`. We then integrate this functionality into a custom layer, whose code mostly addresses book-keeping matters, such as moving data to the right device context, allocating and initializing any required variables, keeping track of moving averages (here for mean and variance), and so on. This pattern enables a clean separation of mathematics from boilerplate code. Also note that for the sake of convenience we did not worry about automatically inferring the input shape here; thus we need to specify the number of features throughout. By now all modern deep learning frameworks offer automatic detection of size and shape in the high-level batch normalization APIs (in practice we will use this instead).

```
class BatchNorm(nn.Module):
    # num_features: the number of outputs for a fully connected layer or the
    # number of output channels for a convolutional layer. num_dims: 2 for a
    # fully connected layer and 4 for a convolutional layer
    def __init__(self, num_features, num_dims):
        super().__init__()
        if num_dims == 2:
            shape = (1, num_features)
        else:
            shape = (1, num_features, 1, 1)
        # The scale parameter and the shift parameter (model parameters) are
        # initialized to 1 and 0, respectively
        self.gamma = nn.Parameter(torch.ones(shape))
        self.beta = nn.Parameter(torch.zeros(shape))
        # The variables that are not model parameters are initialized to 0 and
        # 1
        self.moving_mean = torch.zeros(shape)
```

(continues on next page)

(continued from previous page)

```
        self.moving_var = torch.ones(shape)

    def forward(self, X):
        # If X is not on the main memory, copy moving_mean and moving_var to
        # the device where X is located
        if self.moving_mean.device != X.device:
            self.moving_mean = self.moving_mean.to(X.device)
            self.moving_var = self.moving_var.to(X.device)
        # Save the updated moving_mean and moving_var
        Y, self.moving_mean, self.moving_var = batch_norm(
            X, self.gamma, self.beta, self.moving_mean,
            self.moving_var, eps=1e-5, momentum=0.1)
        return Y
```

We used momentum to govern the aggregation over past mean and variance estimates. This is somewhat of a misnomer as it has nothing whatsoever to do with the *momentum* term of optimization. Nonetheless, it is the commonly adopted name for this term and in deference to API naming convention we use the same variable name in our code.

8.5.4 LeNet with Batch Normalization

To see how to apply `BatchNorm` in context, below we apply it to a traditional LeNet model (Section 7.6). Recall that batch normalization is applied after the convolutional layers or fully connected layers but before the corresponding activation functions.

```
class BNLeNetScratch(d2l.Classifier):
    def __init__(self, lr=0.1, num_classes=10):
        super().__init__()
        self.save_hyperparameters()
        self.net = nn.Sequential(
            nn.LazyConv2d(6, kernel_size=5), BatchNorm(6, num_dims=4),
            nn.Sigmoid(), nn.AvgPool2d(kernel_size=2, stride=2),
            nn.LazyConv2d(16, kernel_size=5), BatchNorm(16, num_dims=4),
            nn.Sigmoid(), nn.AvgPool2d(kernel_size=2, stride=2),
            nn.Flatten(), nn.LazyLinear(120),
            BatchNorm(120, num_dims=2), nn.Sigmoid(), nn.LazyLinear(84),
            BatchNorm(84, num_dims=2), nn.Sigmoid(),
            nn.LazyLinear(num_classes))
```

As before, we will train our network on the Fashion-MNIST dataset. This code is virtually identical to that when we first trained LeNet.

```
trainer = d2l.Trainer(max_epochs=10, num_gpus=1)
data = d2l.FashionMNIST(batch_size=128)
model = BNLeNetScratch(lr=0.1)
model.apply_init([next(iter(data.get_dataloader(True)))[0]], d2l.init_cnn)
trainer.fit(model, data)
```

Let's have a look at the scale parameter gamma and the shift parameter beta learned from the first batch normalization layer.

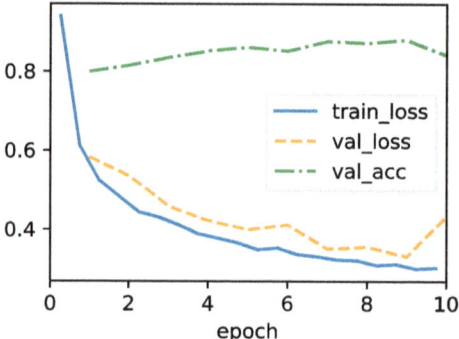

```
model.net[1].gamma.reshape((-1,)), model.net[1].beta.reshape((-1,))
```

```
(tensor([1.2930, 2.2096, 1.7528, 2.1382, 1.7916, 1.8372], device='cuda:0',
        grad_fn=<ReshapeAliasBackward0>),
 tensor([-0.4592, -1.5245,  1.7853, -1.2108,  0.3265, -0.4423], device='cuda:0
↪',
        grad_fn=<ReshapeAliasBackward0>))
```

8.5.5 Concise Implementation

Compared with the BatchNorm class, which we just defined ourselves, we can use the BatchNorm class defined in high-level APIs from the deep learning framework directly. The code looks virtually identical to our implementation above, except that we no longer need to provide additional arguments for it to get the dimensions right.

```
class BNLeNet(d2l.Classifier):
    def __init__(self, lr=0.1, num_classes=10):
        super().__init__()
        self.save_hyperparameters()
        self.net = nn.Sequential(
            nn.LazyConv2d(6, kernel_size=5), nn.LazyBatchNorm2d(),
            nn.Sigmoid(), nn.AvgPool2d(kernel_size=2, stride=2),
            nn.LazyConv2d(16, kernel_size=5), nn.LazyBatchNorm2d(),
            nn.Sigmoid(), nn.AvgPool2d(kernel_size=2, stride=2),
            nn.Flatten(), nn.LazyLinear(120), nn.LazyBatchNorm1d(),
            nn.Sigmoid(), nn.LazyLinear(84), nn.LazyBatchNorm1d(),
            nn.Sigmoid(), nn.LazyLinear(num_classes))
```

Below, we use the same hyperparameters to train our model. Note that as usual, the high-level API variant runs much faster because its code has been compiled to C++ or CUDA while our custom implementation must be interpreted by Python.

```
trainer = d2l.Trainer(max_epochs=10, num_gpus=1)
data = d2l.FashionMNIST(batch_size=128)
```

(continues on next page)

(continued from previous page)

```
model = BNLeNet(lr=0.1)
model.apply_init([next(iter(data.get_dataloader(True)))[0]], d2l.init_cnn)
trainer.fit(model, data)
```

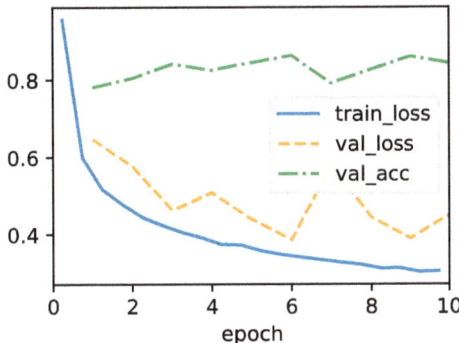

8.5.6 Discussion

Intuitively, batch normalization is thought to make the optimization landscape smoother. However, we must be careful to distinguish between speculative intuitions and true explanations for the phenomena that we observe when training deep models. Recall that we do not even know why simpler deep neural networks (MLPs and conventional CNNs) generalize well in the first place. Even with dropout and weight decay, they remain so flexible that their ability to generalize to unseen data likely needs significantly more refined learning-theoretic generalization guarantees.

The original paper proposing batch normalization (Ioffe and Szegedy, 2015), in addition to introducing a powerful and useful tool, offered an explanation for why it works: by reducing *internal covariate shift*. Presumably by *internal covariate shift* they meant something like the intuition expressed above—the notion that the distribution of variable values changes over the course of training. However, there were two problems with this explanation: i) This drift is very different from *covariate shift*, rendering the name a misnomer. If anything, it is closer to concept drift. ii) The explanation offers an under-specified intuition but leaves the question of *why precisely this technique works* an open question wanting for a rigorous explanation. Throughout this book, we aim to convey the intuitions that practitioners use to guide their development of deep neural networks. However, we believe that it is important to separate these guiding intuitions from established scientific fact. Eventually, when you master this material and start writing your own research papers you will want to be clear to delineate between technical claims and hunches.

Following the success of batch normalization, its explanation in terms of *internal covariate shift* has repeatedly surfaced in debates in the technical literature and broader discourse about how to present machine learning research. In a memorable speech given while accepting a Test of Time Award at the 2017 NeurIPS conference, Ali Rahimi used *internal covariate shift* as a focal point in an argument likening the modern practice of deep learning to alchemy. Subsequently, the example was revisited in detail in a position paper outlining troubling trends in machine learning (Lipton and Steinhardt, 2018). Other authors have proposed alternative explanations for the success of batch normalization, some (Santurkar *et al.*, 2018) claiming that batch normalization's success comes despite exhibiting behavior that is in some ways opposite to those claimed in the original paper.

We note that the *internal covariate shift* is no more worthy of criticism than any of thousands of similarly vague claims made every year in the technical machine learning literature. Likely, its resonance as a focal point of these debates owes to its broad recognizability for the target audience. Batch normalization has proven an indispensable method, applied in nearly all deployed image classifiers, earning the paper that introduced the technique tens of thousands of citations. We conjecture, though, that the guiding principles of regularization through noise injection, acceleration through rescaling and lastly preprocessing may well lead to further inventions of layers and techniques in the future.

On a more practical note, there are a number of aspects worth remembering about batch normalization:

- During model training, batch normalization continuously adjusts the intermediate output of the network by utilizing the mean and standard deviation of the minibatch, so that the values of the intermediate output in each layer throughout the neural network are more stable.

- Batch normalization is slightly different for fully connected layers than for convolutional layers. In fact, for convolutional layers, layer normalization can sometimes be used as an alternative.

- Like a dropout layer, batch normalization layers have different behaviors in training mode than in prediction mode.

- Batch normalization is useful for regularization and improving convergence in optimization. By contrast, the original motivation of reducing internal covariate shift seems not to be a valid explanation.

- For more robust models that are less sensitive to input perturbations, consider removing batch normalization (Wang *et al.*, 2022).

8.5.7 Exercises

1. Should we remove the bias parameter from the fully connected layer or the convolutional layer before the batch normalization? Why?

2. Compare the learning rates for LeNet with and without batch normalization.

   1. Plot the increase in validation accuracy.

   2. How large can you make the learning rate before the optimization fails in both cases?

3. Do we need batch normalization in every layer? Experiment with it.

4. Implement a "lite" version of batch normalization that only removes the mean, or alternatively one that only removes the variance. How does it behave?

5. Fix the parameters beta and gamma. Observe and analyze the results.

6. Can you replace dropout by batch normalization? How does the behavior change?

7. Research ideas: think of other normalization transforms that you can apply:

    1. Can you apply the probability integral transform?

    2. Can you use a full-rank covariance estimate? Why should you probably not do that?

    3. Can you use other compact matrix variants (block-diagonal, low-displacement rank, Monarch, etc.)?

    4. Does a sparsification compression act as a regularizer?

    5. Are there other projections (e.g., convex cone, symmetry group-specific transforms) that you can use?

Discussions [132].

132

8.6 Residual Networks (ResNet) and ResNeXt

As we design ever deeper networks it becomes imperative to understand how adding layers can increase the complexity and expressiveness of the network. Even more important is the ability to design networks where adding layers makes networks strictly more expressive rather than just different. To make some progress we need a bit of mathematics.

```python
import torch
from torch import nn
from torch.nn import functional as F
from d2l import torch as d2l
```

8.6.1 Function Classes

Consider $\mathcal{F}$, the class of functions that a specific network architecture (together with learning rates and other hyperparameter settings) can reach. That is, for all $f \in \mathcal{F}$ there exists some set of parameters (e.g., weights and biases) that can be obtained through training on a suitable dataset. Let's assume that f^* is the "truth" function that we really would like to find. If it is in $\mathcal{F}$, we are in good shape but typically we will not be quite so lucky. Instead, we will try to find some $f_{\mathcal{F}}^*$ which is our best bet within $\mathcal{F}$. For instance, given a dataset with features $\mathbf{X}$ and labels $\mathbf{y}$, we might try finding it by solving the following optimization problem:

$$f_{\mathcal{F}}^* \stackrel{\text{def}}{=} \operatorname*{argmin}_{f} L(\mathbf{X}, \mathbf{y}, f) \text{ subject to } f \in \mathcal{F}. \tag{8.6.1}$$

We know that regularization (Morozov, 1984, Tikhonov and Arsenin, 1977) may control complexity of $\mathcal{F}$ and achieve consistency, so a larger size of training data generally leads to better $f_{\mathcal{F}}^*$. It is only reasonable to assume that if we design a different and more powerful architecture $\mathcal{F}'$ we should arrive at a better outcome. In other words, we would expect that $f_{\mathcal{F}'}^*$ is "better" than $f_{\mathcal{F}}^*$. However, if $\mathcal{F} \not\subseteq \mathcal{F}'$ there is no guarantee that this should even happen. In fact, $f_{\mathcal{F}'}^*$ might well be worse. As illustrated by Fig. 8.6.1, for non-nested function classes, a larger function class does not always move closer to the "truth" function f^*. For instance, on the left of Fig. 8.6.1, though $\mathcal{F}_3$ is closer to f^* than $\mathcal{F}_1$, $\mathcal{F}_6$ moves away and there is no guarantee that further increasing the complexity can reduce the distance from f^*. With nested function classes where $\mathcal{F}_1 \subseteq \cdots \subseteq \mathcal{F}_6$ on the right of Fig. 8.6.1, we can avoid the aforementioned issue from the non-nested function classes.

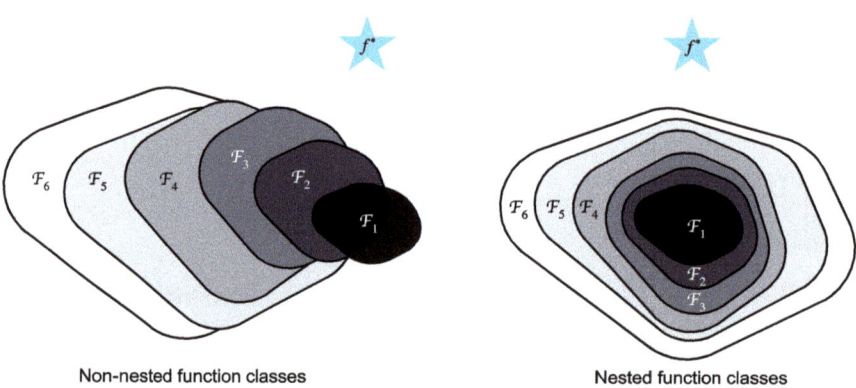

Non-nested function classes   Nested function classes

Fig. 8.6.1   For non-nested function classes, a larger (indicated by area) function class does not guarantee we will get closer to the "truth" function (f^*). This does not happen in nested function classes.

Thus, only if larger function classes contain the smaller ones are we guaranteed that increasing them strictly increases the expressive power of the network. For deep neural networks, if we can train the newly-added layer into an identity function $f(\mathbf{x}) = \mathbf{x}$, the new model will be as effective as the original model. As the new model may get a better solution to fit the training dataset, the added layer might make it easier to reduce training errors.

This is the question that He *et al.* (2016) considered when working on very deep computer vision models. At the heart of their proposed *residual network* (*ResNet*) is the idea that every additional layer should more easily contain the identity function as one of its elements. These considerations are rather profound but they led to a surprisingly simple solution, a *residual block*. With it, ResNet won the ImageNet Large Scale Visual Recognition Challenge in 2015. The design had a profound influence on how to build deep neural networks. For instance, residual blocks have been added to recurrent networks (Kim *et al.*, 2017, Prakash *et al.*, 2016). Likewise, Transformers (Vaswani *et al.*, 2017) use them to stack many layers of networks efficiently. It is also used in graph neural networks (Kipf and Welling, 2016) and, as a basic concept, it has been used extensively in computer vision (Redmon and Farhadi, 2018, Ren *et al.*, 2015). Note that residual networks are predated by highway networks (Srivastava *et al.*, 2015) that share some of the motivation, albeit without the elegant parametrization around the identity function.

8.6.2 Residual Blocks

Let's focus on a local part of a neural network, as depicted in Fig. 8.6.2. Denote the input by $\mathbf{x}$. We assume that $f(\mathbf{x})$, the desired underlying mapping we want to obtain by learning, is to be used as input to the activation function on the top. On the left, the portion within the dotted-line box must directly learn $f(\mathbf{x})$. On the right, the portion within the dotted-line box needs to learn the *residual mapping* $g(\mathbf{x}) = f(\mathbf{x}) - \mathbf{x}$, which is how the residual block derives its name. If the identity mapping $f(\mathbf{x}) = \mathbf{x}$ is the desired underlying mapping, the residual mapping amounts to $g(\mathbf{x}) = 0$ and it is thus easier to learn: we only need to push the weights and biases of the upper weight layer (e.g., fully connected layer and convolutional layer) within the dotted-line box to zero. The right figure illustrates the *residual block* of ResNet, where the solid line carrying the layer input $\mathbf{x}$ to the addition operator is called a *residual connection* (or *shortcut connection*). With residual blocks, inputs can forward propagate faster through the residual connections across layers. In fact, the residual block can be thought of as a special case of the multi-branch Inception block: it has two branches one of which is the identity mapping.

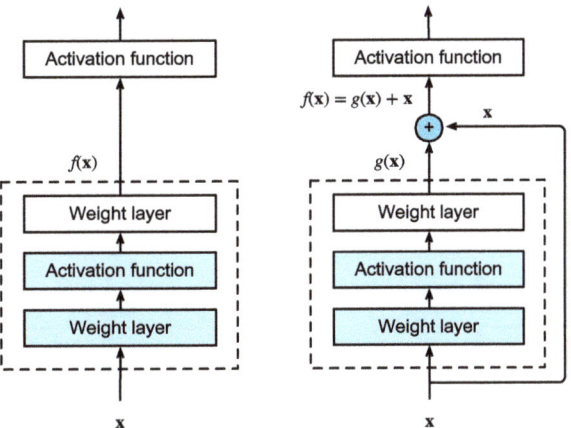

Fig. 8.6.2 In a regular block (left), the portion within the dotted-line box must directly learn the mapping $f(\mathbf{x})$. In a residual block (right), the portion within the dotted-line box needs to learn the residual mapping $g(\mathbf{x}) = f(\mathbf{x}) - \mathbf{x}$, making the identity mapping $f(\mathbf{x}) = \mathbf{x}$ easier to learn.

ResNet has VGG's full 3×3 convolutional layer design. The residual block has two 3×3 convolutional layers with the same number of output channels. Each convolutional layer is followed by a batch normalization layer and a ReLU activation function. Then, we skip these two convolution operations and add the input directly before the final ReLU activation function. This kind of design requires that the output of the two convolutional layers has to be of the same shape as the input, so that they can be added together. If we want to change the number of channels, we need to introduce an additional 1×1 convolutional layer to transform the input into the desired shape for the addition operation. Let's have a look at the code below.

```python
class Residual(nn.Module):  #@save
    """The Residual block of ResNet models."""
    def __init__(self, num_channels, use_1x1conv=False, strides=1):
        super().__init__()
        self.conv1 = nn.LazyConv2d(num_channels, kernel_size=3, padding=1,
                                   stride=strides)
        self.conv2 = nn.LazyConv2d(num_channels, kernel_size=3, padding=1)
        if use_1x1conv:
            self.conv3 = nn.LazyConv2d(num_channels, kernel_size=1,
                                       stride=strides)
        else:
            self.conv3 = None
        self.bn1 = nn.LazyBatchNorm2d()
        self.bn2 = nn.LazyBatchNorm2d()

    def forward(self, X):
        Y = F.relu(self.bn1(self.conv1(X)))
        Y = self.bn2(self.conv2(Y))
        if self.conv3:
            X = self.conv3(X)
        Y += X
        return F.relu(Y)
```

This code generates two types of networks: one where we add the input to the output before applying the ReLU nonlinearity whenever use_1x1conv=False; and one where we adjust channels and resolution by means of a 1×1 convolution before adding. Fig. 8.6.3 illustrates this.

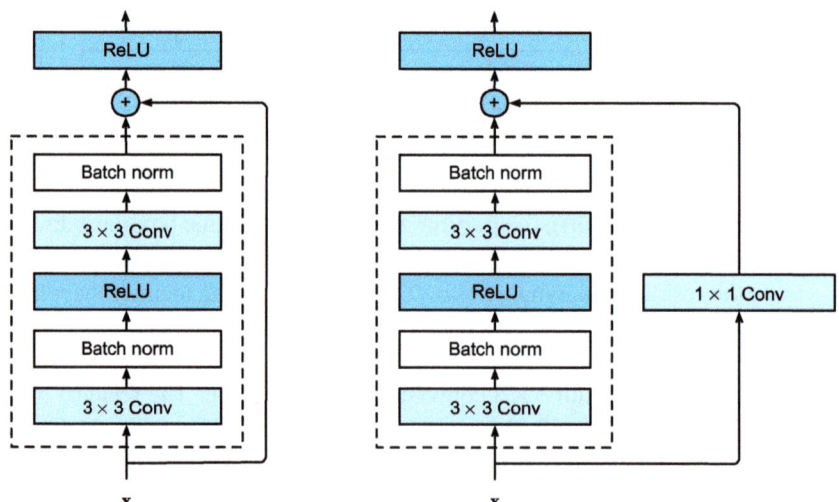

Fig. 8.6.3   ResNet block with and without 1×1 convolution, which transforms the input into the desired shape for the addition operation.

Now let's look at a situation where the input and output are of the same shape, where 1×1 convolution is not needed.

```
blk = Residual(3)
X = torch.randn(4, 3, 6, 6)
blk(X).shape
```

```
torch.Size([4, 3, 6, 6])
```

We also have the option to halve the output height and width while increasing the number of output channels. In this case we use 1×1 convolutions via use_1x1conv=True. This comes in handy at the beginning of each ResNet block to reduce the spatial dimensionality via strides=2.

```
blk = Residual(6, use_1x1conv=True, strides=2)
blk(X).shape
```

```
torch.Size([4, 6, 3, 3])
```

8.6.3 ResNet Model

The first two layers of ResNet are the same as those of the GoogLeNet we described before: the 7×7 convolutional layer with 64 output channels and a stride of 2 is followed by the 3×3 max-pooling layer with a stride of 2. The difference is the batch normalization layer added after each convolutional layer in ResNet.

```
class ResNet(d2l.Classifier):
    def b1(self):
        return nn.Sequential(
            nn.LazyConv2d(64, kernel_size=7, stride=2, padding=3),
            nn.LazyBatchNorm2d(), nn.ReLU(),
            nn.MaxPool2d(kernel_size=3, stride=2, padding=1))
```

GoogLeNet uses four modules made up of Inception blocks. However, ResNet uses four modules made up of residual blocks, each of which uses several residual blocks with the same number of output channels. The number of channels in the first module is the same as the number of input channels. Since a max-pooling layer with a stride of 2 has already been used, it is not necessary to reduce the height and width. In the first residual block for each of the subsequent modules, the number of channels is doubled compared with that of the previous module, and the height and width are halved.

```
@d2l.add_to_class(ResNet)
def block(self, num_residuals, num_channels, first_block=False):
    blk = []
    for i in range(num_residuals):
        if i == 0 and not first_block:
```

(continues on next page)

(continued from previous page)

```
        blk.append(Residual(num_channels, use_1x1conv=True, strides=2))
    else:
        blk.append(Residual(num_channels))
return nn.Sequential(*blk)
```

Then, we add all the modules to ResNet. Here, two residual blocks are used for each module. Lastly, just like GoogLeNet, we add a global average pooling layer, followed by the fully connected layer output.

```
@d2l.add_to_class(ResNet)
def __init__(self, arch, lr=0.1, num_classes=10):
    super(ResNet, self).__init__()
    self.save_hyperparameters()
    self.net = nn.Sequential(self.b1())
    for i, b in enumerate(arch):
        self.net.add_module(f'b{i+2}', self.block(*b, first_block=(i==0)))
    self.net.add_module('last', nn.Sequential(
        nn.AdaptiveAvgPool2d((1, 1)), nn.Flatten(),
        nn.LazyLinear(num_classes)))
    self.net.apply(d2l.init_cnn)
```

There are four convolutional layers in each module (excluding the 1×1 convolutional layer). Together with the first 7×7 convolutional layer and the final fully connected layer, there are 18 layers in total. Therefore, this model is commonly known as ResNet-18. By configuring different numbers of channels and residual blocks in the module, we can create different ResNet models, such as the deeper 152-layer ResNet-152. Although the main architecture of ResNet is similar to that of GoogLeNet, ResNet's structure is simpler and easier to modify. All these factors have resulted in the rapid and widespread use of ResNet. Fig. 8.6.4 depicts the full ResNet-18.

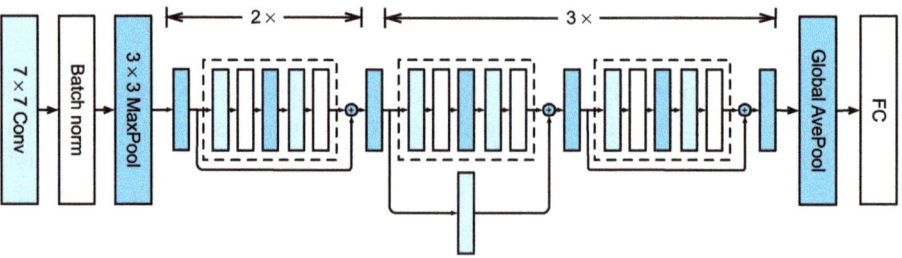

Fig. 8.6.4   The ResNet-18 architecture.

Before training ResNet, let's observe how the input shape changes across different modules in ResNet. As in all the previous architectures, the resolution decreases while the number of channels increases up until the point where a global average pooling layer aggregates all features.

```
class ResNet18(ResNet):
    def __init__(self, lr=0.1, num_classes=10):
        super().__init__(((2, 64), (2, 128), (2, 256), (2, 512)),
                         lr, num_classes)
```

```
ResNet18().layer_summary((1, 1, 96, 96))
```

```
Sequential output shape:        torch.Size([1, 64, 24, 24])
Sequential output shape:        torch.Size([1, 64, 24, 24])
Sequential output shape:        torch.Size([1, 128, 12, 12])
Sequential output shape:        torch.Size([1, 256, 6, 6])
Sequential output shape:        torch.Size([1, 512, 3, 3])
Sequential output shape:        torch.Size([1, 10])
```

8.6.4 Training

We train ResNet on the Fashion-MNIST dataset, just like before. ResNet is quite a powerful and flexible architecture. The plot capturing training and validation loss illustrates a significant gap between both graphs, with the training loss being considerably lower. For a network of this flexibility, more training data would offer distinct benefit in closing the gap and improving accuracy.

```
model = ResNet18(lr=0.01)
trainer = d2l.Trainer(max_epochs=10, num_gpus=1)
data = d2l.FashionMNIST(batch_size=128, resize=(96, 96))
model.apply_init([next(iter(data.get_dataloader(True)))[0]], d2l.init_cnn)
trainer.fit(model, data)
```

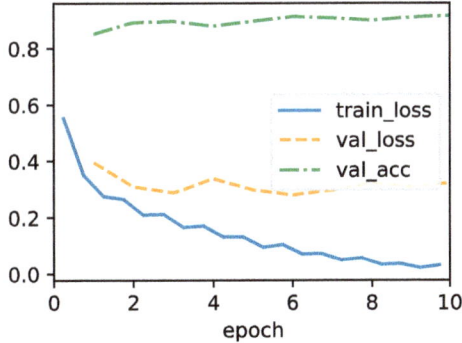

8.6.5 ResNeXt

One of the challenges one encounters in the design of ResNet is the trade-off between non-linearity and dimensionality within a given block. That is, we could add more nonlinearity by increasing the number of layers, or by increasing the width of the convolutions. An alternative strategy is to increase the number of channels that can carry information between blocks. Unfortunately, the latter comes with a quadratic penalty since the computational cost of ingesting c_i channels and emitting c_o channels is proportional to $O(c_i \cdot c_o)$ (see our discussion in Section 7.4).

We can take some inspiration from the Inception block of Fig. 8.4.1 which has information flowing through the block in separate groups. Applying the idea of multiple independent groups to the ResNet block of Fig. 8.6.3 led to the design of ResNeXt (Xie *et al.*, 2017). Different from the smorgasbord of transformations in Inception, ResNeXt adopts the *same* transformation in all branches, thus minimizing the need for manual tuning of each branch.

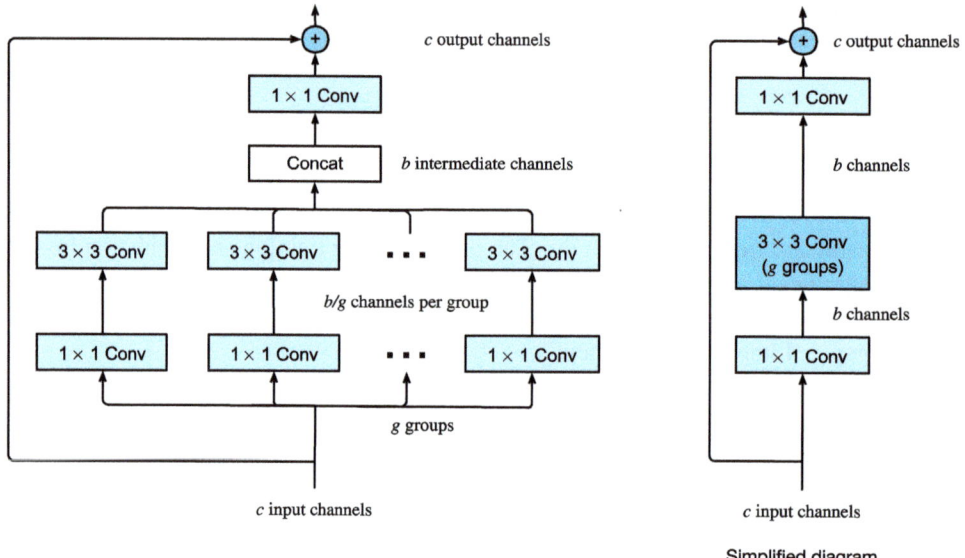

Fig. 8.6.5   The ResNeXt block. The use of grouped convolution with g groups is g times faster than a dense convolution. It is a bottleneck residual block when the number of intermediate channels b is less than c.

Breaking up a convolution from c_i to c_o channels into one of g groups of size c_i/g generating g outputs of size c_o/g is called, quite fittingly, a *grouped convolution*. The computational cost (proportionally) is reduced from $O(c_i \cdot c_o)$ to $O(g \cdot (c_i/g) \cdot (c_o/g)) = O(c_i \cdot c_o/g)$, i.e., it is g times faster. Even better, the number of parameters needed to generate the output is also reduced from a $c_i \times c_o$ matrix to g smaller matrices of size $(c_i/g) \times (c_o/g)$, again a g times reduction. In what follows we assume that both c_i and c_o are divisible by g.

The only challenge in this design is that no information is exchanged between the g groups. The ResNeXt block of Fig. 8.6.5 amends this in two ways: the grouped convolution with a 3×3 kernel is sandwiched in between two 1×1 convolutions. The second one serves double duty in changing the number of channels back. The benefit is that we only pay the $O(c \cdot b)$ cost for 1×1 kernels and can make do with an $O(b^2/g)$ cost for 3×3 kernels. Similar to the residual block implementation in Section 8.6.2, the residual connection is replaced (thus generalized) by a 1×1 convolution.

The right-hand figure in Fig. 8.6.5 provides a much more concise summary of the resulting network block. It will also play a major role in the design of generic modern CNNs in Section 8.8. Note that the idea of grouped convolutions dates back to the implementation of AlexNet (Krizhevsky *et al.*, 2012). When distributing the network across two GPUs with limited memory, the implementation treated each GPU as its own channel with no ill effects.

The following implementation of the ResNeXtBlock class takes as argument groups (g), with bot_channels (b) intermediate (bottleneck) channels. Lastly, when we need to reduce the height and width of the representation, we add a stride of 2 by setting use_1x1conv=True, strides=2.

```
class ResNeXtBlock(nn.Module):  #@save
    """The ResNeXt block."""
    def __init__(self, num_channels, groups, bot_mul, use_1x1conv=False,
                 strides=1):
        super().__init__()
        bot_channels = int(round(num_channels * bot_mul))
        self.conv1 = nn.LazyConv2d(bot_channels, kernel_size=1, stride=1)
        self.conv2 = nn.LazyConv2d(bot_channels, kernel_size=3,
                                   stride=strides, padding=1,
                                   groups=bot_channels//groups)
        self.conv3 = nn.LazyConv2d(num_channels, kernel_size=1, stride=1)
        self.bn1 = nn.LazyBatchNorm2d()
        self.bn2 = nn.LazyBatchNorm2d()
        self.bn3 = nn.LazyBatchNorm2d()
        if use_1x1conv:
            self.conv4 = nn.LazyConv2d(num_channels, kernel_size=1,
                                       stride=strides)
            self.bn4 = nn.LazyBatchNorm2d()
        else:
            self.conv4 = None

    def forward(self, X):
        Y = F.relu(self.bn1(self.conv1(X)))
        Y = F.relu(self.bn2(self.conv2(Y)))
        Y = self.bn3(self.conv3(Y))
        if self.conv4:
            X = self.bn4(self.conv4(X))
        return F.relu(Y + X)
```

Its use is entirely analogous to that of the ResNetBlock discussed previously. For instance, when using (use_1x1conv=False, strides=1), the input and output are of the same shape. Alternatively, setting use_1x1conv=True, strides=2 halves the output height and width.

```
blk = ResNeXtBlock(32, 16, 1)
X = torch.randn(4, 32, 96, 96)
blk(X).shape
```

```
torch.Size([4, 32, 96, 96])
```

8.6.6 Summary and Discussion

Nested function classes are desirable since they allow us to obtain strictly *more power-ful* rather than also subtly *different* function classes when adding capacity. One way of accomplishing this is by letting additional layers to simply pass through the input to the output. Residual connections allow for this. As a consequence, this changes the inductive bias from simple functions being of the form $f(\mathbf{x}) = 0$ to simple functions looking like $f(\mathbf{x}) = \mathbf{x}$.

The residual mapping can learn the identity function more easily, such as pushing param-eters in the weight layer to zero. We can train an effective *deep* neural network by having residual blocks. Inputs can forward propagate faster through the residual connections across layers. As a consequence, we can thus train much deeper networks. For instance, the origi-nal ResNet paper (He *et al.*, 2016) allowed for up to 152 layers. Another benefit of residual networks is that it allows us to add layers, initialized as the identity function, *during* the training process. After all, the default behavior of a layer is to let the data pass through unchanged. This can accelerate the training of very large networks in some cases.

Prior to residual connections, bypassing paths with gating units were introduced to effec-tively train highway networks with over 100 layers (Srivastava *et al.*, 2015). Using identity functions as bypassing paths, ResNet performed remarkably well on multiple computer vi-sion tasks. Residual connections had a major influence on the design of subsequent deep neural networks, of either convolutional or sequential nature. As we will introduce later, the Transformer architecture (Vaswani *et al.*, 2017) adopts residual connections (together with other design choices) and is pervasive in areas as diverse as language, vision, speech, and reinforcement learning.

ResNeXt is an example for how the design of convolutional neural networks has evolved over time: by being more frugal with computation and trading it off against the size of the activations (number of channels), it allows for faster and more accurate networks at lower cost. An alternative way of viewing grouped convolutions is to think of a block-diagonal matrix for the convolutional weights. Note that there are quite a few such "tricks" that lead to more efficient networks. For instance, ShiftNet (Wu *et al.*, 2018) mimics the effects of a 3×3 convolution, simply by adding shifted activations to the channels, offering increased function complexity, this time without any computational cost.

A common feature of the designs we have discussed so far is that the network design is fairly manual, primarily relying on the ingenuity of the designer to find the "right" network hyperparameters. While clearly feasible, it is also very costly in terms of human time and there is no guarantee that the outcome is optimal in any sense. In Section 8.8 we will discuss a number of strategies for obtaining high quality networks in a more automated fashion. In particular, we will review the notion of *network design spaces* that led to the RegNetX/Y models (Radosavovic *et al.*, 2020).

8.6.7 Exercises

1. What are the major differences between the Inception block in Fig. 8.4.1 and the residual block? How do they compare in terms of computation, accuracy, and the classes of functions they can describe?

2. Refer to Table 1 in the ResNet paper (He *et al.*, 2016) to implement different variants of the network.

3. For deeper networks, ResNet introduces a "bottleneck" architecture to reduce model complexity. Try to implement it.

4. In subsequent versions of ResNet, the authors changed the "convolution, batch normalization, and activation" structure to the "batch normalization, activation, and convolution" structure. Make this improvement yourself. See Figure 1 in He *et al.* (2016) for details.

5. Why can't we just increase the complexity of functions without bound, even if the function classes are nested?

 Discussions[133].

8.7 Densely Connected Networks (DenseNet)

ResNet significantly changed the view of how to parametrize the functions in deep networks. *DenseNet* (dense convolutional network) is to some extent the logical extension of this (Huang *et al.*, 2017). DenseNet is characterized by both the connectivity pattern where each layer connects to all the preceding layers and the concatenation operation (rather than the addition operator in ResNet) to preserve and reuse features from earlier layers. To understand how to arrive at it, let's take a small detour to mathematics.

```
import torch
from torch import nn
from d2l import torch as d2l
```

8.7.1 From ResNet to DenseNet

Recall the Taylor expansion for functions. At the point $x = 0$ it can be written as

$$f(x) = f(0) + x \cdot \left[f'(0) + x \cdot \left[\frac{f''(0)}{2!} + x \cdot \left[\frac{f'''(0)}{3!} + \cdots \right] \right] \right]. \qquad (8.7.1)$$

The key point is that it decomposes a function into terms of increasingly higher order. In a similar vein, ResNet decomposes functions into

$$f(\mathbf{x}) = \mathbf{x} + g(\mathbf{x}). \qquad (8.7.2)$$

That is, ResNet decomposes f into a simple linear term and a more complex nonlinear one. What if we wanted to capture (not necessarily add) information beyond two terms? One such solution is DenseNet (Huang *et al.*, 2017).

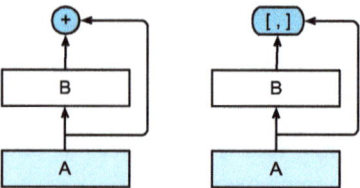

Fig. 8.7.1 The main difference between ResNet (left) and DenseNet (right) in cross-layer connections: use of addition and use of concatenation.

As shown in Fig. 8.7.1, the key difference between ResNet and DenseNet is that in the latter case outputs are *concatenated* (denoted by $[,]$) rather than added. As a result, we perform a mapping from $\mathbf{x}$ to its values after applying an increasingly complex sequence of functions:

$$\mathbf{x} \to [\mathbf{x}, f_1(\mathbf{x}), f_2([\mathbf{x}, f_1(\mathbf{x})]), f_3([\mathbf{x}, f_1(\mathbf{x}), f_2([\mathbf{x}, f_1(\mathbf{x})])]), \ldots]. \qquad (8.7.3)$$

In the end, all these functions are combined in MLP to reduce the number of features again. In terms of implementation this is quite simple: rather than adding terms, we concatenate them. The name DenseNet arises from the fact that the dependency graph between variables becomes quite dense. The final layer of such a chain is densely connected to all previous layers. The dense connections are shown in Fig. 8.7.2.

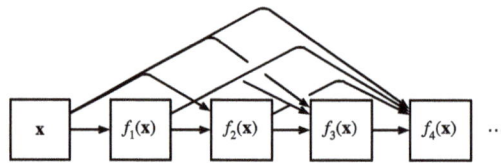

Fig. 8.7.2 Dense connections in DenseNet. Note how the dimensionality increases with depth.

The main components that comprise a DenseNet are *dense blocks* and *transition layers*. The former define how the inputs and outputs are concatenated, while the latter control the number of channels so that it is not too large, since the expansion

$$\mathbf{x} \to [\mathbf{x}, f_1(\mathbf{x}), f_2([\mathbf{x}, f_1(\mathbf{x})]), \ldots]$$

can be quite high-dimensional.

8.7.2 Dense Blocks

DenseNet uses the modified "batch normalization, activation, and convolution" structure of ResNet (see the exercise in Section 8.6). First, we implement this convolution block structure.

```
def conv_block(num_channels):
    return nn.Sequential(
        nn.LazyBatchNorm2d(), nn.ReLU(),
        nn.LazyConv2d(num_channels, kernel_size=3, padding=1))
```

A *dense block* consists of multiple convolution blocks, each using the same number of output channels. In the forward propagation, however, we concatenate the input and output of each convolution block on the channel dimension. Lazy evaluation allows us to adjust the dimensionality automatically.

```
class DenseBlock(nn.Module):
    def __init__(self, num_convs, num_channels):
        super(DenseBlock, self).__init__()
        layer = []
        for i in range(num_convs):
            layer.append(conv_block(num_channels))
        self.net = nn.Sequential(*layer)

    def forward(self, X):
        for blk in self.net:
            Y = blk(X)
            # Concatenate input and output of each block along the channels
            X = torch.cat((X, Y), dim=1)
        return X
```

In the following example, we define a DenseBlock instance with two convolution blocks of 10 output channels. When using an input with three channels, we will get an output with $3 + 10 + 10 = 23$ channels. The number of convolution block channels controls the growth in the number of output channels relative to the number of input channels. This is also referred to as the *growth rate*.

```
blk = DenseBlock(2, 10)
X = torch.randn(4, 3, 8, 8)
Y = blk(X)
Y.shape
```

```
torch.Size([4, 23, 8, 8])
```

8.7.3 Transition Layers

Since each dense block will increase the number of channels, adding too many of them will lead to an excessively complex model. A *transition layer* is used to control the complexity of the model. It reduces the number of channels by using a 1×1 convolution. Moreover, it halves the height and width via average pooling with a stride of 2.

```
def transition_block(num_channels):
    return nn.Sequential(
```

(continues on next page)

(continued from previous page)

```
        nn.LazyBatchNorm2d(), nn.ReLU(),
        nn.LazyConv2d(num_channels, kernel_size=1),
        nn.AvgPool2d(kernel_size=2, stride=2))
```

Apply a transition layer with 10 channels to the output of the dense block in the previous example. This reduces the number of output channels to 10, and halves the height and width.

```
blk = transition_block(10)
blk(Y).shape
```

```
torch.Size([4, 10, 4, 4])
```

8.7.4 DenseNet Model

Next, we will construct a DenseNet model. DenseNet first uses the same single convolutional layer and max-pooling layer as in ResNet.

```
class DenseNet(d2l.Classifier):
    def b1(self):
        return nn.Sequential(
            nn.LazyConv2d(64, kernel_size=7, stride=2, padding=3),
            nn.LazyBatchNorm2d(), nn.ReLU(),
            nn.MaxPool2d(kernel_size=3, stride=2, padding=1))
```

Then, similar to the four modules made up of residual blocks that ResNet uses, DenseNet uses four dense blocks. As with ResNet, we can set the number of convolutional layers used in each dense block. Here, we set it to 4, consistent with the ResNet-18 model in Section 8.6. Furthermore, we set the number of channels (i.e., growth rate) for the convolutional layers in the dense block to 32, so 128 channels will be added to each dense block.

In ResNet, the height and width are reduced between each module by a residual block with a stride of 2. Here, we use the transition layer to halve the height and width and halve the number of channels. Similar to ResNet, a global pooling layer and a fully connected layer are connected at the end to produce the output.

```
@d2l.add_to_class(DenseNet)
def __init__(self, num_channels=64, growth_rate=32, arch=(4, 4, 4, 4),
             lr=0.1, num_classes=10):
    super(DenseNet, self).__init__()
    self.save_hyperparameters()
    self.net = nn.Sequential(self.b1())
    for i, num_convs in enumerate(arch):
        self.net.add_module(f'dense_blk{i+1}', DenseBlock(num_convs,
                                                          growth_rate))
        # The number of output channels in the previous dense block
        num_channels += num_convs * growth_rate
```

(continues on next page)

(continued from previous page)

```
        # A transition layer that halves the number of channels is added
        # between the dense blocks
        if i != len(arch) - 1:
            num_channels //= 2
            self.net.add_module(f'tran_blk{i+1}', transition_block(
                num_channels))
    self.net.add_module('last', nn.Sequential(
        nn.LazyBatchNorm2d(), nn.ReLU(),
        nn.AdaptiveAvgPool2d((1, 1)), nn.Flatten(),
        nn.LazyLinear(num_classes)))
    self.net.apply(d2l.init_cnn)
```

8.7.5 Training

Since we are using a deeper network here, in this section, we will reduce the input height and width from 224 to 96 to simplify the computation.

```
model = DenseNet(lr=0.01)
trainer = d2l.Trainer(max_epochs=10, num_gpus=1)
data = d2l.FashionMNIST(batch_size=128, resize=(96, 96))
trainer.fit(model, data)
```

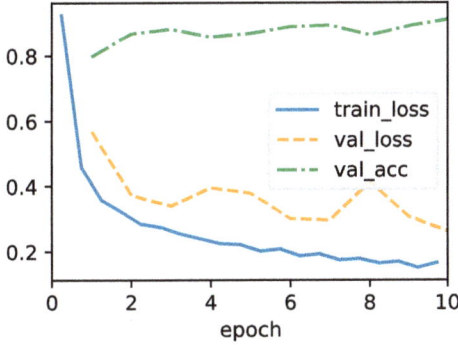

8.7.6 Summary and Discussion

The main components that comprise DenseNet are dense blocks and transition layers. For the latter, we need to keep the dimensionality under control when composing the network by adding transition layers that shrink the number of channels again. In terms of cross-layer connections, in contrast to ResNet, where inputs and outputs are added together, DenseNet concatenates inputs and outputs on the channel dimension. Although these concatenation operations reuse features to achieve computational efficiency, unfortunately they lead to heavy GPU memory consumption. As a result, applying DenseNet may require more memory-efficient implementations that may increase training time (Pleiss *et al.*, 2017).

8.7.7 Exercises

1. Why do we use average pooling rather than max-pooling in the transition layer?

2. One of the advantages mentioned in the DenseNet paper is that its model parameters are smaller than those of ResNet. Why is this the case?

3. One problem for which DenseNet has been criticized is its high memory consumption.

    1. Is this really the case? Try to change the input shape to 224×224 to compare the actual GPU memory consumption empirically.

    2. Can you think of an alternative means of reducing the memory consumption? How would you need to change the framework?

4. Implement the various DenseNet versions presented in Table 1 of the DenseNet paper (Huang *et al.*, 2017).

5. Design an MLP-based model by applying the DenseNet idea. Apply it to the housing price prediction task in Section 5.7.

Discussions [134].

8.8 Designing Convolution Network Architectures

The previous sections have taken us on a tour of modern network design for computer vision. Common to all the work we covered was that it greatly relied on the intuition of scientists. Many of the architectures are heavily informed by human creativity and to a much lesser extent by systematic exploration of the design space that deep networks offer. Nonetheless, this *network engineering* approach has been tremendously successful.

Ever since AlexNet (Section 8.1) beat conventional computer vision models on ImageNet, it has become popular to construct very deep networks by stacking blocks of convolutions, all designed according to the same pattern. In particular, 3×3 convolutions were popularized by VGG networks (Section 8.2). NiN (Section 8.3) showed that even 1×1 convolutions could be beneficial by adding local nonlinearities. Moreover, NiN solved the problem of aggregating information at the head of a network by aggregating across all locations. GoogLeNet (Section 8.4) added multiple branches of different convolution width, combining the advantages of VGG and NiN in its Inception block. ResNets (Section 8.6) changed the inductive bias towards the identity mapping (from $f(x) = 0$). This allowed for very deep networks. Almost a decade later, the ResNet design is still popular, a testament to its design. Lastly, ResNeXt (Section 8.6.5) added grouped convolutions, offering a better trade-off between parameters and computation. A precursor to Transformers for vision, the Squeeze-and-Excitation Networks (SENets) allow for efficient information transfer between locations (Hu *et al.*, 2018). This was accomplished by computing a per-channel global attention function.

Up to now we have omitted networks obtained via *neural architecture search* (NAS) (Liu *et al.*, 2018, Zoph and Le, 2016). We chose to do so since their cost is usually enormous, relying on brute-force search, genetic algorithms, reinforcement learning, or some other form of hyperparameter optimization. Given a fixed search space, NAS uses a search strategy to automatically select an architecture based on the returned performance estimation. The outcome of NAS is a single network instance. EfficientNets are a notable outcome of this search (Tan and Le, 2019).

In the following we discuss an idea that is quite different to the quest for the *single best network*. It is computationally relatively inexpensive, it leads to scientific insights on the way, and it is quite effective in terms of the quality of outcomes. Let's review the strategy by Radosavovic *et al.* (2020) to *design network design spaces*. The strategy combines the strength of manual design and NAS. It accomplishes this by operating on *distributions of networks* and optimizing the distributions in a way to obtain good performance for entire families of networks. The outcome of it are *RegNets*, specifically RegNetX and RegNetY, plus a range of guiding principles for the design of performant CNNs.

```
import torch
from torch import nn
from torch.nn import functional as F
from d2l import torch as d2l
```

8.8.1 The AnyNet Design Space

The description below closely follows the reasoning in Radosavovic *et al.* (2020) with some abbreviations to make it fit in the scope of the book. To begin, we need a template for the family of networks to explore. One of the commonalities of the designs in this chapter is that the networks consist of a *stem*, a *body* and a *head*. The stem performs initial image processing, often through convolutions with a larger window size. The body consists of multiple blocks, carrying out the bulk of the transformations needed to go from raw images to object representations. Lastly, the head converts this into the desired outputs, such as via a softmax regressor for multiclass classification. The body, in turn, consists of multiple stages, operating on the image at decreasing resolutions. In fact, both the stem and each subsequent stage quarter the spatial resolution. Lastly, each stage consists of one or more blocks. This pattern is common to all networks, from VGG to ResNeXt. Indeed, for the design of generic AnyNet networks, Radosavovic *et al.* (2020) used the ResNeXt block of Fig. 8.6.5.

Let's review the structure outlined in Fig. 8.8.1 in detail. As mentioned, an AnyNet consists of a stem, body, and head. The stem takes as its input RGB images (3 channels), using a 3×3 convolution with a stride of 2, followed by a batch norm, to halve the resolution from $r \times r$ to $r/2 \times r/2$. Moreover, it generates c_0 channels that serve as input to the body.

Since the network is designed to work well with ImageNet images of shape $224 \times 224 \times 3$, the body serves to reduce this to $7 \times 7 \times c_4$ through 4 stages (recall that $224/2^{1+4} = 7$), each with an eventual stride of 2. Lastly, the head employs an entirely standard design via

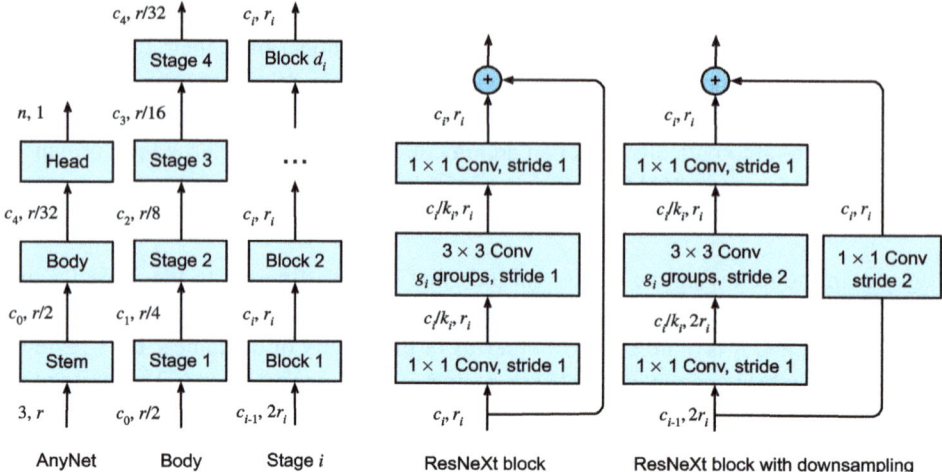

The AnyNet design space. The numbers (c, r) along each arrow indicate the number of channels c and the resolution $r \times r$ of the images at that point. From left to right: generic network structure composed of stem, body, and head; body composed of four stages; detailed structure of a stage; two alternative structures for blocks, one without downsampling and one that halves the resolution in each dimension. Design choices include depth d_i, the number of output channels c_i, the number of groups g_i, and bottleneck ratio k_i for any stage i.

global average pooling, similar to NiN (Section 8.3), followed by a fully connected layer to emit an n-dimensional vector for n-class classification.

Most of the relevant design decisions are inherent to the body of the network. It proceeds in stages, where each stage is composed of the same type of ResNeXt blocks as we discussed in Section 8.6.5. The design there is again entirely generic: we begin with a block that halves the resolution by using a stride of 2 (the rightmost in Fig. 8.8.1). To match this, the residual branch of the ResNeXt block needs to pass through a 1×1 convolution. This block is followed by a variable number of additional ResNeXt blocks that leave both resolution and the number of channels unchanged. Note that a common design practice is to add a slight bottleneck in the design of convolutional blocks. As such, with bottleneck ratio $k_i \geq 1$ we afford some number of channels, c_i/k_i, within each block for stage i (as the experiments show, this is not really effective and should be skipped). Lastly, since we are dealing with ResNeXt blocks, we also need to pick the number of groups g_i for grouped convolutions at stage i.

This seemingly generic design space provides us nonetheless with many parameters: we can set the block width (number of channels) $c_0, \ldots c_4$, the depth (number of blocks) per stage $d_1, \ldots d_4$, the bottleneck ratios $k_1, \ldots k_4$, and the group widths (numbers of groups) $g_1, \ldots g_4$. In total this adds up to 17 parameters, resulting in an unreasonably large number of configurations that would warrant exploring. We need some tools to reduce this huge design space effectively. This is where the conceptual beauty of design spaces comes in. Before we do so, let's implement the generic design first.

```
class AnyNet(d2l.Classifier):
    def stem(self, num_channels):
        return nn.Sequential(
            nn.LazyConv2d(num_channels, kernel_size=3, stride=2, padding=1),
            nn.LazyBatchNorm2d(), nn.ReLU())
```

Each stage consists of depth ResNeXt blocks, where num_channels specifies the block width. Note that the first block halves the height and width of input images.

```
@d2l.add_to_class(AnyNet)
def stage(self, depth, num_channels, groups, bot_mul):
    blk = []
    for i in range(depth):
        if i == 0:
            blk.append(d2l.ResNeXtBlock(num_channels, groups, bot_mul,
                use_1x1conv=True, strides=2))
        else:
            blk.append(d2l.ResNeXtBlock(num_channels, groups, bot_mul))
    return nn.Sequential(*blk)
```

Putting the network stem, body, and head together, we complete the implementation of AnyNet.

```
@d2l.add_to_class(AnyNet)
def __init__(self, arch, stem_channels, lr=0.1, num_classes=10):
    super(AnyNet, self).__init__()
    self.save_hyperparameters()
    self.net = nn.Sequential(self.stem(stem_channels))
    for i, s in enumerate(arch):
        self.net.add_module(f'stage{i+1}', self.stage(*s))
    self.net.add_module('head', nn.Sequential(
        nn.AdaptiveAvgPool2d((1, 1)), nn.Flatten(),
        nn.LazyLinear(num_classes)))
    self.net.apply(d2l.init_cnn)
```

8.8.2 Distributions and Parameters of Design Spaces

As just discussed in Section 8.8.1, parameters of a design space are hyperparameters of networks in that design space. Consider the problem of identifying good parameters in the AnyNet design space. We could try finding the *single best* parameter choice for a given amount of computation (e.g., FLOPs and compute time). If we allowed for even only *two* possible choices for each parameter, we would have to explore $2^{17} = 131072$ combinations to find the best solution. This is clearly infeasible because of its exorbitant cost. Even worse, we do not really learn anything from this exercise in terms of how one should design a network. Next time we add, say, an X-stage, or a shift operation, or similar, we would need to start from scratch. Even worse, due to the stochasticity in training (rounding, shuffling, bit errors), no two runs are likely to produce exactly the same results. A better strategy would be to try to determine general guidelines of how the choices of parameters should be related. For instance, the bottleneck ratio, the number of channels, blocks, groups, or their change between layers should ideally be governed by a collection of simple rules. The approach in Radosavovic *et al.* (2019) relies on the following four assumptions:

1. We assume that general design principles actually exist, so that many networks satisfying these requirements should offer good performance. Consequently, identifying a *distribution* over networks can be a sensible strategy. In other words, we assume that there are many good needles in the haystack.

2. We need not train networks to convergence before we can assess whether a network is good. Instead, it is sufficient to use the intermediate results as reliable guidance for final accuracy. Using (approximate) proxies to optimize an objective is referred to as multi-fidelity optimization (Forrester *et al.*, 2007). Consequently, design optimization is carried out, based on the accuracy achieved after only a few passes through the dataset, reducing the cost significantly.

3. Results obtained at a smaller scale (for smaller networks) generalize to larger ones. Consequently, optimization is carried out for networks that are structurally similar, but with a smaller number of blocks, fewer channels, etc. Only in the end will we need to verify that the so-found networks also offer good performance at scale.

4. Aspects of the design can be approximately factorized so that it is possible to infer their effect on the quality of the outcome somewhat independently. In other words, the optimization problem is moderately easy.

These assumptions allow us to test many networks cheaply. In particular, we can *sample* uniformly from the space of configurations and evaluate their performance. Subsequently, we can evaluate the quality of the choice of parameters by reviewing the *distribution* of error/accuracy that can be achieved with said networks. Denote by $F(e)$ the cumulative distribution function (CDF) for errors committed by networks of a given design space, drawn using probability disribution p. That is,

$$F(e, p) \stackrel{\text{def}}{=} P_{\text{net} \sim p}\{e(\text{net}) \le e\}. \tag{8.8.1}$$

Our goal is now to find a distribution p over *networks* such that most networks have a very low error rate and where the support of p is concise. Of course, this is computationally infeasible to perform accurately. We resort to a sample of networks $\mathcal{Z} \stackrel{\text{def}}{=} \{\text{net}_1, \ldots \text{net}_n\}$ (with errors $e_1, \ldots, e_n$, respectively) from p and use the empirical CDF $\hat{F}(e, \mathcal{Z})$ instead:

$$\hat{F}(e, \mathcal{Z}) = \frac{1}{n} \sum_{i=1}^{n} \mathbf{1}(e_i \le e). \tag{8.8.2}$$

Whenever the CDF for one set of choices majorizes (or matches) another CDF it follows that its choice of parameters is superior (or indifferent). Accordingly Radosavovic *et al.* (2020) experimented with a shared network bottleneck ratio $k_i = k$ for all stages i of the network. This gets rid of three of the four parameters governing the bottleneck ratio. To assess whether this (negatively) affects the performance one can draw networks from the constrained and from the unconstrained distribution and compare the corresonding CDFs.

It turns out that this constraint does not affect the accuracy of the distribution of networks at all, as can be seen in the first panel of Fig. 8.8.2. Likewise, we could choose to pick the same group width $g_i = g$ occurring at the various stages of the network. Again, this does not affect performance, as can be seen in the second panel of Fig. 8.8.2. Both steps combined reduce the number of free parameters by six.

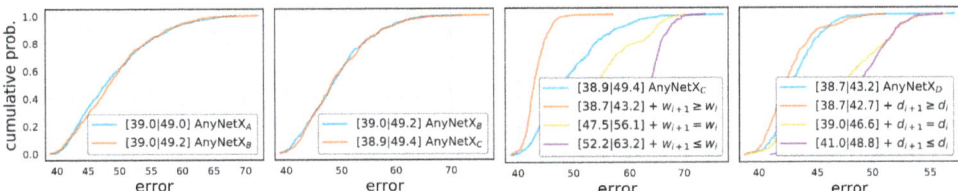

Fig. 8.8.2 Comparing error empirical distribution functions of design spaces. AnyNet$_A$ is the original design space; AnyNet$_B$ ties the bottleneck ratios, AnyNet$_C$ also ties group widths, AnyNet$_D$ increases the network depth across stages. From left to right: (i) tying bottleneck ratios has no effect on performance; (ii) tying group widths has no effect on performance; (iii) increasing network widths (channels) across stages improves performance; (iv) increasing network depths across stages improves performance. Figure courtesy of Radosavovic et al. (2020).

Next we look for ways to reduce the multitude of potential choices for width and depth of the stages. It is a reasonable assumption that, as we go deeper, the number of channels should increase, i.e., $c_i \geq c_{i-1}$ ($w_{i+1} \geq w_i$ per their notation in Fig. 8.8.2), yielding AnyNetX$_D$. Likewise, it is equally reasonable to assume that as the stages progress, they should become deeper, i.e., $d_i \geq d_{i-1}$, yielding AnyNetX$_E$. This can be experimentally verified in the third and fourth panel of Fig. 8.8.2, respectively.

8.8.3 RegNet

The resulting AnyNetX$_E$ design space consists of simple networks following easy-to-interpret design principles:

- Share the bottleneck ratio $k_i = k$ for all stages i;

- Share the group width $g_i = g$ for all stages i;

- Increase network width across stages: $c_i \leq c_{i+1}$;

- Increase network depth across stages: $d_i \leq d_{i+1}$.

This leaves us with a final set of choices: how to pick the specific values for the above parameters of the eventual AnyNetX$_E$ design space. By studying the best-performing networks from the distribution in AnyNetX$_E$ one can observe the following: the width of the network ideally increases linearly with the block index across the network, i.e., $c_j \approx c_0 + c_a j$, where j is the block index and slope $c_a > 0$. Given that we get to choose a different block width only per stage, we arrive at a piecewise constant function, engineered to match this dependence. Furthermore, experiments also show that a bottleneck ratio of $k = 1$ performs best, i.e., we are advised not to use bottlenecks at all.

We recommend the interested reader reviews further details in the design of specific networks for different amounts of computation by perusing Radosavovic *et al.* (2020). For instance, an effective 32-layer RegNetX variant is given by $k = 1$ (no bottleneck), $g = 16$ (group width is 16), $c_1 = 32$ and $c_2 = 80$ channels for the first and second stage, respectively, chosen to be $d_1 = 4$ and $d_2 = 6$ blocks deep. The astonishing insight from the design is that it still applies, even when investigating networks at a larger scale. Even better, it even holds for Squeeze-and-Excitation (SE) network designs (RegNetY) that have a global channel activation (Hu *et al.*, 2018).

```python
class RegNetX32(AnyNet):
    def __init__(self, lr=0.1, num_classes=10):
        stem_channels, groups, bot_mul = 32, 16, 1
        depths, channels = (4, 6), (32, 80)
        super().__init__(
            ((depths[0], channels[0], groups, bot_mul),
             (depths[1], channels[1], groups, bot_mul)),
            stem_channels, lr, num_classes)
```

We can see that each RegNetX stage progressively reduces resolution and increases output channels.

```python
RegNetX32().layer_summary((1, 1, 96, 96))
```

```
Sequential output shape:     torch.Size([1, 32, 48, 48])
Sequential output shape:     torch.Size([1, 32, 24, 24])
Sequential output shape:     torch.Size([1, 80, 12, 12])
Sequential output shape:     torch.Size([1, 10])
```

8.8.4 Training

Training the 32-layer RegNetX on the Fashion-MNIST dataset is just like before.

```python
model = RegNetX32(lr=0.05)
trainer = d2l.Trainer(max_epochs=10, num_gpus=1)
data = d2l.FashionMNIST(batch_size=128, resize=(96, 96))
trainer.fit(model, data)
```

8.8.5 Discussion

With desirable inductive biases (assumptions or preferences) like locality and translation invariance (Section 7.1) for vision, CNNs have been the dominant architectures in this area. This remained the case from LeNet up until Transformers (Section 11.7) (Dosovitskiy *et al.*, 2021, Touvron *et al.*, 2021) started surpassing CNNs in terms of accuracy. While much of the recent progress in terms of vision Transformers *can* be backported into CNNs (Liu *et al.*, 2022), it is only possible at a higher computational cost. Just as importantly, recent hardware optimizations (NVIDIA Ampere and Hopper) have only widened the gap in favor of Transformers.

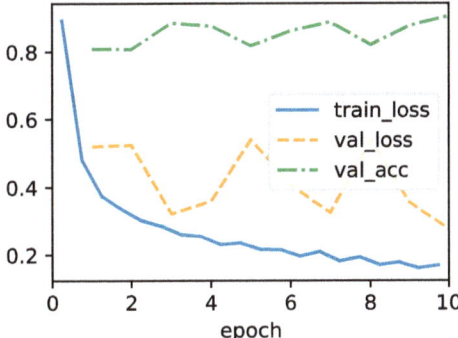

It is worth noting that Transformers have a significantly lower degree of inductive bias towards locality and translation invariance than CNNs. That learned structures prevailed is due, not least, to the availability of large image collections, such as LAION-400m and LAION-5B (Schuhmann *et al.*, 2022) with up to 5 billion images. Quite surprisingly, some of the more relevant work in this context even includes MLPs (Tolstikhin *et al.*, 2021).

In sum, vision Transformers (Section 11.8) by now lead in terms of state-of-the-art performance in large-scale image classification, showing that *scalability trumps inductive biases* (Dosovitskiy *et al.*, 2021). This includes pretraining large-scale Transformers (Section 11.9) with multi-head self-attention (Section 11.5). We invite the readers to dive into these chapters for a much more detailed discussion.

8.8.6 Exercises

1. Increase the number of stages to four. Can you design a deeper RegNetX that performs better?

2. De-ResNeXt-ify RegNets by replacing the ResNeXt block with the ResNet block. How does your new model perform?

3. Implement multiple instances of a "VioNet" family by *violating* the design principles of RegNetX. How do they perform? Which of (d_i, c_i, g_i, b_i) is the most important factor?

4. Your goal is to design the "perfect" MLP. Can you use the design principles introduced above to find good architectures? Is it possible to extrapolate from small to large networks?

Discussions[135].

9 Recurrent Neural Networks

Up until now, we have focused primarily on fixed-length data. When introducing linear and logistic regression in Chapter 3 and Chapter 4 and multilayer perceptrons in Chapter 5, we were happy to assume that each feature vector $\mathbf{x}_i$ consisted of a fixed number of components $x_1, \ldots, x_d$, where each numerical feature x_j corresponded to a particular attribute. These datasets are sometimes called *tabular*, because they can be arranged in tables, where each example i gets its own row, and each attribute gets its own column. Crucially, with tabular data, we seldom assume any particular structure over the columns.

Subsequently, in Chapter 7, we moved on to image data, where inputs consist of the raw pixel values at each coordinate in an image. Image data hardly fitted the bill of a protypical tabular dataset. There, we needed to call upon convolutional neural networks (CNNs) to handle the hierarchical structure and invariances. However, our data were still of fixed length. Every Fashion-MNIST image is represented as a 28×28 grid of pixel values. Moreover, our goal was to develop a model that looked at just one image and then outputted a single prediction. But what should we do when faced with a sequence of images, as in a video, or when tasked with producing a sequentially structured prediction, as in the case of image captioning?

A great many learning tasks require dealing with sequential data. Image captioning, speech synthesis, and music generation all require that models produce outputs consisting of sequences. In other domains, such as time series prediction, video analysis, and musical information retrieval, a model must learn from inputs that are sequences. These demands often arise simultaneously: tasks such as translating passages of text from one natural language to another, engaging in dialogue, or controlling a robot, demand that models both ingest and output sequentially structured data.

Recurrent neural networks (RNNs) are deep learning models that capture the dynamics of sequences via *recurrent* connections, which can be thought of as cycles in the network of nodes. This might seem counterintuitive at first. After all, it is the feedforward nature of neural networks that makes the order of computation unambiguous. However, recurrent edges are defined in a precise way that ensures that no such ambiguity can arise. Recurrent neural networks are *unrolled* across time steps (or sequence steps), with the *same* underlying parameters applied at each step. While the standard connections are applied *synchronously* to propagate each layer's activations to the subsequent layer *at the same time step*, the recurrent connections are *dynamic*, passing information across adjacent time steps. As the unfolded view in Fig. 9.1 reveals, RNNs can be thought of as feedforward neural networks where each layer's parameters (both conventional and recurrent) are shared across time steps.

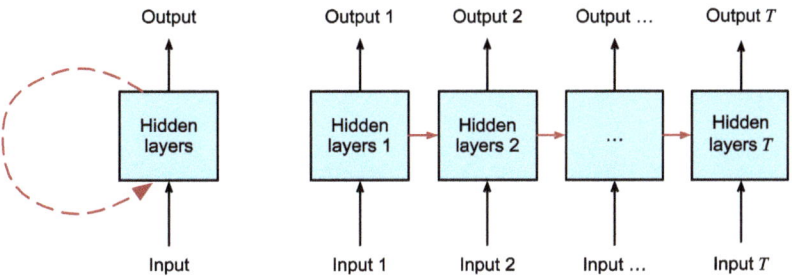

On the left recurrent connections are depicted via cyclic edges. On the right, we unfold the RNN over time steps. Here, recurrent edges span adjacent time steps, while conventional connections are computed synchronously.

Like neural networks more broadly, RNNs have a long discipline-spanning history, originating as models of the brain popularized by cognitive scientists and subsequently adopted as practical modeling tools employed by the machine learning community. As we do for deep learning more broadly, in this book we adopt the machine learning perspective, focusing on RNNs as practical tools that rose to popularity in the 2010s owing to breakthrough results on such diverse tasks as handwriting recognition (Graves *et al.*, 2008), machine translation (Sutskever *et al.*, 2014), and recognizing medical diagnoses (Lipton *et al.*, 2016). We point the reader interested in more background material to a publicly available comprehensive review (Lipton *et al.*, 2015). We also note that sequentiality is not unique to RNNs. For example, the CNNs that we already introduced can be adapted to handle data of varying length, e.g., images of varying resolution. Moreover, RNNs have recently ceded considerable market share to Transformer models, which will be covered in Chapter 11. However, RNNs rose to prominence as the default models for handling complex sequential structure in deep learning, and remain staple models for sequential modeling to this day. The stories of RNNs and of sequence modeling are inextricably linked, and this is as much a chapter about the ABCs of sequence modeling problems as it is a chapter about RNNs.

One key insight paved the way for a revolution in sequence modeling. While the inputs and targets for many fundamental tasks in machine learning cannot easily be represented as fixed-length vectors, they can often nevertheless be represented as varying-length sequences of fixed-length vectors. For example, documents can be represented as sequences of words; medical records can often be represented as sequences of events (encounters, medications, procedures, lab tests, diagnoses); videos can be represented as varying-length sequences of still images.

While sequence models have popped up in numerous application areas, basic research in the area has been driven predominantly by advances on core tasks in natural language processing. Thus, throughout this chapter, we will focus our exposition and examples on text data. If you get the hang of these examples, then applying the models to other data modalities should be relatively straightforward. In the next few sections, we introduce basic notation for sequences and some evaluation measures for assessing the quality of sequentially structured model outputs. After that, we discuss basic concepts of a language model and use this discussion to motivate our first RNN models. Finally, we describe the method for calculating gradients when backpropagating through RNNs and explore some challenges that are often encountered when training such networks, motivating the modern RNN architectures that will follow in Chapter 10.

9.1 Working with Sequences

Up until now, we have focused on models whose inputs consisted of a single feature vector $\mathbf{x} \in \mathbb{R}^d$. The main change of perspective when developing models capable of processing sequences is that we now focus on inputs that consist of an ordered list of feature vectors $\mathbf{x}_1, \ldots, \mathbf{x}_T$, where each feature vector $\mathbf{x}_t$ is indexed by a time step $t \in \mathbb{Z}^+$ lying in $\mathbb{R}^d$.

Some datasets consist of a single massive sequence. Consider, for example, the extremely long streams of sensor readings that might be available to climate scientists. In such cases, we might create training datasets by randomly sampling subsequences of some predetermined length. More often, our data arrives as a collection of sequences. Consider the following examples: (i) a collection of documents, each represented as its own sequence of words, and each having its own length T_i; (ii) sequence representation of patient stays in the hospital, where each stay consists of a number of events and the sequence length depends roughly on the length of the stay.

Previously, when dealing with individual inputs, we assumed that they were sampled independently from the same underlying distribution $P(X)$. While we still assume that entire sequences (e.g., entire documents or patient trajectories) are sampled independently, we cannot assume that the data arriving at each time step are independent of each other. For example, the words that likely to appear later in a document depend heavily on words occurring earlier in the document. The medicine a patient is likely to receive on the 10th day of a hospital visit depends heavily on what transpired in the previous nine days.

This should come as no surprise. If we did not believe that the elements in a sequence were related, we would not have bothered to model them as a sequence in the first place. Consider the usefulness of the auto-fill features that are popular on search tools and modern email clients. They are useful precisely because it is often possible to predict (imperfectly, but better than random guessing) what the likely continuations of a sequence might be, given some initial prefix. For most sequence models, we do not require independence, or even stationarity, of our sequences. Instead, we require only that the sequences themselves are sampled from some fixed underlying distribution over entire sequences.

This flexible approach allows for such phenomena as (i) documents looking significantly different at the beginning than at the end; or (ii) patient status evolving either towards recovery or towards death over the course of a hospital stay; or (iii) customer taste evolving in predictable ways over the course of continued interaction with a recommender system.

We sometimes wish to predict a fixed target y given sequentially structured input (e.g., sentiment classification based on a movie review). At other times, we wish to predict a sequentially structured target $(y_1, \ldots, y_T)$ given a fixed input (e.g., image captioning). Still other times, our goal is to predict sequentially structured targets based on sequentially structured inputs (e.g., machine translation or video captioning). Such sequence-to-sequence tasks take two forms: (i) *aligned*: where the input at each time step aligns with a corresponding target (e.g., part of speech tagging); (ii) *unaligned*: where the input and target do not necessarily exhibit a step-for-step correspondence (e.g., machine translation).

Before we worry about handling targets of any kind, we can tackle the most straightforward problem: unsupervised density modeling (also called *sequence modeling*). Here, given a collection of sequences, our goal is to estimate the probability mass function that tells us how likely we are to see any given sequence, i.e., $p(\mathbf{x}_1, \ldots, \mathbf{x}_T)$.

```
%matplotlib inline
import torch
from torch import nn
from d2l import torch as d2l
```

9.1.1 Autoregressive Models

Before introducing specialized neural networks designed to handle sequentially structured data, let's take a look at some actual sequence data and build up some basic intuitions and statistical tools. In particular, we will focus on stock price data from the FTSE 100 index (Fig. 9.1.1). At each *time step* $t \in \mathbb{Z}^+$, we observe the price, x_t, of the index at that time.

Fig. 9.1.1   FTSE 100 index over about 30 years.

Now suppose that a trader would like to make short-term trades, strategically getting into or out of the index, depending on whether they believe that it will rise or decline in the subsequent time step. Absent any other features (news, financial reporting data, etc.), the only available signal for predicting the subsequent value is the history of prices to date. The trader is thus interested in knowing the probability distribution

$$P(x_t \mid x_{t-1}, \ldots, x_1) \qquad (9.1.1)$$

over prices that the index might take in the subsequent time step. While estimating the entire distribution over a continuously valued random variable can be difficult, the trader would be happy to focus on a few key statistics of the distribution, particularly the expected value and the variance. One simple strategy for estimating the conditional expectation

$$\mathbb{E}[(x_t \mid x_{t-1}, \ldots, x_1)], \qquad (9.1.2)$$

would be to apply a linear regression model (recall Section 3.1). Such models that regress the value of a signal on the previous values of that same signal are naturally called *autoregressive models*. There is just one major problem: the number of inputs, $x_{t-1}, \ldots, x_1$ varies, depending on t. In other words, the number of inputs increases with the amount of data that we encounter. Thus if we want to treat our historical data as a training set, we are left with the problem that each example has a different number of features. Much of what follows in this chapter will revolve around techniques for overcoming these challenges when engaging in such *autoregressive* modeling problems where the object of interest is $P(x_t \mid x_{t-1}, \ldots, x_1)$ or some statistic(s) of this distribution.

A few strategies recur frequently. First of all, we might believe that although long sequences $x_{t-1}, \ldots, x_1$ are available, it may not be necessary to look back so far in the history when predicting the near future. In this case we might content ourselves to condition on some window of length τ and only use $x_{t-1}, \ldots, x_{t-\tau}$ observations. The immediate benefit is that now the number of arguments is always the same, at least for $t > \tau$. This allows us to train any linear model or deep network that requires fixed-length vectors as inputs. Second, we might develop models that maintain some summary h_t of the past observations (see Fig. 9.1.2) and at the same time update h_t in addition to the prediction $\hat{x}_t$. This leads to models that estimate not only x_t with $\hat{x}_t = P(x_t \mid h_t)$ but also updates of the form $h_t = g(h_{t-1}, x_{t-1})$. Since h_t is never observed, these models are also called *latent autoregressive models*.

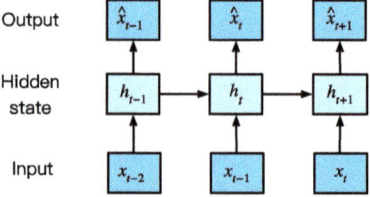

Input

Fig. 9.1.2   A latent autoregressive model.

To construct training data from historical data, one typically creates examples by sampling windows randomly. In general, we do not expect time to stand still. However, we often assume that while the specific values of x_t might change, the dynamics according to which each subsequent observation is generated given the previous observations do not. Statisticians call dynamics that do not change *stationary*.

9.1.2 Sequence Models

Sometimes, especially when working with language, we wish to estimate the joint probability of an entire sequence. This is a common task when working with sequences composed of discrete *tokens*, such as words. Generally, these estimated functions are called *sequence models* and for natural language data, they are called *language models*. The field of sequence modeling has been driven so much by natural language processing, that we often describe sequence models as "language models", even when dealing with non-language data. Language models prove useful for all sorts of reasons. Sometimes we want to evaluate the likelihood of sentences. For example, we might wish to compare the naturalness of two candidate outputs generated by a machine translation system or by a speech recognition system. But language modeling gives us not only the capacity to *evaluate* likelihood, but the ability to *sample* sequences, and even to optimize for the most likely sequences.

While language modeling might not, at first glance, look like an autoregressive problem, we can reduce language modeling to autoregressive prediction by decomposing the joint density of a sequence $p(x_1, \ldots, x_T)$ into the product of conditional densities in a left-to-right fashion by applying the chain rule of probability:

$$P(x_1, \ldots, x_T) = P(x_1) \prod_{t=2}^{T} P(x_t \mid x_{t-1}, \ldots, x_1). \qquad (9.1.3)$$

Note that if we are working with discrete signals such as words, then the autoregressive model must be a probabilistic classifier, outputting a full probability distribution over the vocabulary for whatever word will come next, given the leftwards context.

Markov Models

Now suppose that we wish to employ the strategy mentioned above, where we condition only on the τ previous time steps, i.e., $x_{t-1}, \ldots, x_{t-\tau}$, rather than the entire sequence history $x_{t-1}, \ldots, x_1$. Whenever we can throw away the history beyond the previous τ steps without any loss in predictive power, we say that the sequence satisfies a *Markov condition*, i.e., *that the future is conditionally independent of the past, given the recent history*. When $\tau = 1$, we say that the data is characterized by a *first-order Markov model*, and when $\tau = k$, we say that the data is characterized by a kth-order Markov model. For when the first-order Markov condition holds ($\tau = 1$) the factorization of our joint probability becomes a product of probabilities of each word given the previous *word*:

$$P(x_1, \ldots, x_T) = P(x_1) \prod_{t=2}^{T} P(x_t \mid x_{t-1}). \qquad (9.1.4)$$

We often find it useful to work with models that proceed as though a Markov condition were satisfied, even when we know that this is only *approximately* true. With real text documents we continue to gain information as we include more and more leftwards context. But these gains diminish rapidly. Thus, sometimes we compromise, obviating computational and statistical difficulties by training models whose validity depends on a kth-order Markov condition. Even today's massive RNN- and Transformer-based language models seldom incorporate more than thousands of words of context.

With discrete data, a true Markov model simply counts the number of times that each word has occurred in each context, producing the relative frequency estimate of $P(x_t \mid x_{t-1})$. Whenever the data assumes only discrete values (as in language), the most likely sequence of words can be computed efficiently using dynamic programming.

The Order of Decoding

You may be wondering why we represented the factorization of a text sequence $P(x_1, \ldots, x_T)$ as a left-to-right chain of conditional probabilities. Why not right-to-left or some other, seemingly random order? In principle, there is nothing wrong with unfolding $P(x_1, \ldots, x_T)$ in reverse order. The result is a valid factorization:

$$P(x_1, \ldots, x_T) = P(x_T) \prod_{t=T-1}^{1} P(x_t \mid x_{t+1}, \ldots, x_T). \tag{9.1.5}$$

However, there are many reasons why factorizing text in the same direction in which we read it (left-to-right for most languages, but right-to-left for Arabic and Hebrew) is preferred for the task of language modeling. First, this is just a more natural direction for us to think about. After all we all read text every day, and this process is guided by our ability to anticipate which words and phrases are likely to come next. Just think of how many times you have completed someone else's sentence. Thus, even if we had no other reason to prefer such in-order decodings, they would be useful if only because we have better intuitions for what should be likely when predicting in this order.

Second, by factorizing in order, we can assign probabilities to arbitrarily long sequences using the same language model. To convert a probability over steps 1 through t into one that extends to word $t + 1$ we simply multiply by the conditional probability of the additional token given the previous ones: $P(x_{t+1}, \ldots, x_1) = P(x_t, \ldots, x_1) \cdot P(x_{t+1} \mid x_t, \ldots, x_1)$.

Third, we have stronger predictive models for predicting adjacent words than words at arbitrary other locations. While all orders of factorization are valid, they do not necessarily all represent equally easy predictive modeling problems. This is true not only for language, but for other kinds of data as well, e.g., when the data is causally structured. For example, we believe that future events cannot influence the past. Hence, if we change x_t, we may be able to influence what happens for x_{t+1} going forward but not the converse. That is, if we change x_t, the distribution over past events will not change. In some contexts, this makes it easier to predict $P(x_{t+1} \mid x_t)$ than to predict $P(x_t \mid x_{t+1})$. For instance, in some cases, we can find $x_{t+1} = f(x_t) + \epsilon$ for some additive noise ϵ, whereas the converse is not true (Hoyer *et al.*, 2009). This is great news, since it is typically the forward direction that we are interested in estimating. The book by Peters *et al.* (2017) contains more on this topic. We barely scratch the surface of it.

9.1.3 Training

Before we focus our attention on text data, let's first try this out with some continuous-valued synthetic data.

Here, our 1000 synthetic data will follow the trigonometric sin function, applied to 0.01 times the time step. To make the problem a little more interesting, we corrupt each sample with additive noise. From this sequence we extract training examples, each consisting of features and a label.

```
class Data(d2l.DataModule):
    def __init__(self, batch_size=16, T=1000, num_train=600, tau=4):
        self.save_hyperparameters()
        self.time = torch.arange(1, T + 1, dtype=torch.float32)
        self.x = torch.sin(0.01 * self.time) + torch.randn(T) * 0.2
```

```
data = Data()
d2l.plot(data.time, data.x, 'time', 'x', xlim=[1, 1000], figsize=(6, 3))
```

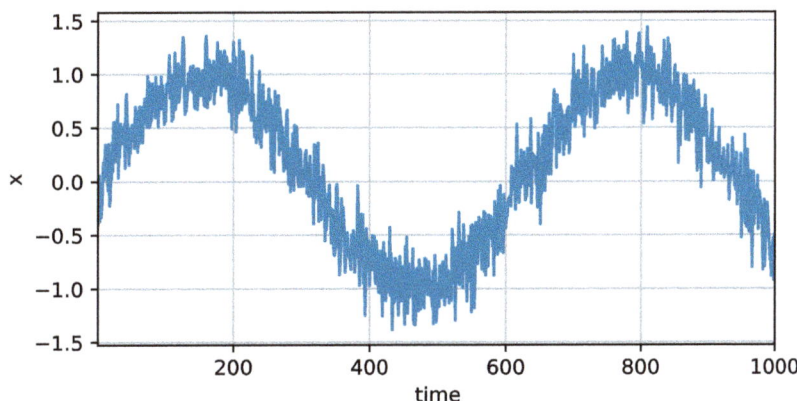

To begin, we try a model that acts as if the data satisfied a τth-order Markov condition, and thus predicts x_t using only the past τ observations. Thus for each time step we have an example with label $y = x_t$ and features $\mathbf{x}_t = [x_{t-\tau}, \ldots, x_{t-1}]$. The astute reader might have noticed that this results in $1000 - \tau$ examples, since we lack sufficient history for $y_1, \ldots, y_\tau$. While we could pad the first τ sequences with zeros, to keep things simple, we drop them for now. The resulting dataset contains $T - \tau$ examples, where each input to the model has sequence length τ. We create a data iterator on the first 600 examples, covering a period of the sin function.

```
@d2l.add_to_class(Data)
def get_dataloader(self, train):
    features = [self.x[i : self.T-self.tau+i] for i in range(self.tau)]
    self.features = torch.stack(features, 1)
    self.labels = self.x[self.tau:].reshape((-1, 1))
```

(continues on next page)

(continued from previous page)

```
    i = slice(0, self.num_train) if train else slice(self.num_train, None)
    return self.get_tensorloader([self.features, self.labels], train, i)
```

In this example our model will be a standard linear regression.

```
model = d2l.LinearRegression(lr=0.01)
trainer = d2l.Trainer(max_epochs=5)
trainer.fit(model, data)
```

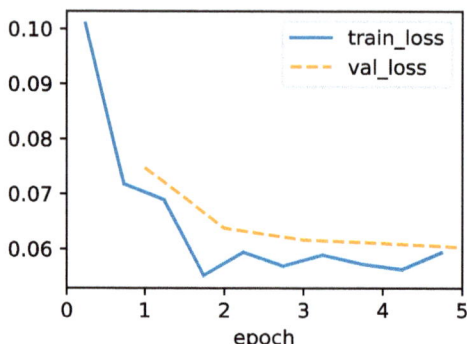

9.1.4 Prediction

To evaluate our model, we first check how well it performs at one-step-ahead prediction.

```
onestep_preds = model(data.features).detach().numpy()
d2l.plot(data.time[data.tau:], [data.labels, onestep_preds], 'time', 'x',
        legend=['labels', '1-step preds'], figsize=(6, 3))
```

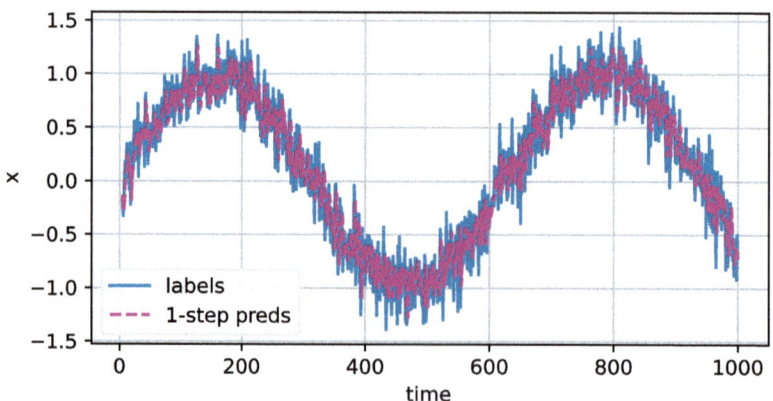

These predictions look good, even near the end at $t = 1000$.

But what if we only observed sequence data up until time step 604 (n_train + tau) and wished to make predictions several steps into the future? Unfortunately, we cannot directly compute the one-step-ahead prediction for time step 609, because we do not know the corresponding inputs, having seen only up to x_{604}. We can address this problem by plugging in our earlier predictions as inputs to our model for making subsequent predictions, projecting forward, one step at a time, until reaching the desired time step:

$$
\begin{aligned}
\hat{x}_{605} &= f(x_{601}, x_{602}, x_{603}, x_{604}), \\
\hat{x}_{606} &= f(x_{602}, x_{603}, x_{604}, \hat{x}_{605}), \\
\hat{x}_{607} &= f(x_{603}, x_{604}, \hat{x}_{605}, \hat{x}_{606}), \\
\hat{x}_{608} &= f(x_{604}, \hat{x}_{605}, \hat{x}_{606}, \hat{x}_{607}), \\
\hat{x}_{609} &= f(\hat{x}_{605}, \hat{x}_{606}, \hat{x}_{607}, \hat{x}_{608}), \\
&\vdots
\end{aligned}
\tag{9.1.6}
$$

Generally, for an observed sequence $x_1, \ldots, x_t$, its predicted output $\hat{x}_{t+k}$ at time step $t + k$ is called the *k-step-ahead prediction*. Since we have observed up to x_{604}, its k-step-ahead prediction is $\hat{x}_{604+k}$. In other words, we will have to keep on using our own predictions to make multistep-ahead predictions. Let's see how well this goes.

```
multistep_preds = torch.zeros(data.T)
multistep_preds[:] = data.x
for i in range(data.num_train + data.tau, data.T):
    multistep_preds[i] = model(
        multistep_preds[i - data.tau:i].reshape((1, -1)))
multistep_preds = multistep_preds.detach().numpy()
```

```
d2l.plot([data.time[data.tau:], data.time[data.num_train+data.tau:]],
         [onestep_preds, multistep_preds[data.num_train+data.tau:]], 'time',
         'x', legend=['1-step preds', 'multistep preds'], figsize=(6, 3))
```

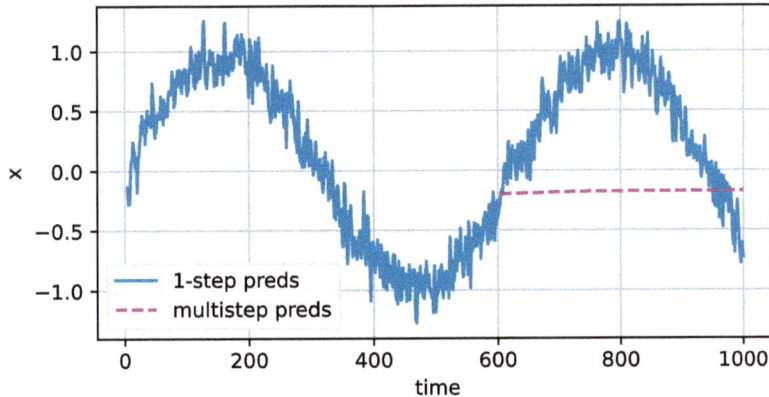

Unfortunately, in this case we fail spectacularly. The predictions decay to a constant pretty quickly after a few steps. Why did the algorithm perform so much worse when predicting further into the future? Ultimately, this is down to the fact that errors build up. Let's say that after step 1 we have some error $\epsilon_1 = \bar{\epsilon}$. Now the *input* for step 2 is perturbed by ϵ_1, hence we suffer some error in the order of $\epsilon_2 = \bar{\epsilon} + c\epsilon_1$ for some constant c, and so on. The predictions can diverge rapidly from the true observations. You may already be familiar with this common phenomenon. For instance, weather forecasts for the next 24 hours tend to be pretty accurate but beyond that, accuracy declines rapidly. We will discuss methods for improving this throughout this chapter and beyond.

Let's take a closer look at the difficulties in k-step-ahead predictions by computing predictions on the entire sequence for $k = 1, 4, 16, 64$.

```python
def k_step_pred(k):
    features = []
    for i in range(data.tau):
        features.append(data.x[i : i+data.T-data.tau-k+1])
    # The (i+tau)-th element stores the (i+1)-step-ahead predictions
    for i in range(k):
        preds = model(torch.stack(features[i : i+data.tau], 1))
        features.append(preds.reshape(-1))
    return features[data.tau:]
```

```python
steps = (1, 4, 16, 64)
preds = k_step_pred(steps[-1])
d2l.plot(data.time[data.tau+steps[-1]-1:],
         [preds[k - 1].detach().numpy() for k in steps], 'time', 'x',
         legend=[f'{k}-step preds' for k in steps], figsize=(6, 3))
```

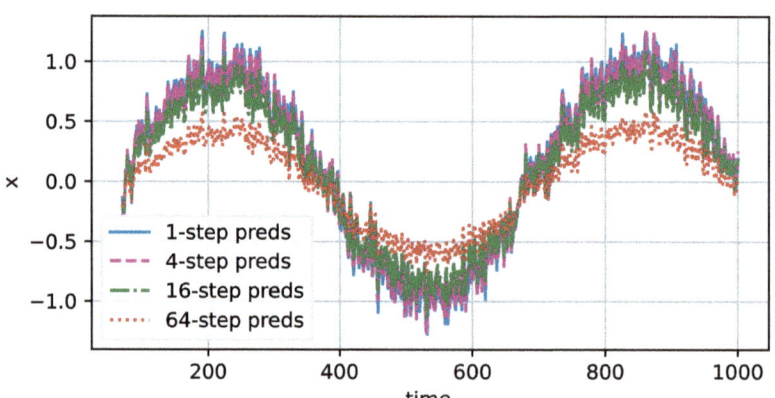

This clearly illustrates how the quality of the prediction changes as we try to predict further into the future. While the 4-step-ahead predictions still look good, anything beyond that is almost useless.

9.1.5 Summary

There is quite a difference in difficulty between interpolation and extrapolation. Consequently, if you have a sequence, always respect the temporal order of the data when training, i.e., never train on future data. Given this kind of data, sequence models require specialized statistical tools for estimation. Two popular choices are autoregressive models and latent-variable autoregressive models. For causal models (e.g., time going forward), estimating the forward direction is typically a lot easier than the reverse direction. For an observed sequence up to time step t, its predicted output at time step $t + k$ is the k-*step-ahead prediction*. As we predict further in time by increasing k, the errors accumulate and the quality of the prediction degrades, often dramatically.

9.1.6 Exercises

1. Improve the model in the experiment of this section.

   1. Incorporate more than the past four observations? How many do you really need?

   2. How many past observations would you need if there was no noise? Hint: you can write sin and cos as a differential equation.

   3. Can you incorporate older observations while keeping the total number of features constant? Does this improve accuracy? Why?

   4. Change the neural network architecture and evaluate the performance. You may train the new model with more epochs. What do you observe?

2. An investor wants to find a good security to buy. They look at past returns to decide which one is likely to do well. What could possibly go wrong with this strategy?

3. Does causality also apply to text? To which extent?

4. Give an example for when a latent autoregressive model might be needed to capture the dynamic of the data.

Discussions[136].

9.2 Converting Raw Text into Sequence Data

Throughout this book, we will often work with text data represented as sequences of words, characters, or word pieces. To get going, we will need some basic tools for converting raw text into sequences of the appropriate form. Typical preprocessing pipelines execute the following steps:

1. Load text as strings into memory.

2. Split the strings into tokens (e.g., words or characters).

3. Build a vocabulary dictionary to associate each vocabulary element with a numerical index.

4. Convert the text into sequences of numerical indices.

```
import collections
import random
import re
import torch
from d2l import torch as d2l
```

9.2.1 Reading the Dataset

137

Here, we will work with H. G. Wells' The Time Machine [137], a book containing just over 30,000 words. While real applications will typically involve significantly larger datasets, this is sufficient to demonstrate the preprocessing pipeline. The following _download method reads the raw text into a string.

```
class TimeMachine(d2l.DataModule): #@save
    """The Time Machine dataset."""
    def _download(self):
        fname = d2l.download(d2l.DATA_URL + 'timemachine.txt', self.root,
                             '090b5e7e70c295757f55df93cb0a180b9691891a')
        with open(fname) as f:
            return f.read()

data = TimeMachine()
raw_text = data._download()
raw_text[:60]
```

```
'The Time Machine, by H. G. Wells [1898]nnnnnInnnThe Time Tra'
```

For simplicity, we ignore punctuation and capitalization when preprocessing the raw text.

```
@d2l.add_to_class(TimeMachine)   #@save
def _preprocess(self, text):
    return re.sub('[^A-Za-z]+', ' ', text).lower()

text = data._preprocess(raw_text)
text[:60]
```

```
'the time machine by h g wells i the time traveller for so it'
```

9.2.2 Tokenization

Tokens are the atomic (indivisible) units of text. Each time step corresponds to 1 token, but what precisely constitutes a token is a design choice. For example, we could represent the sentence "Baby needs a new pair of shoes" as a sequence of 7 words, where the set of all words comprise a large vocabulary (typically tens or hundreds of thousands of words). Or we would represent the same sentence as a much longer sequence of 30 characters, using a much smaller vocabulary (there are only 256 distinct ASCII characters). Below, we tokenize our preprocessed text into a sequence of characters.

```
@d2l.add_to_class(TimeMachine)    #@save
def _tokenize(self, text):
    return list(text)

tokens = data._tokenize(text)
','.join(tokens[:30])
```

```
't,h,e, ,t,i,m,e, ,m,a,c,h,i,n,e, ,b,y, ,h, ,g, ,w,e,l,l,s, '
```

9.2.3 Vocabulary

These tokens are still strings. However, the inputs to our models must ultimately consist of numerical inputs. Next, we introduce a class for constructing *vocabularies*, i.e., objects that associate each distinct token value with a unique index. First, we determine the set of unique tokens in our training *corpus*. We then assign a numerical index to each unique token. Rare vocabulary elements are often dropped for convenience. Whenever we encounter a token at training or test time that had not been previously seen or was dropped from the vocabulary, we represent it by a special "<unk>" token, signifying that this is an *unknown* value.

```
class Vocab:   #@save
    """Vocabulary for text."""
    def __init__(self, tokens=[], min_freq=0, reserved_tokens=[]):
        # Flatten a 2D list if needed
        if tokens and isinstance(tokens[0], list):
            tokens = [token for line in tokens for token in line]
        # Count token frequencies
        counter = collections.Counter(tokens)
        self.token_freqs = sorted(counter.items(), key=lambda x: x[1],
                                  reverse=True)
        # The list of unique tokens
        self.idx_to_token = list(sorted(set(['<unk>'] + reserved_tokens + [
            token for token, freq in self.token_freqs if freq >= min_freq])))
        self.token_to_idx = {token: idx
                             for idx, token in enumerate(self.idx_to_token)}

    def __len__(self):
        return len(self.idx_to_token)

    def __getitem__(self, tokens):
        if not isinstance(tokens, (list, tuple)):
            return self.token_to_idx.get(tokens, self.unk)
        return [self.__getitem__(token) for token in tokens]

    def to_tokens(self, indices):
        if hasattr(indices, '__len__') and len(indices) > 1:
            return [self.idx_to_token[int(index)] for index in indices]
        return self.idx_to_token[indices]

    @property
    def unk(self):   # Index for the unknown token
        return self.token_to_idx['<unk>']
```

We now construct a vocabulary for our dataset, converting the sequence of strings into a list of numerical indices. Note that we have not lost any information and can easily convert our dataset back to its original (string) representation.

```
vocab = Vocab(tokens)
indices = vocab[tokens[:10]]
print('indices:', indices)
print('words:', vocab.to_tokens(indices))
```

```
indices: [21, 9, 6, 0, 21, 10, 14, 6, 0, 14]
words: ['t', 'h', 'e', ' ', 't', 'i', 'm', 'e', ' ', 'm']
```

9.2.4 Putting It All Together

Using the above classes and methods, we package everything into the following build method of the TimeMachine class, which returns corpus, a list of token indices, and vocab, the vocabulary of *The Time Machine* corpus. The modifications we did here are: (i) we tokenize text into characters, not words, to simplify the training in later sections; (ii) corpus is a single list, not a list of token lists, since each text line in *The Time Machine* dataset is not necessarily a sentence or paragraph.

```
@d2l.add_to_class(TimeMachine)  #@save
def build(self, raw_text, vocab=None):
    tokens = self._tokenize(self._preprocess(raw_text))
    if vocab is None: vocab = Vocab(tokens)
    corpus = [vocab[token] for token in tokens]
    return corpus, vocab

corpus, vocab = data.build(raw_text)
len(corpus), len(vocab)
```

```
(173428, 28)
```

9.2.5 Exploratory Language Statistics

Using the real corpus and the Vocab class defined over words, we can inspect basic statistics concerning word use in our corpus. Below, we construct a vocabulary from words used in *The Time Machine* and print the ten most frequently occurring of them.

```
words = text.split()
vocab = Vocab(words)
vocab.token_freqs[:10]
```

```
[('the', 2261),
 ('i', 1267),
 ('and', 1245),
```

(continues on next page)

(continued from previous page)

```
('of', 1155),
('a', 816),
('to', 695),
('was', 552),
('in', 541),
('that', 443),
('my', 440)]
```

Note that the ten most frequent words are not all that descriptive. You might even imagine that we might see a very similar list if we had chosen any book at random. Articles like "the" and "a", pronouns like "i" and "my", and prepositions like "of", "to", and "in" occur often because they serve common syntactic roles. Such words that are common but not particularly descriptive are often called *stop words* and, in previous generations of text classifiers based on so-called bag-of-words representations, they were most often filtered out. However, they carry meaning and it is not necessary to filter them out when working with modern RNN- and Transformer-based neural models. If you look further down the list, you will notice that word frequency decays quickly. The 10th most frequent word is less than $1/5$ as common as the most popular. Word frequency tends to follow a power law distribution (specifically the Zipfian) as we go down the ranks. To get a better idea, we plot the figure of the word frequency.

```
freqs = [freq for token, freq in vocab.token_freqs]
d2l.plot(freqs, xlabel='token: x', ylabel='frequency: n(x)',
         xscale='log', yscale='log')
```

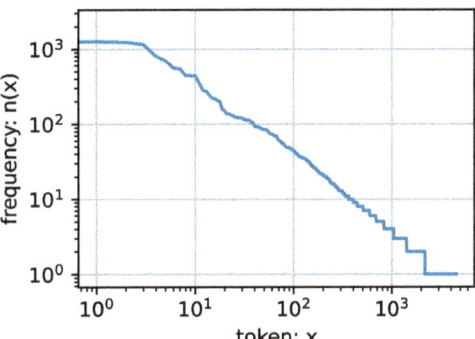

After dealing with the first few words as exceptions, all the remaining words roughly follow a straight line on a log–log plot. This phenomenon is captured by *Zipf's law*, which states that the frequency n_i of the ith most frequent word is:

$$n_i \propto \frac{1}{i^\alpha}, \tag{9.2.1}$$

which is equivalent to

$$\log n_i = -\alpha \log i + c, \tag{9.2.2}$$

where α is the exponent that characterizes the distribution and c is a constant. This should already give us pause for thought if we want to model words by counting statistics. After all, we will significantly overestimate the frequency of the tail, also known as the infrequent words. But what about the other word combinations, such as two consecutive words (bigrams), three consecutive words (trigrams), and beyond? Let's see whether the bigram frequency behaves in the same manner as the single word (unigram) frequency.

```python
bigram_tokens = ['--'.join(pair) for pair in zip(words[:-1], words[1:])]
bigram_vocab = Vocab(bigram_tokens)
bigram_vocab.token_freqs[:10]
```

```
[('of--the', 309),
 ('in--the', 169),
 ('i--had', 130),
 ('i--was', 112),
 ('and--the', 109),
 ('the--time', 102),
 ('it--was', 99),
 ('to--the', 85),
 ('as--i', 78),
 ('of--a', 73)]
```

One thing is notable here. Out of the ten most frequent word pairs, nine are composed of both stop words and only one is relevant to the actual book—"the time". Furthermore, let's see whether the trigram frequency behaves in the same manner.

```python
trigram_tokens = ['--'.join(triple) for triple in zip(
    words[:-2], words[1:-1], words[2:])]
trigram_vocab = Vocab(trigram_tokens)
trigram_vocab.token_freqs[:10]
```

```
[('the--time--traveller', 59),
 ('the--time--machine', 30),
 ('the--medical--man', 24),
 ('it--seemed--to', 16),
 ('it--was--a', 15),
 ('here--and--there', 15),
 ('seemed--to--me', 14),
 ('i--did--not', 14),
 ('i--saw--the', 13),
 ('i--began--to', 13)]
```

Now, let's visualize the token frequency among these three models: unigrams, bigrams, and trigrams.

```python
bigram_freqs = [freq for token, freq in bigram_vocab.token_freqs]
trigram_freqs = [freq for token, freq in trigram_vocab.token_freqs]
d2l.plot([freqs, bigram_freqs, trigram_freqs], xlabel='token: x',
         ylabel='frequency: n(x)', xscale='log', yscale='log',
         legend=['unigram', 'bigram', 'trigram'])
```

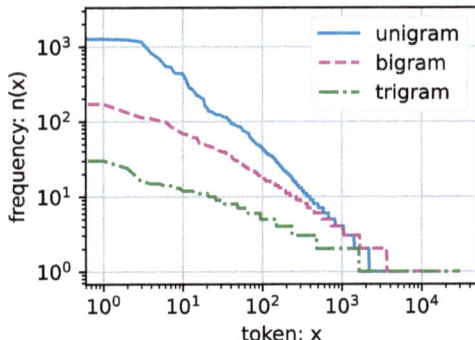

This figure is quite exciting. First, beyond unigram words, sequences of words also appear to be following Zipf's law, albeit with a smaller exponent α in (9.2.1), depending on the sequence length. Second, the number of distinct n-grams is not that large. This gives us hope that there is quite a lot of structure in language. Third, many n-grams occur very rarely. This makes certain methods unsuitable for language modeling and motivates the use of deep learning models. We will discuss this in the next section.

9.2.6 Summary

Text is among the most common forms of sequence data encountered in deep learning. Common choices for what constitutes a token are characters, words, and word pieces. To preprocess text, we usually (i) split text into tokens; (ii) build a vocabulary to map token strings to numerical indices; and (iii) convert text data into token indices for models to manipulate. In practice, the frequency of words tends to follow Zipf's law. This is true not just for individual words (unigrams), but also for n-grams.

9.2.7 Exercises

1. In the experiment of this section, tokenize text into words and vary the min_freq argument value of the Vocab instance. Qualitatively characterize how changes in min_freq impact the size of the resulting vocabulary.

2. Estimate the exponent of Zipfian distribution for unigrams, bigrams, and trigrams in this corpus.

3. Find some other sources of data (download a standard machine learning dataset, pick another public domain book, scrape a website, etc). For each, tokenize the data at both the word and character levels. How do the vocabulary sizes compare with *The Time Machine* corpus at equivalent values of min_freq. Estimate the exponent of the Zipfian distribution corresponding to the unigram and bigram distributions for these corpora. How do they compare with the values that you observed for *The Time Machine* corpus?

138

Discussions [138].

9.3 Language Models

In Section 9.2, we saw how to map text sequences into tokens, where these tokens can be viewed as a sequence of discrete observations such as words or characters. Assume that the tokens in a text sequence of length T are in turn $x_1, x_2, \ldots, x_T$. The goal of *language models* is to estimate the joint probability of the whole sequence:

$$P(x_1, x_2, \ldots, x_T), \tag{9.3.1}$$

where statistical tools in Section 9.1 can be applied.

Language models are incredibly useful. For instance, an ideal language model should generate natural text on its own, simply by drawing one token at a time $x_t \sim P(x_t \mid x_{t-1}, \ldots, x_1)$. Quite unlike the monkey using a typewriter, all text emerging from such a model would pass as natural language, e.g., English text. Furthermore, it would be sufficient for generating a meaningful dialog, simply by conditioning the text on previous dialog fragments. Clearly we are still very far from designing such a system, since it would need to *understand* the text rather than just generate grammatically sensible content.

Nonetheless, language models are of great service even in their limited form. For instance, the phrases "to recognize speech" and "to wreck a nice beach" sound very similar. This can cause ambiguity in speech recognition, which is easily resolved through a language model that rejects the second translation as outlandish. Likewise, in a document summarization algorithm it is worthwhile knowing that "dog bites man" is much more frequent than "man bites dog", or that "I want to eat grandma" is a rather disturbing statement, whereas "I want to eat, grandma" is much more benign.

```
import torch
from d2l import torch as d2l
```

9.3.1 Learning Language Models

The obvious question is how we should model a document, or even a sequence of tokens. Suppose that we tokenize text data at the word level. Let's start by applying basic probability rules:

$$P(x_1, x_2, \ldots, x_T) = \prod_{t=1}^{T} P(x_t \mid x_1, \ldots, x_{t-1}). \tag{9.3.2}$$

For example, the probability of a text sequence containing four words would be given as:

$$
\begin{aligned}
&P(\text{deep, learning, is, fun}) \\
&= P(\text{deep})P(\text{learning} \mid \text{deep})P(\text{is} \mid \text{deep, learning})P(\text{fun} \mid \text{deep, learning, is}).
\end{aligned}
\tag{9.3.3}
$$

Markov Models and n-grams

Among those sequence model analyses in Section 9.1, let's apply Markov models to language modeling. A distribution over sequences satisfies the Markov property of first order if $P(x_{t+1} \mid x_t, \ldots, x_1) = P(x_{t+1} \mid x_t)$. Higher orders correspond to longer dependencies. This leads to a number of approximations that we could apply to model a sequence:

$$
\begin{aligned}
P(x_1, x_2, x_3, x_4) &= P(x_1)P(x_2)P(x_3)P(x_4), \\
P(x_1, x_2, x_3, x_4) &= P(x_1)P(x_2 \mid x_1)P(x_3 \mid x_2)P(x_4 \mid x_3), \\
P(x_1, x_2, x_3, x_4) &= P(x_1)P(x_2 \mid x_1)P(x_3 \mid x_1, x_2)P(x_4 \mid x_2, x_3).
\end{aligned}
\tag{9.3.4}
$$

The probability formulae that involve one, two, and three variables are typically referred to as *unigram*, *bigram*, and *trigram* models, respectively. In order to compute the language model, we need to calculate the probability of words and the conditional probability of a word given the previous few words. Note that such probabilities are language model parameters.

Word Frequency

139

Here, we assume that the training dataset is a large text corpus, such as all Wikipedia entries, Project Gutenberg [139], and all text posted on the web. The probability of words can be calculated from the relative word frequency of a given word in the training dataset. For example, the estimate $\hat{P}(\text{deep})$ can be calculated as the probability of any sentence starting with the word "deep". A slightly less accurate approach would be to count all occurrences of the word "deep" and divide it by the total number of words in the corpus. This works fairly well, particularly for frequent words. Moving on, we could attempt to estimate

$$
\hat{P}(\text{learning} \mid \text{deep}) = \frac{n(\text{deep, learning})}{n(\text{deep})},
\tag{9.3.5}
$$

where $n(x)$ and $n(x, x')$ are the number of occurrences of singletons and consecutive word pairs, respectively. Unfortunately, estimating the probability of a word pair is somewhat more difficult, since the occurrences of "deep learning" are a lot less frequent. In particular, for some unusual word combinations it may be tricky to find enough occurrences to get accurate estimates. As suggested by the empirical results in Section 9.2.5, things take a turn for the worse for three-word combinations and beyond. There will be many plausible three-word combinations that we likely will not see in our dataset. Unless we provide some solution to assign such word combinations a nonzero count, we will not be able to use them in a language model. If the dataset is small or if the words are very rare, we might not find even a single one of them.

Laplace Smoothing

A common strategy is to perform some form of *Laplace smoothing*. The solution is to add a small constant to all counts. Denote by n the total number of words in the training set and *m* the number of unique words. This solution helps with singletons, e.g., via

$$\hat{P}(x) = \frac{n(x) + \epsilon_1/m}{n + \epsilon_1},$$

$$\hat{P}(x' \mid x) = \frac{n(x, x') + \epsilon_2 \hat{P}(x')}{n(x) + \epsilon_2}, \tag{9.3.6}$$

$$\hat{P}(x'' \mid x, x') = \frac{n(x, x', x'') + \epsilon_3 \hat{P}(x'')}{n(x, x') + \epsilon_3}.$$

Here ϵ_1, ϵ_2, and ϵ_3 are hyperparameters. Take ϵ_1 as an example: when $\epsilon_1 = 0$, no smoothing is applied; when ϵ_1 approaches positive infinity, $\hat{P}(x)$ approaches the uniform probability $1/m$. The above is a rather primitive variant of what other techniques can accomplish (Wood *et al.*, 2011).

Unfortunately, models like this get unwieldy rather quickly for the following reasons. First, as discussed in Section 9.2.5, many *n*-grams occur very rarely, making Laplace smoothing rather unsuitable for language modeling. Second, we need to store all counts. Third, this entirely ignores the meaning of the words. For instance, "cat" and "feline" should occur in related contexts. It is quite difficult to adjust such models to additional contexts, whereas, deep learning based language models are well suited to take this into account. Last, long word sequences are almost certain to be novel, hence a model that simply counts the frequency of previously seen word sequences is bound to perform poorly there. Therefore, we focus on using neural networks for language modeling in the rest of the chapter.

9.3.2 Perplexity

Next, let's discuss about how to measure the quality of the language model, which we will then use to evaluate our models in the subsequent sections. One way is to check how surprising the text is. A good language model is able to predict, with high accuracy, the tokens that come next. Consider the following continuations of the phrase "It is raining", as proposed by different language models:

1. "It is raining outside"

2. "It is raining banana tree"

3. "It is raining piouw;kcj pwepoiut"

In terms of quality, Example 1 is clearly the best. The words are sensible and logically coherent. While it might not quite accurately reflect which word follows semantically ("in San Francisco" and "in winter" would have been perfectly reasonable extensions), the model is able to capture which kind of word follows. Example 2 is considerably worse by producing a nonsensical extension. Nonetheless, at least the model has learned how to spell words and some degree of correlation between words. Last, Example 3 indicates a poorly trained model that does not fit data properly.

We might measure the quality of the model by computing the likelihood of the sequence. Unfortunately this is a number that is hard to understand and difficult to compare. After all, shorter sequences are much more likely to occur than the longer ones, hence evaluating the model on Tolstoy's magnum opus *War and Peace* will inevitably produce a much smaller likelihood than, say, on Saint-Exupery's novella *The Little Prince*. What is missing is the equivalent of an average.

Information theory comes handy here. We defined entropy, surprisal, and cross-entropy when we introduced the softmax regression (Section 4.1.3). If we want to compress text, we can ask about predicting the next token given the current set of tokens. A better language model should allow us to predict the next token more accurately. Thus, it should allow us to spend fewer bits in compressing the sequence. So we can measure it by the cross-entropy loss averaged over all the n tokens of a sequence:

$$\frac{1}{n} \sum_{t=1}^{n} -\log P(x_t \mid x_{t-1}, \ldots, x_1), \tag{9.3.7}$$

where P is given by a language model and x_t is the actual token observed at time step t from the sequence. This makes the performance on documents of different lengths comparable. For historical reasons, scientists in natural language processing prefer to use a quantity called *perplexity*. In a nutshell, it is the exponential of (9.3.7):

$$\exp\left(-\frac{1}{n} \sum_{t=1}^{n} \log P(x_t \mid x_{t-1}, \ldots, x_1)\right). \tag{9.3.8}$$

Perplexity can be best understood as the reciprocal of the geometric mean of the number of real choices that we have when deciding which token to pick next. Let's look at a number of cases:

- In the best case scenario, the model always perfectly estimates the probability of the target token as 1. In this case the perplexity of the model is 1.

- In the worst case scenario, the model always predicts the probability of the target token as 0. In this situation, the perplexity is positive infinity.

- At the baseline, the model predicts a uniform distribution over all the available tokens of the vocabulary. In this case, the perplexity equals the number of unique tokens of the vocabulary. In fact, if we were to store the sequence without any compression, this would be the best we could do for encoding it. Hence, this provides a nontrivial upper bound that any useful model must beat.

9.3.3 Partitioning Sequences

We will design language models using neural networks and use perplexity to evaluate how good the model is at predicting the next token given the current set of tokens in text sequences. Before introducing the model, let's assume that it processes a minibatch of sequences with predefined length at a time. Now the question is how to read minibatches of input sequences and target sequences at random.

Suppose that the dataset takes the form of a sequence of T token indices in corpus. We will partition it into subsequences, where each subsequence has n tokens (time steps). To iterate over (almost) all the tokens of the entire dataset for each epoch and obtain all possible length-n subsequences, we can introduce randomness. More concretely, at the beginning of each epoch, discard the first d tokens, where $d \in [0, n)$ is uniformly sampled at random. The rest of the sequence is then partitioned into $m = \lfloor (T - d)/n \rfloor$ subsequences. Denote by $\mathbf{x}_t = [x_t, \ldots, x_{t+n-1}]$ the length-n subsequence starting from token x_t at time step t. The resulting m partitioned subsequences are $\mathbf{x}_d, \mathbf{x}_{d+n}, \ldots, \mathbf{x}_{d+n(m-1)}$. Each subsequence will be used as an input sequence into the language model.

For language modeling, the goal is to predict the next token based on the tokens we have seen so far; hence the targets (labels) are the original sequence, shifted by one token. The target sequence for any input sequence $\mathbf{x}_t$ is $\mathbf{x}_{t+1}$ with length n.

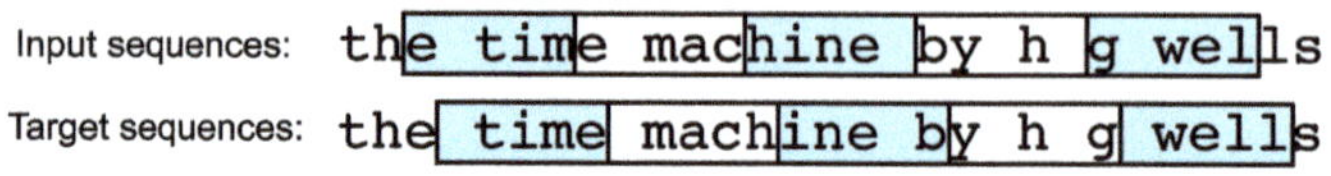

Input sequences: the time machine by h g wells

Target sequences: the time machine by h g wells

Fig. 9.3.1 Obtaining five pairs of input sequences and target sequences from partitioned length-5 subsequences.

Fig. 9.3.1 shows an example of obtaining five pairs of input sequences and target sequences with $n = 5$ and $d = 2$.

```
@d2l.add_to_class(d2l.TimeMachine)  #@save
def __init__(self, batch_size, num_steps, num_train=10000, num_val=5000):
    super(d2l.TimeMachine, self).__init__()
    self.save_hyperparameters()
    corpus, self.vocab = self.build(self._download())
    array = torch.tensor([corpus[i:i+num_steps+1]
                        for i in range(len(corpus)-num_steps)])
    self.X, self.Y = array[:,:-1], array[:,1:]
```

To train language models, we will randomly sample pairs of input sequences and target sequences in minibatches. The following data loader randomly generates a minibatch from the dataset each time. The argument `batch_size` specifies the number of subsequence examples in each minibatch and `num_steps` is the subsequence length in tokens.

```
@d2l.add_to_class(d2l.TimeMachine)  #@save
def get_dataloader(self, train):
    idx = slice(0, self.num_train) if train else slice(
        self.num_train, self.num_train + self.num_val)
    return self.get_tensorloader([self.X, self.Y], train, idx)
```

As we can see in the following, a minibatch of target sequences can be obtained by shifting the input sequences by one token.

```
data = d2l.TimeMachine(batch_size=2, num_steps=10)
for X, Y in data.train_dataloader():
    print('X:', X, '\nY:', Y)
    break
```

9.3.4 Summary and Discussion

Language models estimate the joint probability of a text sequence. For long sequences, *n*-grams provide a convenient model by truncating the dependence. However, there is a lot of structure but not enough frequency to deal efficiently with infrequent word combinations via Laplace smoothing. Thus, we will focus on neural language modeling in subsequent sections. To train language models, we can randomly sample pairs of input sequences and target sequences in minibatches. After training, we will use perplexity to measure the language model quality.

Language models can be scaled up with increased data size, model size, and amount in training compute. Large language models can perform desired tasks by predicting output text given input text instructions. As we will discuss later (e.g., Section 11.9), at the present moment large language models form the basis of state-of-the-art systems across diverse tasks.

9.3.5 Exercises

1. Suppose there are 100,000 words in the training dataset. How much word frequency and multi-word adjacent frequency does a four-gram need to store?

2. How would you model a dialogue?

3. What other methods can you think of for reading long sequence data?

4. Consider our method for discarding a uniformly random number of the first few tokens at the beginning of each epoch.

    1. Does it really lead to a perfectly uniform distribution over the sequences on the document?

    2. What would you have to do to make things even more uniform?

5. If we want a sequence example to be a complete sentence, what kind of problem does this introduce in minibatch sampling? How can we fix it?

Discussions[140].

9.4 Recurrent Neural Networks

In Section 9.3 we described Markov models and n-grams for language modeling, where the conditional probability of token x_t at time step t only depends on the $n-1$ previous tokens. If we want to incorporate the possible effect of tokens earlier than time step $t - (n-1)$ on x_t, we need to increase n. However, the number of model parameters would also increase exponentially with it, as we need to store $|\mathcal{V}|^n$ numbers for a vocabulary set $\mathcal{V}$. Hence, rather than modeling $P(x_t \mid x_{t-1}, \ldots, x_{t-n+1})$ it is preferable to use a latent variable model,

$$P(x_t \mid x_{t-1}, \ldots, x_1) \approx P(x_t \mid h_{t-1}), \tag{9.4.1}$$

where h_{t-1} is a *hidden state* that stores the sequence information up to time step $t-1$. In general, the hidden state at any time step t could be computed based on both the current input x_t and the previous hidden state h_{t-1}:

$$h_t = f(x_t, h_{t-1}). \tag{9.4.2}$$

For a sufficiently powerful function f in (9.4.2), the latent variable model is not an approximation. After all, h_t may simply store all the data it has observed so far. However, it could potentially make both computation and storage expensive.

Recall that we have discussed hidden layers with hidden units in Chapter 5. It is noteworthy that hidden layers and hidden states refer to two very different concepts. Hidden layers are, as explained, layers that are hidden from view on the path from input to output. Hidden states are technically speaking *inputs* to whatever we do at a given step, and they can only be computed by looking at data at previous time steps.

Recurrent neural networks (RNNs) are neural networks with hidden states. Before introducing the RNN model, we first revisit the MLP model introduced in Section 5.1.

```
import torch
from d2l import torch as d2l
```

9.4.1 Neural Networks without Hidden States

Let's take a look at an MLP with a single hidden layer. Let the hidden layer's activation function be ϕ. Given a minibatch of examples $\mathbf{X} \in \mathbb{R}^{n \times d}$ with batch size n and d inputs, the hidden layer output $\mathbf{H} \in \mathbb{R}^{n \times h}$ is calculated as

$$\mathbf{H} = \phi(\mathbf{X}\mathbf{W}_{xh} + \mathbf{b}_h). \tag{9.4.3}$$

In (9.4.3), we have the weight parameter $\mathbf{W}_{xh} \in \mathbb{R}^{d \times h}$, the bias parameter $\mathbf{b}_h \in \mathbb{R}^{1 \times h}$, and the number of hidden units h, for the hidden layer. So armed, we apply broadcasting (see Section 2.1.4) during the summation. Next, the hidden layer output $\mathbf{H}$ is used as input of the output layer, which is given by

$$\mathbf{O} = \mathbf{H}\mathbf{W}_{hq} + \mathbf{b}_q, \qquad\qquad (9.4.4)$$

where $\mathbf{O} \in \mathbb{R}^{n \times q}$ is the output variable, $\mathbf{W}_{hq} \in \mathbb{R}^{h \times q}$ is the weight parameter, and $\mathbf{b}_q \in \mathbb{R}^{1 \times q}$ is the bias parameter of the output layer. If it is a classification problem, we can use softmax($\mathbf{O}$) to compute the probability distribution of the output categories.

This is entirely analogous to the regression problem we solved previously in Section 9.1, hence we omit details. Suffice it to say that we can pick feature-label pairs at random and learn the parameters of our network via automatic differentiation and stochastic gradient descent.

9.4.2 Recurrent Neural Networks with Hidden States

Matters are entirely different when we have hidden states. Let's look at the structure in some more detail.

Assume that we have a minibatch of inputs $\mathbf{X}_t \in \mathbb{R}^{n \times d}$ at time step t. In other words, for a minibatch of n sequence examples, each row of $\mathbf{X}_t$ corresponds to one example at time step t from the sequence. Next, denote by $\mathbf{H}_t \in \mathbb{R}^{n \times h}$ the hidden layer output of time step t. Unlike with MLP, here we save the hidden layer output $\mathbf{H}_{t-1}$ from the previous time step and introduce a new weight parameter $\mathbf{W}_{hh} \in \mathbb{R}^{h \times h}$ to describe how to use the hidden layer output of the previous time step in the current time step. Specifically, the calculation of the hidden layer output of the current time step is determined by the input of the current time step together with the hidden layer output of the previous time step:

$$\mathbf{H}_t = \phi(\mathbf{X}_t \mathbf{W}_{xh} + \mathbf{H}_{t-1} \mathbf{W}_{hh} + \mathbf{b}_h). \qquad\qquad (9.4.5)$$

Compared with (9.4.3), (9.4.5) adds one more term $\mathbf{H}_{t-1} \mathbf{W}_{hh}$ and thus instantiates (9.4.2). From the relationship between hidden layer outputs $\mathbf{H}_t$ and $\mathbf{H}_{t-1}$ of adjacent time steps, we know that these variables captured and retained the sequence's historical information up to their current time step, just like the state or memory of the neural network's current time step. Therefore, such a hidden layer output is called a *hidden state*. Since the hidden state uses the same definition of the previous time step in the current time step, the computation of (9.4.5) is *recurrent*. Hence, as we said, neural networks with hidden states based on recurrent computation are named *recurrent neural networks*. Layers that perform the computation of (9.4.5) in RNNs are called *recurrent layers*.

There are many different ways for constructing RNNs. Those with a hidden state defined by (9.4.5) are very common. For time step t, the output of the output layer is similar to the computation in the MLP:

$$\mathbf{O}_t = \mathbf{H}_t \mathbf{W}_{hq} + \mathbf{b}_q. \qquad\qquad (9.4.6)$$

Parameters of the RNN include the weights $\mathbf{W}_{xh} \in \mathbb{R}^{d \times h}$, $\mathbf{W}_{hh} \in \mathbb{R}^{h \times h}$, and the bias $\mathbf{b}_h \in \mathbb{R}^{1 \times h}$ of the hidden layer, together with the weights $\mathbf{W}_{hq} \in \mathbb{R}^{h \times q}$ and the bias $\mathbf{b}_q \in \mathbb{R}^{1 \times q}$ of the output layer. It is worth mentioning that even at different time steps, RNNs always use these model parameters. Therefore, the parametrization cost of an RNN does not grow as the number of time steps increases.

Fig. 9.4.1 illustrates the computational logic of an RNN at three adjacent time steps. At any time step t, the computation of the hidden state can be treated as: (i) concatenating the input $\mathbf{X}_t$ at the current time step t and the hidden state $\mathbf{H}_{t-1}$ at the previous time step $t - 1$; (ii) feeding the concatenation result into a fully connected layer with the activation function ϕ. The output of such a fully connected layer is the hidden state $\mathbf{H}_t$ of the current time step t. In this case, the model parameters are the concatenation of $\mathbf{W}_{xh}$ and $\mathbf{W}_{hh}$, and a bias of $\mathbf{b}_h$, all from (9.4.5). The hidden state of the current time step t, $\mathbf{H}_t$, will participate in computing the hidden state $\mathbf{H}_{t+1}$ of the next time step $t + 1$. What is more, $\mathbf{H}_t$ will also be fed into the fully connected output layer to compute the output $\mathbf{O}_t$ of the current time step t.

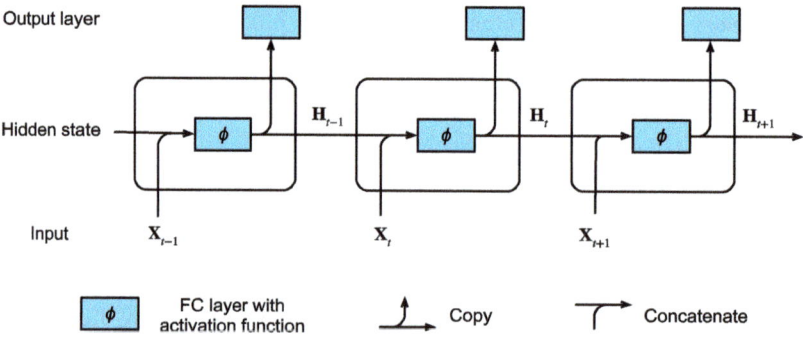

Fig. 9.4.1   An RNN with a hidden state.

We just mentioned that the calculation of $\mathbf{X}_t \mathbf{W}_{xh} + \mathbf{H}_{t-1} \mathbf{W}_{hh}$ for the hidden state is equivalent to matrix multiplication of the concatenation of $\mathbf{X}_t$ and $\mathbf{H}_{t-1}$ and the concatenation of $\mathbf{W}_{xh}$ and $\mathbf{W}_{hh}$. Though this can be proven mathematically, in the following we just use a simple code snippet as a demonstration. To begin with, we define matrices X, W_xh, H, and W_hh, whose shapes are (3, 1), (1, 4), (3, 4), and (4, 4), respectively. Multiplying X by W_xh, and H by W_hh, and then adding these two products, we obtain a matrix of shape (3, 4).

```
X, W_xh = torch.randn(3, 1), torch.randn(1, 4)
H, W_hh = torch.randn(3, 4), torch.randn(4, 4)
torch.matmul(X, W_xh) + torch.matmul(H, W_hh)
```

```
tensor([[ 0.8087, -0.0097, -1.4436,  0.6615],
        [-3.5964,  4.6303, -1.1824, -1.1187],
        [-0.1704,  0.9537, -1.4047, -0.7128]])
```

Now we concatenate the matrices X and H along columns (axis 1), and the matrices W_xh and W_hh along rows (axis 0). These two concatenations result in matrices of shape (3, 5)

and of shape (5, 4), respectively. Multiplying these two concatenated matrices, we obtain
the same output matrix of shape (3, 4) as above.

```
torch.matmul(torch.cat((X, H), 1), torch.cat((W_xh, W_hh), 0))
```

```
tensor([[ 0.8087, -0.0097, -1.4436,  0.6615],
        [-3.5964,  4.6303, -1.1824, -1.1187],
        [-0.1704,  0.9537, -1.4047, -0.7128]])
```

9.4.3 RNN-Based Character-Level Language Models

Recall that for language modeling in Section 9.3, we aim to predict the next token based on
the current and past tokens; thus we shift the original sequence by one token as the targets
(labels). Bengio *et al.* (2003) first proposed to use a neural network for language modeling.
In the following we illustrate how RNNs can be used to build a language model. Let the
minibatch size be one, and the sequence of the text be "machine". To simplify training
in subsequent sections, we tokenize text into characters rather than words and consider a
character-level language model. Fig. 9.4.2 demonstrates how to predict the next character based on the current and previous characters via an RNN for character-level language
modeling.

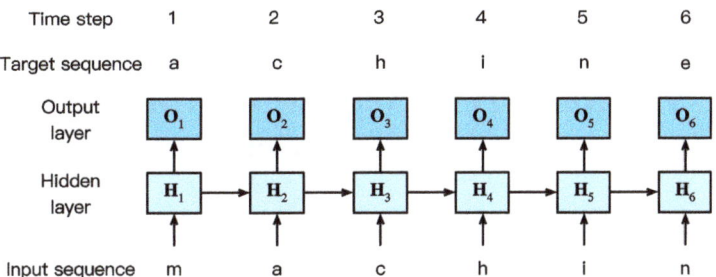

Fig. 9.4.2 A character-level language model based on the RNN. The input and target sequences are
"machin" and "achine", respectively.

During the training process, we run a softmax operation on the output from the output layer
for each time step, and then use the cross-entropy loss to compute the error between the
model output and the target. Because of the recurrent computation of the hidden state in the
hidden layer, the output, O_3, of time step 3 in Fig. 9.4.2 is determined by the text sequence
"m", "a", and "c". Since the next character of the sequence in the training data is "h", the
loss of time step 3 will depend on the probability distribution of the next character generated
based on the feature sequence "m", "a", "c" and the target "h" of this time step.

In practice, each token is represented by a d-dimensional vector, and we use a batch size
$n > 1$. Therefore, the input $\mathbf{X}_t$ at time step t will be an $n \times d$ matrix, which is identical to
what we discussed in Section 9.4.2.

In the following sections, we will implement RNNs for character-level language models.

9.4.4 Summary

A neural network that uses recurrent computation for hidden states is called a recurrent neural network (RNN). The hidden state of an RNN can capture historical information of the sequence up to the current time step. With recurrent computation, the number of RNN model parameters does not grow as the number of time steps increases. As for applications, an RNN can be used to create character-level language models.

9.4.5 Exercises

1. If we use an RNN to predict the next character in a text sequence, what is the required dimension for any output?

2. Why can RNNs express the conditional probability of a token at some time step based on all the previous tokens in the text sequence?

3. What happens to the gradient if you backpropagate through a long sequence?

4. What are some of the problems associated with the language model described in this section?

Discussions [141].

141

9.5 Recurrent Neural Network Implementation from Scratch

We are now ready to implement an RNN from scratch. In particular, we will train this RNN to function as a character-level language model (see Section 9.4) and train it on a corpus consisting of the entire text of H. G. Wells' *The Time Machine*, following the data processing steps outlined in Section 9.2. We start by loading the dataset.

```
%matplotlib inline
import math
import torch
from torch import nn
from torch.nn import functional as F
from d2l import torch as d2l
```

9.5.1 RNN Model

We begin by defining a class to implement the RNN model (Section 9.4.2). Note that the number of hidden units num_hiddens is a tunable hyperparameter.

```
class RNNScratch(d2l.Module):  #@save
    """The RNN model implemented from scratch."""
    def __init__(self, num_inputs, num_hiddens, sigma=0.01):
        super().__init__()
        self.save_hyperparameters()
        self.W_xh = nn.Parameter(
            torch.randn(num_inputs, num_hiddens) * sigma)
        self.W_hh = nn.Parameter(
            torch.randn(num_hiddens, num_hiddens) * sigma)
        self.b_h = nn.Parameter(torch.zeros(num_hiddens))
```

The forward method below defines how to compute the output and hidden state at any time step, given the current input and the state of the model at the previous time step. Note that the RNN model loops through the outermost dimension of inputs, updating the hidden state one time step at a time. The model here uses a tanh activation function (Section 5.1.2).

```
@d2l.add_to_class(RNNScratch)  #@save
def forward(self, inputs, state=None):
    if state is None:
        # Initial state with shape: (batch_size, num_hiddens)
        state = torch.zeros((inputs.shape[1], self.num_hiddens),
                            device=inputs.device)
    else:
        state, = state
    outputs = []
    for X in inputs:  # Shape of inputs: (num_steps, batch_size, num_inputs)
        state = torch.tanh(torch.matmul(X, self.W_xh) +
                           torch.matmul(state, self.W_hh) + self.b_h)
        outputs.append(state)
    return outputs, state
```

We can feed a minibatch of input sequences into an RNN model as follows.

```
batch_size, num_inputs, num_hiddens, num_steps = 2, 16, 32, 100
rnn = RNNScratch(num_inputs, num_hiddens)
X = torch.ones((num_steps, batch_size, num_inputs))
outputs, state = rnn(X)
```

Let's check whether the RNN model produces results of the correct shapes to ensure that the dimensionality of the hidden state remains unchanged.

```
def check_len(a, n):  #@save
    """Check the length of a list."""
    assert len(a) == n, f'list\'s length {len(a)} != expected length {n}'

def check_shape(a, shape):  #@save
    """Check the shape of a tensor."""
    assert a.shape == shape, \
            f'tensor\'s shape {a.shape} != expected shape {shape}'
```

(continues on next page)

(continued from previous page)

```
check_len(outputs, num_steps)
check_shape(outputs[0], (batch_size, num_hiddens))
check_shape(state, (batch_size, num_hiddens))
```

9.5.2 RNN-Based Language Model

The following RNNLMScratch class defines an RNN-based language model, where we pass
in our RNN via the rnn argument of the __init__ method. When training language models, the inputs and outputs are from the same vocabulary. Hence, they have the same dimension, which is equal to the vocabulary size. Note that we use perplexity to evaluate the
model. As discussed in Section 9.3.2, this ensures that sequences of different length are
comparable.

```
class RNNLMScratch(d2l.Classifier):  #@save
    """The RNN-based language model implemented from scratch."""
    def __init__(self, rnn, vocab_size, lr=0.01):
        super().__init__()
        self.save_hyperparameters()
        self.init_params()

    def init_params(self):
        self.W_hq = nn.Parameter(
            torch.randn(
                self.rnn.num_hiddens, self.vocab_size) * self.rnn.sigma)
        self.b_q = nn.Parameter(torch.zeros(self.vocab_size))

    def training_step(self, batch):
        l = self.loss(self(*batch[:-1]), batch[-1])
        self.plot('ppl', torch.exp(l), train=True)
        return l

    def validation_step(self, batch):
        l = self.loss(self(*batch[:-1]), batch[-1])
        self.plot('ppl', torch.exp(l), train=False)
```

One-Hot Encoding

Recall that each token is represented by a numerical index indicating the position in the
vocabulary of the corresponding word/character/word piece. You might be tempted to build
a neural network with a single input node (at each time step), where the index could be fed
in as a scalar value. This works when we are dealing with numerical inputs like price or
temperature, where any two values sufficiently close together should be treated similarly.
But this does not quite make sense. The 45th and 46th words in our vocabulary happen to
be "their" and "said", whose meanings are not remotely similar.

When dealing with such categorical data, the most common strategy is to represent each
item by a *one-hot encoding* (recall from Section 4.1.1). A one-hot encoding is a vector
whose length is given by the size of the vocabulary N, where all entries are set to 0, except

for the entry corresponding to our token, which is set to 1. For example, if the vocabulary had five elements, then the one-hot vectors corresponding to indices 0 and 2 would be the following.

```
F.one_hot(torch.tensor([0, 2]), 5)
```

```
tensor([[1, 0, 0, 0, 0],
        [0, 0, 1, 0, 0]])
```

The minibatches that we sample at each iteration will take the shape (batch size, number of time steps). Once representing each input as a one-hot vector, we can think of each minibatch as a three-dimensional tensor, where the length along the third axis is given by the vocabulary size (len(vocab)). We often transpose the input so that we will obtain an output of shape (number of time steps, batch size, vocabulary size). This will allow us to loop more conveniently through the outermost dimension for updating hidden states of a minibatch, time step by time step (e.g., in the above forward method).

```
@d2l.add_to_class(RNNLMScratch)  #@save
def one_hot(self, X):
    # Output shape: (num_steps, batch_size, vocab_size)
    return F.one_hot(X.T, self.vocab_size).type(torch.float32)
```

Transforming RNN Outputs

The language model uses a fully connected output layer to transform RNN outputs into token predictions at each time step.

```
@d2l.add_to_class(RNNLMScratch)  #@save
def output_layer(self, rnn_outputs):
    outputs = [torch.matmul(H, self.W_hq) + self.b_q for H in rnn_outputs]
    return torch.stack(outputs, 1)

@d2l.add_to_class(RNNLMScratch)  #@save
def forward(self, X, state=None):
    embs = self.one_hot(X)
    rnn_outputs, _ = self.rnn(embs, state)
    return self.output_layer(rnn_outputs)
```

Let's check whether the forward computation produces outputs with the correct shape.

```
model = RNNLMScratch(rnn, num_inputs)
outputs = model(torch.ones((batch_size, num_steps), dtype=torch.int64))
check_shape(outputs, (batch_size, num_steps, num_inputs))
```

9.5.3 Gradient Clipping

While you are already used to thinking of neural networks as "deep" in the sense that many layers separate the input and output even within a single time step, the length of the sequence introduces a new notion of depth. In addition to the passing through the network in the input-to-output direction, inputs at the first time step must pass through a chain of T layers along the time steps in order to influence the output of the model at the final time step. Taking the backwards view, in each iteration, we backpropagate gradients through time, resulting in a chain of matrix-products of length $O(T)$. As mentioned in Section 5.4, this can result in numerical instability, causing the gradients either to explode or vanish, depending on the properties of the weight matrices.

Dealing with vanishing and exploding gradients is a fundamental problem when designing RNNs and has inspired some of the biggest advances in modern neural network architectures. In the next chapter, we will talk about specialized architectures that were designed in hopes of mitigating the vanishing gradient problem. However, even modern RNNs often suffer from exploding gradients. One inelegant but ubiquitous solution is to simply clip the gradients forcing the resulting "clipped" gradients to take smaller values.

Generally speaking, when optimizing some objective by gradient descent, we iteratively update the parameter of interest, say a vector $\mathbf{x}$, but pushing it in the direction of the negative gradient $\mathbf{g}$ (in stochastic gradient descent, we calculate this gradient on a randomly sampled minibatch). For example, with learning rate $\eta > 0$, each update takes the form $\mathbf{x} \leftarrow \mathbf{x} - \eta \mathbf{g}$. Let's further assume that the objective function f is sufficiently smooth. Formally, we say that the objective is *Lipschitz continuous* with constant L, meaning that for any $\mathbf{x}$ and $\mathbf{y}$, we have

$$|f(\mathbf{x}) - f(\mathbf{y})| \leq L\|\mathbf{x} - \mathbf{y}\|. \tag{9.5.1}$$

As you can see, when we update the parameter vector by subtracting $\eta \mathbf{g}$, the change in the value of the objective depends on the learning rate, the norm of the gradient and L as follows:

$$|f(\mathbf{x}) - f(\mathbf{x} - \eta \mathbf{g})| \leq L\eta\|\mathbf{g}\|. \tag{9.5.2}$$

In other words, the objective cannot change by more than $L\eta\|\mathbf{g}\|$. Having a small value for this upper bound might be viewed as good or bad. On the downside, we are limiting the speed at which we can reduce the value of the objective. On the bright side, this limits by just how much we can go wrong in any one gradient step.

When we say that gradients explode, we mean that $\|\mathbf{g}\|$ becomes excessively large. In this worst case, we might do so much damage in a single gradient step that we could undo all of the progress made over the course of thousands of training iterations. When gradients can be so large, neural network training often diverges, failing to reduce the value of the objective. At other times, training eventually converges but is unstable owing to massive spikes in the loss.

One way to limit the size of $L\eta\|\mathbf{g}\|$ is to shrink the learning rate η to tiny values. This has the advantage that we do not bias the updates. But what if we only *rarely* get large gradients? This drastic move slows down our progress at all steps, just to deal with the rare exploding gradient events. A popular alternative is to adopt a *gradient clipping* heuristic projecting the gradients **g** onto a ball of some given radius θ as follows:

$$\mathbf{g} \leftarrow \min\left(1, \frac{\theta}{\|\mathbf{g}\|}\right)\mathbf{g}. \qquad (9.5.3)$$

This ensures that the gradient norm never exceeds θ and that the updated gradient is entirely aligned with the original direction of **g**. It also has the desirable side-effect of limiting the influence any given minibatch (and within it any given sample) can exert on the parameter vector. This bestows a certain degree of robustness to the model. To be clear, it is a hack. Gradient clipping means that we are not always following the true gradient and it is hard to reason analytically about the possible side effects. However, it is a very useful hack, and is widely adopted in RNN implementations in most deep learning frameworks.

Below we define a method to clip gradients, which is invoked by the `fit_epoch` method of the `d2l.Trainer` class (see Section 3.4). Note that when computing the gradient norm, we are concatenating all model parameters, treating them as a single giant parameter vector.

```
@d2l.add_to_class(d2l.Trainer)  #@save
def clip_gradients(self, grad_clip_val, model):
    params = [p for p in model.parameters() if p.requires_grad]
    norm = torch.sqrt(sum(torch.sum((p.grad ** 2)) for p in params))
    if norm > grad_clip_val:
        for param in params:
            param.grad[:] *= grad_clip_val / norm
```

9.5.4 Training

Using *The Time Machine* dataset (`data`), we train a character-level language model (`model`) based on the RNN (`rnn`) implemented from scratch. Note that we first calculate the gradients, then clip them, and finally update the model parameters using the clipped gradients.

```
data = d2l.TimeMachine(batch_size=1024, num_steps=32)
rnn = RNNScratch(num_inputs=len(data.vocab), num_hiddens=32)
model = RNNLMScratch(rnn, vocab_size=len(data.vocab), lr=1)
trainer = d2l.Trainer(max_epochs=100, gradient_clip_val=1, num_gpus=1)
trainer.fit(model, data)
```

9.5.5 Decoding

Once a language model has been learned, we can use it not only to predict the next token but to continue predicting each subsequent one, treating the previously predicted token as though it were the next in the input. Sometimes we will just want to generate text as though we were starting at the beginning of a document. However, it is often useful to condition the language model on a user-supplied prefix. For example, if we were developing an autocom-

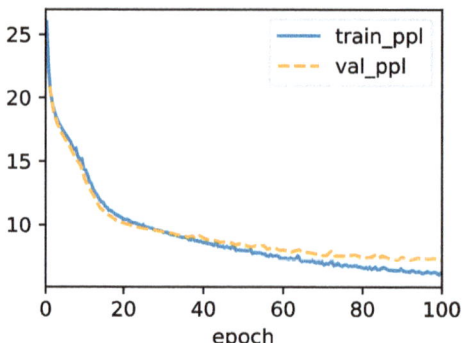

plete feature for a search engine or to assist users in writing emails, we would want to feed in what they had written so far (the prefix), and then generate a likely continuation.

The following `predict` method generates a continuation, one character at a time, after ingesting a user-provided `prefix`. When looping through the characters in `prefix`, we keep passing the hidden state to the next time step but do not generate any output. This is called the *warm-up* period. After ingesting the prefix, we are now ready to begin emitting the subsequent characters, each of which will be fed back into the model as the input at the next time step.

```python
@d2l.add_to_class(RNNLMScratch)  #@save
def predict(self, prefix, num_preds, vocab, device=None):
    state, outputs = None, [vocab[prefix[0]]]
    for i in range(len(prefix) + num_preds - 1):
        X = torch.tensor([[outputs[-1]]], device=device)
        embs = self.one_hot(X)
        rnn_outputs, state = self.rnn(embs, state)
        if i < len(prefix) - 1:  # Warm-up period
            outputs.append(vocab[prefix[i + 1]])
        else:  # Predict num_preds steps
            Y = self.output_layer(rnn_outputs)
            outputs.append(int(Y.argmax(axis=2).reshape(1)))
    return ''.join([vocab.idx_to_token[i] for i in outputs])
```

In the following, we specify the prefix and have it generate 20 additional characters.

```python
model.predict('it has', 20, data.vocab, d2l.try_gpu())
```

```
'it has it treat in the tim'
```

While implementing the above RNN model from scratch is instructive, it is not convenient. In the next section, we will see how to leverage deep learning frameworks to whip up RNNs using standard architectures, and to reap performance gains by relying on highly optimized library functions.

9.5.6 Summary

We can train RNN-based language models to generate text following the user-provided text prefix. A simple RNN language model consists of input encoding, RNN modeling, and output generation. During training, gradient clipping can mitigate the problem of exploding gradients but does not address the problem of vanishing gradients. In the experiment, we implemented a simple RNN language model and trained it with gradient clipping on sequences of text, tokenized at the character level. By conditioning on a prefix, we can use a language model to generate likely continuations, which proves useful in many applications, e.g., autocomplete features.

9.5.7 Exercises

1. Does the implemented language model predict the next token based on all the past tokens up to the very first token in *The Time Machine*?

2. Which hyperparameter controls the length of history used for prediction?

3. Show that one-hot encoding is equivalent to picking a different embedding for each object.

4. Adjust the hyperparameters (e.g., number of epochs, number of hidden units, number of time steps in a minibatch, and learning rate) to improve the perplexity. How low can you go while sticking with this simple architecture?

5. Replace one-hot encoding with learnable embeddings. Does this lead to better performance?

6. Conduct an experiment to determine how well this language model trained on *The Time Machine* works on other books by H. G. Wells, e.g., *The War of the Worlds*.

7. Conduct another experiment to evaluate the perplexity of this model on books written by other authors.

8. Modify the prediction method so as to use sampling rather than picking the most likely next character.

   - What happens?

   - Bias the model towards likelier outputs, e.g., by sampling from $q(x_t \mid x_{t-1}, \ldots, x_1) \propto P(x_t \mid x_{t-1}, \ldots, x_1)^\alpha$ for $\alpha > 1$.

9. Run the code in this section without clipping the gradient. What happens?

10. Replace the activation function used in this section with ReLU and repeat the experiments in this section. Do we still need gradient clipping? Why?

[142]

Discussions[142].

9.6 Concise Implementation of Recurrent Neural Networks

Like most of our from-scratch implementations, Section 9.5 was designed to provide insight into how each component works. But when you are using RNNs every day or writing production code, you will want to rely more on libraries that cut down on both implementation time (by supplying library code for common models and functions) and computation time (by optimizing the heck out of these library implementations). This section will show you how to implement the same language model more efficiently using the high-level API provided by your deep learning framework. We begin, as before, by loading *The Time Machine* dataset.

```python
import torch
from torch import nn
from torch.nn import functional as F
from d2l import torch as d2l
```

9.6.1 Defining the Model

We define the following class using the RNN implemented by high-level APIs.

```python
class RNN(d2l.Module):  #@save
    """The RNN model implemented with high-level APIs."""
    def __init__(self, num_inputs, num_hiddens):
        super().__init__()
        self.save_hyperparameters()
        self.rnn = nn.RNN(num_inputs, num_hiddens)

    def forward(self, inputs, H=None):
        return self.rnn(inputs, H)
```

Inheriting from the RNNLMScratch class in Section 9.5, the following RNNLM class defines a complete RNN-based language model. Note that we need to create a separate fully connected output layer.

```python
class RNNLM(d2l.RNNLMScratch):  #@save
    """The RNN-based language model implemented with high-level APIs."""
    def init_params(self):
        self.linear = nn.LazyLinear(self.vocab_size)

    def output_layer(self, hiddens):
        return self.linear(hiddens).swapaxes(0, 1)
```

9.6.2 Training and Predicting

Before training the model, let's make a prediction with a model initialized with random weights. Given that we have not trained the network, it will generate nonsensical predictions.

```
data = d2l.TimeMachine(batch_size=1024, num_steps=32)
rnn = RNN(num_inputs=len(data.vocab), num_hiddens=32)
model = RNNLM(rnn, vocab_size=len(data.vocab), lr=1)
model.predict('it has', 20, data.vocab)
```

```
'it haspllllllllllllllllllll'
```

Next, we train our model, leveraging the high-level API.

```
trainer = d2l.Trainer(max_epochs=100, gradient_clip_val=1, num_gpus=1)
trainer.fit(model, data)
```

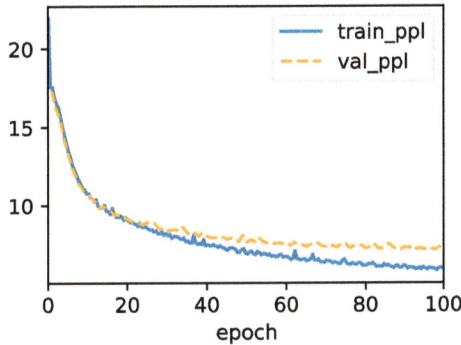

Compared with Section 9.5, this model achieves comparable perplexity, but runs faster due to the optimized implementations. As before, we can generate predicted tokens following the specified prefix string.

```
model.predict('it has', 20, data.vocab, d2l.try_gpu())
```

```
'it has so for the the the '
```

9.6.3 Summary

High-level APIs in deep learning frameworks provide implementations of standard RNNs. These libraries help you to avoid wasting time reimplementing standard models. Moreover, framework implementations are often highly optimized, leading to significant (computational) performance gains when compared with implementations from scratch.

9.6.4 Exercises

1. Can you make the RNN model overfit using the high-level APIs?

2. Implement the autoregressive model of Section 9.1 using an RNN.

Discussions [143].

143

9.7 Backpropagation Through Time

If you completed the exercises in Section 9.5, you would have seen that gradient clipping is vital for preventing the occasional massive gradients from destabilizing training. We hinted that the exploding gradients stem from backpropagating across long sequences. Before introducing a slew of modern RNN architectures, let's take a closer look at how *backpropagation* works in sequence models in mathematical detail. Hopefully, this discussion will bring some precision to the notion of *vanishing* and *exploding* gradients. If you recall our discussion of forward and backward propagation through computational graphs when we introduced MLPs in Section 5.3, then forward propagation in RNNs should be relatively straightforward. Applying backpropagation in RNNs is called *backpropagation through time* (Werbos, 1990). This procedure requires us to expand (or unroll) the computational graph of an RNN one time step at a time. The unrolled RNN is essentially a feedforward neural network with the special property that the same parameters are repeated throughout the unrolled network, appearing at each time step. Then, just as in any feedforward neural network, we can apply the chain rule, backpropagating gradients through the unrolled net. The gradient with respect to each parameter must be summed across all places that the parameter occurs in the unrolled net. Handling such weight tying should be familiar from our chapters on convolutional neural networks.

Complications arise because sequences can be rather long. It is not unusual to work with text sequences consisting of over a thousand tokens. Note that this poses problems both from a computational (too much memory) and optimization (numerical instability) standpoint. Input from the first step passes through over 1000 matrix products before arriving at the output, and another 1000 matrix products are required to compute the gradient. We now analyze what can go wrong and how to address it in practice.

9.7.1 Analysis of Gradients in RNNs

We start with a simplified model of how an RNN works. This model ignores details about the specifics of the hidden state and how it is updated. The mathematical notation here does not explicitly distinguish scalars, vectors, and matrices. We are just trying to develop some intuition. In this simplified model, we denote h_t as the hidden state, x_t as input, and o_t as output at time step t. Recall our discussions in Section 9.4.2 that the input and the hidden state can be concatenated before being multiplied by one weight variable in the hidden layer. Thus, we use w_h and w_o to indicate the weights of the hidden layer and the output layer, respectively. As a result, the hidden states and outputs at each time step are

$$
\begin{aligned}
h_t &= f(x_t, h_{t-1}, w_{\mathrm{h}}), \\
o_t &= g(h_t, w_{\mathrm{o}}),
\end{aligned}
\tag{9.7.1}
$$

where f and g are transformations of the hidden layer and the output layer, respectively. Hence, we have a chain of values $\{\ldots, (x_{t-1}, h_{t-1}, o_{t-1}), (x_t, h_t, o_t), \ldots\}$ that depend on each other via recurrent computation. The forward propagation is fairly straightforward. All we need is to loop through the (x_t, h_t, o_t) triples one time step at a time. The discrepancy between output o_t and the desired target y_t is then evaluated by an objective function across all the T time steps as

$$
L(x_1, \ldots, x_T, y_1, \ldots, y_T, w_{\mathrm{h}}, w_{\mathrm{o}}) = \frac{1}{T} \sum_{t=1}^{T} l(y_t, o_t).
\tag{9.7.2}
$$

For backpropagation, matters are a bit trickier, especially when we compute the gradients with regard to the parameters w_{h} of the objective function L. To be specific, by the chain rule,

$$
\begin{aligned}
\frac{\partial L}{\partial w_{\mathrm{h}}} &= \frac{1}{T} \sum_{t=1}^{T} \frac{\partial l(y_t, o_t)}{\partial w_{\mathrm{h}}} \\
&= \frac{1}{T} \sum_{t=1}^{T} \frac{\partial l(y_t, o_t)}{\partial o_t} \frac{\partial g(h_t, w_{\mathrm{o}})}{\partial h_t} \frac{\partial h_t}{\partial w_{\mathrm{h}}}.
\end{aligned}
\tag{9.7.3}
$$

The first and the second factors of the product in (9.7.3) are easy to compute. The third factor $\partial h_t / \partial w_{\mathrm{h}}$ is where things get tricky, since we need to recurrently compute the effect of the parameter w_{h} on h_t. According to the recurrent computation in (9.7.1), h_t depends on both h_{t-1} and w_{h}, where computation of h_{t-1} also depends on w_{h}. Thus, evaluating the total derivate of h_t with respect to w_{h} using the chain rule yields

$$
\frac{\partial h_t}{\partial w_{\mathrm{h}}} = \frac{\partial f(x_t, h_{t-1}, w_{\mathrm{h}})}{\partial w_{\mathrm{h}}} + \frac{\partial f(x_t, h_{t-1}, w_{\mathrm{h}})}{\partial h_{t-1}} \frac{\partial h_{t-1}}{\partial w_{\mathrm{h}}}.
\tag{9.7.4}
$$

To derive the above gradient, assume that we have three sequences $\{a_t\}, \{b_t\}, \{c_t\}$ satisfying $a_0 = 0$ and $a_t = b_t + c_t a_{t-1}$ for $t = 1, 2, \ldots$. Then for $t \geq 1$, it is easy to show

$$
a_t = b_t + \sum_{i=1}^{t-1} \left(\prod_{j=i+1}^{t} c_j \right) b_i.
\tag{9.7.5}
$$

By substituting a_t, b_t, and c_t according to

$$
\begin{aligned}
a_t &= \frac{\partial h_t}{\partial w_{\mathrm{h}}}, \\
b_t &= \frac{\partial f(x_t, h_{t-1}, w_{\mathrm{h}})}{\partial w_{\mathrm{h}}}, \\
c_t &= \frac{\partial f(x_t, h_{t-1}, w_{\mathrm{h}})}{\partial h_{t-1}},
\end{aligned}
\tag{9.7.6}
$$

the gradient computation in (9.7.4) satisfies $a_t = b_t + c_t a_{t-1}$. Thus, per (9.7.5), we can remove the recurrent computation in (9.7.4) with

$$\frac{\partial h_t}{\partial w_{\text{h}}} = \frac{\partial f(x_t, h_{t-1}, w_{\text{h}})}{\partial w_{\text{h}}} + \sum_{i=1}^{t-1} \left(\prod_{j=i+1}^{t} \frac{\partial f(x_j, h_{j-1}, w_{\text{h}})}{\partial h_{j-1}} \right) \frac{\partial f(x_i, h_{i-1}, w_{\text{h}})}{\partial w_{\text{h}}}. \tag{9.7.7}$$

While we can use the chain rule to compute $\partial h_t / \partial w_{\text{h}}$ recursively, this chain can get very long whenever t is large. Let's discuss a number of strategies for dealing with this problem.

Full Computation

One idea might be to compute the full sum in (9.7.7). However, this is very slow and gradients can blow up, since subtle changes in the initial conditions can potentially affect the outcome a lot. That is, we could see things similar to the butterfly effect, where minimal changes in the initial conditions lead to disproportionate changes in the outcome. This is generally undesirable. After all, we are looking for robust estimators that generalize well. Hence this strategy is almost never used in practice.

Truncating Time Steps

Alternatively, we can truncate the sum in (9.7.7) after τ steps. This is what we have been discussing so far. This leads to an *approximation* of the true gradient, simply by terminating the sum at $\partial h_{t-\tau} / \partial w_{\text{h}}$. In practice this works quite well. It is what is commonly referred to as truncated backpropgation through time (Jaeger, 2002). One of the consequences of this is that the model focuses primarily on short-term influence rather than long-term consequences. This is actually *desirable*, since it biases the estimate towards simpler and more stable models.

Randomized Truncation

Last, we can replace $\partial h_t / \partial w_{\text{h}}$ by a random variable which is correct in expectation but truncates the sequence. This is achieved by using a sequence of ξ_t with predefined $0 \leq \pi_t \leq 1$, where $P(\xi_t = 0) = 1 - \pi_t$ and $P(\xi_t = \pi_t^{-1}) = \pi_t$, thus $E[\xi_t] = 1$. We use this to replace the gradient $\partial h_t / \partial w_{\text{h}}$ in (9.7.4) with

$$z_t = \frac{\partial f(x_t, h_{t-1}, w_{\text{h}})}{\partial w_{\text{h}}} + \xi_t \frac{\partial f(x_t, h_{t-1}, w_{\text{h}})}{\partial h_{t-1}} \frac{\partial h_{t-1}}{\partial w_{\text{h}}}. \tag{9.7.8}$$

It follows from the definition of ξ_t that $E[z_t] = \partial h_t / \partial w_{\text{h}}$. Whenever $\xi_t = 0$ the recurrent computation terminates at that time step t. This leads to a weighted sum of sequences of varying lengths, where long sequences are rare but appropriately overweighted. This idea was proposed by Tallec and Ollivier (2017).

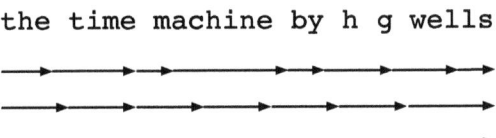

the time machine by h g wells

Fig. 9.7.1   Comparing strategies for computing gradients in RNNs. From top to bottom: randomized truncation, regular truncation, and full computation.

Comparing Strategies

Fig. 9.7.1 illustrates the three strategies when analyzing the first few characters of *The Time Machine* using backpropagation through time for RNNs:

- The first row is the randomized truncation that partitions the text into segments of varying lengths.

- The second row is the regular truncation that breaks the text into subsequences of the same length. This is what we have been doing in RNN experiments.

- The third row is the full backpropagation through time that leads to a computationally infeasible expression.

Unfortunately, while appealing in theory, randomized truncation does not work much better than regular truncation, most likely due to a number of factors. First, the effect of an observation after a number of backpropagation steps into the past is quite sufficient to capture dependencies in practice. Second, the increased variance counteracts the fact that the gradient is more accurate with more steps. Third, we actually *want* models that have only a short range of interactions. Hence, regularly truncated backpropagation through time has a slight regularizing effect that can be desirable.

9.7.2 Backpropagation Through Time in Detail

After discussing the general principle, let's discuss backpropagation through time in detail. In contrast to the analysis in Section 9.7.1, in the following we will show how to compute the gradients of the objective function with respect to all the decomposed model parameters. To keep things simple, we consider an RNN without bias parameters, whose activation function in the hidden layer uses the identity mapping ($\phi(x) = x$). For time step t, let the single example input and the target be $\mathbf{x}_t \in \mathbb{R}^d$ and y_t, respectively. The hidden state $\mathbf{h}_t \in \mathbb{R}^h$ and the output $\mathbf{o}_t \in \mathbb{R}^q$ are computed as

$$\begin{aligned}
\mathbf{h}_t &= \mathbf{W}_{hx}\mathbf{x}_t + \mathbf{W}_{hh}\mathbf{h}_{t-1}, \\
\mathbf{o}_t &= \mathbf{W}_{qh}\mathbf{h}_t,
\end{aligned} \tag{9.7.9}$$

where $\mathbf{W}_{hx} \in \mathbb{R}^{h \times d}$, $\mathbf{W}_{hh} \in \mathbb{R}^{h \times h}$, and $\mathbf{W}_{qh} \in \mathbb{R}^{q \times h}$ are the weight parameters. Denote by $l(\mathbf{o}_t, y_t)$ the loss at time step t. Our objective function, the loss over T time steps from the beginning of the sequence is thus

$$L = \frac{1}{T} \sum_{t=1}^{T} l(\mathbf{o}_t, y_t). \tag{9.7.10}$$

In order to visualize the dependencies among model variables and parameters during computation of the RNN, we can draw a computational graph for the model, as shown in Fig. 9.7.2. For example, the computation of the hidden states of time step 3, $\mathbf{h}_3$, depends on the model parameters $\mathbf{W}_{hx}$ and $\mathbf{W}_{hh}$, the hidden state of the previous time step $\mathbf{h}_2$, and the input of the current time step $\mathbf{x}_3$.

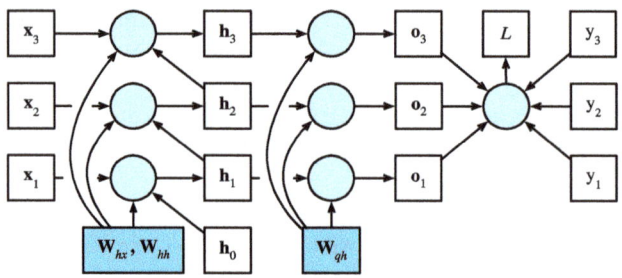

Fig. 9.7.2 Computational graph showing dependencies for an RNN model with three time steps. Boxes represent variables (not shaded) or parameters (shaded) and circles represent operators.

As just mentioned, the model parameters in Fig. 9.7.2 are $\mathbf{W}_{hx}$, $\mathbf{W}_{hh}$, and $\mathbf{W}_{qh}$. Generally, training this model requires gradient computation with respect to these parameters $\partial L/\partial \mathbf{W}_{hx}$, $\partial L/\partial \mathbf{W}_{hh}$, and $\partial L/\partial \mathbf{W}_{qh}$. According to the dependencies in Fig. 9.7.2, we can traverse in the opposite direction of the arrows to calculate and store the gradients in turn. To flexibly express the multiplication of matrices, vectors, and scalars of different shapes in the chain rule, we continue to use the prod operator as described in Section 5.3.

First of all, differentiating the objective function with respect to the model output at any time step t is fairly straightforward:

$$\frac{\partial L}{\partial \mathbf{o}_t} = \frac{\partial l(\mathbf{o}_t, y_t)}{T \cdot \partial \mathbf{o}_t} \in \mathbb{R}^q. \tag{9.7.11}$$

Now we can calculate the gradient of the objective with respect to the parameter $\mathbf{W}_{qh}$ in the output layer: $\partial L/\partial \mathbf{W}_{qh} \in \mathbb{R}^{q \times h}$. Based on Fig. 9.7.2, the objective L depends on $\mathbf{W}_{qh}$ via $\mathbf{o}_1, \ldots, \mathbf{o}_T$. Using the chain rule yields

$$\frac{\partial L}{\partial \mathbf{W}_{qh}} = \sum_{t=1}^{T} \text{prod}\left(\frac{\partial L}{\partial \mathbf{o}_t}, \frac{\partial \mathbf{o}_t}{\partial \mathbf{W}_{qh}}\right) = \sum_{t=1}^{T} \frac{\partial L}{\partial \mathbf{o}_t} \mathbf{h}_t^\top, \tag{9.7.12}$$

where $\partial L/\partial \mathbf{o}_t$ is given by (9.7.11).

Next, as shown in Fig. 9.7.2, at the final time step T, the objective function L depends on the hidden state $\mathbf{h}_T$ only via $\mathbf{o}_T$. Therefore, we can easily find the gradient $\partial L/\partial \mathbf{h}_T \in \mathbb{R}^h$

using the chain rule:

$$\frac{\partial L}{\partial \mathbf{h}_T} = \text{prod}\left(\frac{\partial L}{\partial \mathbf{o}_T}, \frac{\partial \mathbf{o}_T}{\partial \mathbf{h}_T}\right) = \mathbf{W}_{qh}^{\top} \frac{\partial L}{\partial \mathbf{o}_T}. \tag{9.7.13}$$

It gets trickier for any time step $t < T$, where the objective function L depends on $\mathbf{h}_t$ via $\mathbf{h}_{t+1}$ and $\mathbf{o}_t$. According to the chain rule, the gradient of the hidden state $\partial L/\partial \mathbf{h}_t \in \mathbb{R}^h$ at any time step $t < T$ can be recurrently computed as:

$$\frac{\partial L}{\partial \mathbf{h}_t} = \text{prod}\left(\frac{\partial L}{\partial \mathbf{h}_{t+1}}, \frac{\partial \mathbf{h}_{t+1}}{\partial \mathbf{h}_t}\right) + \text{prod}\left(\frac{\partial L}{\partial \mathbf{o}_t}, \frac{\partial \mathbf{o}_t}{\partial \mathbf{h}_t}\right) = \mathbf{W}_{hh}^{\top} \frac{\partial L}{\partial \mathbf{h}_{t+1}} + \mathbf{W}_{qh}^{\top} \frac{\partial L}{\partial \mathbf{o}_t}. \tag{9.7.14}$$

For analysis, expanding the recurrent computation for any time step $1 \leq t \leq T$ gives

$$\frac{\partial L}{\partial \mathbf{h}_t} = \sum_{i=t}^{T} \left(\mathbf{W}_{hh}^{\top}\right)^{T-i} \mathbf{W}_{qh}^{\top} \frac{\partial L}{\partial \mathbf{o}_{T+t-i}}. \tag{9.7.15}$$

We can see from (9.7.15) that this simple linear example already exhibits some key problems of long sequence models: it involves potentially very large powers of $\mathbf{W}_{hh}^{\top}$. In it, eigenvalues smaller than 1 vanish and eigenvalues larger than 1 diverge. This is numerically unstable, which manifests itself in the form of vanishing and exploding gradients. One way to address this is to truncate the time steps at a computationally convenient size as discussed in Section 9.7.1. In practice, this truncation can also be effected by detaching the gradient after a given number of time steps. Later on, we will see how more sophisticated sequence models such as long short-term memory can alleviate this further.

Finally, Fig. 9.7.2 shows that the objective function L depends on model parameters $\mathbf{W}_{hx}$ and $\mathbf{W}_{hh}$ in the hidden layer via hidden states $\mathbf{h}_1, \ldots, \mathbf{h}_T$. To compute gradients with respect to such parameters $\partial L/\partial \mathbf{W}_{hx} \in \mathbb{R}^{h \times d}$ and $\partial L/\partial \mathbf{W}_{hh} \in \mathbb{R}^{h \times h}$, we apply the chain rule giving

$$\begin{aligned}
\frac{\partial L}{\partial \mathbf{W}_{hx}} &= \sum_{t=1}^{T} \text{prod}\left(\frac{\partial L}{\partial \mathbf{h}_t}, \frac{\partial \mathbf{h}_t}{\partial \mathbf{W}_{hx}}\right) = \sum_{t=1}^{T} \frac{\partial L}{\partial \mathbf{h}_t} \mathbf{x}_t^{\top}, \\
\frac{\partial L}{\partial \mathbf{W}_{hh}} &= \sum_{t=1}^{T} \text{prod}\left(\frac{\partial L}{\partial \mathbf{h}_t}, \frac{\partial \mathbf{h}_t}{\partial \mathbf{W}_{hh}}\right) = \sum_{t=1}^{T} \frac{\partial L}{\partial \mathbf{h}_t} \mathbf{h}_{t-1}^{\top},
\end{aligned} \tag{9.7.16}$$

where $\partial L/\partial \mathbf{h}_t$ which is recurrently computed by (9.7.13) and (9.7.14) is the key quantity that affects the numerical stability.

Since backpropagation through time is the application of backpropagation in RNNs, as we have explained in Section 5.3, training RNNs alternates forward propagation with backpropagation through time. Moreover, backpropagation through time computes and stores the above gradients in turn. Specifically, stored intermediate values are reused to avoid duplicate calculations, such as storing $\partial L/\partial \mathbf{h}_t$ to be used in computation of both $\partial L/\partial \mathbf{W}_{hx}$ and $\partial L/\partial \mathbf{W}_{hh}$.

9.7.3 Summary

Backpropagation through time is merely an application of backpropagation to sequence models with a hidden state. Truncation, such as regular or randomized, is needed for computational convenience and numerical stability. High powers of matrices can lead to divergent or vanishing eigenvalues. This manifests itself in the form of exploding or vanishing gradients. For efficient computation, intermediate values are cached during backpropagation through time.

9.7.4 Exercises

1. Assume that we have a symmetric matrix $M \in \mathbb{R}^{n \times n}$ with eigenvalues λ_i whose corresponding eigenvectors are v_i $(i = 1, \ldots, n)$. Without loss of generality, assume that they are ordered in the order $|\lambda_i| \geq |\lambda_{i+1}|$.

2. Show that M^k has eigenvalues λ_i^k.

3. Prove that for a random vector $x \in \mathbb{R}^n$, with high probability $M^k x$ will be very much aligned with the eigenvector v_1 of M. Formalize this statement.

4. What does the above result mean for gradients in RNNs?

5. Besides gradient clipping, can you think of any other methods to cope with gradient explosion in recurrent neural networks?

Discussions[144].

144

Modern Recurrent Neural Networks

The previous chapter introduced the key ideas behind recurrent neural networks (RNNs). However, just as with convolutional neural networks, there has been a tremendous amount of innovation in RNN architectures, culminating in several complex designs that have proven successful in practice. In particular, the most popular designs feature mechanisms for mitigating the notorious numerical instability faced by RNNs, as typified by vanishing and exploding gradients. Recall that in Chapter 9 we dealt with exploding gradients by applying a blunt gradient clipping heuristic. Despite the efficacy of this hack, it leaves open the problem of vanishing gradients.

In this chapter, we introduce the key ideas behind the most successful RNN architectures for sequences, which stem from two papers. The first, *Long Short-Term Memory* (Hochreiter and Schmidhuber, 1997), introduces the *memory cell*, a unit of computation that replaces traditional nodes in the hidden layer of a network. With these memory cells, networks are able to overcome difficulties with training encountered by earlier recurrent networks. Intuitively, the memory cell avoids the vanishing gradient problem by keeping values in each memory cell's internal state cascading along a recurrent edge with weight 1 across many successive time steps. A set of multiplicative gates help the network to determine not only the inputs to allow into the memory state, but when the content of the memory state should influence the model's output.

The second paper, *Bidirectional Recurrent Neural Networks* (Schuster and Paliwal, 1997), introduces an architecture in which information from both the future (subsequent time steps) and the past (preceding time steps) are used to determine the output at any point in the sequence. This is in contrast to previous networks, in which only past input can affect the output. Bidirectional RNNs have become a mainstay for sequence labeling tasks in natural language processing, among a myriad of other tasks. Fortunately, the two innovations are not mutually exclusive, and have been successfully combined for phoneme classification (Graves and Schmidhuber, 2005) and handwriting recognition (Graves *et al.*, 2008).

The first sections in this chapter will explain the LSTM architecture, a lighter-weight version called the gated recurrent unit (GRU), the key ideas behind bidirectional RNNs and a brief explanation of how RNN layers are stacked together to form deep RNNs. Subsequently, we will explore the application of RNNs in sequence-to-sequence tasks, introducing machine translation along with key ideas such as *encoder–decoder* architectures and *beam search*.

10.1 Long Short-Term Memory (LSTM)

Shortly after the first Elman-style RNNs were trained using backpropagation (Elman, 1990), the problems of learning long-term dependencies (owing to vanishing and exploding gradients) became salient, with Bengio and Hochreiter discussing the problem (Bengio *et al.*, 1994, Hochreiter *et al.*, 2001). Hochreiter had articulated this problem as early as 1991 in his Master's thesis, although the results were not widely known because the thesis was written in German. While gradient clipping helps with exploding gradients, handling vanishing gradients appears to require a more elaborate solution. One of the first and most successful techniques for addressing vanishing gradients came in the form of the long short-term memory (LSTM) model due to Hochreiter and Schmidhuber (1997). LSTMs resemble standard recurrent neural networks but here each ordinary recurrent node is replaced by a *memory cell*. Each memory cell contains an *internal state*, i.e., a node with a self-connected recurrent edge of fixed weight 1, ensuring that the gradient can pass across many time steps without vanishing or exploding.

The term "long short-term memory" comes from the following intuition. Simple recurrent neural networks have *long-term memory* in the form of weights. The weights change slowly during training, encoding general knowledge about the data. They also have *short-term memory* in the form of ephemeral activations, which pass from each node to successive nodes. The LSTM model introduces an intermediate type of storage via the memory cell. A memory cell is a composite unit, built from simpler nodes in a specific connectivity pattern, with the novel inclusion of multiplicative nodes.

```
import torch
from torch import nn
from d2l import torch as d2l
```

10.1.1 Gated Memory Cell

Each memory cell is equipped with an *internal state* and a number of multiplicative gates that determine whether (i) a given input should impact the internal state (the *input gate*), (ii) the internal state should be flushed to 0 (the *forget gate*), and (iii) the internal state of a given neuron should be allowed to impact the cell's output (the *output* gate).

Gated Hidden State

The key distinction between vanilla RNNs and LSTMs is that the latter support gating of the hidden state. This means that we have dedicated mechanisms for when a hidden state should be *updated* and also for when it should be *reset*. These mechanisms are learned and they address the concerns listed above. For instance, if the first token is of great importance we will learn not to update the hidden state after the first observation. Likewise, we will learn to skip irrelevant temporary observations. Last, we will learn to reset the latent state whenever needed. We discuss this in detail below.

Input Gate, Forget Gate, and Output Gate

The data feeding into the LSTM gates are the input at the current time step and the hidden state of the previous time step, as illustrated in Fig. 10.1.1. Three fully connected layers with sigmoid activation functions compute the values of the input, forget, and output gates. As a result of the sigmoid activation, all values of the three gates are in the range of $(0, 1)$. Additionally, we require an *input node*, typically computed with a *tanh* activation function. Intuitively, the *input gate* determines how much of the input node's value should be added to the current memory cell internal state. The *forget gate* determines whether to keep the current value of the memory or flush it. And the *output gate* determines whether the memory cell should influence the output at the current time step.

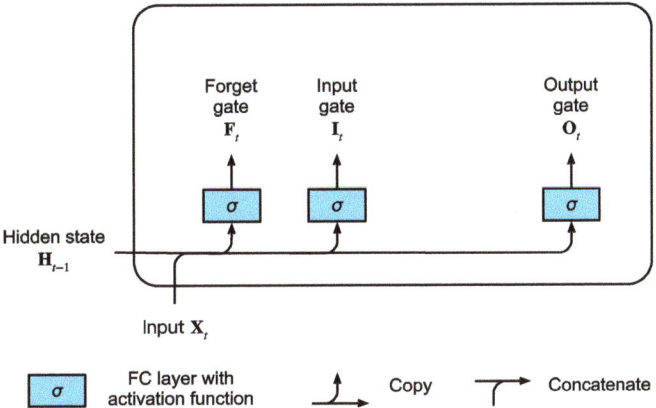

Fig. 10.1.1   Computing the input gate, the forget gate, and the output gate in an LSTM model.

Mathematically, suppose that there are h hidden units, the batch size is n, and the number of inputs is d. Thus, the input is $\mathbf{X}_t \in \mathbb{R}^{n \times d}$ and the hidden state of the previous time step is $\mathbf{H}_{t-1} \in \mathbb{R}^{n \times h}$. Correspondingly, the gates at time step t are defined as follows: the input gate is $\mathbf{I}_t \in \mathbb{R}^{n \times h}$, the forget gate is $\mathbf{F}_t \in \mathbb{R}^{n \times h}$, and the output gate is $\mathbf{O}_t \in \mathbb{R}^{n \times h}$. They are calculated as follows:

$$\begin{aligned}
\mathbf{I}_t &= \sigma(\mathbf{X}_t \mathbf{W}_{\mathrm{xi}} + \mathbf{H}_{t-1} \mathbf{W}_{\mathrm{hi}} + \mathbf{b}_{\mathrm{i}}), \\
\mathbf{F}_t &= \sigma(\mathbf{X}_t \mathbf{W}_{\mathrm{xf}} + \mathbf{H}_{t-1} \mathbf{W}_{\mathrm{hf}} + \mathbf{b}_{\mathrm{f}}), \\
\mathbf{O}_t &= \sigma(\mathbf{X}_t \mathbf{W}_{\mathrm{xo}} + \mathbf{H}_{t-1} \mathbf{W}_{\mathrm{ho}} + \mathbf{b}_{\mathrm{o}}),
\end{aligned} \tag{10.1.1}$$

where $\mathbf{W}_{\mathrm{xi}}, \mathbf{W}_{\mathrm{xf}}, \mathbf{W}_{\mathrm{xo}} \in \mathbb{R}^{d \times h}$ and $\mathbf{W}_{\mathrm{hi}}, \mathbf{W}_{\mathrm{hf}}, \mathbf{W}_{\mathrm{ho}} \in \mathbb{R}^{h \times h}$ are weight parameters and $\mathbf{b}_{\mathrm{i}}, \mathbf{b}_{\mathrm{f}}, \mathbf{b}_{\mathrm{o}} \in \mathbb{R}^{1 \times h}$ are bias parameters. Note that broadcasting (see Section 2.1.4) is triggered during the summation. We use sigmoid functions (as introduced in Section 5.1) to map the input values to the interval $(0, 1)$.

Input Node

Next we design the memory cell. Since we have not specified the action of the various gates yet, we first introduce the *input node* $\tilde{\mathbf{C}}_t \in \mathbb{R}^{n \times h}$. Its computation is similar to that of the three gates described above, but uses a tanh function with a value range for $(-1, 1)$ as the activation function. This leads to the following equation at time step t:

$$\tilde{\mathbf{C}}_t = \tanh(\mathbf{X}_t\mathbf{W}_{xc} + \mathbf{H}_{t-1}\mathbf{W}_{hc} + \mathbf{b}_c), \tag{10.1.2}$$

where $\mathbf{W}_{xc} \in \mathbb{R}^{d \times h}$ and $\mathbf{W}_{hc} \in \mathbb{R}^{h \times h}$ are weight parameters and $\mathbf{b}_c \in \mathbb{R}^{1 \times h}$ is a bias parameter.

A quick illustration of the input node is shown in Fig. 10.1.2.

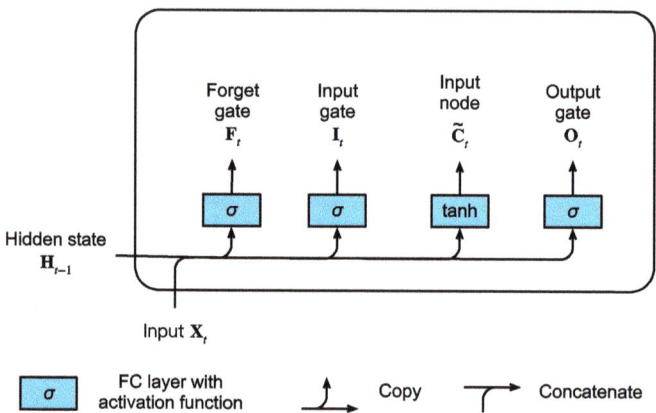

Fig. 10.1.2   Computing the input node in an LSTM model.

Memory Cell Internal State

In LSTMs, the input gate $\mathbf{I}_t$ governs how much we take new data into account via $\tilde{\mathbf{C}}_t$ and the forget gate $\mathbf{F}_t$ addresses how much of the old cell internal state $\mathbf{C}_{t-1} \in \mathbb{R}^{n \times h}$ we retain. Using the Hadamard (elementwise) product operator $\odot$ we arrive at the following update equation:

$$\mathbf{C}_t = \mathbf{F}_t \odot \mathbf{C}_{t-1} + \mathbf{I}_t \odot \tilde{\mathbf{C}}_t. \tag{10.1.3}$$

If the forget gate is always 1 and the input gate is always 0, the memory cell internal state $\mathbf{C}_{t-1}$ will remain constant forever, passing unchanged to each subsequent time step. However, input gates and forget gates give the model the flexibility of being able to learn when to keep this value unchanged and when to perturb it in response to subsequent inputs. In practice, this design alleviates the vanishing gradient problem, resulting in models that are much easier to train, especially when facing datasets with long sequence lengths.

We thus arrive at the flow diagram in Fig. 10.1.3.

Hidden State

Last, we need to define how to compute the output of the memory cell, i.e., the hidden state $\mathbf{H}_t \in \mathbb{R}^{n \times h}$, as seen by other layers. This is where the output gate comes into play. In

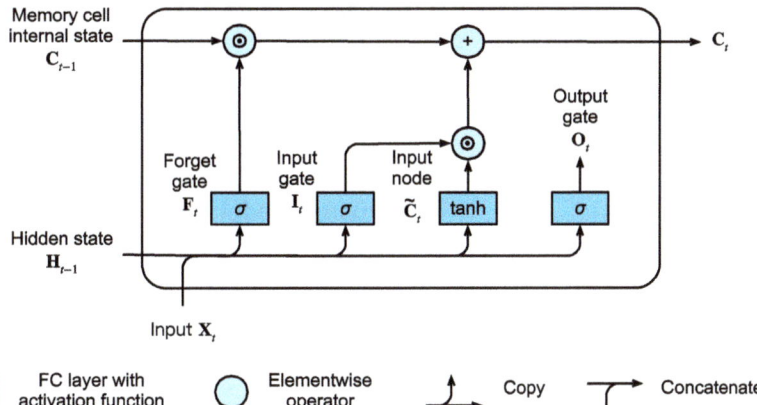

Computing the memory cell internal state in an LSTM model.

LSTMs, we first apply tanh to the memory cell internal state and then apply another point-wise multiplication, this time with the output gate. This ensures that the values of $\mathbf{H}_t$ are always in the interval $(-1, 1)$:

$$\mathbf{H}_t = \mathbf{O}_t \odot \tanh(\mathbf{C}_t). \tag{10.1.4}$$

Whenever the output gate is close to 1, we allow the memory cell internal state to impact the subsequent layers uninhibited, whereas for output gate values close to 0, we prevent the current memory from impacting other layers of the network at the current time step. Note that a memory cell can accrue information across many time steps without impacting the rest of the network (as long as the output gate takes values close to 0), and then suddenly impact the network at a subsequent time step as soon as the output gate flips from values close to 0 to values close to 1. Fig. 10.1.4 has a graphical illustration of the data flow.

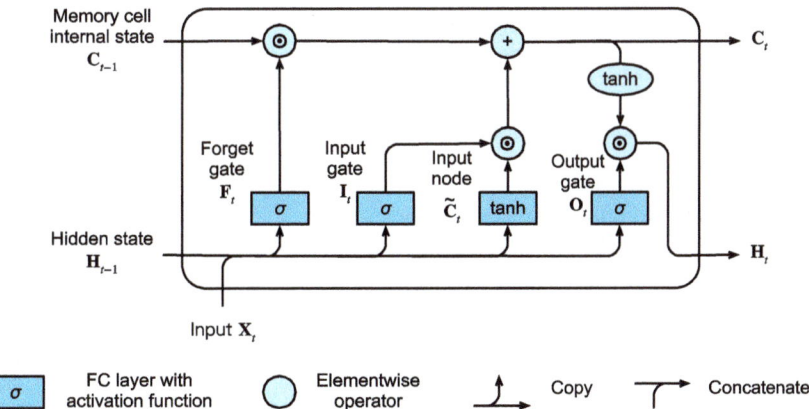

Computing the hidden state in an LSTM model.

10.1.2 Implementation from Scratch

Now let's implement an LSTM from scratch. As same as the experiments in Section 9.5, we first load *The Time Machine* dataset.

Initializing Model Parameters

Next, we need to define and initialize the model parameters. As previously, the hyperparameter num_hiddens dictates the number of hidden units. We initialize weights following a Gaussian distribution with 0.01 standard deviation, and we set the biases to 0.

```python
class LSTMScratch(d2l.Module):
    def __init__(self, num_inputs, num_hiddens, sigma=0.01):
        super().__init__()
        self.save_hyperparameters()

        init_weight = lambda *shape: nn.Parameter(torch.randn(*shape) * sigma)
        triple = lambda: (init_weight(num_inputs, num_hiddens),
                          init_weight(num_hiddens, num_hiddens),
                          nn.Parameter(torch.zeros(num_hiddens)))
        self.W_xi, self.W_hi, self.b_i = triple()  # Input gate
        self.W_xf, self.W_hf, self.b_f = triple()  # Forget gate
        self.W_xo, self.W_ho, self.b_o = triple()  # Output gate
        self.W_xc, self.W_hc, self.b_c = triple()  # Input node
```

The actual model is defined as described above, consisting of three gates and an input node. Note that only the hidden state is passed to the output layer.

```python
@d2l.add_to_class(LSTMScratch)
def forward(self, inputs, H_C=None):
    if H_C is None:
        # Initial state with shape: (batch_size, num_hiddens)
        H = torch.zeros((inputs.shape[1], self.num_hiddens),
                        device=inputs.device)
        C = torch.zeros((inputs.shape[1], self.num_hiddens),
                        device=inputs.device)
    else:
        H, C = H_C
    outputs = []
    for X in inputs:
        I = torch.sigmoid(torch.matmul(X, self.W_xi) +
                          torch.matmul(H, self.W_hi) + self.b_i)
        F = torch.sigmoid(torch.matmul(X, self.W_xf) +
                          torch.matmul(H, self.W_hf) + self.b_f)
        O = torch.sigmoid(torch.matmul(X, self.W_xo) +
                          torch.matmul(H, self.W_ho) + self.b_o)
        C_tilde = torch.tanh(torch.matmul(X, self.W_xc) +
                             torch.matmul(H, self.W_hc) + self.b_c)
        C = F * C + I * C_tilde
        H = O * torch.tanh(C)
        outputs.append(H)
    return outputs, (H, C)
```

Training and Prediction

Let's train an LSTM model by instantiating the RNNLMScratch class from Section 9.5.

```
data = d2l.TimeMachine(batch_size=1024, num_steps=32)
lstm = LSTMScratch(num_inputs=len(data.vocab), num_hiddens=32)
model = d2l.RNNLMScratch(lstm, vocab_size=len(data.vocab), lr=4)
trainer = d2l.Trainer(max_epochs=50, gradient_clip_val=1, num_gpus=1)
trainer.fit(model, data)
```

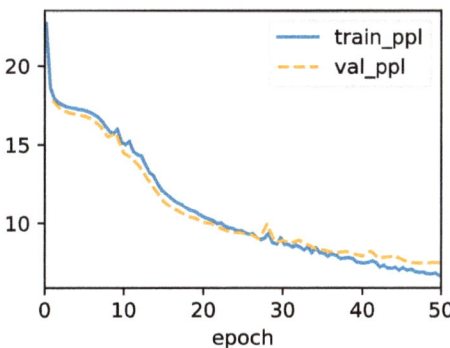

10.1.3 Concise Implementation

Using high-level APIs, we can directly instantiate an LSTM model. This encapsulates all the configuration details that we made explicit above. The code is significantly faster as it uses compiled operators rather than Python for many details that we spelled out before.

```
class LSTM(d2l.RNN):
    def __init__(self, num_inputs, num_hiddens):
        d2l.Module.__init__(self)
        self.save_hyperparameters()
        self.rnn = nn.LSTM(num_inputs, num_hiddens)

    def forward(self, inputs, H_C=None):
        return self.rnn(inputs, H_C)
```

```
lstm = LSTM(num_inputs=len(data.vocab), num_hiddens=32)
model = d2l.RNNLM(lstm, vocab_size=len(data.vocab), lr=4)
trainer.fit(model, data)
```

```
model.predict('it has', 20, data.vocab, d2l.try_gpu())
```

```
'it has in the time travell'
```

LSTMs are the prototypical latent variable autoregressive model with nontrivial state control. Many variants thereof have been proposed over the years, e.g., multiple layers, residual connections, different types of regularization. However, training LSTMs and other

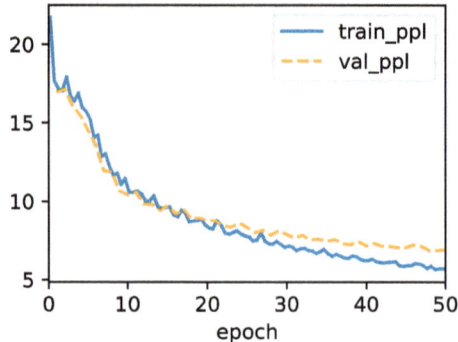

sequence models (such as GRUs) is quite costly because of the long range dependency of the sequence. Later we will encounter alternative models such as Transformers that can be used in some cases.

10.1.4 Summary

While LSTMs were published in 1997, they rose to great prominence with some victories in prediction competitions in the mid-2000s, and became the dominant models for sequence learning from 2011 until the rise of Transformer models, starting in 2017. Even Tranformers owe some of their key ideas to architecture design innovations introduced by the LSTM.

LSTMs have three types of gates: input gates, forget gates, and output gates that control the flow of information. The hidden layer output of LSTM includes the hidden state and the memory cell internal state. Only the hidden state is passed into the output layer while the memory cell internal state remains entirely internal. LSTMs can alleviate vanishing and exploding gradients.

10.1.5 Exercises

1. Adjust the hyperparameters and analyze their influence on running time, perplexity, and the output sequence.

2. How would you need to change the model to generate proper words rather than just sequences of characters?

3. Compare the computational cost for GRUs, LSTMs, and regular RNNs for a given hidden dimension. Pay special attention to the training and inference cost.

4. Since the candidate memory cell ensures that the value range is between -1 and 1 by using the tanh function, why does the hidden state need to use the tanh function again to ensure that the output value range is between -1 and 1?

5. Implement an LSTM model for time series prediction rather than character sequence prediction.

145

Discussions[145].

10.2 Gated Recurrent Units (GRU)

As RNNs and particularly the LSTM architecture (Section 10.1) rapidly gained popularity during the 2010s, a number of researchers began to experiment with simplified architectures in hopes of retaining the key idea of incorporating an internal state and multiplicative gating mechanisms but with the aim of speeding up computation. The gated recurrent unit (GRU) (Cho *et al.*, 2014) offered a streamlined version of the LSTM memory cell that often achieves comparable performance but with the advantage of being faster to compute (Chung *et al.*, 2014).

```
import torch
from torch import nn
from d2l import torch as d2l
```

10.2.1 Reset Gate and Update Gate

Here, the LSTM's three gates are replaced by two: the *reset gate* and the *update gate*. As with LSTMs, these gates are given sigmoid activations, forcing their values to lie in the interval $(0, 1)$. Intuitively, the reset gate controls how much of the previous state we might still want to remember. Likewise, an update gate would allow us to control how much of the new state is just a copy of the old one. Fig. 10.2.1 illustrates the inputs for both the reset and update gates in a GRU, given the input of the current time step and the hidden state of the previous time step. The outputs of the gates are given by two fully connected layers with a sigmoid activation function.

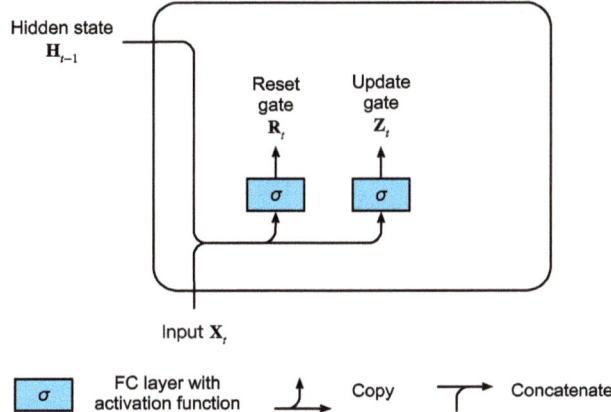

Fig. 10.2.1   Computing the reset gate and the update gate in a GRU model.

Mathematically, for a given time step t, suppose that the input is a minibatch $\mathbf{X}_t \in \mathbb{R}^{n \times d}$ (number of examples = n; number of inputs = d) and the hidden state of the previous time step is $\mathbf{H}_{t-1} \in \mathbb{R}^{n \times h}$ (number of hidden units = h). Then the reset gate $\mathbf{R}_t \in \mathbb{R}^{n \times h}$ and update gate $\mathbf{Z}_t \in \mathbb{R}^{n \times h}$ are computed as follows:

$$
\begin{aligned}
\mathbf{R}_t &= \sigma(\mathbf{X}_t \mathbf{W}_{xr} + \mathbf{H}_{t-1} \mathbf{W}_{hr} + \mathbf{b}_r), \\
\mathbf{Z}_t &= \sigma(\mathbf{X}_t \mathbf{W}_{xz} + \mathbf{H}_{t-1} \mathbf{W}_{hz} + \mathbf{b}_z),
\end{aligned}
\tag{10.2.1}
$$

where $\mathbf{W}_{xr}, \mathbf{W}_{xz} \in \mathbb{R}^{d \times h}$ and $\mathbf{W}_{hr}, \mathbf{W}_{hz} \in \mathbb{R}^{h \times h}$ are weight parameters and $\mathbf{b}_r, \mathbf{b}_z \in \mathbb{R}^{1 \times h}$ are bias parameters.

10.2.2 Candidate Hidden State

Next, we integrate the reset gate $\mathbf{R}_t$ with the regular updating mechanism in (9.4.5), leading to the following *candidate hidden state* $\tilde{\mathbf{H}}_t \in \mathbb{R}^{n \times h}$ at time step t:

$$
\tilde{\mathbf{H}}_t = \tanh(\mathbf{X}_t \mathbf{W}_{xh} + (\mathbf{R}_t \odot \mathbf{H}_{t-1}) \mathbf{W}_{hh} + \mathbf{b}_h),
\tag{10.2.2}
$$

where $\mathbf{W}_{xh} \in \mathbb{R}^{d \times h}$ and $\mathbf{W}_{hh} \in \mathbb{R}^{h \times h}$ are weight parameters, $\mathbf{b}_h \in \mathbb{R}^{1 \times h}$ is the bias, and the symbol $\odot$ is the Hadamard (elementwise) product operator. Here we use a tanh activation function.

The result is a *candidate*, since we still need to incorporate the action of the update gate. Comparing with (9.4.5), the influence of the previous states can now be reduced with the elementwise multiplication of $\mathbf{R}_t$ and $\mathbf{H}_{t-1}$ in (10.2.2). Whenever the entries in the reset gate $\mathbf{R}_t$ are close to 1, we recover a vanilla RNN such as that in (9.4.5). For all entries of the reset gate $\mathbf{R}_t$ that are close to 0, the candidate hidden state is the result of an MLP with $\mathbf{X}_t$ as input. Any pre-existing hidden state is thus *reset* to defaults.

Fig. 10.2.2 illustrates the computational flow after applying the reset gate.

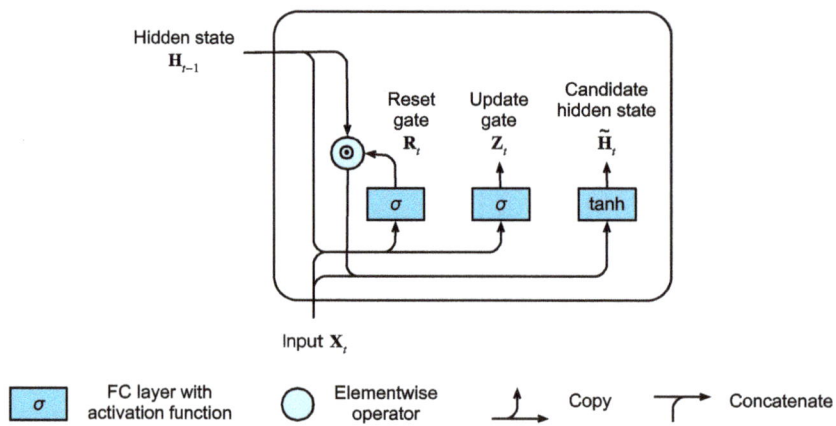

Fig. 10.2.2 Computing the candidate hidden state in a GRU model.

10.2.3 Hidden State

Finally, we need to incorporate the effect of the update gate $\mathbf{Z}_t$. This determines the extent to which the new hidden state $\mathbf{H}_t \in \mathbb{R}^{n \times h}$ matches the old state $\mathbf{H}_{t-1}$ compared with how much it resembles the new candidate state $\tilde{\mathbf{H}}_t$. The update gate $\mathbf{Z}_t$ can be used for this purpose, simply by taking elementwise convex combinations of $\mathbf{H}_{t-1}$ and $\tilde{\mathbf{H}}_t$. This leads to the final update equation for the GRU:

$$\mathbf{H}_t = \mathbf{Z}_t \odot \mathbf{H}_{t-1} + (1 - \mathbf{Z}_t) \odot \tilde{\mathbf{H}}_t. \qquad (10.2.3)$$

Whenever the update gate $\mathbf{Z}_t$ is close to 1, we simply retain the old state. In this case the information from $\mathbf{X}_t$ is ignored, effectively skipping time step t in the dependency chain. By contrast, whenever $\mathbf{Z}_t$ is close to 0, the new latent state $\mathbf{H}_t$ approaches the candidate latent state $\tilde{\mathbf{H}}_t$. Fig. 10.2.3 shows the computational flow after the update gate is in action.

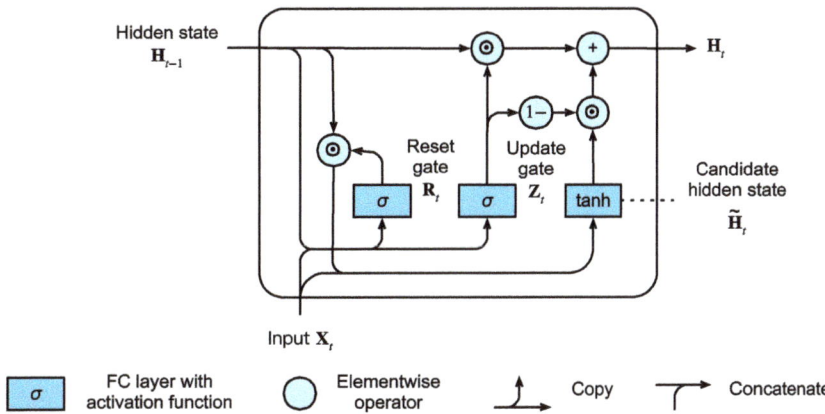

Fig. 10.2.3   Computing the hidden state in a GRU model.

In summary, GRUs have the following two distinguishing features:

• Reset gates help capture short-term dependencies in sequences.

• Update gates help capture long-term dependencies in sequences.

10.2.4 Implementation from Scratch

To gain a better understanding of the GRU model, let's implement it from scratch.

Initializing Model Parameters

The first step is to initialize the model parameters. We draw the weights from a Gaussian distribution with standard deviation to be `sigma` and set the bias to 0. The hyperparameter `num_hiddens` defines the number of hidden units. We instantiate all weights and biases relating to the update gate, the reset gate, and the candidate hidden state.

```
class GRUScratch(d2l.Module):
    def __init__(self, num_inputs, num_hiddens, sigma=0.01):
        super().__init__()
        self.save_hyperparameters()
```

(continues on next page)

(continued from previous page)

```
    init_weight = lambda *shape: nn.Parameter(torch.randn(*shape) * sigma)
    triple = lambda: (init_weight(num_inputs, num_hiddens),
                      init_weight(num_hiddens, num_hiddens),
                      nn.Parameter(torch.zeros(num_hiddens)))
    self.W_xz, self.W_hz, self.b_z = triple()  # Update gate
    self.W_xr, self.W_hr, self.b_r = triple()  # Reset gate
    self.W_xh, self.W_hh, self.b_h = triple()  # Candidate hidden state
```

Defining the Model

Now we are ready to define the GRU forward computation. Its structure is the same as that of the basic RNN cell, except that the update equations are more complex.

```
@d2l.add_to_class(GRUScratch)
def forward(self, inputs, H=None):
    if H is None:
        # Initial state with shape: (batch_size, num_hiddens)
        H = torch.zeros((inputs.shape[1], self.num_hiddens),
                        device=inputs.device)
    outputs = []
    for X in inputs:
        Z = torch.sigmoid(torch.matmul(X, self.W_xz) +
                          torch.matmul(H, self.W_hz) + self.b_z)
        R = torch.sigmoid(torch.matmul(X, self.W_xr) +
                          torch.matmul(H, self.W_hr) + self.b_r)
        H_tilde = torch.tanh(torch.matmul(X, self.W_xh) +
                            torch.matmul(R * H, self.W_hh) + self.b_h)
        H = Z * H + (1 - Z) * H_tilde
        outputs.append(H)
    return outputs, H
```

Training

Training a language model on *The Time Machine* dataset works in exactly the same manner as in Section 9.5.

```
data = d2l.TimeMachine(batch_size=1024, num_steps=32)
gru = GRUScratch(num_inputs=len(data.vocab), num_hiddens=32)
model = d2l.RNNLMScratch(gru, vocab_size=len(data.vocab), lr=4)
trainer = d2l.Trainer(max_epochs=50, gradient_clip_val=1, num_gpus=1)
trainer.fit(model, data)
```

10.2.5 Concise Implementation

In high-level APIs, we can directly instantiate a GRU model. This encapsulates all the configuration detail that we made explicit above.

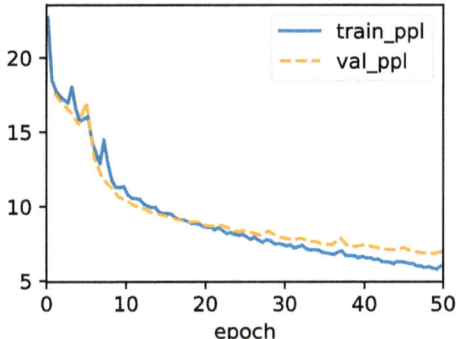

```
class GRU(d2l.RNN):
    def __init__(self, num_inputs, num_hiddens):
        d2l.Module.__init__(self)
        self.save_hyperparameters()
        self.rnn = nn.GRU(num_inputs, num_hiddens)
```

The code is significantly faster in training as it uses compiled operators rather than Python.

```
gru = GRU(num_inputs=len(data.vocab), num_hiddens=32)
model = d2l.RNNLM(gru, vocab_size=len(data.vocab), lr=4)
trainer.fit(model, data)
```

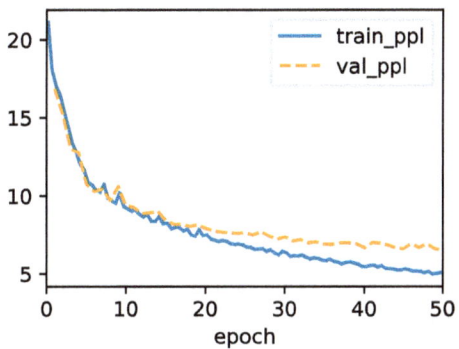

After training, we print out the perplexity on the training set and the predicted sequence following the provided prefix.

```
model.predict('it has', 20, data.vocab, d2l.try_gpu())
```

```
'it has the time traveller '
```

10.2.6 Summary

Compared with LSTMs, GRUs achieve similar performance but tend to be lighter computationally. Generally, compared with simple RNNs, gated RNNS, just like LSTMs and GRUs, can better capture dependencies for sequences with large time step distances. GRUs contain basic RNNs as their extreme case whenever the reset gate is switched on. They can also skip subsequences by turning on the update gate.

10.2.7 Exercises

1. Assume that we only want to use the input at time step t' to predict the output at time step $t > t'$. What are the best values for the reset and update gates for each time step?

2. Adjust the hyperparameters and analyze their influence on running time, perplexity, and the output sequence.

3. Compare runtime, perplexity, and the output strings for `rnn.RNN` and `rnn.GRU` implementations with each other.

4. What happens if you implement only parts of a GRU, e.g., with only a reset gate or only an update gate?

146

Discussions[146].

10.3 Deep Recurrent Neural Networks

Up until now, we have focused on defining networks consisting of a sequence input, a single hidden RNN layer, and an output layer. Despite having just one hidden layer between the input at any time step and the corresponding output, there is a sense in which these networks are deep. Inputs from the first time step can influence the outputs at the final time step T (often 100s or 1000s of steps later). These inputs pass through T applications of the recurrent layer before reaching the final output. However, we often also wish to retain the ability to express complex relationships between the inputs at a given time step and the outputs at that same time step. Thus we often construct RNNs that are deep not only in the time direction but also in the input-to-output direction. This is precisely the notion of depth that we have already encountered in our development of MLPs and deep CNNs.

The standard method for building this sort of deep RNN is strikingly simple: we stack the RNNs on top of each other. Given a sequence of length T, the first RNN produces a sequence of outputs, also of length T. These, in turn, constitute the inputs to the next RNN layer. In this short section, we illustrate this design pattern and present a simple example for how to code up such stacked RNNs. Below, in Fig. 10.3.1, we illustrate a deep RNN with L hidden layers. Each hidden state operates on a sequential input and produces a sequential output. Moreover, any RNN cell (white box in Fig. 10.3.1) at each time step depends on both the same layer's value at the previous time step and the previous layer's value at the same time step.

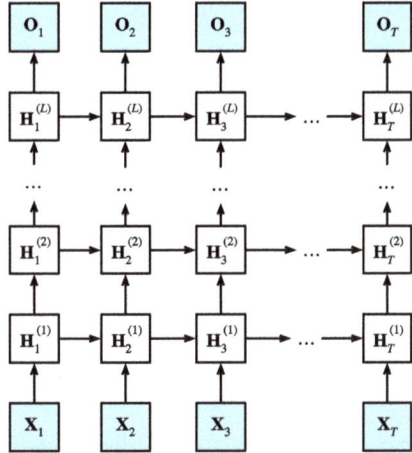

Fig. 10.3.1 Architecture of a deep RNN.

Formally, suppose that we have a minibatch input $\mathbf{X}_t \in \mathbb{R}^{n \times d}$ (number of examples $= n$; number of inputs in each example $= d$) at time step t. At the same time step, let the hidden state of the lth hidden layer ($l = 1, \ldots, L$) be $\mathbf{H}_t^{(l)} \in \mathbb{R}^{n \times h}$ (number of hidden units $= h$) and the output layer variable be $\mathbf{O}_t \in \mathbb{R}^{n \times q}$ (number of outputs: q). Setting $\mathbf{H}_t^{(0)} = \mathbf{X}_t$, the hidden state of the lth hidden layer that uses the activation function ϕ_l is calculated as follows:

$$\mathbf{H}_t^{(l)} = \phi_l(\mathbf{H}_t^{(l-1)}\mathbf{W}_{\text{xh}}^{(l)} + \mathbf{H}_{t-1}^{(l)}\mathbf{W}_{\text{hh}}^{(l)} + \mathbf{b}_{\text{h}}^{(l)}), \tag{10.3.1}$$

where the weights $\mathbf{W}_{\text{xh}}^{(l)} \in \mathbb{R}^{h \times h}$ and $\mathbf{W}_{\text{hh}}^{(l)} \in \mathbb{R}^{h \times h}$, together with the bias $\mathbf{b}_{\text{h}}^{(l)} \in \mathbb{R}^{1 \times h}$, are the model parameters of the lth hidden layer.

At the end, the calculation of the output layer is only based on the hidden state of the final L^{th} hidden layer:

$$\mathbf{O}_t = \mathbf{H}_t^{(L)}\mathbf{W}_{\text{hq}} + \mathbf{b}_{\text{q}}, \tag{10.3.2}$$

where the weight $\mathbf{W}_{\text{hq}} \in \mathbb{R}^{h \times q}$ and the bias $\mathbf{b}_{\text{q}} \in \mathbb{R}^{1 \times q}$ are the model parameters of the output layer.

Just as with MLPs, the number of hidden layers L and the number of hidden units h are hyperparameters that we can tune. Common RNN layer widths (h) are in the range $(64, 2056)$, and common depths (L) are in the range $(1, 8)$. In addition, we can easily get a deep-gated RNN by replacing the hidden state computation in (10.3.1) with that from an LSTM or a GRU.

```python
import torch
from torch import nn
from d2l import torch as d2l
```

10.3.1 Implementation from Scratch

To implement a multilayer RNN from scratch, we can treat each layer as an `RNNScratch` instance with its own learnable parameters.

```
class StackedRNNScratch(d2l.Module):
    def __init__(self, num_inputs, num_hiddens, num_layers, sigma=0.01):
        super().__init__()
        self.save_hyperparameters()
        self.rnns = nn.Sequential(*[d2l.RNNScratch(
            num_inputs if i==0 else num_hiddens, num_hiddens, sigma)
                                     for i in range(num_layers)])
```

The multilayer forward computation simply performs forward computation layer by layer.

```
@d2l.add_to_class(StackedRNNScratch)
def forward(self, inputs, Hs=None):
    outputs = inputs
    if Hs is None: Hs = [None] * self.num_layers
    for i in range(self.num_layers):
        outputs, Hs[i] = self.rnns[i](outputs, Hs[i])
        outputs = torch.stack(outputs, 0)
    return outputs, Hs
```

As an example, we train a deep GRU model on *The Time Machine* dataset (same as in Section 9.5). To keep things simple we set the number of layers to 2.

```
data = d2l.TimeMachine(batch_size=1024, num_steps=32)
rnn_block = StackedRNNScratch(num_inputs=len(data.vocab),
                               num_hiddens=32, num_layers=2)
model = d2l.RNNLMScratch(rnn_block, vocab_size=len(data.vocab), lr=2)
trainer = d2l.Trainer(max_epochs=100, gradient_clip_val=1, num_gpus=1)
trainer.fit(model, data)
```

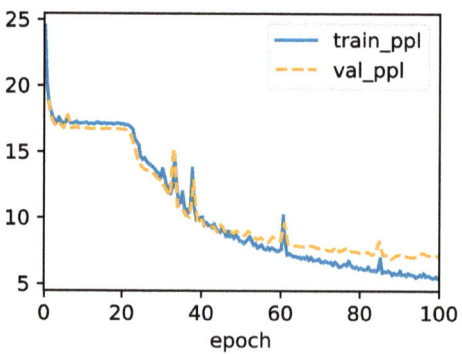

10.3.2 Concise Implementation

Fortunately many of the logistical details required to implement multiple layers of an RNN are readily available in high-level APIs. Our concise implementation will use such built-in functionalities. The code generalizes the one we used previously in Section 10.2, letting us specify the number of layers explicitly rather than picking the default of only one layer.

```
class GRU(d2l.RNN):  #@save
    """The multilayer GRU model."""
    def __init__(self, num_inputs, num_hiddens, num_layers, dropout=0):
        d2l.Module.__init__(self)
        self.save_hyperparameters()
        self.rnn = nn.GRU(num_inputs, num_hiddens, num_layers,
                          dropout=dropout)
```

The architectural decisions such as choosing hyperparameters are very similar to those of Section 10.2. We pick the same number of inputs and outputs as we have distinct tokens, i.e., vocab_size. The number of hidden units is still 32. The only difference is that we now select a nontrivial number of hidden layers by specifying the value of num_layers.

```
gru = GRU(num_inputs=len(data.vocab), num_hiddens=32, num_layers=2)
model = d2l.RNNLM(gru, vocab_size=len(data.vocab), lr=2)
trainer.fit(model, data)
```

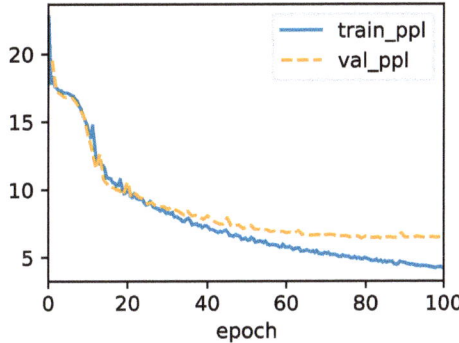

```
model.predict('it has', 20, data.vocab, d2l.try_gpu())
```

```
'it has so it is all and th'
```

10.3.3 Summary

In deep RNNs, the hidden state information is passed to the next time step of the current layer and the current time step of the next layer. There exist many different flavors of deep RNNs, such as LSTMs, GRUs, or vanilla RNNs. Conveniently, these models are all available as parts of the high-level APIs of deep learning frameworks. Initialization of models requires care. Overall, deep RNNs require considerable amount of work (such as learning rate and clipping) to ensure proper convergence.

10.3.4 Exercises

1. Replace the GRU by an LSTM and compare the accuracy and training speed.

2. Increase the training data to include multiple books. How low can you go on the perplexity scale?

3. Would you want to combine sources of different authors when modeling text? Why is this a good idea? What could go wrong?

Discussions[147].

147

10.4 Bidirectional Recurrent Neural Networks

So far, our working example of a sequence learning task has been language modeling, where we aim to predict the next token given all previous tokens in a sequence. In this scenario, we wish only to condition upon the leftward context, and thus the unidirectional chaining of a standard RNN seems appropriate. However, there are many other sequence learning tasks contexts where it is perfectly fine to condition the prediction at every time step on both the leftward and the rightward context. Consider, for example, part of speech detection. Why shouldn't we take the context in both directions into account when assessing the part of speech associated with a given word?

Another common task—often useful as a pretraining exercise prior to fine-tuning a model on an actual task of interest—is to mask out random tokens in a text document and then to train a sequence model to predict the values of the missing tokens. Note that depending on what comes after the blank, the likely value of the missing token changes dramatically:

- I am ___.

- I am ___ hungry.

- I am ___ hungry, and I can eat half a pig.

In the first sentence "happy" seems to be a likely candidate. The words "not" and "very" seem plausible in the second sentence, but "not" seems incompatible with the third sentences.

Fortunately, a simple technique transforms any unidirectional RNN into a bidirectional RNN (Schuster and Paliwal, 1997). We simply implement two unidirectional RNN layers chained together in opposite directions and acting on the same input (Fig. 10.4.1). For the first RNN layer, the first input is x_1 and the last input is x_T, but for the second RNN layer, the first input is x_T and the last input is x_1. To produce the output of this bidirectional RNN layer, we simply concatenate together the corresponding outputs of the two underlying unidirectional RNN layers.

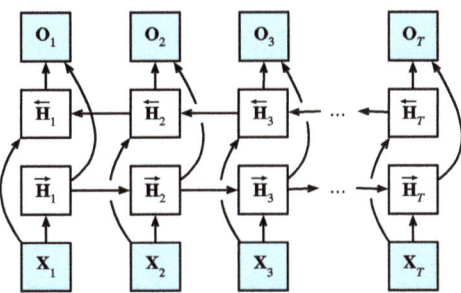

Fig. 10.4.1 Architecture of a bidirectional RNN.

Formally for any time step t, we consider a minibatch input $\mathbf{X}_t \in \mathbb{R}^{n \times d}$ (number of examples = n; number of inputs in each example = d) and let the hidden layer activation function be ϕ. In the bidirectional architecture, the forward and backward hidden states for this time step are $\overrightarrow{\mathbf{H}}_t \in \mathbb{R}^{n \times h}$ and $\overleftarrow{\mathbf{H}}_t \in \mathbb{R}^{n \times h}$, respectively, where h is the number of hidden units. The forward and backward hidden state updates are as follows:

$$\begin{aligned}
\overrightarrow{\mathbf{H}}_t &= \phi(\mathbf{X}_t \mathbf{W}_{xh}^{(f)} + \overrightarrow{\mathbf{H}}_{t-1} \mathbf{W}_{hh}^{(f)} + \mathbf{b}_h^{(f)}), \\
\overleftarrow{\mathbf{H}}_t &= \phi(\mathbf{X}_t \mathbf{W}_{xh}^{(b)} + \overleftarrow{\mathbf{H}}_{t+1} \mathbf{W}_{hh}^{(b)} + \mathbf{b}_h^{(b)}),
\end{aligned} \tag{10.4.1}$$

where the weights $\mathbf{W}_{xh}^{(f)} \in \mathbb{R}^{d \times h}, \mathbf{W}_{hh}^{(f)} \in \mathbb{R}^{h \times h}, \mathbf{W}_{xh}^{(b)} \in \mathbb{R}^{d \times h}$, and $\mathbf{W}_{hh}^{(b)} \in \mathbb{R}^{h \times h}$, and the biases $\mathbf{b}_h^{(f)} \in \mathbb{R}^{1 \times h}$ and $\mathbf{b}_h^{(b)} \in \mathbb{R}^{1 \times h}$ are all the model parameters.

Next, we concatenate the forward and backward hidden states $\overrightarrow{\mathbf{H}}_t$ and $\overleftarrow{\mathbf{H}}_t$ to obtain the hidden state $\mathbf{H}_t \in \mathbb{R}^{n \times 2h}$ for feeding into the output layer. In deep bidirectional RNNs with multiple hidden layers, such information is passed on as *input* to the next bidirectional layer. Last, the output layer computes the output $\mathbf{O}_t \in \mathbb{R}^{n \times q}$ (number of outputs = q):

$$\mathbf{O}_t = \mathbf{H}_t \mathbf{W}_{hq} + \mathbf{b}_q. \tag{10.4.2}$$

Here, the weight matrix $\mathbf{W}_{hq} \in \mathbb{R}^{2h \times q}$ and the bias $\mathbf{b}_q \in \mathbb{R}^{1 \times q}$ are the model parameters of the output layer. While technically, the two directions can have different numbers of hidden units, this design choice is seldom made in practice. We now demonstrate a simple implementation of a bidirectional RNN.

```python
import torch
from torch import nn
from d2l import torch as d2l
```

10.4.1 Implementation from Scratch

If we want to implement a bidirectional RNN from scratch, we can include two unidirectional RNNScratch instances with separate learnable parameters.

```
class BiRNNScratch(d2l.Module):
    def __init__(self, num_inputs, num_hiddens, sigma=0.01):
        super().__init__()
        self.save_hyperparameters()
        self.f_rnn = d2l.RNNScratch(num_inputs, num_hiddens, sigma)
        self.b_rnn = d2l.RNNScratch(num_inputs, num_hiddens, sigma)
        self.num_hiddens *= 2  # The output dimension will be doubled
```

States of forward and backward RNNs are updated separately, while outputs of these two RNNs are concatenated.

```
@d2l.add_to_class(BiRNNScratch)
def forward(self, inputs, Hs=None):
    f_H, b_H = Hs if Hs is not None else (None, None)
    f_outputs, f_H = self.f_rnn(inputs, f_H)
    b_outputs, b_H = self.b_rnn(reversed(inputs), b_H)
    outputs = [torch.cat((f, b), -1) for f, b in zip(
        f_outputs, reversed(b_outputs))]
    return outputs, (f_H, b_H)
```

10.4.2 Concise Implementation

Using the high-level APIs, we can implement bidirectional RNNs more concisely. Here we take a GRU model as an example.

```
class BiGRU(d2l.RNN):
    def __init__(self, num_inputs, num_hiddens):
        d2l.Module.__init__(self)
        self.save_hyperparameters()
        self.rnn = nn.GRU(num_inputs, num_hiddens, bidirectional=True)
        self.num_hiddens *= 2
```

10.4.3 Summary

In bidirectional RNNs, the hidden state for each time step is simultaneously determined by the data prior to and after the current time step. Bidirectional RNNs are mostly useful for sequence encoding and the estimation of observations given bidirectional context. Bidirectional RNNs are very costly to train due to long gradient chains.

10.4.4 Exercises

1. If the different directions use a different number of hidden units, how will the shape of $\mathbf{H}_t$ change?

2. Design a bidirectional RNN with multiple hidden layers.

3. Polysemy is common in natural languages. For example, the word "bank" has different meanings in contexts "i went to the bank to deposit cash" and "i went to the bank to sit down". How can we design a neural network model such that given a context sequence

and a word, a vector representation of the word in the correct context will be returned? What type of neural architectures is preferred for handling polysemy?

148

Discussions[148].

10.5 Machine Translation and the Dataset

Among the major breakthroughs that prompted widespread interest in modern RNNs was a major advance in the applied field of statistical *machine translation*. Here, the model is presented with a sentence in one language and must predict the corresponding sentence in another. Note that here the sentences may be of different lengths, and that corresponding words in the two sentences may not occur in the same order, owing to differences in the two language's grammatical structure.

Many problems have this flavor of mapping between two such "unaligned" sequences. Examples include mapping from dialog prompts to replies or from questions to answers. Broadly, such problems are called *sequence-to-sequence* (seq2seq) problems and they are our focus for both the remainder of this chapter and much of Chapter 11.

In this section, we introduce the machine translation problem and an example dataset that we will use in the subsequent examples. For decades, statistical formulations of translation between languages had been popular (Brown *et al.*, 1990, Brown *et al.*, 1988), even before researchers got neural network approaches working (methods were often lumped together under the term *neural machine translation*).

First we will need some new code to process our data. Unlike the language modeling that we saw in Section 9.3, here each example consists of two separate text sequences, one in the source language and another (the translation) in the target language. The following code snippets will show how to load the preprocessed data into minibatches for training.

```
import os
import torch
from d2l import torch as d2l
```

149

10.5.1 Downloading and Preprocessing the Dataset

To begin, we download an English–French dataset that consists of bilingual sentence pairs from the Tatoeba Project[149]. Each line in the dataset is a tab-delimited pair consisting of an English text sequence (the *source*) and the translated French text sequence (the *target*). Note that each text sequence can be just one sentence, or a paragraph of multiple sentences.

```
class MTFraEng(d2l.DataModule):  #@save
    """The English-French dataset."""
    def _download(self):
        d2l.extract(d2l.download(
            d2l.DATA_URL+'fra-eng.zip', self.root,
            '94646ad1522d915e7b0f9296181140edcf86a4f5'))
        with open(self.root + '/fra-eng/fra.txt', encoding='utf-8') as f:
            return f.read()
```

```
data = MTFraEng()
raw_text = data._download()
print(raw_text[:75])
```

```
Downloading ../data/fra-eng.zip from http://d2l-data.s3-accelerate.amazonaws.
↪com/fra-eng.zip...
Go.     Va !
Hi.     Salut !
Run!    Cours !
Run!    Courez !
Who?    Qui ?
Wow!    Ça alors !
```

After downloading the dataset, we proceed with several preprocessing steps for the raw text data. For instance, we replace non-breaking space with space, convert uppercase letters to lowercase ones, and insert space between words and punctuation marks.

```
@d2l.add_to_class(MTFraEng)  #@save
def _preprocess(self, text):
    # Replace non-breaking space with space
    text = text.replace('\u202f', ' ').replace('\xa0', ' ')
    # Insert space between words and punctuation marks
    no_space = lambda char, prev_char: char in ',.!?' and prev_char != ' '
    out = [' ' + char if i > 0 and no_space(char, text[i - 1]) else char
           for i, char in enumerate(text.lower())]
    return ''.join(out)
```

```
text = data._preprocess(raw_text)
print(text[:80])
```

```
go .    va !
hi .    salut !
run !   cours !
run !   courez !
who ?   qui ?
wow !   ça alors !
```

10.5.2 Tokenization

Unlike the character-level tokenization in Section 9.3, for machine translation we prefer word-level tokenization here (today's state-of-the-art models use more complex tokenization techniques). The following _tokenize method tokenizes the first max_examples text sequence pairs, where each token is either a word or a punctuation mark. We append the special "<eos>" token to the end of every sequence to indicate the end of the sequence. When a model is predicting by generating a sequence token after token, the generation of the "<eos>" token can suggest that the output sequence is complete. In the end, the method below returns two lists of token lists: src and tgt. Specifically, src[i] is a list of tokens from the ith text sequence in the source language (English here) and tgt[i] is that in the target language (French here).

```python
@d2l.add_to_class(MTFraEng)  #@save
def _tokenize(self, text, max_examples=None):
    src, tgt = [], []
    for i, line in enumerate(text.split('\n')):
        if max_examples and i > max_examples: break
        parts = line.split('\t')
        if len(parts) == 2:
            # Skip empty tokens
            src.append([t for t in f'{parts[0]} <eos>'.split(' ') if t])
            tgt.append([t for t in f'{parts[1]} <eos>'.split(' ') if t])
    return src, tgt
```

```python
src, tgt = data._tokenize(text)
src[:6], tgt[:6]
```

```
([['go', '.', '<eos>'],
  ['hi', '.', '<eos>'],
  ['run', '!', '<eos>'],
  ['run', '!', '<eos>'],
  ['who', '?', '<eos>'],
  ['wow', '!', '<eos>']],
 [['va', '!', '<eos>'],
  ['salut', '!', '<eos>'],
  ['cours', '!', '<eos>'],
  ['courez', '!', '<eos>'],
  ['qui', '?', '<eos>'],
  ['ça', 'alors', '!', '<eos>']])
```

Let's plot the histogram of the number of tokens per text sequence. In this simple English–French dataset, most of the text sequences have fewer than 20 tokens.

```python
#@save
def show_list_len_pair_hist(legend, xlabel, ylabel, xlist, ylist):
    """Plot the histogram for list length pairs."""
    d2l.set_figsize()
    _, _, patches = d2l.plt.hist(
        [[len(l) for l in xlist], [len(l) for l in ylist]])
```

(continues on next page)

(continued from previous page)

```
d21.plt.xlabel(xlabel)
d21.plt.ylabel(ylabel)
for patch in patches[1].patches:
    patch.set_hatch('/')
d21.plt.legend(legend)
```

```
show_list_len_pair_hist(['source', 'target'], '# tokens per sequence',
                        'count', src, tgt);
```

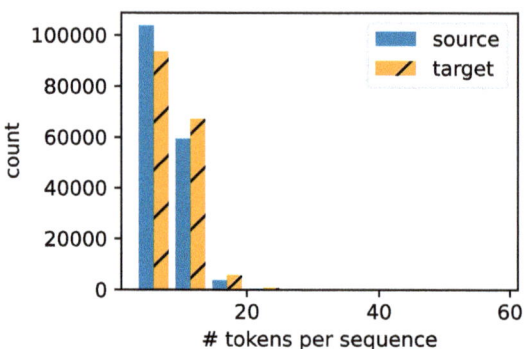

10.5.3 Loading Sequences of Fixed Length

Recall that in language modeling each example sequence, either a segment of one sentence or a span over multiple sentences, had a fixed length. This was specified by the num_steps (number of time steps or tokens) argument from Section 9.3. In machine translation, each example is a pair of source and target text sequences, where the two text sequences may have different lengths.

For computational efficiency, we can still process a minibatch of text sequences at one time by *truncation* and *padding*. Suppose that every sequence in the same minibatch should have the same length num_steps. If a text sequence has fewer than num_steps tokens, we will keep appending the special "<pad>" token to its end until its length reaches num_steps. Otherwise, we will truncate the text sequence by only taking its first num_steps tokens and discarding the remaining. In this way, every text sequence will have the same length to be loaded in minibatches of the same shape. Furthermore, we also record length of the source sequence excluding padding tokens. This information will be needed by some models that we will cover later.

Since the machine translation dataset consists of pairs of languages, we can build two vocabularies for both the source language and the target language separately. With word-level tokenization, the vocabulary size will be significantly larger than that using character-level tokenization. To alleviate this, here we treat infrequent tokens that appear less than twice as the same unknown ("<unk>") token. As we will explain later (Fig. 10.7.1), when training with target sequences, the decoder output (label tokens) can be the same decoder input (target tokens), shifted by one token; and the special beginning-of-sequence "<bos>" token will be used as the first input token for predicting the target sequence (Fig. 10.7.3).

```python
@d2l.add_to_class(MTFraEng)  #@save
def __init__(self, batch_size, num_steps=9, num_train=512, num_val=128):
    super(MTFraEng, self).__init__()
    self.save_hyperparameters()
    self.arrays, self.src_vocab, self.tgt_vocab = self._build_arrays(
        self._download())
```

```python
@d2l.add_to_class(MTFraEng)  #@save
def _build_arrays(self, raw_text, src_vocab=None, tgt_vocab=None):
    def _build_array(sentences, vocab, is_tgt=False):
        pad_or_trim = lambda seq, t: (
            seq[:t] if len(seq) > t else seq + ['<pad>'] * (t - len(seq)))
        sentences = [pad_or_trim(s, self.num_steps) for s in sentences]
        if is_tgt:
            sentences = [['<bos>'] + s for s in sentences]
        if vocab is None:
            vocab = d2l.Vocab(sentences, min_freq=2)
        array = torch.tensor([vocab[s] for s in sentences])
        valid_len = (array != vocab['<pad>']).type(torch.int32).sum(1)
        return array, vocab, valid_len
    src, tgt = self._tokenize(self._preprocess(raw_text),
                              self.num_train + self.num_val)
    src_array, src_vocab, src_valid_len = _build_array(src, src_vocab)
    tgt_array, tgt_vocab, _ = _build_array(tgt, tgt_vocab, True)
    return ((src_array, tgt_array[:,:-1], src_valid_len, tgt_array[:,1:]),
            src_vocab, tgt_vocab)
```

10.5.4 Reading the Dataset

Finally, we define the `get_dataloader` method to return the data iterator.

```python
@d2l.add_to_class(MTFraEng)  #@save
def get_dataloader(self, train):
    idx = slice(0, self.num_train) if train else slice(self.num_train, None)
    return self.get_tensorloader(self.arrays, train, idx)
```

Let's read the first minibatch from the English–French dataset.

```python
data = MTFraEng(batch_size=3)
src, tgt, src_valid_len, label = next(iter(data.train_dataloader()))
print('source:', src.type(torch.int32))
print('decoder input:', tgt.type(torch.int32))
print('source len excluding pad:', src_valid_len.type(torch.int32))
print('label:', label.type(torch.int32))
```

```
source: tensor([[ 84, 190,   2,   3,   4,   4,   4,   4,   4],
        [ 28, 122,   2,   3,   4,   4,   4,   4,   4],
```

(continues on next page)

(continued from previous page)

```
        [ 59, 118,    2,    3,    4,    4,    4,    4,    4]], dtype=torch.int32)
decoder input: tensor([[    3, 111,  94,    2,    4,    5,    5,    5,    5],
        [    3, 206,    0,    4,    5,    5,    5,    5,    5],
        [    3, 203, 124,    2,    4,    5,    5,    5,    5]], dtype=torch.int32)
source len excluding pad: tensor([4, 4, 4], dtype=torch.int32)
label: tensor([[111,  94,    2,    4,    5,    5,    5,    5,    5],
        [206,    0,    4,    5,    5,    5,    5,    5,    5],
        [203, 124,    2,    4,    5,    5,    5,    5,    5]], dtype=torch.int32)
```

We show a pair of source and target sequences processed by the above _build_arrays method (in the string format).

```
@d21.add_to_class(MTFraEng)  #@save
def build(self, src_sentences, tgt_sentences):
    raw_text = '\n'.join([src + '\t' + tgt for src, tgt in zip(
        src_sentences, tgt_sentences)])
    arrays, _, _ = self._build_arrays(
        raw_text, self.src_vocab, self.tgt_vocab)
    return arrays
```

```
src, tgt, _, _ = data.build(['hi .'], ['salut .'])
print('source:', data.src_vocab.to_tokens(src[0].type(torch.int32)))
print('target:', data.tgt_vocab.to_tokens(tgt[0].type(torch.int32)))
```

```
source: ['hi', '.', '<eos>', '<pad>', '<pad>', '<pad>', '<pad>', '<pad>', '
↪<pad>']
target: ['<bos>', 'salut', '.', '<eos>', '<pad>', '<pad>', '<pad>', '<pad>', '
↪<pad>']
```

10.5.5 Summary

In natural language processing, *machine translation* refers to the task of automatically mapping from a sequence representing a string of text in a *source* language to a string representing a plausible translation in a *target* language. Using word-level tokenization, the vocabulary size will be significantly larger than that using character-level tokenization, but the sequence lengths will be much shorter. To mitigate the large vocabulary size, we can treat infrequent tokens as some "unknown" token. We can truncate and pad text sequences so that all of them will have the same length to be loaded in minibatches. Modern implementations often bucket sequences with similar lengths to avoid wasting excessive computation on padding.

10.5.6 Exercises

1. Try different values of the max_examples argument in the _tokenize method. How does this affect the vocabulary sizes of the source language and the target language?

2. Text in some languages such as Chinese and Japanese does not have word boundary

indicators (e.g., space). Is word-level tokenization still a good idea for such cases? Why or why not?

Discussions[150].

150

10.6 The Encoder–Decoder Architecture

In general sequence-to-sequence problems like machine translation (Section 10.5), inputs and outputs are of varying lengths that are unaligned. The standard approach to handling this sort of data is to design an *encoder–decoder* architecture (Fig. 10.6.1) consisting of two major components: an *encoder* that takes a variable-length sequence as input, and a *decoder* that acts as a conditional language model, taking in the encoded input and the leftwards context of the target sequence and predicting the subsequent token in the target sequence.

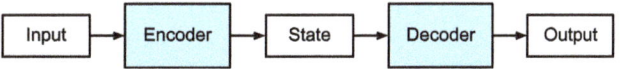

Fig. 10.6.1   The encoder–decoder architecture.

Let's take machine translation from English to French as an example. Given an input sequence in English: "They", "are", "watching", ".", this encoder–decoder architecture first encodes the variable-length input into a state, then decodes the state to generate the translated sequence, token by token, as output: "Ils", "regardent", ".". Since the encoder–decoder architecture forms the basis of different sequence-to-sequence models in subsequent sections, this section will convert this architecture into an interface that will be implemented later.

```
from torch import nn
from d2l import torch as d2l
```

10.6.1 Encoder

In the encoder interface, we just specify that the encoder takes variable-length sequences as input X. The implementation will be provided by any model that inherits this base Encoder class.

```
class Encoder(nn.Module):  #@save
    """The base encoder interface for the encoder--decoder architecture."""
    def __init__(self):
        super().__init__()

    # Later there can be additional arguments (e.g., length excluding padding)
    def forward(self, X, *args):
        raise NotImplementedError
```

10.6.2 Decoder

In the following decoder interface, we add an additional `init_state` method to convert the encoder output (`enc_all_outputs`) into the encoded state. Note that this step may require extra inputs, such as the valid length of the input, which was explained in Section 10.5. To generate a variable-length sequence token by token, every time the decoder may map an input (e.g., the generated token at the previous time step) and the encoded state into an output token at the current time step.

```python
class Decoder(nn.Module):  #@save
    """The base decoder interface for the encoder--decoder architecture."""
    def __init__(self):
        super().__init__()

    # Later there can be additional arguments (e.g., length excluding padding)
    def init_state(self, enc_all_outputs, *args):
        raise NotImplementedError

    def forward(self, X, state):
        raise NotImplementedError
```

10.6.3 Putting the Encoder and Decoder Together

In the forward propagation, the output of the encoder is used to produce the encoded state, and this state will be further used by the decoder as one of its input.

```python
class EncoderDecoder(d2l.Classifier):  #@save
    """The base class for the encoder--decoder architecture."""
    def __init__(self, encoder, decoder):
        super().__init__()
        self.encoder = encoder
        self.decoder = decoder

    def forward(self, enc_X, dec_X, *args):
        enc_all_outputs = self.encoder(enc_X, *args)
        dec_state = self.decoder.init_state(enc_all_outputs, *args)
        # Return decoder output only
        return self.decoder(dec_X, dec_state)[0]
```

In the next section, we will see how to apply RNNs to design sequence-to-sequence models based on this encoder–decoder architecture.

10.6.4 Summary

Encoder-decoder architectures can handle inputs and outputs that both consist of variable-length sequences and thus are suitable for sequence-to-sequence problems such as machine translation. The encoder takes a variable-length sequence as input and transforms it into a state with a fixed shape. The decoder maps the encoded state of a fixed shape to a variable-length sequence.

10.6.5 Exercises

1. Suppose that we use neural networks to implement the encoder–decoder architecture. Do the encoder and the decoder have to be the same type of neural network?

2. Besides machine translation, can you think of another application where the encoder–decoder architecture can be applied?

Discussions[151].

151

10.7 Sequence-to-Sequence Learning for Machine Translation

In so-called sequence-to-sequence problems such as machine translation (as discussed in Section 10.5), where inputs and outputs each consist of variable-length unaligned sequences, we generally rely on encoder–decoder architectures (Section 10.6). In this section, we will demonstrate the application of an encoder–decoder architecture, where both the encoder and decoder are implemented as RNNs, to the task of machine translation (Cho *et al.*, 2014, Sutskever *et al.*, 2014).

Here, the encoder RNN will take a variable-length sequence as input and transform it into a fixed-shape hidden state. Later, in Chapter 11, we will introduce attention mechanisms, which allow us to access encoded inputs without having to compress the entire input into a single fixed-length representation.

Then to generate the output sequence, one token at a time, the decoder model, consisting of a separate RNN, will predict each successive target token given both the input sequence and the preceding tokens in the output. During training, the decoder will typically be conditioned upon the preceding tokens in the official "ground truth" label. However, at test time, we will want to condition each output of the decoder on the tokens already predicted. Note that if we ignore the encoder, the decoder in a sequence-to-sequence architecture behaves just like a normal language model. Fig. 10.7.1 illustrates how to use two RNNs for sequence-to-sequence learning in machine translation.

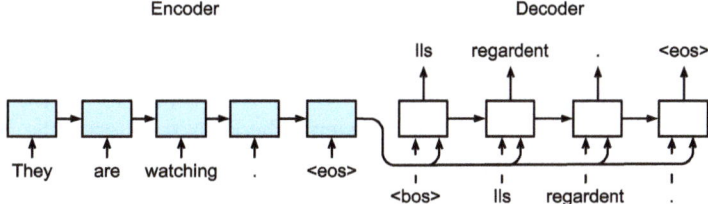

Fig. 10.7.1   Sequence-to-sequence learning with an RNN encoder and an RNN decoder.

In Fig. 10.7.1, the special "<eos>" token marks the end of the sequence. Our model can stop making predictions once this token is generated. At the initial time step of the RNN decoder, there are two special design decisions to be aware of: First, we begin every input with a special beginning-of-sequence "<bos>" token. Second, we may feed the final hidden state of the encoder into the decoder at every single decoding time step (Cho *et al.*, 2014). In some other designs, such as that of Sutskever *et al.* (2014), the final hidden state of the RNN encoder is used to initiate the hidden state of the decoder only at the first decoding step.

```
import collections
import math
import torch
from torch import nn
from torch.nn import functional as F
from d2l import torch as d2l
```

10.7.1 Teacher Forcing

While running the encoder on the input sequence is relatively straightforward, handling the input and output of the decoder requires more care. The most common approach is sometimes called *teacher forcing*. Here, the original target sequence (token labels) is fed into the decoder as input. More concretely, the special beginning-of-sequence token and the original target sequence, excluding the final token, are concatenated as input to the decoder, while the decoder output (labels for training) is the original target sequence, shifted by one token: "<bos>", "Ils", "regardent", "." → "Ils", "regardent", ".", "<eos>" (Fig. 10.7.1).

Our implementation in Section 10.5.3 prepared training data for teacher forcing, where shifting tokens for self-supervised learning is similar to the training of language models in Section 9.3. An alternative approach is to feed the *predicted* token from the previous time step as the current input to the decoder.

In the following, we explain the design depicted in Fig. 10.7.1 in greater detail. We will train this model for machine translation on the English–French dataset as introduced in Section 10.5.

10.7.2 Encoder

Recall that the encoder transforms an input sequence of variable length into a fixed-shape *context variable* **c** (see Fig. 10.7.1).

Consider a single sequence example (batch size 1). Suppose the input sequence is $x_1, \ldots, x_T$, such that x_t is the tth token. At time step t, the RNN transforms the input feature vector $\mathbf{x}_t$ for x_t and the hidden state $\mathbf{h}_{t-1}$ from the previous time step into the current hidden state $\mathbf{h}_t$. We can use a function f to express the transformation of the RNN's recurrent layer:

$$\mathbf{h}_t = f(\mathbf{x}_t, \mathbf{h}_{t-1}). \tag{10.7.1}$$

In general, the encoder transforms the hidden states at all time steps into a context variable through a customized function q:

$$\mathbf{c} = q(\mathbf{h}_1, \ldots, \mathbf{h}_T). \tag{10.7.2}$$

For example, in Fig. 10.7.1, the context variable is just the hidden state $\mathbf{h}_T$ corresponding to the encoder RNN's representation after processing the final token of the input sequence.

In this example, we have used a unidirectional RNN to design the encoder, where the hidden state only depends on the input subsequence at and before the time step of the hidden state. We can also construct encoders using bidirectional RNNs. In this case, a hidden state depends on the subsequence before and after the time step (including the input at the current time step), which encodes the information of the entire sequence.

Now let's implement the RNN encoder. Note that we use an *embedding layer* to obtain the feature vector for each token in the input sequence. The weight of an embedding layer is a matrix, where the number of rows corresponds to the size of the input vocabulary (vocab_size) and number of columns corresponds to the feature vector's dimension (embed_size). For any input token index i, the embedding layer fetches the ith row (starting from 0) of the weight matrix to return its feature vector. Here we implement the encoder with a multilayer GRU.

```
def init_seq2seq(module):  #@save
    """Initialize weights for sequence-to-sequence learning."""
    if type(module) == nn.Linear:
        nn.init.xavier_uniform_(module.weight)
    if type(module) == nn.GRU:
        for param in module._flat_weights_names:
            if "weight" in param:
                nn.init.xavier_uniform_(module._parameters[param])
```

```
class Seq2SeqEncoder(d2l.Encoder):  #@save
    """The RNN encoder for sequence-to-sequence learning."""
    def __init__(self, vocab_size, embed_size, num_hiddens, num_layers,
                 dropout=0):
        super().__init__()
        self.embedding = nn.Embedding(vocab_size, embed_size)
        self.rnn = d2l.GRU(embed_size, num_hiddens, num_layers, dropout)
        self.apply(init_seq2seq)

    def forward(self, X, *args):
        # X shape: (batch_size, num_steps)
        embs = self.embedding(X.t().type(torch.int64))
        # embs shape: (num_steps, batch_size, embed_size)
        outputs, state = self.rnn(embs)
        # outputs shape: (num_steps, batch_size, num_hiddens)
        # state shape: (num_layers, batch_size, num_hiddens)
        return outputs, state
```

Let's use a concrete example to illustrate the above encoder implementation. Below, we instantiate a two-layer GRU encoder whose number of hidden units is 16. Given a minibatch of sequence inputs X (batch size = 4; number of time steps = 9), the hidden states of the final layer at all the time steps (enc_outputs returned by the encoder's recurrent layers) are a tensor of shape (number of time steps, batch size, number of hidden units).

```
vocab_size, embed_size, num_hiddens, num_layers = 10, 8, 16, 2
batch_size, num_steps = 4, 9
encoder = Seq2SeqEncoder(vocab_size, embed_size, num_hiddens, num_layers)
X = torch.zeros((batch_size, num_steps))
enc_outputs, enc_state = encoder(X)
d2l.check_shape(enc_outputs, (num_steps, batch_size, num_hiddens))
```

Since we are using a GRU here, the shape of the multilayer hidden states at the final time step is (number of hidden layers, batch size, number of hidden units).

```
d2l.check_shape(enc_state, (num_layers, batch_size, num_hiddens))
```

10.7.3 Decoder

Given a target output sequence $y_1, y_2, \ldots, y_{T'}$ for each time step t' (we use t' to differentiate from the input sequence time steps), the decoder assigns a predicted probability to each possible token occurring at step $y_{t'+1}$ conditioned upon the previous tokens in the target $y_1, \ldots, y_{t'}$ and the context variable $\mathbf{c}$, i.e., $P(y_{t'+1} \mid y_1, \ldots, y_{t'}, \mathbf{c})$.

To predict the subsequent token $t' + 1$ in the target sequence, the RNN decoder takes the previous step's target token $y_{t'}$, the hidden RNN state from the previous time step $\mathbf{s}_{t'-1}$, and the context variable $\mathbf{c}$ as its input, and transforms them into the hidden state $\mathbf{s}_{t'}$ at the current time step. We can use a function g to express the transformation of the decoder's hidden layer:

$$\mathbf{s}_{t'} = g(y_{t'-1}, \mathbf{c}, \mathbf{s}_{t'-1}). \tag{10.7.3}$$

After obtaining the hidden state of the decoder, we can use an output layer and the softmax operation to compute the predictive distribution $p(y_{t'+1} \mid y_1, \ldots, y_{t'}, \mathbf{c})$ over the subsequent output token $t' + 1$.

Following Fig. 10.7.1, when implementing the decoder as follows, we directly use the hidden state at the final time step of the encoder to initialize the hidden state of the decoder. This requires that the RNN encoder and the RNN decoder have the same number of layers and hidden units. To further incorporate the encoded input sequence information, the context variable is concatenated with the decoder input at all the time steps. To predict the probability distribution of the output token, we use a fully connected layer to transform the hidden state at the final layer of the RNN decoder.

```
class Seq2SeqDecoder(d2l.Decoder):
    """The RNN decoder for sequence to sequence learning."""
    def __init__(self, vocab_size, embed_size, num_hiddens, num_layers,
                 dropout=0):
        super().__init__()
```

(continues on next page)

(continued from previous page)

```python
        self.embedding = nn.Embedding(vocab_size, embed_size)
        self.rnn = d2l.GRU(embed_size+num_hiddens, num_hiddens,
                           num_layers, dropout)
        self.dense = nn.LazyLinear(vocab_size)
        self.apply(init_seq2seq)

    def init_state(self, enc_all_outputs, *args):
        return enc_all_outputs

    def forward(self, X, state):
        # X shape: (batch_size, num_steps)
        # embs shape: (num_steps, batch_size, embed_size)
        embs = self.embedding(X.t().type(torch.int32))
        enc_output, hidden_state = state
        # context shape: (batch_size, num_hiddens)
        context = enc_output[-1]
        # Broadcast context to (num_steps, batch_size, num_hiddens)
        context = context.repeat(embs.shape[0], 1, 1)
        # Concat at the feature dimension
        embs_and_context = torch.cat((embs, context), -1)
        outputs, hidden_state = self.rnn(embs_and_context, hidden_state)
        outputs = self.dense(outputs).swapaxes(0, 1)
        # outputs shape: (batch_size, num_steps, vocab_size)
        # hidden_state shape: (num_layers, batch_size, num_hiddens)
        return outputs, [enc_output, hidden_state]
```

To illustrate the implemented decoder, below we instantiate it with the same hyperparameters from the aforementioned encoder. As we can see, the output shape of the decoder becomes (batch size, number of time steps, vocabulary size), where the final dimension of the tensor stores the predicted token distribution.

```python
decoder = Seq2SeqDecoder(vocab_size, embed_size, num_hiddens, num_layers)
state = decoder.init_state(encoder(X))
dec_outputs, state = decoder(X, state)
d2l.check_shape(dec_outputs, (batch_size, num_steps, vocab_size))
d2l.check_shape(state[1], (num_layers, batch_size, num_hiddens))
```

The layers in the above RNN encoder–decoder model are summarized in Fig. 10.7.2.

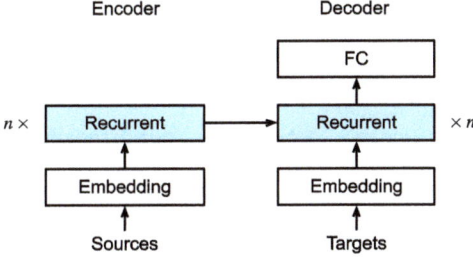

Fig. 10.7.2   Layers in an RNN encoder–decoder model.

10.7.4 Encoder–Decoder for Sequence-to-Sequence Learning

Putting it all together in code yields the following:

```
class Seq2Seq(d2l.EncoderDecoder):  #@save
    """The RNN encoder--decoder for sequence to sequence learning."""
    def __init__(self, encoder, decoder, tgt_pad, lr):
        super().__init__(encoder, decoder)
        self.save_hyperparameters()

    def validation_step(self, batch):
        Y_hat = self(*batch[:-1])
        self.plot('loss', self.loss(Y_hat, batch[-1]), train=False)

    def configure_optimizers(self):
        # Adam optimizer is used here
        return torch.optim.Adam(self.parameters(), lr=self.lr)
```

10.7.5 Loss Function with Masking

At each time step, the decoder predicts a probability distribution for the output tokens. As with language modeling, we can apply softmax to obtain the distribution and calculate the cross-entropy loss for optimization. Recall from Section 10.5 that the special padding tokens are appended to the end of sequences and so sequences of varying lengths can be efficiently loaded in minibatches of the same shape. However, prediction of padding tokens should be excluded from loss calculations. To this end, we can mask irrelevant entries with zero values so that multiplication of any irrelevant prediction with zero equates to zero.

```
@d2l.add_to_class(Seq2Seq)
def loss(self, Y_hat, Y):
    l = super(Seq2Seq, self).loss(Y_hat, Y, averaged=False)
    mask = (Y.reshape(-1) != self.tgt_pad).type(torch.float32)
    return (l * mask).sum() / mask.sum()
```

10.7.6 Training

Now we can create and train an RNN encoder–decoder model for sequence-to-sequence learning on the machine translation dataset.

```
data = d2l.MTFraEng(batch_size=128)
embed_size, num_hiddens, num_layers, dropout = 256, 256, 2, 0.2
encoder = Seq2SeqEncoder(
    len(data.src_vocab), embed_size, num_hiddens, num_layers, dropout)
decoder = Seq2SeqDecoder(
    len(data.tgt_vocab), embed_size, num_hiddens, num_layers, dropout)
model = Seq2Seq(encoder, decoder, tgt_pad=data.tgt_vocab['<pad>'],
                lr=0.005)
trainer = d2l.Trainer(max_epochs=30, gradient_clip_val=1, num_gpus=1)
trainer.fit(model, data)
```

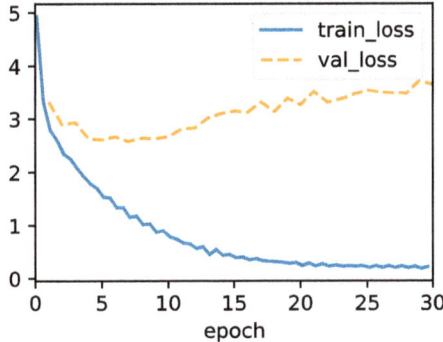

10.7.7 Prediction

To predict the output sequence at each step, the predicted token from the previous time step is fed into the decoder as an input. One simple strategy is to sample whichever token that has been assigned by the decoder the highest probability when predicting at each step. As in training, at the initial time step the beginning-of-sequence ("<bos>") token is fed into the decoder. This prediction process is illustrated in Fig. 10.7.3. When the end-of-sequence ("<eos>") token is predicted, the prediction of the output sequence is complete.

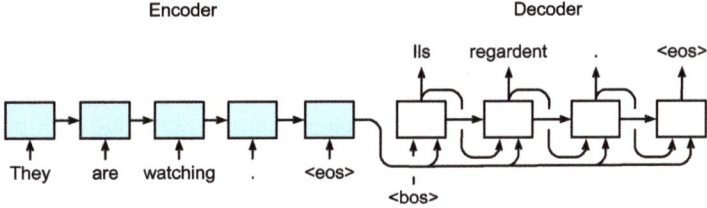

Fig. 10.7.3   Predicting the output sequence token by token using an RNN encoder–decoder.

In the next section, we will introduce more sophisticated strategies based on beam search (Section 10.8).

```python
@d2l.add_to_class(d2l.EncoderDecoder)  #@save
def predict_step(self, batch, device, num_steps,
                 save_attention_weights=False):
    batch = [a.to(device) for a in batch]
    src, tgt, src_valid_len, _ = batch
    enc_all_outputs = self.encoder(src, src_valid_len)
    dec_state = self.decoder.init_state(enc_all_outputs, src_valid_len)
    outputs, attention_weights = [tgt[:, 0].unsqueeze(1), ], []
    for _ in range(num_steps):
        Y, dec_state = self.decoder(outputs[-1], dec_state)
        outputs.append(Y.argmax(2))
        # Save attention weights (to be covered later)
        if save_attention_weights:
            attention_weights.append(self.decoder.attention_weights)
    return torch.cat(outputs[1:], 1), attention_weights
```

10.7.8 Evaluation of Predicted Sequences

We can evaluate a predicted sequence by comparing it with the target sequence (the ground truth). But what precisely is the appropriate measure for comparing similarity between two sequences?

Bilingual Evaluation Understudy (BLEU), though originally proposed for evaluating machine translation results (Papineni *et al.*, 2002), has been extensively used in measuring the quality of output sequences for different applications. In principle, for any n-gram (Section 9.3.1) in the predicted sequence, BLEU evaluates whether this n-gram appears in the target sequence.

Denote by p_n the precision of an n-gram, defined as the ratio of the number of matched n-grams in the predicted and target sequences to the number of n-grams in the predicted sequence. To explain, given a target sequence A, B, C, D, E, F, and a predicted sequence A, B, B, C, D, we have $p_1 = 4/5$, $p_2 = 3/4$, $p_3 = 1/3$, and $p_4 = 0$. Now let $\text{len}_{\text{label}}$ and len_{pred} be the numbers of tokens in the target sequence and the predicted sequence, respectively. Then, BLEU is defined as

$$\exp\left(\min\left(0, 1 - \frac{\text{len}_{\text{label}}}{\text{len}_{\text{pred}}}\right)\right) \prod_{n=1}^{k} p_n^{1/2^n}, \tag{10.7.4}$$

where k is the longest n-gram for matching.

Based on the definition of BLEU in (10.7.4), whenever the predicted sequence is the same as the target sequence, BLEU is 1. Moreover, since matching longer n-grams is more difficult, BLEU assigns a greater weight when a longer n-gram has high precision. Specifically, when p_n is fixed, $p_n^{1/2^n}$ increases as n grows (the original paper uses $p_n^{1/n}$). Furthermore, since predicting shorter sequences tends to yield a higher p_n value, the coefficient before the multiplication term in (10.7.4) penalizes shorter predicted sequences. For example, when $k = 2$, given the target sequence A, B, C, D, E, F and the predicted sequence A, B, although $p_1 = p_2 = 1$, the penalty factor $\exp(1 - 6/2) \approx 0.14$ lowers the BLEU.

We implement the BLEU measure as follows.

```python
def bleu(pred_seq, label_seq, k):  #@save
    """Compute the BLEU."""
    pred_tokens, label_tokens = pred_seq.split(' '), label_seq.split(' ')
    len_pred, len_label = len(pred_tokens), len(label_tokens)
    score = math.exp(min(0, 1 - len_label / len_pred))
    for n in range(1, min(k, len_pred) + 1):
        num_matches, label_subs = 0, collections.defaultdict(int)
        for i in range(len_label - n + 1):
            label_subs[' '.join(label_tokens[i: i + n])] += 1
        for i in range(len_pred - n + 1):
            if label_subs[' '.join(pred_tokens[i: i + n])] > 0:
                num_matches += 1
                label_subs[' '.join(pred_tokens[i: i + n])] -= 1
        score *= math.pow(num_matches / (len_pred - n + 1), math.pow(0.5, n))
    return score
```

In the end, we use the trained RNN encoder–decoder to translate a few English sentences into French and compute the BLEU of the results.

```
engs = ['go .', 'i lost .', 'he\'s calm .', 'i\'m home .']
fras = ['va !', 'j\'ai perdu .', 'il est calme .', 'je suis chez moi .']
preds, _ = model.predict_step(
    data.build(engs, fras), d2l.try_gpu(), data.num_steps)
for en, fr, p in zip(engs, fras, preds):
    translation = []
    for token in data.tgt_vocab.to_tokens(p):
        if token == '<eos>':
            break
        translation.append(token)
    print(f'{en} => {translation}, bleu,'
          f'{bleu(" ".join(translation), fr, k=2):.3f}')
```

```
go . => ['va', '!'], bleu,1.000
i lost . => ["j'ai", 'perdu', '.'], bleu,1.000
he's calm . => ['sois', 'calme', '.'], bleu,0.492
i'm home . => ['je', 'suis', 'chez', 'moi', '.'], bleu,1.000
```

10.7.9 Summary

Following the design of the encoder–decoder architecture, we can use two RNNs to design a model for sequence-to-sequence learning. In encoder–decoder training, the teacher forcing approach feeds original output sequences (in contrast to predictions) into the decoder. When implementing the encoder and the decoder, we can use multilayer RNNs. We can use masks to filter out irrelevant computations, such as when calculating the loss. For evaluating output sequences, BLEU is a popular measure that matches n-grams between the predicted sequence and the target sequence.

10.7.10 Exercises

1. Can you adjust the hyperparameters to improve the translation results?

2. Rerun the experiment without using masks in the loss calculation. What results do you observe? Why?

3. If the encoder and the decoder differ in the number of layers or the number of hidden units, how can we initialize the hidden state of the decoder?

4. In training, replace teacher forcing with feeding the prediction at the previous time step into the decoder. How does this influence the performance?

5. Rerun the experiment by replacing GRU with LSTM.

6. Are there any other ways to design the output layer of the decoder?

Discussions[152].

10.8 Beam Search

In Section 10.7, we introduced the encoder–decoder architecture, and the standard techniques for training them end-to-end. However, when it came to test-time prediction, we mentioned only the *greedy* strategy, where we select at each time step the token given the highest predicted probability of coming next, until, at some time step, we find that we have predicted the special end-of-sequence "<eos>" token. In this section, we will begin by formalizing this *greedy search* strategy and identifying some problems that practitioners tend to run into. Subsequently, we compare this strategy with two alternatives: *exhaustive search* (illustrative but not practical) and *beam search* (the standard method in practice).

Let's begin by setting up our mathematical notation, borrowing conventions from Section 10.7. At any time step t', the decoder outputs predictions representing the probability of each token in the vocabulary coming next in the sequence (the likely value of $y_{t'+1}$), conditioned on the previous tokens $y_1, \ldots, y_{t'}$ and the context variable $\mathbf{c}$, produced by the encoder to represent the input sequence. To quantify computational cost, denote by $\mathcal{Y}$ the output vocabulary (including the special end-of-sequence token "<eos>"). Let's also specify the maximum number of tokens of an output sequence as T'. Our goal is to search for an ideal output from all $O(|\mathcal{Y}|^{T'})$ possible output sequences. Note that this slightly overestimates the number of distinct outputs because there are no subsequent tokens once the "<eos>" token occurs. However, for our purposes, this number roughly captures the size of the search space.

10.8.1 Greedy Search

Consider the simple *greedy search* strategy from Section 10.7. Here, at any time step t', we simply select the token with the highest conditional probability from $\mathcal{Y}$, i.e.,

$$y_{t'} = \underset{y \in \mathcal{Y}}{\operatorname{argmax}} \, P(y \mid y_1, \ldots, y_{t'-1}, \mathbf{c}). \tag{10.8.1}$$

Once our model outputs "<eos>" (or we reach the maximum length T') the output sequence is completed.

This strategy might look reasonable, and in fact it is not so bad! Considering how computationally undemanding it is, you'd be hard pressed to get more bang for your buck. However, if we put aside efficiency for a minute, it might seem more reasonable to search for the *most likely sequence*, not the sequence of (greedily selected) *most likely tokens*. It turns out that these two objects can be quite different. The most likely sequence is the one that maximizes the expression $\prod_{t'=1}^{T'} P(y_{t'} \mid y_1, \ldots, y_{t'-1}, \mathbf{c})$. In our machine translation example, if the decoder truly recovered the probabilities of the underlying generative process, then this would give us the most likely translation. Unfortunately, there is no guarantee that greedy search will give us this sequence.

Let's illustrate it with an example. Suppose that there are four tokens "A", "B", "C", and

"<eos>" in the output dictionary. In Fig. 10.8.1, the four numbers under each time step represent the conditional probabilities of generating "A", "B", "C", and "<eos>" respectively, at that time step.

Time step	1	2	3	4
A	0.5	0.1	0.2	0.0
B	0.2	0.4	0.2	0.2
C	0.2	0.3	0.4	0.2
<eos>	0.1	0.2	0.2	0.6

Fig. 10.8.1 At each time step, greedy search selects the token with the highest conditional probability.

At each time step, greedy search selects the token with the highest conditional probability. Therefore, the output sequence "A", "B", "C", and "<eos>" will be predicted (Fig. 10.8.1). The conditional probability of this output sequence is $0.5 \times 0.4 \times 0.4 \times 0.6 = 0.048$.

Next, let's look at another example in Fig. 10.8.2. Unlike in Fig. 10.8.1, at time step 2 we select the token "C", which has the *second* highest conditional probability.

Time step	1	2	3	4
A	0.5	0.1	0.1	0.1
B	0.2	0.4	0.6	0.2
C	0.2	0.3	0.2	0.1
<eos>	0.1	0.2	0.1	0.6

Fig. 10.8.2 The four numbers under each time step represent the conditional probabilities of generating "A", "B", "C", and "<eos>" at that time step. At time step 2, the token "C", which has the second highest conditional probability, is selected.

Since the output subsequences at time steps 1 and 2, on which time step 3 is based, have changed from "A" and "B" in Fig. 10.8.1 to "A" and "C" in Fig. 10.8.2, the conditional probability of each token at time step 3 has also changed in Fig. 10.8.2. Suppose that we choose the token "B" at time step 3. Now time step 4 is conditional on the output subsequence at the first three time steps "A", "C", and "B", which has changed from "A", "B", and "C" in Fig. 10.8.1. Therefore, the conditional probability of generating each token at time step 4 in Fig. 10.8.2 is also different from that in Fig. 10.8.1. As a result, the conditional probability of the output sequence "A", "C", "B", and "<eos>" in Fig. 10.8.2 is $0.5 \times 0.3 \times 0.6 \times 0.6 = 0.054$, which is greater than that of greedy search in Fig. 10.8.1. In this example, the output sequence "A", "B", "C", and "<eos>" obtained by the greedy search is not optimal.

10.8.2 Exhaustive Search

If the goal is to obtain the most likely sequence, we may consider using *exhaustive search*: enumerate all the possible output sequences with their conditional probabilities, and then output the one that scores the highest predicted probability.

While this would certainly give us what we desire, it would come at a prohibitive computational cost of $O(|\mathcal{Y}|^{T'})$, exponential in the sequence length and with an enormous base given by the vocabulary size. For example, when $|\mathcal{Y}| = 10000$ and $T' = 10$, both small numbers when compared with ones in real applications, we will need to evaluate $10000^{10} = 10^{40}$ sequences, which is already beyond the capabilities of any foreseeable computers. On the other hand, the computational cost of greedy search is $O(|\mathcal{Y}| T')$: miraculously cheap but far from optimal. For example, when $|\mathcal{Y}| = 10000$ and $T' = 10$, we only need to evaluate $10000 \times 10 = 10^5$ sequences.

10.8.3 Beam Search

You could view sequence decoding strategies as lying on a spectrum, with *beam search* striking a compromise between the efficiency of greedy search and the optimality of exhaustive search. The most straightforward version of beam search is characterized by a single hyperparameter, the *beam size*, k. Let's explain this terminology. At time step 1, we select the k tokens with the highest predicted probabilities. Each of them will be the first token of k candidate output sequences, respectively. At each subsequent time step, based on the k candidate output sequences at the previous time step, we continue to select k candidate output sequences with the highest predicted probabilities from $k |\mathcal{Y}|$ possible choices.

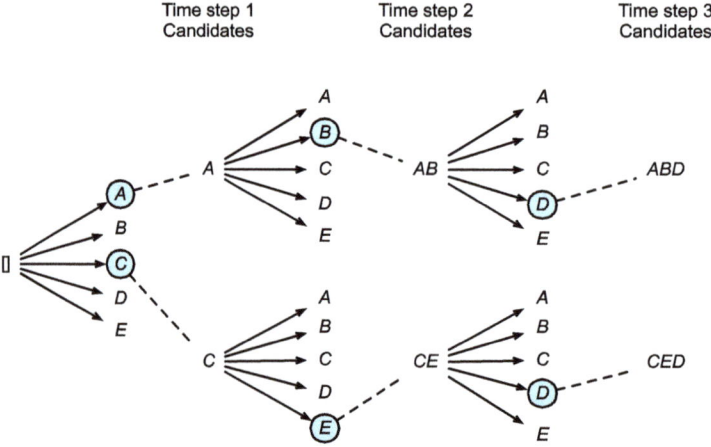

Fig. 10.8.3   The process of beam search (beam size = 2; maximum length of an output sequence = 3). The candidate output sequences are A, C, AB, CE, ABD, and CED.

Fig. 10.8.3 demonstrates the process of beam search with an example. Suppose that the output vocabulary contains only five elements: $\mathcal{Y} = \{A, B, C, D, E\}$, where one of them is "<eos>". Let the beam size be two and the maximum length of an output sequence be three. At time step 1, suppose that the tokens with the highest conditional probabilities $P(y_1 \mid \mathbf{c})$ are A and C. At time step 2, for all $y_2 \in \mathcal{Y}$, we compute

$$P(A, y_2 \mid \mathbf{c}) = P(A \mid \mathbf{c})P(y_2 \mid A, \mathbf{c}),$$
$$P(C, y_2 \mid \mathbf{c}) = P(C \mid \mathbf{c})P(y_2 \mid C, \mathbf{c}), \tag{10.8.2}$$

and pick the largest two among these ten values, say $P(A, B \mid \mathbf{c})$ and $P(C, E \mid \mathbf{c})$. Then at time step 3, for all $y_3 \in \mathcal{Y}$, we compute

$$P(A, B, y_3 \mid \mathbf{c}) = P(A, B \mid \mathbf{c})P(y_3 \mid A, B, \mathbf{c}),$$
$$P(C, E, y_3 \mid \mathbf{c}) = P(C, E \mid \mathbf{c})P(y_3 \mid C, E, \mathbf{c}), \tag{10.8.3}$$

and pick the largest two among these ten values, say $P(A, B, D \mid \mathbf{c})$ and $P(C, E, D \mid \mathbf{c})$. As a result, we get six candidates output sequences: (i) A; (ii) C; (iii) A, B; (iv) C, E; (v) A, B, D; and (vi) C, E, D.

In the end, we obtain the set of final candidate output sequences based on these six sequences (e.g., discard portions including and after "<eos>"). Then we choose the output sequence which maximizes the following score:

$$\frac{1}{L^\alpha} \log P(y_1, \dots, y_L \mid \mathbf{c}) = \frac{1}{L^\alpha} \sum_{t'=1}^{L} \log P(y_{t'} \mid y_1, \dots, y_{t'-1}, \mathbf{c}); \tag{10.8.4}$$

here L is the length of the final candidate sequence and α is usually set to 0.75. Since a longer sequence has more logarithmic terms in the summation of (10.8.4), the term L^α in the denominator penalizes long sequences.

The computational cost of beam search is $O(k \, |\mathcal{Y}| \, T')$. This result is in between that of greedy search and that of exhaustive search. Greedy search can be treated as a special case of beam search arising when the beam size is set to 1.

10.8.4 Summary

Sequence searching strategies include greedy search, exhaustive search, and beam search. Beam search provides a trade-off between accuracy and computational cost via the flexible choice of the beam size.

10.8.5 Exercises

1. Can we treat exhaustive search as a special type of beam search? Why or why not?

2. Apply beam search in the machine translation problem in Section 10.7. How does the beam size affect the translation results and the prediction speed?

3. We used language modeling for generating text following user-provided prefixes in Section 9.5. Which kind of search strategy does it use? Can you improve it?

Discussions[153].

153

The earliest years of the deep learning boom were driven primarily by results produced using the multilayer perceptron, convolutional network, and recurrent network architectures. Remarkably, the model architectures that underpinned many of deep learning's breakthroughs in the 2010s had changed remarkably little relative to their antecedents despite the lapse of nearly 30 years. While plenty of new methodological innovations made their way into most practitioner's toolkits—ReLU activations, residual layers, batch normalization, dropout, and adaptive learning rate schedules come to mind—the core underlying architectures were clearly recognizable as scaled-up implementations of classic ideas. Despite thousands of papers proposing alternative ideas, models resembling classical convolutional neural networks (Chapter 7) retained *state-of-the-art* status in computer vision and models resembling Sepp Hochreiter's original design for the LSTM recurrent neural network (Section 10.1), dominated most applications in natural language processing. Arguably, to that point, the rapid emergence of deep learning appeared to be primarily attributable to shifts in the available computational resources (thanks to innovations in parallel computing with GPUs) and the availability of massive data resources (thanks to cheap storage and Internet services). While these factors may indeed remain the primary drivers behind this technology's increasing power we are also witnessing, at long last, a sea change in the landscape of dominant architectures.

At the present moment, the dominant models for nearly all natural language processing tasks are based on the Transformer architecture. Given any new task in natural language processing, the default first-pass approach is to grab a large Transformer-based pretrained model, (e.g., BERT (Devlin *et al.*, 2018), ELECTRA (Clark *et al.*, 2020), RoBERTa (Liu *et al.*, 2019), or Longformer (Beltagy *et al.*, 2020)) adapting the output layers as necessary, and fine-tuning the model on the available data for the downstream task. If you have been paying attention to the last few years of breathless news coverage centered on OpenAI's large language models, then you have been tracking a conversation centered on the GPT-2 and GPT-3 Transformer-based models (Brown *et al.*, 2020, Radford *et al.*, 2019). Meanwhile, the vision Transformer has emerged as a default model for diverse vision tasks, including image recognition, object detection, semantic segmentation, and superresolution (Dosovitskiy *et al.*, 2021, Liu *et al.*, 2021). Transformers also showed up as competitive methods for speech recognition (Gulati *et al.*, 2020), reinforcement learning (Chen *et al.*, 2021), and graph neural networks (Dwivedi and Bresson, 2020).

The core idea behind the Transformer model is the *attention mechanism*, an innovation that was originally envisioned as an enhancement for encoder–decoder RNNs applied to sequence-to-sequence applications, such as machine translations (Bahdanau *et al.*, 2014). You might recall that in the first sequence-to-sequence models for machine translation (Sutskever *et al.*, 2014), the entire input was compressed by the encoder into a single fixed-length vector to be fed into the decoder. The intuition behind attention is that rather than compressing the input, it might be better for the decoder to revisit the input sequence at every step. Moreover, rather than always seeing the same representation of the input, one might imagine that the decoder should selectively focus on particular parts of the input sequence at particular decoding steps. Bahdanau's attention mechanism provided a simple means by which the decoder could dynamically *attend* to different parts of the input at each decoding step. The high-level idea is that the encoder could produce a representation of length equal to the original input sequence. Then, at decoding time, the decoder can (via some control mechanism) receive as input a context vector consisting of a weighted sum of the representations on the input at each time step. Intuitively, the weights determine the extent to which each step's context "focuses" on each input token, and the key is to make this process for assigning the weights differentiable so that it can be learned along with all of the other neural network parameters.

Initially, the idea was a remarkably successful enhancement to the recurrent neural networks that already dominated machine translation applications. The models performed better than the original encoder–decoder sequence-to-sequence architectures. Furthermore, researchers noted that some nice qualitative insights sometimes emerged from inspecting the pattern of attention weights. In translation tasks, attention models often assigned high attention weights to cross-lingual synonyms when generating the corresponding words in the target language. For example, when translating the sentence "my feet hurt" to "j'ai mal au pieds", the neural network might assign high attention weights to the representation of "feet" when generating the corresponding French word "pieds". These insights spurred claims that attention models confer "interpretability" although what precisely the attention weights mean—i.e., how, if at all, they should be *interpreted* remains a hazy research topic.

However, attention mechanisms soon emerged as more significant concerns, beyond their usefulness as an enhancement for encoder–decoder recurrent neural networks and their putative usefulness for picking out salient inputs. Vaswani *et al.* (2017) proposed the Transformer architecture for machine translation, dispensing with recurrent connections altogether, and instead relying on cleverly arranged attention mechanisms to capture all relationships among input and output tokens. The architecture performed remarkably well, and by 2018 the Transformer began showing up in the majority of state-of-the-art natural language processing systems. Moreover, at the same time, the dominant practice in natural language processing became to pretrain large-scale models on enormous generic background corpora to optimize some self-supervised pretraining objective, and then to fine-tune these models using the available downstream data. The gap between Transformers and traditional architectures grew especially wide when applied in this pretraining paradigm, and thus the ascendance of Transformers coincided with the ascendence of such large-scale pretrained models, now sometimes called *foundation models* (Bommasani *et al.*, 2021).

In this chapter, we introduce attention models, starting with the most basic intuitions and the simplest instantiations of the idea. We then work our way up to the Transformer archi-tecture, the vision Transformer, and the landscape of modern Transformer-based pretrained models.

11.1 Queries, Keys, and Values

So far all the networks we have reviewed crucially relied on the input being of a well-defined size. For instance, the images in ImageNet are of size 224×224 pixels and CNNs are specifically tuned to this size. Even in natural language processing the input size for RNNs is well defined and fixed. Variable size is addressed by sequentially processing one token at a time, or by specially designed convolution kernels (Kalchbrenner *et al.*, 2014). This approach can lead to significant problems when the input is truly of varying size with varying information content, such as in Section 10.7 in the transformation of text (Sutskever *et al.*, 2014). In particular, for long sequences it becomes quite difficult to keep track of everything that has already been generated or even viewed by the network. Even explicit tracking heuristics such as proposed by Yang *et al.* (2016) only offer limited benefit.

Compare this to databases. In their simplest form they are collections of keys (k) and values (v). For instance, our database $\mathcal{D}$ might consist of tuples {("Zhang", "Aston"), ("Lipton", "Zachary"), ("Li", "Mu"), ("Smola", "Alex"), ("Hu", "Rachel"), ("Werness", "Brent")} with the last name being the key and the first name being the value. We can operate on $\mathcal{D}$, for instance with the exact query (q) for "Li" which would return the value "Mu". If ("Li", "Mu") was not a record in $\mathcal{D}$, there would be no valid answer. If we also allowed for approximate matches, we would retrieve ("Lipton", "Zachary") instead. This quite simple and trivial example nonetheless teaches us a number of useful things:

- We can design queries q that operate on (k,v) pairs in such a manner as to be valid regardless of the database size.

- The same query can receive different answers, according to the contents of the database.

- The "code" being executed for operating on a large state space (the database) can be quite simple (e.g., exact match, approximate match, top-k).

- There is no need to compress or simplify the database to make the operations effective.

Clearly we would not have introduced a simple database here if it wasn't for the purpose of explaining deep learning. Indeed, this leads to one of the most exciting concepts introduced in deep learning in the past decade: the *attention mechanism* (Bahdanau *et al.*, 2014). We will cover the specifics of its application to machine translation later. For now, simply consider the following: denote by $\mathcal{D} \overset{\text{def}}{=} \{(\mathbf{k}_1, \mathbf{v}_1), \ldots (\mathbf{k}_m, \mathbf{v}_m)\}$ a database of m tuples of *keys* and *values*. Moreover, denote by $\mathbf{q}$ a *query*. Then we can define the *attention* over $\mathcal{D}$ as

$$\text{Attention}(\mathbf{q}, \mathcal{D}) \overset{\text{def}}{=} \sum_{i=1}^{m} \alpha(\mathbf{q}, \mathbf{k}_i) \mathbf{v}_i, \tag{11.1.1}$$

where $\alpha(\mathbf{q}, \mathbf{k}_i) \in \mathbb{R}$ ($i = 1, \ldots, m$) are scalar attention weights. The operation itself is typically referred to as *attention pooling*. The name *attention* derives from the fact that the operation pays particular attention to the terms for which the weight α is significant (i.e., large). As such, the attention over $\mathcal{D}$ generates a linear combination of values contained in the database. In fact, this contains the above example as a special case where all but one weight is zero. We have a number of special cases:

- The weights $\alpha(\mathbf{q}, \mathbf{k}_i)$ are nonnegative. In this case the output of the attention mechanism is contained in the convex cone spanned by the values $\mathbf{v}_i$.

- The weights $\alpha(\mathbf{q}, \mathbf{k}_i)$ form a convex combination, i.e., $\sum_i \alpha(\mathbf{q}, \mathbf{k}_i) = 1$ and $\alpha(\mathbf{q}, \mathbf{k}_i) \geq 0$ for all i. This is the most common setting in deep learning.

- Exactly one of the weights $\alpha(\mathbf{q}, \mathbf{k}_i)$ is 1, while all others are 0. This is akin to a traditional database query.

- All weights are equal, i.e., $\alpha(\mathbf{q}, \mathbf{k}_i) = \frac{1}{m}$ for all i. This amounts to averaging across the entire database, also called average pooling in deep learning.

A common strategy for ensuring that the weights sum up to 1 is to normalize them via

$$\alpha(\mathbf{q}, \mathbf{k}_i) = \frac{\alpha(\mathbf{q}, \mathbf{k}_i)}{\sum_j \alpha(\mathbf{q}, \mathbf{k}_j)}. \tag{11.1.2}$$

In particular, to ensure that the weights are also nonnegative, one can resort to exponentiation. This means that we can now pick *any* function $a(\mathbf{q}, \mathbf{k})$ and then apply the softmax operation used for multinomial models to it via

$$\alpha(\mathbf{q}, \mathbf{k}_i) = \frac{\exp(a(\mathbf{q}, \mathbf{k}_i))}{\sum_j \exp(a(\mathbf{q}, \mathbf{k}_j))}. \tag{11.1.3}$$

This operation is readily available in all deep learning frameworks. It is differentiable and its gradient never vanishes, all of which are desirable properties in a model. Note though, the attention mechanism introduced above is not the only option. For instance, we can design a non-differentiable attention model that can be trained using reinforcement learning methods (Mnih *et al.*, 2014). As one would expect, training such a model is quite complex. Consequently the bulk of modern attention research follows the framework outlined in Fig. 11.1.1. We thus focus our exposition on this family of differentiable mechanisms.

What is quite remarkable is that the actual "code" for executing on the set of keys and values, namely the query, can be quite concise, even though the space to operate on is significant. This is a desirable property for a network layer as it does not require too many parameters to learn. Just as convenient is the fact that attention can operate on arbitrarily large databases without the need to change the way the attention pooling operation is performed.

```
import torch
from d2l import torch as d2l
```

11.1.1 Visualization

One of the benefits of the attention mechanism is that it can be quite intuitive, particularly when the weights are nonnegative and sum to 1. In this case we might *interpret* large

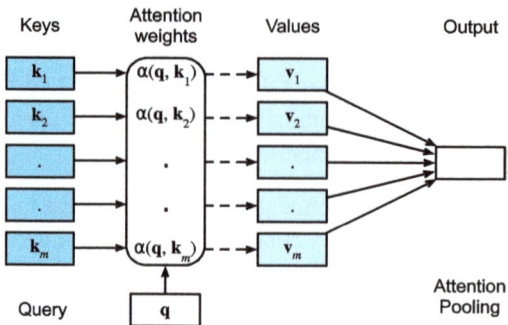

Fig. 11.1.1 The attention mechanism computes a linear combination over values $\mathbf{v}_i$ via attention pooling, where weights are derived according to the compatibility between a query $\mathbf{q}$ and keys $\mathbf{k}_i$.

weights as a way for the model to select components of relevance. While this is a good intuition, it is important to remember that it is just that, an *intuition*. Regardless, we may want to visualize its effect on the given set of keys when applying a variety of different queries. This function will come in handy later.

We thus define the show_heatmaps function. Note that it does not take a matrix (of attention weights) as its input but rather a tensor with four axes, allowing for an array of different queries and weights. Consequently the input matrices has the shape (number of rows for display, number of columns for display, number of queries, number of keys). This will come in handy later on when we want to visualize the workings that are to design Transformers.

```
#@save
def show_heatmaps(matrices, xlabel, ylabel, titles=None, figsize=(2.5, 2.5),
                  cmap='Reds'):
    """Show heatmaps of matrices."""
    d2l.use_svg_display()
    num_rows, num_cols, _, _ = matrices.shape
    fig, axes = d2l.plt.subplots(num_rows, num_cols, figsize=figsize,
                                 sharex=True, sharey=True, squeeze=False)
    for i, (row_axes, row_matrices) in enumerate(zip(axes, matrices)):
        for j, (ax, matrix) in enumerate(zip(row_axes, row_matrices)):
            pcm = ax.imshow(matrix.detach().numpy(), cmap=cmap)
            if i == num_rows - 1:
                ax.set_xlabel(xlabel)
            if j == 0:
                ax.set_ylabel(ylabel)
            if titles:
                ax.set_title(titles[j])
    fig.colorbar(pcm, ax=axes, shrink=0.6);
```

As a quick sanity check let's visualize the identity matrix, representing a case where the attention weight is 1 only when the query and the key are the same.

```
attention_weights = torch.eye(10).reshape((1, 1, 10, 10))
show_heatmaps(attention_weights, xlabel='Keys', ylabel='Queries')
```

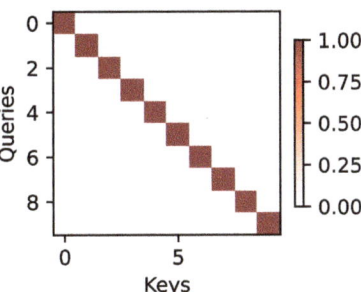

11.1.2 Summary

The attention mechanism allows us to aggregate data from many (key, value) pairs. So far our discussion was quite abstract, simply describing a way to pool data. We have not explained yet where those mysterious queries, keys, and values might arise from. Some intuition might help here: for instance, in a regression setting, the query might correspond to the location where the regression should be carried out. The keys are the locations where past data was observed and the values are the (regression) values themselves. This is the so-called Nadaraya–Watson estimator (Nadaraya, 1964, Watson, 1964) that we will be studying in the next section.

By design, the attention mechanism provides a *differentiable* means of control by which a neural network can select elements from a set and to construct an associated weighted sum over representations.

11.1.3 Exercises

1. Suppose that you wanted to reimplement approximate (key, query) matches as used in classical databases, which attention function would you pick?

2. Suppose that the attention function is given by $a(\mathbf{q}, \mathbf{k}_i) = \mathbf{q}^\top \mathbf{k}_i$ and that $\mathbf{k}_i = \mathbf{v}_i$ for $i = 1, \ldots, m$. Denote by $p(\mathbf{k}_i; \mathbf{q})$ the probability distribution over keys when using the softmax normalization in (11.1.3). Prove that $\nabla_{\mathbf{q}} \text{Attention}(\mathbf{q}, \mathcal{D}) = \text{Cov}_{p(\mathbf{k}_i; \mathbf{q})} [\mathbf{k}_i]$.

3. Design a differentiable search engine using the attention mechanism.

 4. Review the design of the Squeeze and Excitation Networks (Hu *et al.*, 2018) and interpret them through the lens of the attention mechanism.

Discussions[154].

11.2 Attention Pooling by Similarity

Now that we have introduced the primary components of the attention mechanism, let's use them in a rather classical setting, namely regression and classification via kernel density estimation (Nadaraya, 1964, Watson, 1964). This detour simply provides additional background: it is entirely optional and can be skipped if needed. At their core, Nadaraya–Watson estimators rely on some similarity kernel $\alpha(\mathbf{q}, \mathbf{k})$ relating queries $\mathbf{q}$ to keys $\mathbf{k}$. Some common kernels are

$$\alpha(\mathbf{q}, \mathbf{k}) = \exp\left(-\frac{1}{2}\|\mathbf{q} - \mathbf{k}\|^2\right) \qquad \text{Gaussian;}$$

$$\alpha(\mathbf{q}, \mathbf{k}) = 1 \text{ if } \|\mathbf{q} - \mathbf{k}\| \le 1 \qquad \text{Boxcar;} \tag{11.2.1}$$

$$\alpha(\mathbf{q}, \mathbf{k}) = \max\left(0, 1 - \|\mathbf{q} - \mathbf{k}\|\right) \quad \text{Epanechikov.}$$

[155] There are many more choices that we could pick. See a Wikipedia article [155] for a more extensive review and how the choice of kernels is related to kernel density estimation, sometimes also called *Parzen Windows* (Parzen, 1957). All of the kernels are heuristic and can be tuned. For instance, we can adjust the width, not only on a global basis but even on a per-coordinate basis. Regardless, all of them lead to the following equation for regression and classification alike:

$$f(\mathbf{q}) = \sum_i \mathbf{v}_i \frac{\alpha(\mathbf{q}, \mathbf{k}_i)}{\sum_j \alpha(\mathbf{q}, \mathbf{k}_j)}. \tag{11.2.2}$$

In the case of a (scalar) regression with observations $(\mathbf{x}_i, y_i)$ for features and labels respectively, $\mathbf{v}_i = y_i$ are scalars, $\mathbf{k}_i = \mathbf{x}_i$ are vectors, and the query $\mathbf{q}$ denotes the new location where f should be evaluated. In the case of (multiclass) classification, we use one-hot-encoding of y_i to obtain $\mathbf{v}_i$. One of the convenient properties of this estimator is that it requires no training. Even more so, if we suitably narrow the kernel with increasing amounts of data, the approach is consistent (Mack and Silverman, 1982), i.e., it will converge to some statistically optimal solution. Let's start by inspecting some kernels.

```
import numpy as np
import torch
from torch import nn
from torch.nn import functional as F
from d2l import torch as d2l

d2l.use_svg_display()
```

11.2.1 Kernels and Data

All the kernels $\alpha(\mathbf{k}, \mathbf{q})$ defined in this section are *translation and rotation invariant*; that is, if we shift and rotate $\mathbf{k}$ and $\mathbf{q}$ in the same manner, the value of α remains unchanged. For simplicity we thus pick scalar arguments $k, q \in \mathbb{R}$ and pick the key $k = 0$ as the origin. This yields:

```
# Define some kernels
def gaussian(x):
    return torch.exp(-x**2 / 2)

def boxcar(x):
    return torch.abs(x) < 1.0

def constant(x):
    return 1.0 + 0 * x

def epanechikov(x):
    return torch.max(1 - torch.abs(x), torch.zeros_like(x))


fig, axes = d2l.plt.subplots(1, 4, sharey=True, figsize=(12, 3))

kernels = (gaussian, boxcar, constant, epanechikov)
names = ('Gaussian', 'Boxcar', 'Constant', 'Epanechikov')
x = torch.arange(-2.5, 2.5, 0.1)
for kernel, name, ax in zip(kernels, names, axes):
    ax.plot(x.detach().numpy(), kernel(x).detach().numpy())
    ax.set_xlabel(name)

d2l.plt.show()
```

Different kernels correspond to different notions of range and smoothness. For instance, the boxcar kernel only attends to observations within a distance of 1 (or some otherwise defined hyperparameter) and does so indiscriminately.

To see Nadaraya–Watson estimation in action, let's define some training data. In the following we use the dependency

$$y_i = 2\sin(x_i) + x_i + \epsilon, \tag{11.2.3}$$

where ϵ is drawn from a normal distribution with zero mean and unit variance. We draw 40 training examples.

```
def f(x):
    return 2 * torch.sin(x) + x

n = 40
```

(continues on next page)

(continued from previous page)

```
x_train, _ = torch.sort(torch.rand(n) * 5)
y_train = f(x_train) + torch.randn(n)
x_val = torch.arange(0, 5, 0.1)
y_val = f(x_val)
```

11.2.2 Attention Pooling via Nadaraya–Watson Regression

Now that we have data and kernels, all we need is a function that computes the kernel regression estimates. Note that we also want to obtain the relative kernel weights in order to perform some minor diagnostics. Hence we first compute the kernel between all training features (covariates) x_train and all validation features x_val. This yields a matrix, which we subsequently normalize. When multiplied with the training labels y_train we obtain the estimates.

Recall attention pooling in (11.1.1). Let each validation feature be a query, and each training feature–label pair be a key–value pair. As a result, the normalized relative kernel weights (attention_w below) are the *attention weights*.

```
def nadaraya_watson(x_train, y_train, x_val, kernel):
    dists = x_train.reshape((-1, 1)) - x_val.reshape((1, -1))
    # Each column/row corresponds to each query/key
    k = kernel(dists).type(torch.float32)
    # Normalization over keys for each query
    attention_w = k / k.sum(0)
    y_hat = y_train@attention_w
    return y_hat, attention_w
```

Let's have a look at the kind of estimates that the different kernels produce.

```
def plot(x_train, y_train, x_val, y_val, kernels, names, attention=False):
    fig, axes = d2l.plt.subplots(1, 4, sharey=True, figsize=(12, 3))
    for kernel, name, ax in zip(kernels, names, axes):
        y_hat, attention_w = nadaraya_watson(x_train, y_train, x_val, kernel)
        if attention:
            pcm = ax.imshow(attention_w.detach().numpy(), cmap='Reds')
        else:
            ax.plot(x_val, y_hat)
            ax.plot(x_val, y_val, 'm--')
            ax.plot(x_train, y_train, 'o', alpha=0.5);
        ax.set_xlabel(name)
        if not attention:
            ax.legend(['y_hat', 'y'])
    if attention:
        fig.colorbar(pcm, ax=axes, shrink=0.7)
```

```
plot(x_train, y_train, x_val, y_val, kernels, names)
```

The first thing that stands out is that all three nontrivial kernels (Gaussian, Boxcar, and Epanechikov) produce fairly workable estimates that are not too far from the true function.

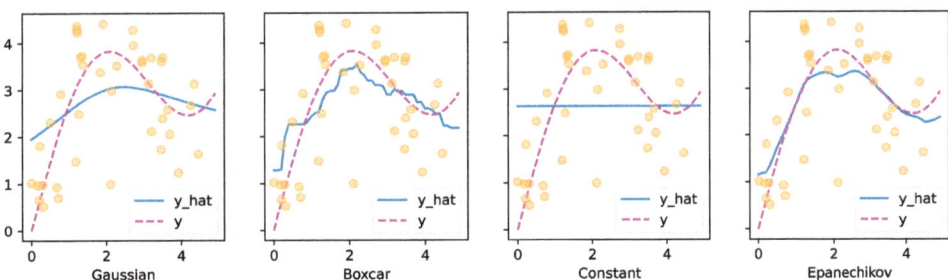

Only the constant kernel that leads to the trivial estimate $f(x) = \frac{1}{n} \sum_i y_i$ produces a rather unrealistic result. Let's inspect the attention weighting a bit more closely:

```
plot(x_train, y_train, x_val, y_val, kernels, names, attention=True)
```

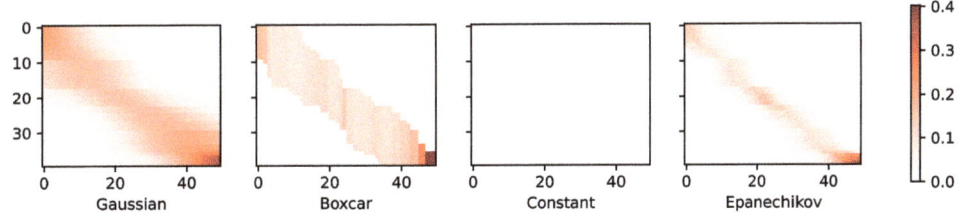

The visualization clearly shows why the estimates for Gaussian, Boxcar, and Epanechikov are very similar: after all, they are derived from very similar attention weights, despite the different functional form of the kernel. This raises the question as to whether this is always the case.

11.2.3 Adapting Attention Pooling

We could replace the Gaussian kernel with one of a different width. That is, we could use $\alpha(\mathbf{q}, \mathbf{k}) = \exp\left(-\frac{1}{2\sigma^2}\|\mathbf{q} - \mathbf{k}\|^2\right)$ where σ^2 determines the width of the kernel. Let's see whether this affects the outcomes.

```
sigmas = (0.1, 0.2, 0.5, 1)
names = ['Sigma ' + str(sigma) for sigma in sigmas]

def gaussian_with_width(sigma):
    return (lambda x: torch.exp(-x**2 / (2*sigma**2)))

kernels = [gaussian_with_width(sigma) for sigma in sigmas]
plot(x_train, y_train, x_val, y_val, kernels, names)
```

Clearly, the narrower the kernel, the less smooth the estimate. At the same time, it adapts better to the local variations. Let's look at the corresponding attention weights.

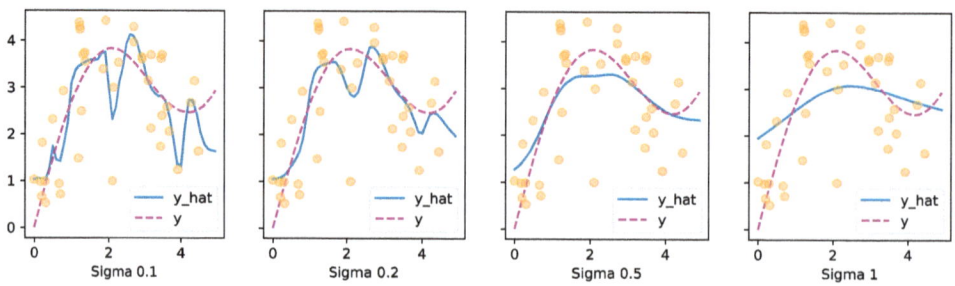

```
plot(x_train, y_train, x_val, y_val, kernels, names, attention=True)
```

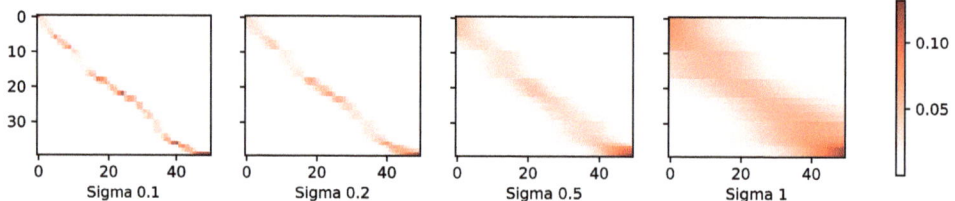

As we would expect, the narrower the kernel, the narrower the range of large attention weights. It is also clear that picking the same width might not be ideal. In fact, Silverman (1986) proposed a heuristic that depends on the local density. Many more such "tricks" have been proposed. For instance, Norelli *et al.* (2022) used a similar nearest-neighbor interpolation technique for designing cross-modal image and text representations.

The astute reader might wonder why we are providing this deep dive for a method that is over half a century old. First, it is one of the earliest precursors of modern attention mechanisms. Second, it is great for visualization. Third, and just as importantly, it demonstrates the limits of hand-crafted attention mechanisms. A much better strategy is to *learn* the mechanism, by learning the representations for queries and keys. This is what we will embark on in the following sections.

11.2.4 Summary

Nadaraya–Watson kernel regression is an early precursor of the current attention mechanisms. It can be used directly with little to no training or tuning, either for classification or regression. The attention weight is assigned according to the similarity (or distance) between query and key, and according to how many similar observations are available.

11.2.5 Exercises

1. Parzen windows density estimates are given by $\hat{p}(\mathbf{x}) = \frac{1}{n} \sum_i k(\mathbf{x}, \mathbf{x}_i)$. Prove that for binary classification the function $\hat{p}(\mathbf{x}, y = 1) - \hat{p}(\mathbf{x}, y = -1)$, as obtained by Parzen windows is equivalent to Nadaraya–Watson classification.

2. Implement stochastic gradient descent to learn a good value for kernel widths in Nadaraya–Watson regression.

   1. What happens if you just use the above estimates to minimize $(f(\mathbf{x_i}) - y_i)^2$ directly? Hint: y_i is part of the terms used to compute f.

   2. Remove $(\mathbf{x}_i, y_i)$ from the estimate for $f(\mathbf{x}_i)$ and optimize over the kernel widths. Do you still observe overfitting?

3. Assume that all $\mathbf{x}$ lie on the unit sphere, i.e., all satisfy $\|\mathbf{x}\| = 1$. Can you simplify the $\|\mathbf{x} - \mathbf{x}_i\|^2$ term in the exponential? Hint: we will later see that this is very closely related to dot product attention.

4. Recall that Mack and Silverman (1982) proved that Nadaraya–Watson estimation is consistent. How quickly should you reduce the scale for the attention mechanism as you get more data? Provide some intuition for your answer. Does it depend on the dimensionality of the data? How?

156

Discussions[156].

11.3 Attention Scoring Functions

In Section 11.2, we used a number of different distance-based kernels, including a Gaussian kernel to model interactions between queries and keys. As it turns out, distance functions are slightly more expensive to compute than dot products. As such, with the softmax operation to ensure nonnegative attention weights, much of the work has gone into *attention scoring functions a* in (11.1.3) and Fig. 11.3.1 that are simpler to compute.

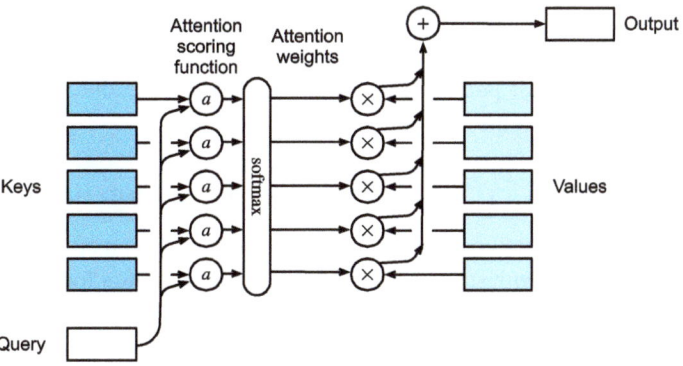

Fig. 11.3.1   Computing the output of attention pooling as a weighted average of values, where weights are computed with the attention scoring function a and the softmax operation.

```
import math
import torch
from torch import nn
from d2l import torch as d2l
```

11.3.1 Dot Product Attention

Let's review the attention function (without exponentiation) from the Gaussian kernel for a moment:

$$a(\mathbf{q}, \mathbf{k}_i) = -\frac{1}{2}\|\mathbf{q} - \mathbf{k}_i\|^2 = \mathbf{q}^\top\mathbf{k}_i - \frac{1}{2}\|\mathbf{k}_i\|^2 - \frac{1}{2}\|\mathbf{q}\|^2. \qquad (11.3.1)$$

First, note that the final term depends on $\mathbf{q}$ only. As such it is identical for all $(\mathbf{q}, \mathbf{k}_i)$ pairs. Normalizing the attention weights to 1, as is done in (11.1.3), ensures that this term disappears entirely. Second, note that both batch and layer normalization (to be discussed later) lead to activations that have well-bounded, and often constant, norms $\|\mathbf{k}_i\|$. This is the case, for instance, whenever the keys $\mathbf{k}_i$ were generated by a layer norm. As such, we can drop it from the definition of a without any major change in the outcome.

Last, we need to keep the order of magnitude of the arguments in the exponential function under control. Assume that all the elements of the query $\mathbf{q} \in \mathbb{R}^d$ and the key $\mathbf{k}_i \in \mathbb{R}^d$ are independent and identically drawn random variables with zero mean and unit variance. The dot product between both vectors has zero mean and a variance of d. To ensure that the variance of the dot product still remains 1 regardless of vector length, we use the *scaled dot product attention* scoring function. That is, we rescale the dot product by $1/\sqrt{d}$. We thus arrive at the first commonly used attention function that is used, e.g., in Transformers (Vaswani *et al.*, 2017):

$$a(\mathbf{q}, \mathbf{k}_i) = \mathbf{q}^\top\mathbf{k}_i/\sqrt{d}. \qquad (11.3.2)$$

Note that attention weights α still need normalizing. We can simplify this further via (11.1.3) by using the softmax operation:

$$\alpha(\mathbf{q}, \mathbf{k}_i) = \mathrm{softmax}(a(\mathbf{q}, \mathbf{k}_i)) = \frac{\exp(\mathbf{q}^\top\mathbf{k}_i/\sqrt{d})}{\sum_{j=1}\exp(\mathbf{q}^\top\mathbf{k}_j/\sqrt{d})}. \qquad (11.3.3)$$

As it turns out, all popular attention mechanisms use the softmax, hence we will limit ourselves to that in the remainder of this chapter.

11.3.2 Convenience Functions

We need a few functions to make the attention mechanism efficient to deploy. This includes tools for dealing with strings of variable lengths (common for natural language processing) and tools for efficient evaluation on minibatches (batch matrix multiplication).

Masked Softmax Operation

One of the most popular applications of the attention mechanism is to sequence models. Hence we need to be able to deal with sequences of different lengths. In some cases, such sequences may end up in the same minibatch, necessitating padding with dummy tokens for shorter sequences (see Section 10.5 for an example). These special tokens do not carry meaning. For instance, assume that we have the following three sentences:

```
Dive  into  Deep    Learning
Learn to    code    <blank>
Hello world <blank> <blank>
```

Since we do not want blanks in our attention model we simply need to limit $\sum_{i=1}^{n} \alpha(\mathbf{q}, \mathbf{k}_i) \mathbf{v}_i$ to $\sum_{i=1}^{l} \alpha(\mathbf{q}, \mathbf{k}_i) \mathbf{v}_i$ for however long, $l \leq n$, the actual sentence is. Since it is such a common problem, it has a name: the *masked softmax operation*.

Let's implement it. Actually, the implementation cheats ever so slightly by setting the values of $\mathbf{v}_i$, for $i > l$, to zero. Moreover, it sets the attention weights to a large negative number, such as -10^6, in order to make their contribution to gradients and values vanish in practice. This is done since linear algebra kernels and operators are heavily optimized for GPUs and it is faster to be slightly wasteful in computation rather than to have code with conditional (if then else) statements.

```python
def masked_softmax(X, valid_lens):  #@save
    """Perform softmax operation by masking elements on the last axis."""
    # X: 3D tensor, valid_lens: 1D or 2D tensor
    def _sequence_mask(X, valid_len, value=0):
        maxlen = X.size(1)
        mask = torch.arange((maxlen), dtype=torch.float32,
                            device=X.device)[None, :] < valid_len[:, None]
        X[~mask] = value
        return X

    if valid_lens is None:
        return nn.functional.softmax(X, dim=-1)
    else:
        shape = X.shape
        if valid_lens.dim() == 1:
            valid_lens = torch.repeat_interleave(valid_lens, shape[1])
        else:
            valid_lens = valid_lens.reshape(-1)
        # On the last axis, replace masked elements with a very large negative
        # value, whose exponentiation outputs 0
        X = _sequence_mask(X.reshape(-1, shape[-1]), valid_lens, value=-1e6)
        return nn.functional.softmax(X.reshape(shape), dim=-1)
```

To illustrate how this function works, consider a minibatch of two examples of size 2×4, where their valid lengths are 2 and 3, respectively. As a result of the masked softmax operation, values beyond the valid lengths for each pair of vectors are all masked as zero.

```python
masked_softmax(torch.rand(2, 2, 4), torch.tensor([2, 3]))
```

```
tensor([[[0.5661, 0.4339, 0.0000, 0.0000],
         [0.3783, 0.6217, 0.0000, 0.0000]],
```

(continues on next page)

(continued from previous page)

```
        [[0.4107, 0.3094, 0.2799, 0.0000],
         [0.2315, 0.4707, 0.2978, 0.0000]]])
```

If we need more fine-grained control to specify the valid length for each of the two vectors of every example, we simply use a two-dimensional tensor of valid lengths. This yields:

```
masked_softmax(torch.rand(2, 2, 4), torch.tensor([[1, 3], [2, 4]]))
```

```
tensor([[[1.0000, 0.0000, 0.0000, 0.0000],
         [0.2053, 0.5043, 0.2904, 0.0000]],

        [[0.6222, 0.3778, 0.0000, 0.0000],
         [0.1884, 0.2884, 0.3161, 0.2072]]])
```

Batch Matrix Multiplication

Another commonly used operation is to multiply batches of matrices by one another. This comes in handy when we have minibatches of queries, keys, and values. More specifically, assume that

$$\mathbf{Q} = [\mathbf{Q}_1, \mathbf{Q}_2, \ldots, \mathbf{Q}_n] \in \mathbb{R}^{n \times a \times b},$$
$$\mathbf{K} = [\mathbf{K}_1, \mathbf{K}_2, \ldots, \mathbf{K}_n] \in \mathbb{R}^{n \times b \times c}. \qquad (11.3.4)$$

Then the batch matrix multiplication (BMM) computes the elementwise product

$$\mathrm{BMM}(\mathbf{Q}, \mathbf{K}) = [\mathbf{Q}_1 \mathbf{K}_1, \mathbf{Q}_2 \mathbf{K}_2, \ldots, \mathbf{Q}_n \mathbf{K}_n] \in \mathbb{R}^{n \times a \times c}. \qquad (11.3.5)$$

Let's see this in action in a deep learning framework.

```
Q = torch.ones((2, 3, 4))
K = torch.ones((2, 4, 6))
d2l.check_shape(torch.bmm(Q, K), (2, 3, 6))
```

11.3.3 Scaled Dot Product Attention

Let's return to the dot product attention introduced in (11.3.2). In general, it requires that both the query and the key have the same vector length, say d, even though this can be addressed easily by replacing $\mathbf{q}^\top \mathbf{k}$ with $\mathbf{q}^\top \mathbf{M} \mathbf{k}$ where $\mathbf{M}$ is a matrix suitably chosen for translating between both spaces. For now assume that the dimensions match.

In practice, we often think of minibatches for efficiency, such as computing attention for n queries and m key-value pairs, where queries and keys are of length d and values are of length v. The scaled dot product attention of queries $\mathbf{Q} \in \mathbb{R}^{n \times d}$, keys $\mathbf{K} \in \mathbb{R}^{m \times d}$, and values $\mathbf{V} \in \mathbb{R}^{m \times v}$ thus can be written as

$$\text{softmax}\left(\frac{\mathbf{Q}\mathbf{K}^{\top}}{\sqrt{d}}\right)\mathbf{V} \in \mathbb{R}^{n \times v}. \tag{11.3.6}$$

Note that when applying this to a minibatch, we need the batch matrix multiplication introduced in (11.3.5). In the following implementation of the scaled dot product attention, we use dropout for model regularization.

```python
class DotProductAttention(nn.Module):  #@save
    """Scaled dot product attention."""
    def __init__(self, dropout):
        super().__init__()
        self.dropout = nn.Dropout(dropout)

    # Shape of queries: (batch_size, no. of queries, d)
    # Shape of keys: (batch_size, no. of key-value pairs, d)
    # Shape of values: (batch_size, no. of key-value pairs, value dimension)
    # Shape of valid_lens: (batch_size,) or (batch_size, no. of queries)
    def forward(self, queries, keys, values, valid_lens=None):
        d = queries.shape[-1]
        # Swap the last two dimensions of keys with keys.transpose(1, 2)
        scores = torch.bmm(queries, keys.transpose(1, 2)) / math.sqrt(d)
        self.attention_weights = masked_softmax(scores, valid_lens)
        return torch.bmm(self.dropout(self.attention_weights), values)
```

To illustrate how the DotProductAttention class works, we use the same keys, values, and valid lengths from the earlier toy example for additive attention. For the purpose of our example we assume that we have a minibatch size of 2, a total of 10 keys and values, and that the dimensionality of the values is 4. Lastly, we assume that the valid length per observation is 2 and 6 respectively. Given that, we expect the output to be a $2 \times 1 \times 4$ tensor, i.e., one row per example of the minibatch.

```python
queries = torch.normal(0, 1, (2, 1, 2))
keys = torch.normal(0, 1, (2, 10, 2))
values = torch.normal(0, 1, (2, 10, 4))
valid_lens = torch.tensor([2, 6])

attention = DotProductAttention(dropout=0.5)
attention.eval()
d2l.check_shape(attention(queries, keys, values, valid_lens), (2, 1, 4))
```

Let's check whether the attention weights actually vanish for anything beyond the second and sixth column respectively (because of setting the valid length to 2 and 6).

```python
d2l.show_heatmaps(attention.attention_weights.reshape((1, 1, 2, 10)),
                  xlabel='Keys', ylabel='Queries')
```

11.3.4 Additive Attention

When queries $\mathbf{q}$ and keys $\mathbf{k}$ are vectors of different dimension, we can either use a matrix to address the mismatch via $\mathbf{q}^{\top}\mathbf{M}\mathbf{k}$, or we can use additive attention as the scoring function.

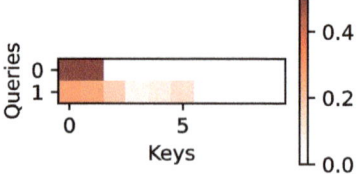

Another benefit is that, as its name indicates, the attention is additive. This can lead to some minor computational savings. Given a query $\mathbf{q} \in \mathbb{R}^q$ and a key $\mathbf{k} \in \mathbb{R}^k$, the *additive attention* scoring function (Bahdanau *et al.*, 2014) is given by

$$a(\mathbf{q}, \mathbf{k}) = \mathbf{w}_v^\top \tanh(\mathbf{W}_q\mathbf{q} + \mathbf{W}_k\mathbf{k}) \in \mathbb{R}, \qquad (11.3.7)$$

where $\mathbf{W}_q \in \mathbb{R}^{h \times q}$, $\mathbf{W}_k \in \mathbb{R}^{h \times k}$, and $\mathbf{w}_v \in \mathbb{R}^h$ are the learnable parameters. This term is then fed into a softmax to ensure both nonnegativity and normalization. An equivalent interpretation of (11.3.7) is that the query and key are concatenated and fed into an MLP with a single hidden layer. Using tanh as the activation function and disabling bias terms, we implement additive attention as follows:

```python
class AdditiveAttention(nn.Module):  #@save
    """Additive attention."""
    def __init__(self, num_hiddens, dropout, **kwargs):
        super(AdditiveAttention, self).__init__(**kwargs)
        self.W_k = nn.LazyLinear(num_hiddens, bias=False)
        self.W_q = nn.LazyLinear(num_hiddens, bias=False)
        self.w_v = nn.LazyLinear(1, bias=False)
        self.dropout = nn.Dropout(dropout)

    def forward(self, queries, keys, values, valid_lens):
        queries, keys = self.W_q(queries), self.W_k(keys)
        # After dimension expansion, shape of queries: (batch_size, no. of
        # queries, 1, num_hiddens) and shape of keys: (batch_size, 1, no. of
        # key-value pairs, num_hiddens). Sum them up with broadcasting
        features = queries.unsqueeze(2) + keys.unsqueeze(1)
        features = torch.tanh(features)
        # There is only one output of self.w_v, so we remove the last
        # one-dimensional entry from the shape. Shape of scores: (batch_size,
        # no. of queries, no. of key-value pairs)
        scores = self.w_v(features).squeeze(-1)
        self.attention_weights = masked_softmax(scores, valid_lens)
        # Shape of values: (batch_size, no. of key-value pairs, value
        # dimension)
        return torch.bmm(self.dropout(self.attention_weights), values)
```

Let's see how `AdditiveAttention` works. In our toy example we pick queries, keys and values of size $(2, 1, 20)$, $(2, 10, 2)$ and $(2, 10, 4)$, respectively. This is identical to our choice for `DotProductAttention`, except that now the queries are 20-dimensional. Likewise, we pick $(2, 6)$ as the valid lengths for the sequences in the minibatch.

```
queries = torch.normal(0, 1, (2, 1, 20))

attention = AdditiveAttention(num_hiddens=8, dropout=0.1)
attention.eval()
d2l.check_shape(attention(queries, keys, values, valid_lens), (2, 1, 4))
```

When reviewing the attention function we see a behavior that is qualitatively quite similar to that of DotProductAttention. That is, only terms within the chosen valid length $(2, 6)$ are nonzero.

```
d2l.show_heatmaps(attention.attention_weights.reshape((1, 1, 2, 10)),
                  xlabel='Keys', ylabel='Queries')
```

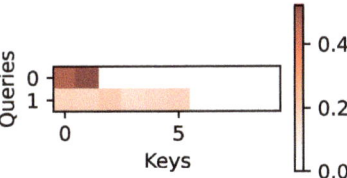

11.3.5 Summary

In this section we introduced the two key attention scoring functions: dot product and additive attention. They are effective tools for aggregating across sequences of variable length. In particular, the dot product attention is the mainstay of modern Transformer architectures. When queries and keys are vectors of different lengths, we can use the additive attention scoring function instead. Optimizing these layers is one of the key areas of advance in recent years. For instance, NVIDIA's Transformer Library[157] and Megatron (Shoeybi *et al.*, 2019) crucially rely on efficient variants of the attention mechanism. We will dive into this in quite a bit more detail as we review Transformers in later sections.

11.3.6 Exercises

1. Implement distance-based attention by modifying the DotProductAttention code. Note that you only need the squared norms of the keys $\|\mathbf{k}_i\|^2$ for an efficient implementation.

2. Modify the dot product attention to allow for queries and keys of different dimensionalities by employing a matrix to adjust dimensions.

3. How does the computational cost scale with the dimensionality of the keys, queries, values, and their number? What about the memory bandwidth requirements?

Discussions[158].

11.4 The Bahdanau Attention Mechanism

When we encountered machine translation in Section 10.7, we designed an encoder–decoder architecture for sequence-to-sequence learning based on two RNNs (Sutskever *et al.*, 2014). Specifically, the RNN encoder transforms a variable-length sequence into a *fixed-shape* context variable. Then, the RNN decoder generates the output (target) sequence token by token based on the generated tokens and the context variable.

Recall Fig. 10.7.2 which we repeat (Fig. 11.4.1) with some additional detail. Conventionally, in an RNN all relevant information about a source sequence is translated into some internal *fixed-dimensional* state representation by the encoder. It is this very state that is used by the decoder as the complete and exclusive source of information for generating the translated sequence. In other words, the sequence-to-sequence mechanism treats the intermediate state as a sufficient statistic of whatever string might have served as input.

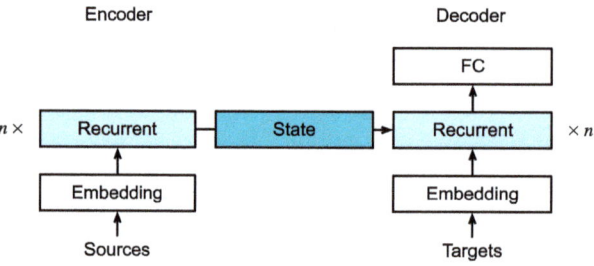

Fig. 11.4.1   Sequence-to-sequence model. The state, as generated by the encoder, is the only piece of information shared between the encoder and the decoder.

While this is quite reasonable for short sequences, it is clear that it is infeasible for long ones, such as a book chapter or even just a very long sentence. After all, before too long there will simply not be enough "space" in the intermediate representation to store all that is important in the source sequence. Consequently the decoder will fail to translate long and complex sentences. One of the first to encounter this was Graves (2013) who tried to design an RNN to generate handwritten text. Since the source text has arbitrary length they designed a differentiable attention model to align text characters with the much longer pen trace, where the alignment moves only in one direction. This, in turn, draws on decoding algorithms in speech recognition, e.g., hidden Markov models (Rabiner and Juang, 1993).

Inspired by the idea of learning to align, Bahdanau *et al.* (2014) proposed a differentiable attention model *without* the unidirectional alignment limitation. When predicting a token, if not all the input tokens are relevant, the model aligns (or attends) only to parts of the input sequence that are deemed relevant to the current prediction. This is then used to update the current state before generating the next token. While quite innocuous in its description, this *Bahdanau attention mechanism* has arguably turned into one of the most influential ideas of the past decade in deep learning, giving rise to Transformers (Vaswani *et al.*, 2017) and many related new architectures.

```
import torch
from torch import nn
from d2l import torch as d2l
```

11.4.1 Model

We follow the notation introduced by the sequence-to-sequence architecture of Section 10.7, in particular (10.7.3). The key idea is that instead of keeping the state, i.e., the context variable $\mathbf{c}$ summarizing the source sentence, as fixed, we dynamically update it, as a function of both the original text (encoder hidden states $\mathbf{h}_t$) and the text that was already generated (decoder hidden states $\mathbf{s}_{t'-1}$). This yields $\mathbf{c}_{t'}$, which is updated after any decoding time step t'. Suppose that the input sequence is of length T. In this case the context variable is the output of attention pooling:

$$\mathbf{c}_{t'} = \sum_{t=1}^{T} \alpha(\mathbf{s}_{t'-1}, \mathbf{h}_t)\mathbf{h}_t. \tag{11.4.1}$$

We used $\mathbf{s}_{t'-1}$ as the query, and $\mathbf{h}_t$ as both the key and the value. Note that $\mathbf{c}_{t'}$ is then used to generate the state $\mathbf{s}_{t'}$ and to generate a new token: see (10.7.3). In particular, the attention weight α is computed as in (11.3.3) using the additive attention scoring function defined by (11.3.7). This RNN encoder–decoder architecture using attention is depicted in Fig. 11.4.2. Note that later this model was modified so as to include the already generated tokens in the decoder as further context (i.e., the attention sum does not stop at T but rather it proceeds up to $t' - 1$). For instance, see Chan *et al.* (2015) for a description of this strategy, as applied to speech recognition.

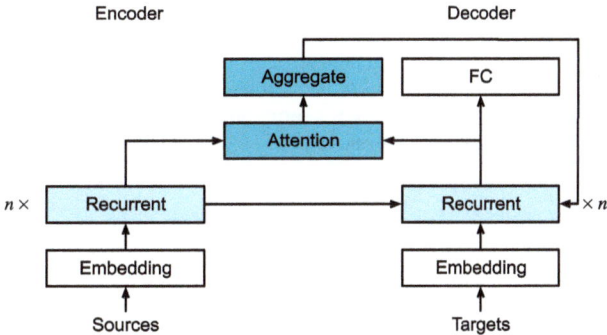

Fig. 11.4.2 Layers in an RNN encoder–decoder model with the Bahdanau attention mechanism.

11.4.2 Defining the Decoder with Attention

To implement the RNN encoder-–decoder with attention, we only need to redefine the decoder (omitting the generated symbols from the attention function simplifies the design). Let's begin with the base interface for decoders with attention by defining the quite unsurprisingly named `AttentionDecoder` class.

```python
class AttentionDecoder(d2l.Decoder):  #@save
    """The base attention-based decoder interface."""
    def __init__(self):
        super().__init__()

    @property
    def attention_weights(self):
        raise NotImplementedError
```

We need to implement the RNN decoder in the Seq2SeqAttentionDecoder class. The state of the decoder is initialized with (i) the hidden states of the last layer of the encoder at all time steps, used as keys and values for attention; (ii) the hidden state of the encoder at all layers at the final time step, which serves to initialize the hidden state of the decoder; and (iii) the valid length of the encoder, to exclude the padding tokens in attention pooling. At each decoding time step, the hidden state of the final layer of the decoder, obtained at the previous time step, is used as the query of the attention mechanism. Both the output of the attention mechanism and the input embedding are concatenated to serve as the input of the RNN decoder.

```python
class Seq2SeqAttentionDecoder(AttentionDecoder):
    def __init__(self, vocab_size, embed_size, num_hiddens, num_layers,
                 dropout=0):
        super().__init__()
        self.attention = d2l.AdditiveAttention(num_hiddens, dropout)
        self.embedding = nn.Embedding(vocab_size, embed_size)
        self.rnn = nn.GRU(
            embed_size + num_hiddens, num_hiddens, num_layers,
            dropout=dropout)
        self.dense = nn.LazyLinear(vocab_size)
        self.apply(d2l.init_seq2seq)

    def init_state(self, enc_outputs, enc_valid_lens):
        # Shape of outputs: (num_steps, batch_size, num_hiddens).
        # Shape of hidden_state: (num_layers, batch_size, num_hiddens)
        outputs, hidden_state = enc_outputs
        return (outputs.permute(1, 0, 2), hidden_state, enc_valid_lens)

    def forward(self, X, state):
        # Shape of enc_outputs: (batch_size, num_steps, num_hiddens).
        # Shape of hidden_state: (num_layers, batch_size, num_hiddens)
        enc_outputs, hidden_state, enc_valid_lens = state
        # Shape of the output X: (num_steps, batch_size, embed_size)
        X = self.embedding(X).permute(1, 0, 2)
        outputs, self._attention_weights = [], []
        for x in X:
            # Shape of query: (batch_size, 1, num_hiddens)
            query = torch.unsqueeze(hidden_state[-1], dim=1)
            # Shape of context: (batch_size, 1, num_hiddens)
            context = self.attention(
                query, enc_outputs, enc_outputs, enc_valid_lens)
            # Concatenate on the feature dimension
            x = torch.cat((context, torch.unsqueeze(x, dim=1)), dim=-1)
            # Reshape x as (1, batch_size, embed_size + num_hiddens)
            out, hidden_state = self.rnn(x.permute(1, 0, 2), hidden_state)
```

(continues on next page)

(continued from previous page)

```
        outputs.append(out)
        self._attention_weights.append(self.attention.attention_weights)
    # After fully connected layer transformation, shape of outputs:
    # (num_steps, batch_size, vocab_size)
    outputs = self.dense(torch.cat(outputs, dim=0))
    return outputs.permute(1, 0, 2), [enc_outputs, hidden_state,
                                      enc_valid_lens]

@property
def attention_weights(self):
    return self._attention_weights
```

In the following, we test the implemented decoder with attention using a minibatch of four sequences, each of which are seven time steps long.

```
vocab_size, embed_size, num_hiddens, num_layers = 10, 8, 16, 2
batch_size, num_steps = 4, 7
encoder = d2l.Seq2SeqEncoder(vocab_size, embed_size, num_hiddens, num_layers)
decoder = Seq2SeqAttentionDecoder(vocab_size, embed_size, num_hiddens,
                                  num_layers)
X = torch.zeros((batch_size, num_steps), dtype=torch.long)
state = decoder.init_state(encoder(X), None)
output, state = decoder(X, state)
d2l.check_shape(output, (batch_size, num_steps, vocab_size))
d2l.check_shape(state[0], (batch_size, num_steps, num_hiddens))
d2l.check_shape(state[1][0], (batch_size, num_hiddens))
```

11.4.3 Training

Now that we specified the new decoder we can proceed analogously to Section 10.7.6: specify the hyperparameters, instantiate a regular encoder and a decoder with attention, and train this model for machine translation.

```
data = d2l.MTFraEng(batch_size=128)
embed_size, num_hiddens, num_layers, dropout = 256, 256, 2, 0.2
encoder = d2l.Seq2SeqEncoder(
    len(data.src_vocab), embed_size, num_hiddens, num_layers, dropout)
decoder = Seq2SeqAttentionDecoder(
    len(data.tgt_vocab), embed_size, num_hiddens, num_layers, dropout)
model = d2l.Seq2Seq(encoder, decoder, tgt_pad=data.tgt_vocab['<pad>'],
                    lr=0.005)
trainer = d2l.Trainer(max_epochs=30, gradient_clip_val=1, num_gpus=1)
trainer.fit(model, data)
```

After the model is trained, we use it to translate a few English sentences into French and compute their BLEU scores.

```
engs = ['go .', 'i lost .', 'he\'s calm .', 'i\'m home .']
fras = ['va !', 'j\'ai perdu .', 'il est calme .', 'je suis chez moi .']
preds, _ = model.predict_step(
    data.build(engs, fras), d2l.try_gpu(), data.num_steps)
```

(continues on next page)

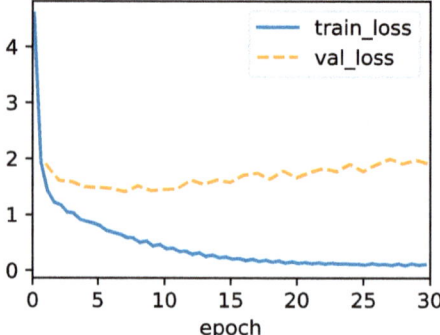

(continued from previous page)

```
for en, fr, p in zip(engs, fras, preds):
    translation = []
    for token in data.tgt_vocab.to_tokens(p):
        if token == '<eos>':
            break
        translation.append(token)
    print(f'{en} => {translation}, bleu,'
          f'{d2l.bleu(" ".join(translation), fr, k=2):.3f}')
```

```
go . => ['va', '!'], bleu,1.000
i lost . => ["j'ai", 'perdu', '.'], bleu,1.000
he's calm . => ['il', 'est', 'bien', '.'], bleu,0.658
i'm home . => ['je', 'suis', 'chez', 'moi', '.'], bleu,1.000
```

Let's visualize the attention weights when translating the last English sentence. We see that each query assigns non-uniform weights over key–value pairs. It shows that at each decoding step, different parts of the input sequences are selectively aggregated in the attention pooling.

```
_, dec_attention_weights = model.predict_step(
    data.build([engs[-1]], [fras[-1]]), d2l.try_gpu(), data.num_steps, True)
attention_weights = torch.cat(
    [step[0][0][0] for step in dec_attention_weights], 0)
attention_weights = attention_weights.reshape((1, 1, -1, data.num_steps))
```

```
# Plus one to include the end-of-sequence token
d2l.show_heatmaps(
    attention_weights[:, :, :, :len(engs[-1].split()) + 1].cpu(),
    xlabel='Key positions', ylabel='Query positions')
```

11.4.4 Summary

When predicting a token, if not all the input tokens are relevant, the RNN encoder–decoder with the Bahdanau attention mechanism selectively aggregates different parts of the input

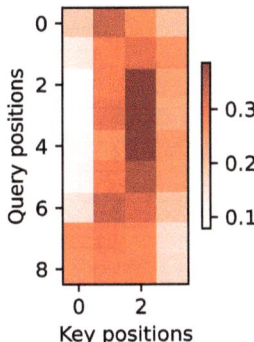

sequence. This is achieved by treating the state (context variable) as an output of additive attention pooling. In the RNN encoder–decoder, the Bahdanau attention mechanism treats the decoder hidden state at the previous time step as the query, and the encoder hidden states at all the time steps as both the keys and values.

11.4.5 Exercises

1. Replace GRU with LSTM in the experiment.

2. Modify the experiment to replace the additive attention scoring function with the scaled dot-product. How does it influence the training efficiency?

Discussions [159].

11.5 Multi-Head Attention

In practice, given the same set of queries, keys, and values we may want our model to combine knowledge from different behaviors of the same attention mechanism, such as capturing dependencies of various ranges (e.g., shorter-range vs. longer-range) within a sequence. Thus, it may be beneficial to allow our attention mechanism to jointly use different representation subspaces of queries, keys, and values.

To this end, instead of performing a single attention pooling, queries, keys, and values can be transformed with h independently learned linear projections. Then these h projected queries, keys, and values are fed into attention pooling in parallel. In the end, h attention-pooling outputs are concatenated and transformed with another learned linear projection to produce the final output. This design is called *multi-head attention*, where each of the h attention pooling outputs is a *head* (Vaswani *et al.*, 2017). Using fully connected layers to perform learnable linear transformations, Fig. 11.5.1 describes multi-head attention.

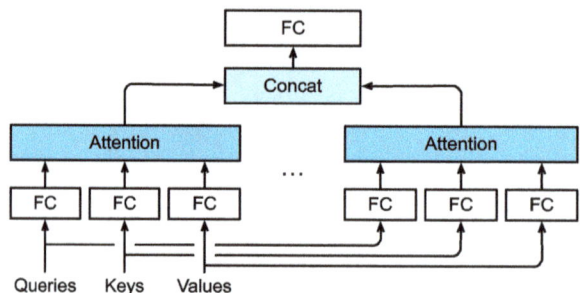

Fig. 11.5.1 Multi-head attention, where multiple heads are concatenated then linearly transformed.

```
import math
import torch
from torch import nn
from d2l import torch as d2l
```

11.5.1 Model

Before providing the implementation of multi-head attention, let's formalize this model mathematically. Given a query $\mathbf{q} \in \mathbb{R}^{d_q}$, a key $\mathbf{k} \in \mathbb{R}^{d_k}$, and a value $\mathbf{v} \in \mathbb{R}^{d_v}$, each attention head $\mathbf{h}_i$ ($i = 1, \ldots, h$) is computed as

$$\mathbf{h}_i = f(\mathbf{W}_i^{(q)}\mathbf{q}, \mathbf{W}_i^{(k)}\mathbf{k}, \mathbf{W}_i^{(v)}\mathbf{v}) \in \mathbb{R}^{p_v}, \tag{11.5.1}$$

where $\mathbf{W}_i^{(q)} \in \mathbb{R}^{p_q \times d_q}$, $\mathbf{W}_i^{(k)} \in \mathbb{R}^{p_k \times d_k}$, and $\mathbf{W}_i^{(v)} \in \mathbb{R}^{p_v \times d_v}$ are learnable parameters and f is attention pooling, such as additive attention and scaled dot product attention in Section 11.3. The multi-head attention output is another linear transformation via learnable parameters $\mathbf{W}_o \in \mathbb{R}^{p_o \times h p_v}$ of the concatenation of h heads:

$$\mathbf{W}_o \begin{bmatrix} \mathbf{h}_1 \\ \vdots \\ \mathbf{h}_h \end{bmatrix} \in \mathbb{R}^{p_o}. \tag{11.5.2}$$

Based on this design, each head may attend to different parts of the input. More sophisticated functions than the simple weighted average can be expressed.

11.5.2 Implementation

In our implementation, we choose the scaled dot product attention for each head of the multi-head attention. To avoid significant growth of computational cost and parametrization cost, we set $p_q = p_k = p_v = p_o/h$. Note that h heads can be computed in parallel if we set the number of outputs of linear transformations for the query, key, and value to $p_q h = p_k h = p_v h = p_o$. In the following implementation, p_o is specified via the argument num_hidden.

```
class MultiHeadAttention(d2l.Module):  #@save
    """Multi-head attention."""
    def __init__(self, num_hiddens, num_heads, dropout, bias=False, **kwargs):
        super().__init__()
        self.num_heads = num_heads
        self.attention = d2l.DotProductAttention(dropout)
        self.W_q = nn.LazyLinear(num_hiddens, bias=bias)
        self.W_k = nn.LazyLinear(num_hiddens, bias=bias)
        self.W_v = nn.LazyLinear(num_hiddens, bias=bias)
        self.W_o = nn.LazyLinear(num_hiddens, bias=bias)

    def forward(self, queries, keys, values, valid_lens):
        # Shape of queries, keys, or values:
        # (batch_size, no. of queries or key-value pairs, num_hiddens)
        # Shape of valid_lens: (batch_size,) or (batch_size, no. of queries)
        # After transposing, shape of output queries, keys, or values:
        # (batch_size * num_heads, no. of queries or key-value pairs,
        # num_hiddens / num_heads)
        queries = self.transpose_qkv(self.W_q(queries))
        keys = self.transpose_qkv(self.W_k(keys))
        values = self.transpose_qkv(self.W_v(values))

        if valid_lens is not None:
            # On axis 0, copy the first item (scalar or vector) for num_heads
            # times, then copy the next item, and so on
            valid_lens = torch.repeat_interleave(
                valid_lens, repeats=self.num_heads, dim=0)

        # Shape of output: (batch_size * num_heads, no. of queries,
        # num_hiddens / num_heads)
        output = self.attention(queries, keys, values, valid_lens)
        # Shape of output_concat: (batch_size, no. of queries, num_hiddens)
        output_concat = self.transpose_output(output)
        return self.W_o(output_concat)
```

To allow for parallel computation of multiple heads, the above `MultiHeadAttention` class uses two transposition methods as defined below. Specifically, the `transpose_output` method reverses the operation of the `transpose_qkv` method.

```
@d2l.add_to_class(MultiHeadAttention)  #@save
def transpose_qkv(self, X):
    """Transposition for parallel computation of multiple attention heads."""
    # Shape of input X: (batch_size, no. of queries or key-value pairs,
    # num_hiddens). Shape of output X: (batch_size, no. of queries or
    # key-value pairs, num_heads, num_hiddens / num_heads)
    X = X.reshape(X.shape[0], X.shape[1], self.num_heads, -1)
    # Shape of output X: (batch_size, num_heads, no. of queries or key-value
    # pairs, num_hiddens / num_heads)
    X = X.permute(0, 2, 1, 3)
    # Shape of output: (batch_size * num_heads, no. of queries or key-value
    # pairs, num_hiddens / num_heads)
    return X.reshape(-1, X.shape[2], X.shape[3])

@d2l.add_to_class(MultiHeadAttention)  #@save
```

(continues on next page)

(continued from previous page)

```
def transpose_output(self, X):
    """Reverse the operation of transpose_qkv."""
    X = X.reshape(-1, self.num_heads, X.shape[1], X.shape[2])
    X = X.permute(0, 2, 1, 3)
    return X.reshape(X.shape[0], X.shape[1], -1)
```

Let's test our implemented `MultiHeadAttention` class using a toy example where keys and values are the same. As a result, the shape of the multi-head attention output is (`batch_size`, `num_queries`, `num_hiddens`).

```
num_hiddens, num_heads = 100, 5
attention = MultiHeadAttention(num_hiddens, num_heads, 0.5)
batch_size, num_queries, num_kvpairs = 2, 4, 6
valid_lens = torch.tensor([3, 2])
X = torch.ones((batch_size, num_queries, num_hiddens))
Y = torch.ones((batch_size, num_kvpairs, num_hiddens))
d2l.check_shape(attention(X, Y, Y, valid_lens),
                (batch_size, num_queries, num_hiddens))
```

11.5.3 Summary

Multi-head attention combines knowledge of the same attention pooling via different representation subspaces of queries, keys, and values. To compute multiple heads of multi-head attention in parallel, proper tensor manipulation is needed.

11.5.4 Exercises

1. Visualize attention weights of multiple heads in this experiment.

2. Suppose that we have a trained model based on multi-head attention and we want to prune less important attention heads to increase the prediction speed. How can we design experiments to measure the importance of an attention head?

Discussions[160].

160

11.6 Self-Attention and Positional Encoding

In deep learning, we often use CNNs or RNNs to encode sequences. Now with attention mechanisms in mind, imagine feeding a sequence of tokens into an attention mechanism such that at every step, each token has its own query, keys, and values. Here, when computing the value of a token's representation at the next layer, the token can attend (via its query vector) to any other's token (matching based on their key vectors). Using the full set of query-key compatibility scores, we can compute, for each token, a representation by building the appropriate weighted sum over the other tokens. Because every token is attending to each other token (unlike the case where decoder steps attend to encoder steps), such architectures are typically described as *self-attention* models (Lin *et al.*, 2017, Vaswani *et al.*, 2017), and elsewhere described as *intra-attention* model (Cheng *et al.*, 2016, Parikh *et al.*, 2016, Paulus *et al.*, 2017). In this section, we will discuss sequence encoding using self-attention, including using additional information for the sequence order.

```
import math
import torch
from torch import nn
from d2l import torch as d2l
```

11.6.1 Self-Attention

Given a sequence of input tokens $\mathbf{x}_1, \ldots, \mathbf{x}_n$ where any $\mathbf{x}_i \in \mathbb{R}^d$ ($1 \le i \le n$), its self-attention outputs a sequence of the same length $\mathbf{y}_1, \ldots, \mathbf{y}_n$, where

$$\mathbf{y}_i = f(\mathbf{x}_i, (\mathbf{x}_1, \mathbf{x}_1), \ldots, (\mathbf{x}_n, \mathbf{x}_n)) \in \mathbb{R}^d \qquad (11.6.1)$$

according to the definition of attention pooling in (11.1.1). Using multi-head attention, the following code snippet computes the self-attention of a tensor with shape (batch size, number of time steps or sequence length in tokens, d). The output tensor has the same shape.

```
num_hiddens, num_heads = 100, 5
attention = d2l.MultiHeadAttention(num_hiddens, num_heads, 0.5)
batch_size, num_queries, valid_lens = 2, 4, torch.tensor([3, 2])
X = torch.ones((batch_size, num_queries, num_hiddens))
d2l.check_shape(attention(X, X, X, valid_lens),
                (batch_size, num_queries, num_hiddens))
```

11.6.2 Comparing CNNs, RNNs, and Self-Attention

Let's compare architectures for mapping a sequence of n tokens to another one of equal length, where each input or output token is represented by a d-dimensional vector. Specifically, we will consider CNNs, RNNs, and self-attention. We will compare their computational complexity, sequential operations, and maximum path lengths. Note that sequential operations prevent parallel computation, while a shorter path between any combination of sequence positions makes it easier to learn long-range dependencies within the sequence (Hochreiter *et al.*, 2001).

Let's regard any text sequence as a "one-dimensional image". Similarly, one-dimensional CNNs can process local features such as n-grams in text. Given a sequence of length n, consider a convolutional layer whose kernel size is k, and whose numbers of input and output channels are both d. The computational complexity of the convolutional layer is $O(knd^2)$. As Fig. 11.6.1 shows, CNNs are hierarchical, so there are $O(1)$ sequential operations and the maximum path length is $O(n/k)$. For example, $\mathbf{x}_1$ and $\mathbf{x}_5$ are within the receptive field of a two-layer CNN with kernel size 3 in Fig. 11.6.1.

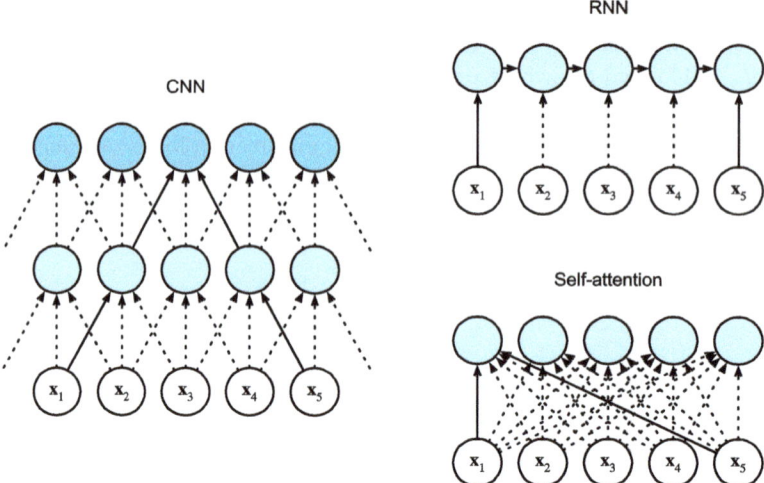

Fig. 11.6.1 Comparing CNN (padding tokens are omitted), RNN, and self-attention architectures.

When updating the hidden state of RNNs, multiplication of the $d \times d$ weight matrix and the d-dimensional hidden state has a computational complexity of $O(d^2)$. Since the sequence length is n, the computational complexity of the recurrent layer is $O(nd^2)$. According to Fig. 11.6.1, there are $O(n)$ sequential operations that cannot be parallelized and the maximum path length is also $O(n)$.

In self-attention, the queries, keys, and values are all $n \times d$ matrices. Consider the scaled dot product attention in (11.3.6), where an $n \times d$ matrix is multiplied by a $d \times n$ matrix, then the output $n \times n$ matrix is multiplied by an $n \times d$ matrix. As a result, the self-attention has a $O(n^2 d)$ computational complexity. As we can see from Fig. 11.6.1, each token is directly connected to any other token via self-attention. Therefore, computation can be parallel with $O(1)$ sequential operations and the maximum path length is also $O(1)$.

All in all, both CNNs and self-attention enjoy parallel computation and self-attention has the shortest maximum path length. However, the quadratic computational complexity with respect to the sequence length makes self-attention prohibitively slow for very long sequences.

11.6.3 Positional Encoding

Unlike RNNs, which recurrently process tokens of a sequence one-by-one, self-attention ditches sequential operations in favor of parallel computation. Note that self-attention by itself does not preserve the order of the sequence. What do we do if it really matters that the model knows in which order the input sequence arrived?

The dominant approach for preserving information about the order of tokens is to represent this to the model as an additional input associated with each token. These inputs are called *positional encodings*, and they can either be learned or fixed *a priori*. We now describe a simple scheme for fixed positional encodings based on sine and cosine functions (Vaswani *et al.*, 2017).

Suppose that the input representation $\mathbf{X} \in \mathbb{R}^{n \times d}$ contains the d-dimensional embeddings for n tokens of a sequence. The positional encoding outputs $\mathbf{X} + \mathbf{P}$ using a positional embedding matrix $\mathbf{P} \in \mathbb{R}^{n \times d}$ of the same shape, whose element on the ith row and the $(2j)^{\text{th}}$ or the $(2j + 1)^{\text{th}}$ column is

$$p_{i,2j} = \sin\left(\frac{i}{10000^{2j/d}}\right),$$

$$p_{i,2j+1} = \cos\left(\frac{i}{10000^{2j/d}}\right).$$

$$(11.6.2)$$

At first glance, this trigonometric function design looks weird. Before we give explanations of this design, let's first implement it in the following `PositionalEncoding` class.

```python
class PositionalEncoding(nn.Module):  #@save
    """Positional encoding."""
    def __init__(self, num_hiddens, dropout, max_len=1000):
        super().__init__()
        self.dropout = nn.Dropout(dropout)
        # Create a long enough P
        self.P = torch.zeros((1, max_len, num_hiddens))
        X = torch.arange(max_len, dtype=torch.float32).reshape(
            -1, 1) / torch.pow(10000, torch.arange(
            0, num_hiddens, 2, dtype=torch.float32) / num_hiddens)
        self.P[:, :, 0::2] = torch.sin(X)
        self.P[:, :, 1::2] = torch.cos(X)

    def forward(self, X):
        X = X + self.P[:, :X.shape[1], :].to(X.device)
        return self.dropout(X)
```

In the positional embedding matrix $\mathbf{P}$, rows correspond to positions within a sequence and columns represent different positional encoding dimensions. In the example below, we can see that the 6th and the 7th columns of the positional embedding matrix have a higher frequency than the 8th and the 9th columns. The offset between the 6th and the 7th (same for the 8th and the 9th) columns is due to the alternation of sine and cosine functions.

```python
encoding_dim, num_steps = 32, 60
pos_encoding = PositionalEncoding(encoding_dim, 0)
X = pos_encoding(torch.zeros((1, num_steps, encoding_dim)))
P = pos_encoding.P[:, :X.shape[1], :]
d2l.plot(torch.arange(num_steps), P[0, :, 6:10].T, xlabel='Row (position)',
         figsize=(6, 2.5), legend=["Col %d" % d for d in torch.arange(6, 10)])
```

Absolute Positional Information

To see how the monotonically decreased frequency along the encoding dimension relates to absolute positional information, let's print out the binary representations of $0, 1, \ldots, 7$.

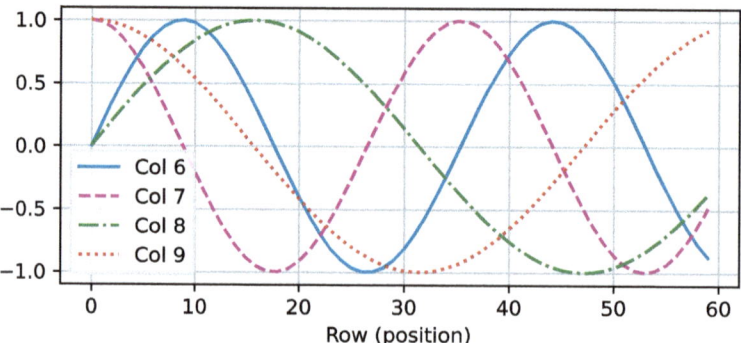

As we can see, the lowest bit, the second-lowest bit, and the third-lowest bit alternate on every number, every two numbers, and every four numbers, respectively.

```python
for i in range(8):
    print(f'{i} in binary is {i:>03b}')
```

```
0 in binary is 000
1 in binary is 001
2 in binary is 010
3 in binary is 011
4 in binary is 100
5 in binary is 101
6 in binary is 110
7 in binary is 111
```

In binary representations, a higher bit has a lower frequency than a lower bit. Similarly, as demonstrated in the heat map below, the positional encoding decreases frequencies along the encoding dimension by using trigonometric functions. Since the outputs are float numbers, such continuous representations are more space-efficient than binary representations.

```python
P = P[0, :, :].unsqueeze(0).unsqueeze(0)
d2l.show_heatmaps(P, xlabel='Column (encoding dimension)',
                  ylabel='Row (position)', figsize=(3.5, 4), cmap='Blues')
```

Relative Positional Information

Besides capturing absolute positional information, the above positional encoding also allows a model to easily learn to attend by relative positions. This is because for any fixed position offset δ, the positional encoding at position $i + \delta$ can be represented by a linear projection of that at position i.

This projection can be explained mathematically. Denoting $\omega_j = 1/10000^{2j/d}$, any pair of $(p_{i,2j}, p_{i,2j+1})$ in (11.6.2) can be linearly projected to $(p_{i+\delta,2j}, p_{i+\delta,2j+1})$ for any fixed

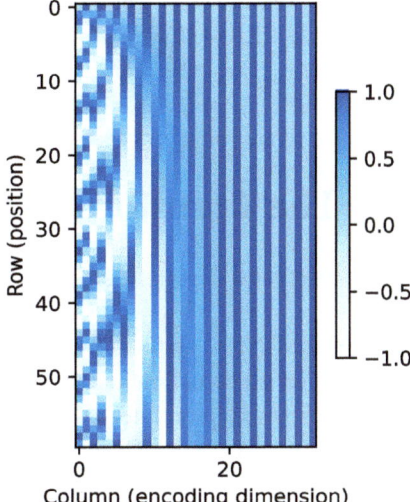

offset δ:

$$\begin{bmatrix} \cos(\delta\omega_j) & \sin(\delta\omega_j) \\ -\sin(\delta\omega_j) & \cos(\delta\omega_j) \end{bmatrix} \begin{bmatrix} p_{i,2j} \\ p_{i,2j+1} \end{bmatrix} = \begin{bmatrix} \cos(\delta\omega_j)\sin(i\omega_j) + \sin(\delta\omega_j)\cos(i\omega_j) \\ -\sin(\delta\omega_j)\sin(i\omega_j) + \cos(\delta\omega_j)\cos(i\omega_j) \end{bmatrix}$$

$$= \begin{bmatrix} \sin\big((i+\delta)\omega_j\big) \\ \cos\big((i+\delta)\omega_j\big) \end{bmatrix}$$

$$= \begin{bmatrix} p_{i+\delta,2j} \\ p_{i+\delta,2j+1} \end{bmatrix},$$

$$(11.6.3)$$

where the 2×2 projection matrix does not depend on any position index i.

11.6.4 Summary

In self-attention, the queries, keys, and values all come from the same place. Both CNNs and self-attention enjoy parallel computation and self-attention has the shortest maximum path length. However, the quadratic computational complexity with respect to the sequence length makes self-attention prohibitively slow for very long sequences. To use the sequence order information, we can inject absolute or relative positional information by adding positional encoding to the input representations.

11.6.5 Exercises

1. Suppose that we design a deep architecture to represent a sequence by stacking self-attention layers with positional encoding. What could the possible issues be?

2. Can you design a learnable positional encoding method?

3. Can we assign different learned embeddings according to different offsets between queries

and keys that are compared in self-attention? Hint: you may refer to relative position embeddings (Huang *et al.*, 2018, Shaw *et al.*, 2018).

Discussions[161].

161

11.7 The Transformer Architecture

We have compared CNNs, RNNs, and self-attention in Section 11.6.2. Notably, self-attention enjoys both parallel computation and the shortest maximum path length. Therefore, it is appealing to design deep architectures by using self-attention. Unlike earlier self-attention models that still rely on RNNs for input representations (Cheng *et al.*, 2016, Lin *et al.*, 2017, Paulus *et al.*, 2017), the Transformer model is solely based on attention mechanisms without any convolutional or recurrent layer (Vaswani *et al.*, 2017). Though originally proposed for sequence-to-sequence learning on text data, Transformers have been pervasive in a wide range of modern deep learning applications, such as in areas to do with language, vision, speech, and reinforcement learning.

```
import math
import pandas as pd
import torch
from torch import nn
from d2l import torch as d2l
```

11.7.1 Model

As an instance of the encoder–decoder architecture, the overall architecture of the Transformer is presented in Fig. 11.7.1. As we can see, the Transformer is composed of an encoder and a decoder. In contrast to Bahdanau attention for sequence-to-sequence learning in Fig. 11.4.2, the input (source) and output (target) sequence embeddings are added with positional encoding before being fed into the encoder and the decoder that stack modules based on self-attention.

Now we provide an overview of the Transformer architecture in Fig. 11.7.1. At a high level, the Transformer encoder is a stack of multiple identical layers, where each layer has two sublayers (either is denoted as sublayer). The first is a multi-head self-attention pooling and the second is a positionwise feed-forward network. Specifically, in the encoder self-attention, queries, keys, and values are all from the outputs of the previous encoder layer. Inspired by the ResNet design of Section 8.6, a residual connection is employed around both sublayers. In the Transformer, for any input $\mathbf{x} \in \mathbb{R}^d$ at any position of the sequence, we require that sublayer$(\mathbf{x}) \in \mathbb{R}^d$ so that the residual connection $\mathbf{x} + \text{sublayer}(\mathbf{x}) \in \mathbb{R}^d$ is feasible. This addition from the residual connection is immediately followed by layer normalization (Ba *et al.*, 2016). As a result, the Transformer encoder outputs a d-dimensional vector representation for each position of the input sequence.

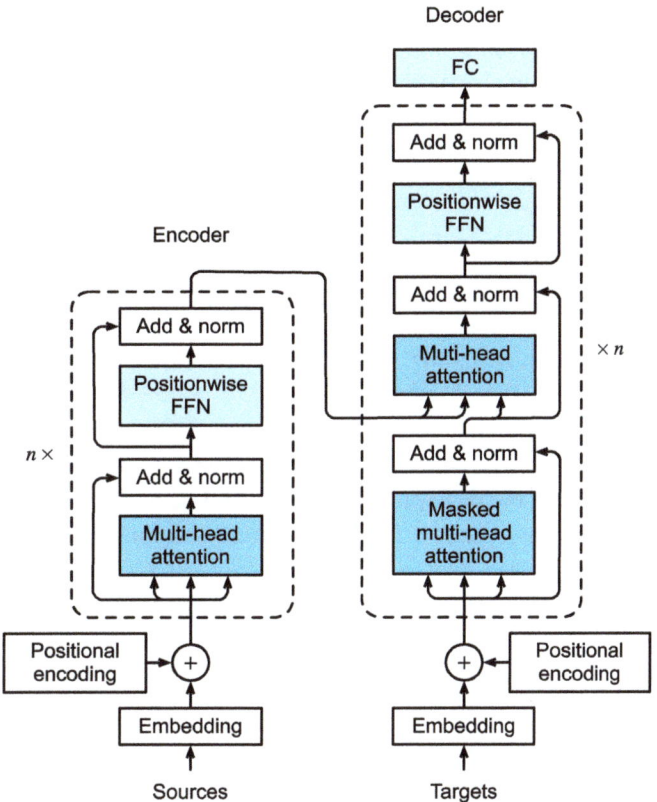

Fig. 11.7.1   The Transformer architecture.

The Transformer decoder is also a stack of multiple identical layers with residual connections and layer normalizations. As well as the two sublayers described in the encoder, the decoder inserts a third sublayer, known as the encoder–decoder attention, between these two. In the encoder–decoder attention, queries are from the outputs of the decoder's self-attention sublayer, and the keys and values are from the Transformer encoder outputs. In the decoder self-attention, queries, keys, and values are all from the outputs of the previous decoder layer. However, each position in the decoder is allowed only to attend to all positions in the decoder up to that position. This *masked* attention preserves the autoregressive property, ensuring that the prediction only depends on those output tokens that have been generated.

We have already described and implemented multi-head attention based on scaled dot products in Section 11.5 and positional encoding in Section 11.6.3. In the following, we will implement the rest of the Transformer model.

11.7.2 Positionwise Feed-Forward Networks

The positionwise feed-forward network transforms the representation at all the sequence positions using the same MLP. This is why we call it *positionwise*. In the implementation below, the input X with shape (batch size, number of time steps or sequence length in tokens, number of hidden units or feature dimension) will be transformed by a two-layer MLP into an output tensor of shape (batch size, number of time steps, ffn_num_outputs).

```python
class PositionWiseFFN(nn.Module):  #@save
    """The positionwise feed-forward network."""
    def __init__(self, ffn_num_hiddens, ffn_num_outputs):
        super().__init__()
        self.dense1 = nn.LazyLinear(ffn_num_hiddens)
        self.relu = nn.ReLU()
        self.dense2 = nn.LazyLinear(ffn_num_outputs)

    def forward(self, X):
        return self.dense2(self.relu(self.dense1(X)))
```

The following example shows that the innermost dimension of a tensor changes to the number of outputs in the positionwise feed-forward network. Since the same MLP transforms at all the positions, when the inputs at all these positions are the same, their outputs are also identical.

```python
ffn = PositionWiseFFN(4, 8)
ffn.eval()
ffn(torch.ones((2, 3, 4)))[0]
```

```
tensor([[ 0.2053,  0.2207,  0.3016, -0.2388,  0.2771,  0.1132,  0.0235,  0.
→0308],
        [ 0.2053,  0.2207,  0.3016, -0.2388,  0.2771,  0.1132,  0.0235,  0.
→0308],
        [ 0.2053,  0.2207,  0.3016, -0.2388,  0.2771,  0.1132,  0.0235,  0.
→0308]],
       grad_fn=<SelectBackward0>)
```

11.7.3 Residual Connection and Layer Normalization

Now let's focus on the "add & norm" component in Fig. 11.7.1. As we described at the beginning of this section, this is a residual connection immediately followed by layer normalization. Both are key to effective deep architectures.

In Section 8.5, we explained how batch normalization recenters and rescales across the examples within a minibatch. As discussed in Section 8.5.2, layer normalization is the same as batch normalization except that the former normalizes across the feature dimension, thus enjoying benefits of scale independence and batch size independence. Despite its pervasive applications in computer vision, batch normalization is usually empirically less effective than layer normalization in natural language processing tasks, where the inputs are often variable-length sequences.

The following code snippet compares the normalization across different dimensions by layer normalization and batch normalization.

```
ln = nn.LayerNorm(2)
bn = nn.LazyBatchNorm1d()
X = torch.tensor([[1, 2], [2, 3]], dtype=torch.float32)
# Compute mean and variance from X in the training mode
print('layer norm:', ln(X), '\nbatch norm:', bn(X))
```

```
layer norm: tensor([[-1.0000,  1.0000],
        [-1.0000,  1.0000]], grad_fn=<NativeLayerNormBackward0>)
batch norm: tensor([[-1.0000, -1.0000],
        [ 1.0000,  1.0000]], grad_fn=<NativeBatchNormBackward0>)
```

Now we can implement the AddNorm class using a residual connection followed by layer normalization. Dropout is also applied for regularization.

```
class AddNorm(nn.Module):  #@save
    """The residual connection followed by layer normalization."""
    def __init__(self, norm_shape, dropout):
        super().__init__()
        self.dropout = nn.Dropout(dropout)
        self.ln = nn.LayerNorm(norm_shape)

    def forward(self, X, Y):
        return self.ln(self.dropout(Y) + X)
```

The residual connection requires that the two inputs are of the same shape so that the output tensor also has the same shape after the addition operation.

```
add_norm = AddNorm(4, 0.5)
shape = (2, 3, 4)
d2l.check_shape(add_norm(torch.ones(shape), torch.ones(shape)), shape)
```

11.7.4 Encoder

With all the essential components to assemble the Transformer encoder, let's start by implementing a single layer within the encoder. The following TransformerEncoderBlock class contains two sublayers: multi-head self-attention and positionwise feed-forward networks, where a residual connection followed by layer normalization is employed around both sublayers.

```
class TransformerEncoderBlock(nn.Module):  #@save
    """The Transformer encoder block."""
    def __init__(self, num_hiddens, ffn_num_hiddens, num_heads, dropout,
                 use_bias=False):
        super().__init__()
        self.attention = d2l.MultiHeadAttention(num_hiddens, num_heads,
                                                dropout, use_bias)
        self.addnorm1 = AddNorm(num_hiddens, dropout)
        self.ffn = PositionWiseFFN(ffn_num_hiddens, num_hiddens)
        self.addnorm2 = AddNorm(num_hiddens, dropout)
```

(continues on next page)

(continued from previous page)

```
def forward(self, X, valid_lens):
    Y = self.addnorm1(X, self.attention(X, X, X, valid_lens))
    return self.addnorm2(Y, self.ffn(Y))
```

As we can see, no layer in the Transformer encoder changes the shape of its input.

```
X = torch.ones((2, 100, 24))
valid_lens = torch.tensor([3, 2])
encoder_blk = TransformerEncoderBlock(24, 48, 8, 0.5)
encoder_blk.eval()
d2l.check_shape(encoder_blk(X, valid_lens), X.shape)
```

In the following Transformer encoder implementation, we stack num_blks instances of the above TransformerEncoderBlock classes. Since we use the fixed positional encoding whose values are always between −1 and 1, we multiply values of the learnable input embeddings by the square root of the embedding dimension to rescale before summing up the input embedding and the positional encoding.

```
class TransformerEncoder(d2l.Encoder):  #@save
    """The Transformer encoder."""
    def __init__(self, vocab_size, num_hiddens, ffn_num_hiddens,
                 num_heads, num_blks, dropout, use_bias=False):
        super().__init__()
        self.num_hiddens = num_hiddens
        self.embedding = nn.Embedding(vocab_size, num_hiddens)
        self.pos_encoding = d2l.PositionalEncoding(num_hiddens, dropout)
        self.blks = nn.Sequential()
        for i in range(num_blks):
            self.blks.add_module("block"+str(i), TransformerEncoderBlock(
                num_hiddens, ffn_num_hiddens, num_heads, dropout, use_bias))

    def forward(self, X, valid_lens):
        # Since positional encoding values are between -1 and 1, the embedding
        # values are multiplied by the square root of the embedding dimension
        # to rescale before they are summed up
        X = self.pos_encoding(self.embedding(X) * math.sqrt(self.num_hiddens))
        self.attention_weights = [None] * len(self.blks)
        for i, blk in enumerate(self.blks):
            X = blk(X, valid_lens)
            self.attention_weights[
                i] = blk.attention.attention.attention_weights
        return X
```

Below we specify hyperparameters to create a two-layer Transformer encoder. The shape of the Transformer encoder output is (batch size, number of time steps, num_hiddens).

```
encoder = TransformerEncoder(200, 24, 48, 8, 2, 0.5)
d2l.check_shape(encoder(torch.ones((2, 100), dtype=torch.long), valid_lens),
                (2, 100, 24))
```

11.7.5 Decoder

As shown in Fig. 11.7.1, the Transformer decoder is composed of multiple identical layers. Each layer is implemented in the following `TransformerDecoderBlock` class, which contains three sublayers: decoder self-attention, encoder–decoder attention, and position-wise feed-forward networks. These sublayers employ a residual connection around them followed by layer normalization.

As we described earlier in this section, in the masked multi-head decoder self-attention (the first sublayer), queries, keys, and values all come from the outputs of the previous decoder layer. When training sequence-to-sequence models, tokens at all the positions (time steps) of the output sequence are known. However, during prediction the output sequence is generated token by token; thus, at any decoder time step only the generated tokens can be used in the decoder self-attention. To preserve autoregression in the decoder, its masked self-attention specifies `dec_valid_lens` so that any query only attends to all positions in the decoder up to the query position.

```python
class TransformerDecoderBlock(nn.Module):
    # The i-th block in the Transformer decoder
    def __init__(self, num_hiddens, ffn_num_hiddens, num_heads, dropout, i):
        super().__init__()
        self.i = i
        self.attention1 = d2l.MultiHeadAttention(num_hiddens, num_heads,
                                                 dropout)
        self.addnorm1 = AddNorm(num_hiddens, dropout)
        self.attention2 = d2l.MultiHeadAttention(num_hiddens, num_heads,
                                                 dropout)
        self.addnorm2 = AddNorm(num_hiddens, dropout)
        self.ffn = PositionWiseFFN(ffn_num_hiddens, num_hiddens)
        self.addnorm3 = AddNorm(num_hiddens, dropout)

    def forward(self, X, state):
        enc_outputs, enc_valid_lens = state[0], state[1]
        # During training, all the tokens of any output sequence are processed
        # at the same time, so state[2][self.i] is None as initialized. When
        # decoding any output sequence token by token during prediction,
        # state[2][self.i] contains representations of the decoded output at
        # the i-th block up to the current time step
        if state[2][self.i] is None:
            key_values = X
        else:
            key_values = torch.cat((state[2][self.i], X), dim=1)
        state[2][self.i] = key_values
        if self.training:
            batch_size, num_steps, _ = X.shape
            # Shape of dec_valid_lens: (batch_size, num_steps), where every
            # row is [1, 2, ..., num_steps]
            dec_valid_lens = torch.arange(
                1, num_steps + 1, device=X.device).repeat(batch_size, 1)
        else:
            dec_valid_lens = None
        # Self-attention
        X2 = self.attention1(X, key_values, key_values, dec_valid_lens)
```

(continues on next page)

(continued from previous page)

```
    Y = self.addnorm1(X, X2)
    # Encoder-decoder attention. Shape of enc_outputs:
    # (batch_size, num_steps, num_hiddens)
    Y2 = self.attention2(Y, enc_outputs, enc_outputs, enc_valid_lens)
    Z = self.addnorm2(Y, Y2)
    return self.addnorm3(Z, self.ffn(Z)), state
```

To facilitate scaled dot product operations in the encoder–decoder attention and addition operations in the residual connections, the feature dimension (num_hiddens) of the decoder is the same as that of the encoder.

```
decoder_blk = TransformerDecoderBlock(24, 48, 8, 0.5, 0)
X = torch.ones((2, 100, 24))
state = [encoder_blk(X, valid_lens), valid_lens, [None]]
d2l.check_shape(decoder_blk(X, state)[0], X.shape)
```

Now we construct the entire Transformer decoder composed of num_blks instances of TransformerDecoderBlock. In the end, a fully connected layer computes the prediction for all the vocab_size possible output tokens. Both of the decoder self-attention weights and the encoder–decoder attention weights are stored for later visualization.

```
class TransformerDecoder(d2l.AttentionDecoder):
    def __init__(self, vocab_size, num_hiddens, ffn_num_hiddens, num_heads,
                 num_blks, dropout):
        super().__init__()
        self.num_hiddens = num_hiddens
        self.num_blks = num_blks
        self.embedding = nn.Embedding(vocab_size, num_hiddens)
        self.pos_encoding = d2l.PositionalEncoding(num_hiddens, dropout)
        self.blks = nn.Sequential()
        for i in range(num_blks):
            self.blks.add_module("block"+str(i), TransformerDecoderBlock(
                num_hiddens, ffn_num_hiddens, num_heads, dropout, i))
        self.dense = nn.LazyLinear(vocab_size)

    def init_state(self, enc_outputs, enc_valid_lens):
        return [enc_outputs, enc_valid_lens, [None] * self.num_blks]

    def forward(self, X, state):
        X = self.pos_encoding(self.embedding(X) * math.sqrt(self.num_hiddens))
        self._attention_weights = [[None] * len(self.blks) for _ in range (2)]
        for i, blk in enumerate(self.blks):
            X, state = blk(X, state)
            # Decoder self-attention weights
            self._attention_weights[0][
                i] = blk.attention1.attention.attention_weights
            # Encoder-decoder attention weights
            self._attention_weights[1][
                i] = blk.attention2.attention.attention_weights
        return self.dense(X), state
```

(continues on next page)

(continued from previous page)

```python
@property
def attention_weights(self):
    return self._attention_weights
```

11.7.6 Training

Let's instantiate an encoder–decoder model by following the Transformer architecture. Here we specify that both the Transformer encoder and the Transformer decoder have two layers using 4-head attention. As in Section 10.7.6, we train the Transformer model for sequence-to-sequence learning on the English–French machine translation dataset.

```python
data = d2l.MTFraEng(batch_size=128)
num_hiddens, num_blks, dropout = 256, 2, 0.2
ffn_num_hiddens, num_heads = 64, 4
encoder = TransformerEncoder(
    len(data.src_vocab), num_hiddens, ffn_num_hiddens, num_heads,
    num_blks, dropout)
decoder = TransformerDecoder(
    len(data.tgt_vocab), num_hiddens, ffn_num_hiddens, num_heads,
    num_blks, dropout)
model = d2l.Seq2Seq(encoder, decoder, tgt_pad=data.tgt_vocab['<pad>'],
                    lr=0.001)
trainer = d2l.Trainer(max_epochs=30, gradient_clip_val=1, num_gpus=1)
trainer.fit(model, data)
```

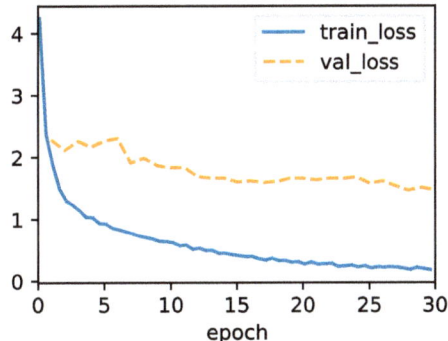

After training, we use the Transformer model to translate a few English sentences into French and compute their BLEU scores.

```python
engs = ['go .', 'i lost .', 'he\'s calm .', 'i\'m home .']
fras = ['va !', 'j\'ai perdu .', 'il est calme .', 'je suis chez moi .']
preds, _ = model.predict_step(
    data.build(engs, fras), d2l.try_gpu(), data.num_steps)
for en, fr, p in zip(engs, fras, preds):
    translation = []
    for token in data.tgt_vocab.to_tokens(p):
```

(continues on next page)

(continued from previous page)

```
        if token == '<eos>':
            break
        translation.append(token)
    print(f'{en} => {translation}, bleu,'
          f'{d21.bleu(" ".join(translation), fr, k=2):.3f}')
```

```
go . => ['va', '!'], bleu,1.000
i lost . => ["j'ai", 'perdu', '.'], bleu,1.000
he's calm . => ['il', 'est', 'mouillé', '.'], bleu,0.658
i'm home . => ['je', 'suis', 'chez', 'moi', '.'], bleu,1.000
```

Let's visualize the Transformer attention weights when translating the final English sentence into French. The shape of the encoder self-attention weights is (number of encoder layers, number of attention heads, num_steps or number of queries, num_steps or number of key-value pairs).

```
_, dec_attention_weights = model.predict_step(
    data.build([engs[-1]], [fras[-1]]), d21.try_gpu(), data.num_steps, True)
enc_attention_weights = torch.cat(model.encoder.attention_weights, 0)
shape = (num_blks, num_heads, -1, data.num_steps)
enc_attention_weights = enc_attention_weights.reshape(shape)
d21.check_shape(enc_attention_weights,
                (num_blks, num_heads, data.num_steps, data.num_steps))
```

In the encoder self-attention, both queries and keys come from the same input sequence. Since padding tokens do not carry meaning, with specified valid length of the input sequence no query attends to positions of padding tokens. In the following, two layers of multi-head attention weights are presented row by row. Each head independently attends based on a separate representation subspace of queries, keys, and values.

```
d21.show_heatmaps(
    enc_attention_weights.cpu(), xlabel='Key positions',
    ylabel='Query positions', titles=['Head %d' % i for i in range(1, 5)],
    figsize=(7, 3.5))
```

To visualize the decoder self-attention weights and the encoder–decoder attention weights, we need more data manipulations. For example, we fill the masked attention weights with zero. Note that the decoder self-attention weights and the encoder–decoder attention weights both have the same queries: the beginning-of-sequence token followed by the output tokens and possibly end-of-sequence tokens.

```
dec_attention_weights_2d = [head[0].tolist()
                            for step in dec_attention_weights
                            for attn in step for blk in attn for head in blk]
dec_attention_weights_filled = torch.tensor(
    pd.DataFrame(dec_attention_weights_2d).fillna(0.0).values)
shape = (-1, 2, num_blks, num_heads, data.num_steps)
```

(continues on next page)

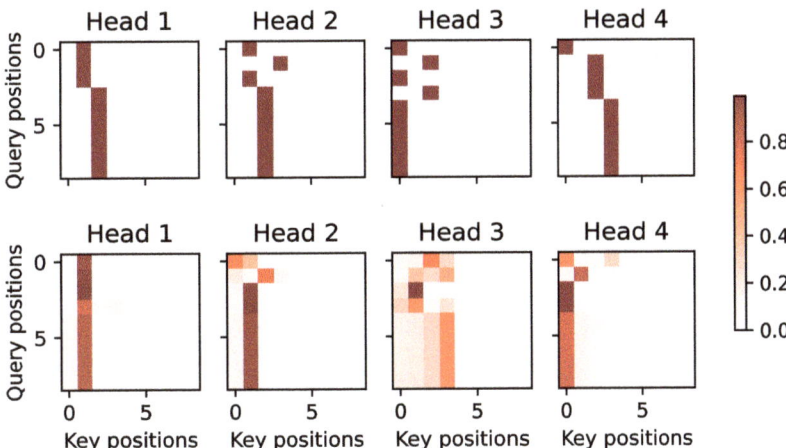

(continued from previous page)

```
dec_attention_weights = dec_attention_weights_filled.reshape(shape)
dec_self_attention_weights, dec_inter_attention_weights = \
    dec_attention_weights.permute(1, 2, 3, 0, 4)
```

```
d2l.check_shape(dec_self_attention_weights,
                (num_blks, num_heads, data.num_steps, data.num_steps))
d2l.check_shape(dec_inter_attention_weights,
                (num_blks, num_heads, data.num_steps, data.num_steps))
```

Because of the autoregressive property of the decoder self-attention, no query attends to key–value pairs after the query position.

```
d2l.show_heatmaps(
    dec_self_attention_weights[:, :, :, :],
    xlabel='Key positions', ylabel='Query positions',
    titles=['Head %d' % i for i in range(1, 5)], figsize=(7, 3.5))
```

Similar to the case in the encoder self-attention, via the specified valid length of the input sequence, no query from the output sequence attends to those padding tokens from the input sequence.

```
d2l.show_heatmaps(
    dec_inter_attention_weights, xlabel='Key positions',
    ylabel='Query positions', titles=['Head %d' % i for i in range(1, 5)],
    figsize=(7, 3.5))
```

Although the Transformer architecture was originally proposed for sequence-to-sequence learning, as we will discover later in the book, either the Transformer encoder or the Transformer decoder is often individually used for different deep learning tasks.

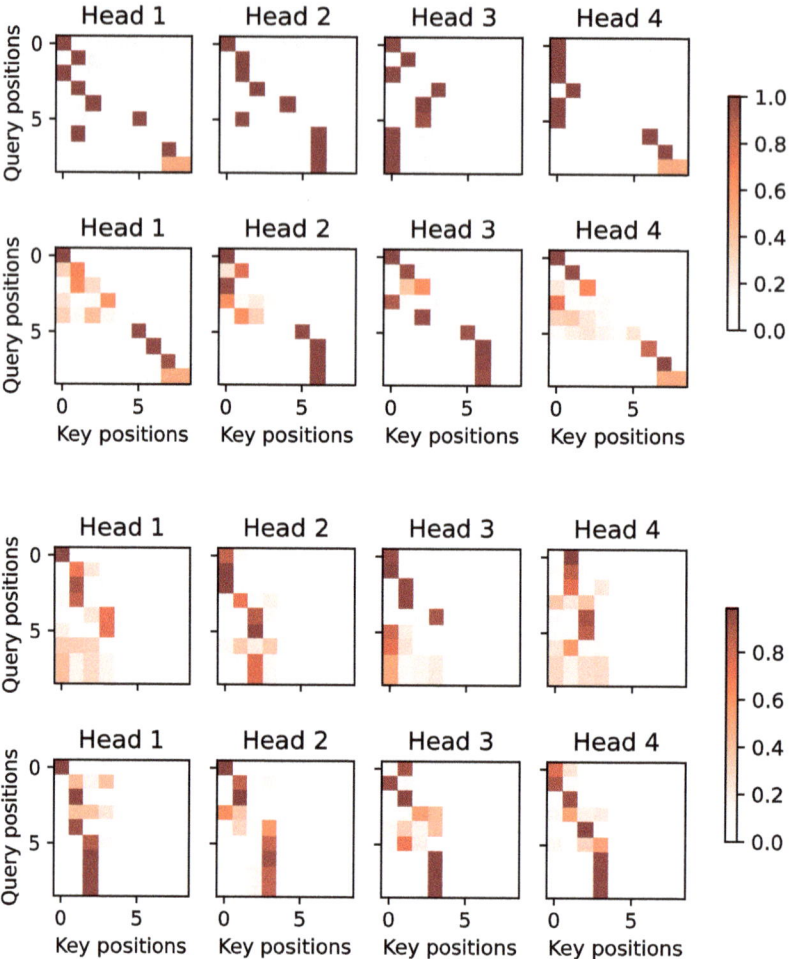

11.7.7 Summary

The Transformer is an instance of the encoder–decoder architecture, though either the encoder or the decoder can be used individually in practice. In the Transformer architecture, multi-head self-attention is used for representing the input sequence and the output sequence, though the decoder has to preserve the autoregressive property via a masked version. Both the residual connections and the layer normalization in the Transformer are important for training a very deep model. The positionwise feed-forward network in the Transformer model transforms the representation at all the sequence positions using the same MLP.

11.7.8 Exercises

1. Train a deeper Transformer in the experiments. How does it affect the training speed and the translation performance?

2. Is it a good idea to replace scaled dot product attention with additive attention in the Transformer? Why?

3. For language modeling, should we use the Transformer encoder, decoder, or both? How would you design this method?

4. What challenges can Transformers face if input sequences are very long? Why?

5. How would you improve the computational and memory efficiency of Transformers? Hint: you may refer to the survey paper by Tay *et al.* (2020).

162

Discussions[162].

11.8 Transformers for Vision

The Transformer architecture was initially proposed for sequence-to-sequence learning, with a focus on machine translation. Subsequently, Transformers emerged as the model of choice in various natural language processing tasks (Brown *et al.*, 2020, Devlin *et al.*, 2018, Radford *et al.*, 2018, Radford *et al.*, 2019, Raffel *et al.*, 2020). However, in the field of computer vision the dominant architecture has remained the CNN (Chapter 8). Naturally, researchers started to wonder if it might be possible to do better by adapting Transformer models to image data. This question sparked immense interest in the computer vision community. Recently, Ramachandran *et al.* (2019) proposed a scheme for replacing convolution with self-attention. However, its use of specialized patterns in attention makes it hard to scale up models on hardware accelerators. Then, Cordonnier *et al.* (2020) theoretically proved that self-attention can learn to behave similarly to convolution. Empirically, 2×2 patches were taken from images as inputs, but the small patch size makes the model only applicable to image data with low resolutions.

Without specific constraints on patch size, *vision Transformers* (ViTs) extract patches from images and feed them into a Transformer encoder to obtain a global representation, which will finally be transformed for classification (Dosovitskiy *et al.*, 2021). Notably, Transformers show better scalability than CNNs: and when training larger models on larger datasets, vision Transformers outperform ResNets by a significant margin. Similar to the landscape of network architecture design in natural language processing, Transformers have also become a game-changer in computer vision.

```
import torch
from torch import nn
from d2l import torch as d2l
```

11.8.1 Model

Fig. 11.8.1 depicts the model architecture of vision Transformers. This architecture consists of a stem that patchifies images, a body based on the multilayer Transformer encoder, and a head that transforms the global representation into the output label.

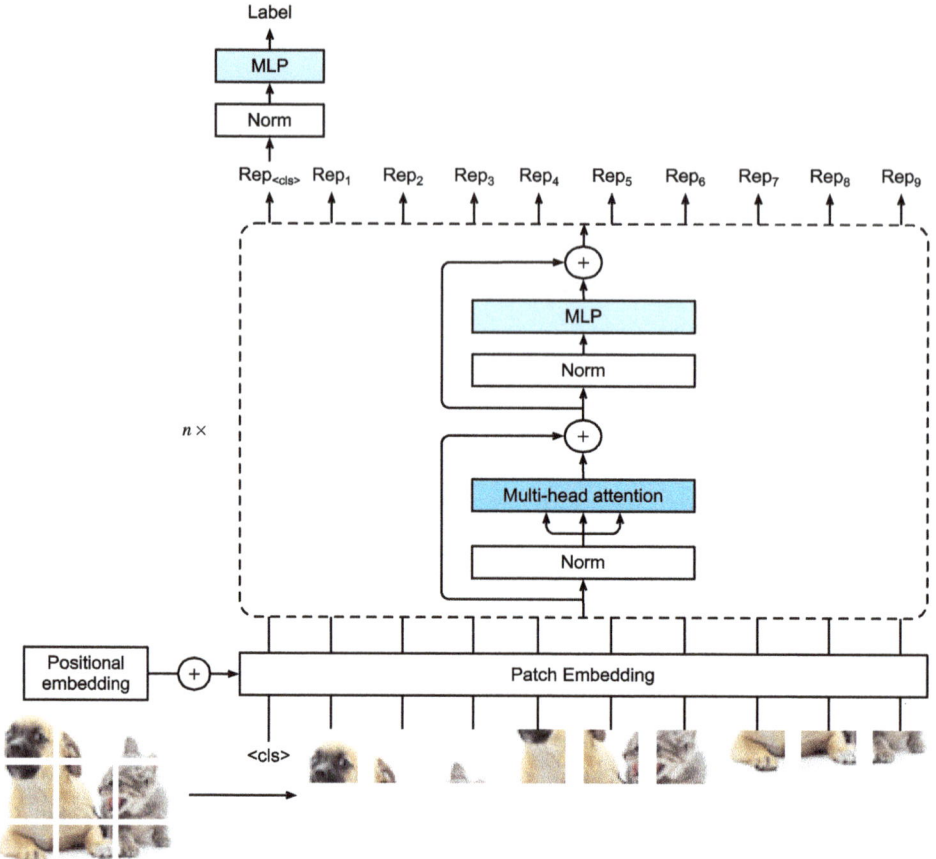

Fig. 11.8.1   The vision Transformer architecture. In this example, an image is split into nine patches. A special "<cls>" token and the nine flattened image patches are transformed via patch embedding and n Transformer encoder blocks into ten representations, respectively. The "<cls>" representation is further transformed into the output label.

Consider an input image with height h, width w, and c channels. Specifying the patch height and width both as p, the image is split into a sequence of $m = hw/p^2$ patches, where each patch is flattened to a vector of length cp^2. In this way, image patches can be treated similarly to tokens in text sequences by Transformer encoders. A special "<cls>" (class) token and the m flattened image patches are linearly projected into a sequence of $m+1$ vectors, summed with learnable positional embeddings. The multilayer Transformer encoder transforms $m+1$ input vectors into the same number of output vector representations of the same length. It works exactly the same way as the original Transformer encoder in Fig. 11.7.1, only differing in the position of normalization. Since the "<cls>" token attends to all the image patches via self-attention (see Fig. 11.6.1), its representation from the Transformer encoder output will be further transformed into the output label.

11.8.2 Patch Embedding

To implement a vision Transformer, let's start with patch embedding in Fig. 11.8.1. Splitting an image into patches and linearly projecting these flattened patches can be simplified as a single convolution operation, where both the kernel size and the stride size are set to the patch size.

```python
class PatchEmbedding(nn.Module):
    def __init__(self, img_size=96, patch_size=16, num_hiddens=512):
        super().__init__()
        def _make_tuple(x):
            if not isinstance(x, (list, tuple)):
                return (x, x)
            return x
        img_size, patch_size = _make_tuple(img_size), _make_tuple(patch_size)
        self.num_patches = (img_size[0] // patch_size[0]) * (
            img_size[1] // patch_size[1])
        self.conv = nn.LazyConv2d(num_hiddens, kernel_size=patch_size,
                                  stride=patch_size)

    def forward(self, X):
        # Output shape: (batch size, no. of patches, no. of channels)
        return self.conv(X).flatten(2).transpose(1, 2)
```

In the following example, taking images with height and width of `img_size` as inputs, the patch embedding outputs `(img_size//patch_size)**2` patches that are linearly projected to vectors of length `num_hiddens`.

```python
img_size, patch_size, num_hiddens, batch_size = 96, 16, 512, 4
patch_emb = PatchEmbedding(img_size, patch_size, num_hiddens)
X = torch.zeros(batch_size, 3, img_size, img_size)
d2l.check_shape(patch_emb(X),
                (batch_size, (img_size//patch_size)**2, num_hiddens))
```

11.8.3 Vision Transformer Encoder

The MLP of the vision Transformer encoder is slightly different from the positionwise FFN of the original Transformer encoder (see Section 11.7.2). First, here the activation function uses the Gaussian error linear unit (GELU), which can be considered as a smoother version of the ReLU (Hendrycks and Gimpel, 2016). Second, dropout is applied to the output of each fully connected layer in the MLP for regularization.

```python
class ViTMLP(nn.Module):
    def __init__(self, mlp_num_hiddens, mlp_num_outputs, dropout=0.5):
        super().__init__()
        self.dense1 = nn.LazyLinear(mlp_num_hiddens)
        self.gelu = nn.GELU()
        self.dropout1 = nn.Dropout(dropout)
        self.dense2 = nn.LazyLinear(mlp_num_outputs)
        self.dropout2 = nn.Dropout(dropout)
```

(continues on next page)

(continued from previous page)

```
def forward(self, x):
    return self.dropout2(self.dense2(self.dropout1(self.gelu(
        self.dense1(x)))))
```

The vision Transformer encoder block implementation just follows the pre-normalization design in Fig. 11.8.1, where normalization is applied right *before* multi-head attention or the MLP. In contrast to post-normalization ("add & norm" in Fig. 11.7.1), where normalization is placed right *after* residual connections, pre-normalization leads to more effective or efficient training for Transformers (Baevski and Auli, 2018, Wang *et al.*, 2019, Xiong *et al.*, 2020).

```
class ViTBlock(nn.Module):
    def __init__(self, num_hiddens, norm_shape, mlp_num_hiddens,
                 num_heads, dropout, use_bias=False):
        super().__init__()
        self.ln1 = nn.LayerNorm(norm_shape)
        self.attention = d2l.MultiHeadAttention(num_hiddens, num_heads,
                                                dropout, use_bias)
        self.ln2 = nn.LayerNorm(norm_shape)
        self.mlp = ViTMLP(mlp_num_hiddens, num_hiddens, dropout)

    def forward(self, X, valid_lens=None):
        X = X + self.attention(*([self.ln1(X)] * 3), valid_lens)
        return X + self.mlp(self.ln2(X))
```

Just as in Section 11.7.4, no vision Transformer encoder block changes its input shape.

```
X = torch.ones((2, 100, 24))
encoder_blk = ViTBlock(24, 24, 48, 8, 0.5)
encoder_blk.eval()
d2l.check_shape(encoder_blk(X), X.shape)
```

11.8.4 Putting It All Together

The forward pass of vision Transformers below is straightforward. First, input images are fed into an PatchEmbedding instance, whose output is concatenated with the "<cls>" token embedding. They are summed with learnable positional embeddings before dropout. Then the output is fed into the Transformer encoder that stacks num_blks instances of the ViT-Block class. Finally, the representation of the "<cls>" token is projected by the network head.

```
class ViT(d2l.Classifier):
    """Vision Transformer."""
    def __init__(self, img_size, patch_size, num_hiddens, mlp_num_hiddens,
                 num_heads, num_blks, emb_dropout, blk_dropout, lr=0.1,
                 use_bias=False, num_classes=10):
        super().__init__()
        self.save_hyperparameters()
```

(continues on next page)

(continued from previous page)

```
    self.patch_embedding = PatchEmbedding(
        img_size, patch_size, num_hiddens)
    self.cls_token = nn.Parameter(torch.zeros(1, 1, num_hiddens))
    num_steps = self.patch_embedding.num_patches + 1  # Add the cls token
    # Positional embeddings are learnable
    self.pos_embedding = nn.Parameter(
        torch.randn(1, num_steps, num_hiddens))
    self.dropout = nn.Dropout(emb_dropout)
    self.blks = nn.Sequential()
    for i in range(num_blks):
        self.blks.add_module(f"{i}", ViTBlock(
            num_hiddens, num_hiddens, mlp_num_hiddens,
            num_heads, blk_dropout, use_bias))
    self.head = nn.Sequential(nn.LayerNorm(num_hiddens),
                              nn.Linear(num_hiddens, num_classes))

def forward(self, X):
    X = self.patch_embedding(X)
    X = torch.cat((self.cls_token.expand(X.shape[0], -1, -1), X), 1)
    X = self.dropout(X + self.pos_embedding)
    for blk in self.blks:
        X = blk(X)
    return self.head(X[:, 0])
```

11.8.5 Training

Training a vision Transformer on the Fashion-MNIST dataset is just like how CNNs were trained in Chapter 8.

```
img_size, patch_size = 96, 16
num_hiddens, mlp_num_hiddens, num_heads, num_blks = 512, 2048, 8, 2
emb_dropout, blk_dropout, lr = 0.1, 0.1, 0.1
model = ViT(img_size, patch_size, num_hiddens, mlp_num_hiddens, num_heads,
            num_blks, emb_dropout, blk_dropout, lr)
trainer = d2l.Trainer(max_epochs=10, num_gpus=1)
data = d2l.FashionMNIST(batch_size=128, resize=(img_size, img_size))
trainer.fit(model, data)
```

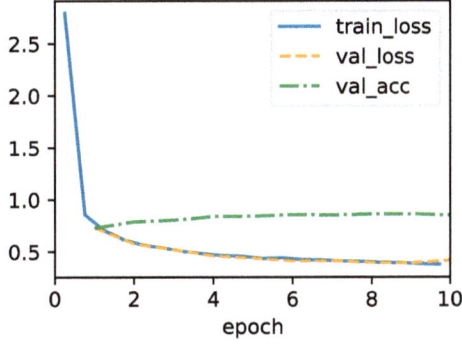

11.8.6 Summary and Discussion

You may have noticed that for small datasets like Fashion-MNIST, our implemented vision Transformer does not outperform the ResNet in Section 8.6. Similar observations can be made even on the ImageNet dataset (1.2 million images). This is because Transformers *lack* those useful principles in convolution, such as translation invariance and locality (Section 7.1). However, the picture changes when training larger models on larger datasets (e.g., 300 million images), where vision Transformers outperform ResNets by a large margin in image classification, demonstrating intrinsic superiority of Transformers in scalability (Dosovitskiy *et al.*, 2021). The introduction of vision Transformers has changed the landscape of network design for modeling image data. They were soon shown to be effective on the ImageNet dataset with data-efficient training strategies of DeiT (Touvron *et al.*, 2021). However, the quadratic complexity of self-attention (Section 11.6) makes the Transformer architecture less suitable for higher-resolution images. Towards a general-purpose backbone network in computer vision, Swin Transformers addressed the quadratic computational complexity with respect to image size (Section 11.6.2) and reinstated convolution-like priors, extending the applicability of Transformers to a range of computer vision tasks beyond image classification with state-of-the-art results (Liu *et al.*, 2021).

11.8.7 Exercises

1. How does the value of `img_size` affect training time?

2. Instead of projecting the "<cls>" token representation to the output, how would you project the averaged patch representations? Implement this change and see how it affects the accuracy.

3. Can you modify hyperparameters to improve the accuracy of the vision Transformer?

Discussions[163].

163

11.9 Large-Scale Pretraining with Transformers

So far in our image classification and machine translation experiments, models have been trained on datasets with input–output examples *from scratch* to perform specific tasks. For example, a Transformer was trained with English–French pairs (Section 11.7) so that this model can translate input English text into French. As a result, each model becomes a *specific expert* that is sensitive to even a slight shift in data distribution (Section 4.7). For better generalized models, or even more competent *generalists* that can perform multiple tasks with or without adaptation, *pretraining* models on large data has been increasingly common.

Given larger data for pretraining, the Transformer architecture performs better with an increased model size and training compute, demonstrating superior *scaling* behavior. Specifically, performance of Transformer-based language models scales as a power law with the amount of model parameters, training tokens, and training compute (Kaplan *et al.*, 2020). The scalability of Transformers is also evidenced by the significantly boosted performance from larger vision Transformers trained on larger data (discussed in Section 11.8). More recent success stories include Gato, a *generalist* model that can play Atari, caption images, chat, and act as a robot (Reed *et al.*, 2022). Gato is a single Transformer that scales well when pretrained on diverse modalities, including text, images, joint torques, and button presses. Notably, all such multimodal data is serialized into a flat sequence of tokens, which can be processed akin to text tokens (Section 11.7) or image patches (Section 11.8) by Transformers.

Prior to the compelling success of pretraining Transformers for multimodal data, Transformers were extensively pretrained with a wealth of text. Originally proposed for machine translation, the Transformer architecture in Fig. 11.7.1 consists of an encoder for representing input sequences and a decoder for generating target sequences. Primarily, Transformers can be used in three different modes: *encoder-only*, *encoder–decoder*, and *decoder-only*. To conclude this chapter, we will review these three modes and explain the scalability in pretraining Transformers.

11.9.1 Encoder-Only

When only the Transformer encoder is used, a sequence of input tokens is converted into the same number of representations that can be further projected into output (e.g., classification). A Transformer encoder consists of self-attention layers, where all input tokens attend to each other. For example, vision Transformers depicted in Fig. 11.8.1 are encoder-only, converting a sequence of input image patches into the representation of a special "<cls>" token. Since this representation depends on all input tokens, it is further projected into classification labels. This design was inspired by an earlier encoder-only Transformer pretrained on text: BERT (Bidirectional Encoder Representations from Transformers) (Devlin *et al.*, 2018).

Pretraining BERT

BERT is pretrained on text sequences using *masked language modeling*: input text with randomly masked tokens is fed into a Transformer encoder to predict the masked tokens. As illustrated in Fig. 11.9.1, an original text sequence "I", "love", "this", "red", "car" is prepended with the "<cls>" token, and the "<mask>" token randomly replaces "love"; then the cross-entropy loss between the masked token "love" and its prediction is to be minimized during pretraining. Note that there is no constraint in the attention pattern of Transformer encoders (right of Fig. 11.9.1) so all tokens can attend to each other. Thus, prediction of "love" depends on input tokens before and after it in the sequence. This is why BERT is a "bidirectional encoder". Without need for manual labeling, large-scale text data from books and Wikipedia can be used for pretraining BERT.

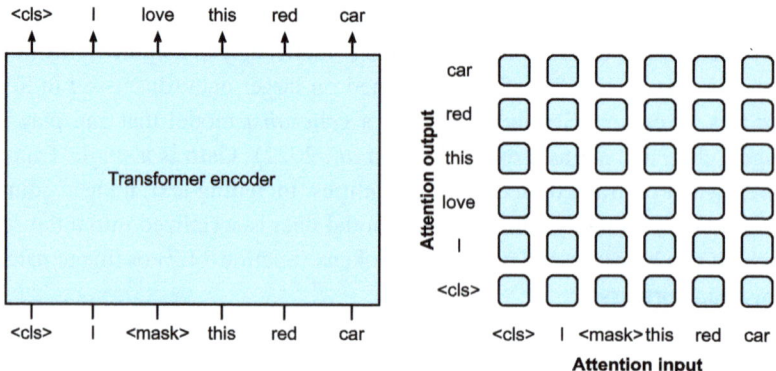

Fig. 11.9.1   Left: Pretraining BERT with masked language modeling. Prediction of the masked "love" token depends on all input tokens before and after "love". Right: Attention pattern in the Transformer encoder. Each token along the vertical axis attends to all input tokens along the horizontal axis.

Fine-Tuning BERT

The pretrained BERT can be *fine-tuned* to downstream encoding tasks involving single text or text pairs. During fine-tuning, additional layers can be added to BERT with randomized parameters: these parameters and those pretrained BERT parameters will be *updated* to fit training data of downstream tasks.

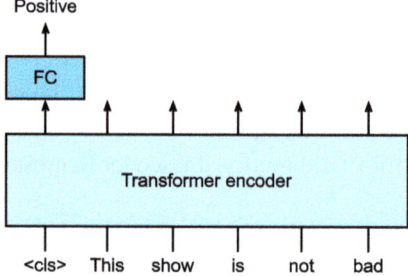

Fig. 11.9.2   Fine-tuning BERT for sentiment analysis.

Fig. 11.9.2 illustrates fine-tuning of BERT for sentiment analysis. The Transformer encoder is a pretrained BERT, which takes a text sequence as input and feeds the "<cls>" representation (global representation of the input) into an additional fully connected layer to predict the sentiment. During fine-tuning, the cross-entropy loss between the prediction and the label on sentiment analysis data is minimized via gradient-based algorithms, where the additional layer is trained from scratch while pretrained parameters of BERT are updated. BERT does more than sentiment analysis. The general language representations learned by the 350-million-parameter BERT from 250 billion training tokens advanced the state of the art for natural language tasks such as single text classification, text pair classification or regression, text tagging, and question answering.

You may note that these downstream tasks include text pair understanding. BERT pretraining has another loss for predicting whether one sentence immediately follows the other. However, this loss was later found to be less useful when pretraining RoBERTa, a BERT variant of the same size, on 2000 billion tokens (Liu *et al.*, 2019). Other derivatives of BERT improved model architectures or pretraining objectives, such as ALBERT (enforcing parameter sharing) (Lan *et al.*, 2019), SpanBERT (representing and predicting spans of text) (Joshi *et al.*, 2020), DistilBERT (lightweight via knowledge distillation) (Sanh *et al.*, 2019), and ELECTRA (replaced token detection) (Clark *et al.*, 2020). Moreover, BERT inspired Transformer pretraining in computer vision, such as with vision Transformers (Dosovitskiy *et al.*, 2021), Swin Transformers (Liu *et al.*, 2021), and MAE (masked autoencoders) (He *et al.*, 2022).

11.9.2 Encoder–Decoder

Since a Transformer encoder converts a sequence of input tokens into the same number of output representations, the encoder-only mode cannot generate a sequence of arbitrary length as in machine translation. As originally proposed for machine translation, the Transformer architecture can be outfitted with a decoder that autoregressively predicts the target sequence of arbitrary length, token by token, conditional on both encoder output and decoder output: (i) for conditioning on encoder output, encoder–decoder cross-attention (multi-head attention of decoder in Fig. 11.7.1) allows target tokens to attend to *all* input tokens; (ii) conditioning on decoder output is achieved by a so-called *causal* attention (this name is common in the literature but is misleading as it has little connection to the proper study of causality) pattern (masked multi-head attention of decoder in Fig. 11.7.1), where any target token can only attend to *past* and *present* tokens in the target sequence.

To pretrain encoder–decoder Transformers beyond human-labeled machine translation data, BART (Lewis *et al.*, 2019) and T5sindexT5 (Raffel *et al.*, 2020) are two concurrently proposed encoder–decoder Transformers pretrained on large-scale text corpora. Both attempt to reconstruct original text in their pretraining objectives, while the former emphasizes noising input (e.g., masking, deletion, permutation, and rotation) and the latter highlights multitask unification with comprehensive ablation studies.

Pretraining T5

As an example of the pretrained Transformer encoder–decoder, T5 (Text-to-Text Transfer Transformer) unifies many tasks as the same text-to-text problem: for any task, the input of the encoder is a task description (e.g., "Summarize", ":") followed by task input (e.g., a sequence of tokens from an article), and the decoder predicts the task output (e.g., a sequence of tokens summarizing the input article). To perform as text-to-text, T5 is trained to generate some target text conditional on input text.

To obtain input and output from any original text, T5 is pretrained to predict consecutive spans. Specifically, tokens from text are randomly replaced by special tokens where each consecutive span is replaced by the same special token. Consider the example in Fig. 11.9.3, where the original text is "I", "love", "this", "red", "car". Tokens "love", "red",

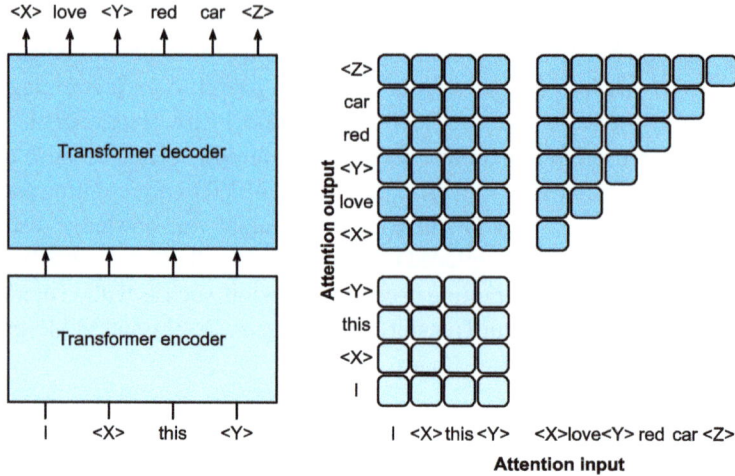

Fig. 11.9.3 Left: Pretraining T5 by predicting consecutive spans. The original sentence is "I", "love", "this", "red", "car", where "love" is replaced by a special "<X>" token, and consecutive "red", "car" are replaced by a special "<Y>" token. The target sequence ends with a special "<Z>" token. Right: Attention pattern in the Transformer encoder–decoder. In the encoder self-attention (lower square), all input tokens attend to each other; In the encoder–decoder cross-attention (upper rectangle), each target token attends to all input tokens; In the decoder self-attention (upper triangle), each target token attends to present and past target tokens only (causal).

"car" are randomly replaced by special tokens. Since "red" and "car" are a consecutive span, they are replaced by the same special token. As a result, the input sequence is "I", "<X>", "this", "<Y>", and the target sequence is "<X>", "love", "<Y>", "red", "car", "<Z>", where "<Z>" is another special token marking the end. As shown in Fig. 11.9.3, the decoder has a causal attention pattern to prevent itself from attending to future tokens during sequence prediction.

In T5, predicting consecutive span is also referred to as reconstructing corrupted text. With this objective, T5 is pretrained with 1000 billion tokens from the C4 (Colossal Clean Crawled Corpus) data, which consists of clean English text from the web (Raffel *et al.*, 2020).

Fine-Tuning T5

Similar to BERT, T5 needs to be fine-tuned (updating T5 parameters) on task-specific training data to perform this task. Major differences from BERT fine-tuning include: (i) T5 input includes task descriptions; (ii) T5 can generate sequences with arbitrary length with its Transformer decoder; (iii) No additional layers are required.

Fig. 11.9.4 explains fine-tuning T5 using text summarization as an example. In this downstream task, the task description tokens "Summarize", ":" followed by the article tokens are input to the encoder.

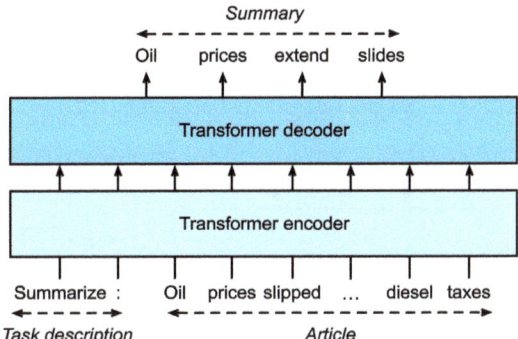

Fig. 11.9.4   Fine-tuning T5 for text summarization. Both the task description and article tokens are fed into the Transformer encoder for predicting the summary.

After fine-tuning, the 11-billion-parameter T5 (T5-11B) achieved state-of-the-art results on multiple encoding (e.g., classification) and generation (e.g., summarization) benchmarks. Since released, T5 has been extensively used in later research. For example, switch Transformers are designed based on T5 to activate a subset of the parameters for better computational efficiency (Fedus *et al.*, 2022). In a text-to-image model called Imagen, text is input to a frozen T5 encoder (T5-XXL) with 4.6 billion parameters (Saharia *et al.*, 2022). The photorealistic text-to-image examples in Fig. 11.9.5 suggest that the T5 encoder alone may effectively represent text even without fine-tuning.

Teddy bears swimming at the Olympics 400m Butter-   A cute corgi lives in a house made out of sushi.   A cute sloth holding a small treasure chest. A bright
fly event.   golden glow is coming from the chest.

Fig. 11.9.5   Text-to-image examples by the Imagen model, whose text encoder is from T5 (figures taken from Saharia et al. (2022)).

11.9.3 Decoder-Only

We have reviewed encoder-only and encoder–decoder Transformers. Alternatively, decoder-only Transformers remove the entire encoder and the decoder sublayer with the encoder–decoder cross-attention from the original encoder–decoder architecture depicted in Fig. 11.7.1. Nowadays, decoder-only Transformers have been the *de facto* architecture in large-scale language modeling (Section 9.3), which leverages the world's abundant unlabeled text corpora via self-supervised learning.

GPT and GPT-2

Using language modeling as the training objective, the GPT (generative pretraining) model chooses a Transformer decoder as its backbone (Radford *et al.*, 2018).

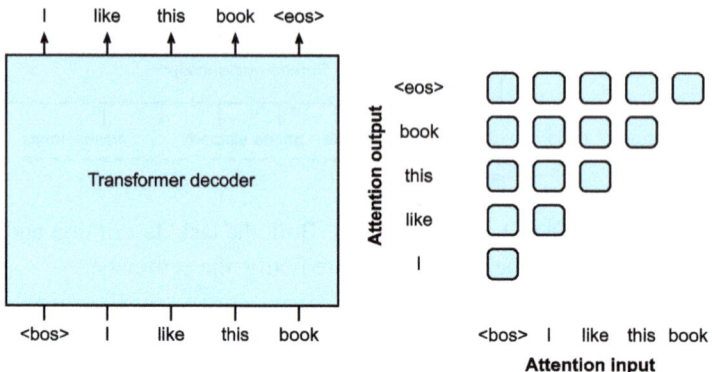

Fig. 11.9.6   Left: Pretraining GPT with language modeling. The target sequence is the input sequence shifted by one token. Both "<bos>" and "<eos>" are special tokens marking the beginning and end of sequences, respectively. Right: Attention pattern in the Transformer decoder. Each token along the vertical axis attends to only its past tokens along the horizontal axis (causal).

Following the autoregressive language model training as described in Section 9.3.3, Fig. 11.9.6 illustrates GPT pretraining with a Transformer encoder, where the target sequence is the input sequence shifted by one token. Note that the attention pattern in the Transformer decoder enforces that each token can only attend to its past tokens (future tokens cannot be attended to because they have not yet been chosen).

GPT has 100 million parameters and needs to be fine-tuned for individual downstream tasks. A much larger Transformer-decoder language model, GPT-2, was introduced one year later (Radford *et al.*, 2019). Compared with the original Transformer decoder in GPT, pre-normalization (discussed in Section 11.8.3) and improved initialization and weight-scaling were adopted in GPT-2. Pretrained on 40 GB of text, the 1.5-billion-parameter GPT-2 obtained the state-of-the-art results on language modeling benchmarks and promising results on multiple other tasks *without updating the parameters or architecture*.

GPT-3 and Beyond

GPT-2 demonstrated potential of using the same language model for multiple tasks without updating the model. This is more computationally efficient than fine-tuning, which requires model updates via gradient computation.

Before explaining the more computationally efficient use of language models without parameter update, recall Section 9.5 that a language model can be trained to generate a text sequence conditional on some prefix text sequence. Thus, a pretrained language model may generate the task output as a sequence *without parameter update*, conditional on an input

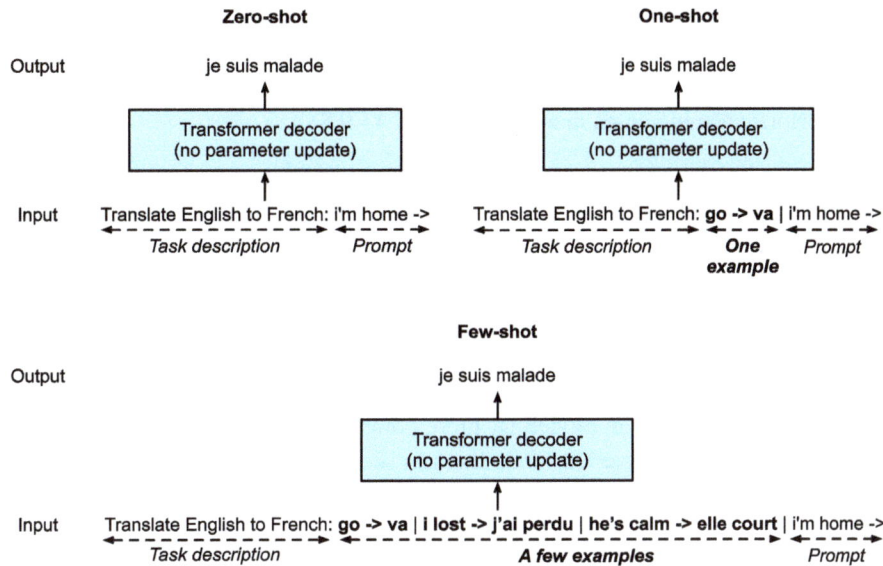

Fig. 11.9.7 Zero-shot, one-shot, few-shot in-context learning with language models (Transformer decoders). No parameter update is needed.

sequence with the task description, task-specific input–output examples, and a prompt (task input). This learning paradigm is called *in-context learning* (Brown *et al.*, 2020), which can be further categorized into *zero-shot*, *one-shot*, and *few-shot*, when there is no, one, and a few task-specific input–output examples (Fig. 11.9.7).

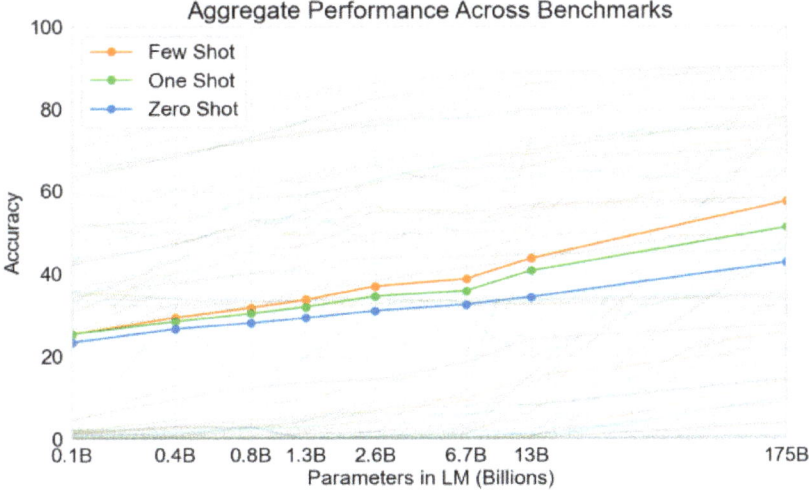

Fig. 11.9.8 Aggregate performance of GPT-3 for all 42 accuracy-denominated benchmarks (caption adapted and figure taken from Brown et al. (2020)).

These three settings were tested in GPT-3 (Brown *et al.*, 2020), whose largest version uses data and model size about two orders of magnitude larger than those in GPT-2. GPT-3 uses the same Transformer decoder architecture as its direct predecessor GPT-2 except that attention patterns (at the right in Fig. 11.9.6) are sparser at alternating layers. Pretrained with 300 billion tokens, GPT-3 performs better with larger model size, where few-shot performance increases most rapidly (Fig. 11.9.8).

The subsequent GPT-4 model did not fully disclose technical details in its report (OpenAI, 2023). By contrast with its predecessors, GPT-4 is a large-scale, multimodal model that can take both text and images as input and generate text output.

11.9.4 Scalability

Fig. 11.9.8 empirically demonstrates scalability of Transformers in the GPT-3 language model. For language modeling, more comprehensive empirical studies on the scalability of Transformers have led researchers to see promise in training larger Transformers with more data and compute (Kaplan *et al.*, 2020).

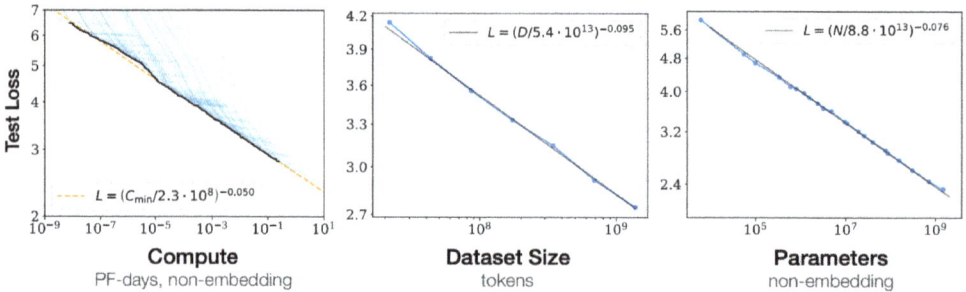

Compute
PF-days, non-embedding

Dataset Size
tokens

Parameters
non-embedding

Fig. 11.9.9 Transformer language model performance improves smoothly as we increase the model size, dataset size, and amount of compute used for training. For optimal performance all three factors must be scaled up in tandem. Empirical performance has a power-law relationship with each individual factor when not bottlenecked by the other two. (Caption adapted and figure taken from Kaplan et al. (2020).)

As shown in Fig. 11.9.9, *power-law scaling* can be observed in the performance with respect to the model size (number of parameters, excluding embedding layers), dataset size (number of training tokens), and amount of training compute (PetaFLOP/s-days, excluding embedding layers). In general, increasing all these three factors in tandem leads to better performance. However, *how* to increase them in tandem still remains a matter of debate (Hoffmann *et al.*, 2022).

As well as increased performance, large models also enjoy better sample efficiency than small models. Fig. 11.9.10 shows that large models need fewer training samples (tokens processed) to perform at the same level achieved by small models, and performance is scaled smoothly with compute.

The empirical scaling behaviors in Kaplan *et al.* (2020) have been tested in subsequent large Transformer models. For example, GPT-3 supported this hypothesis with two more orders of magnitude in Fig. 11.9.11.

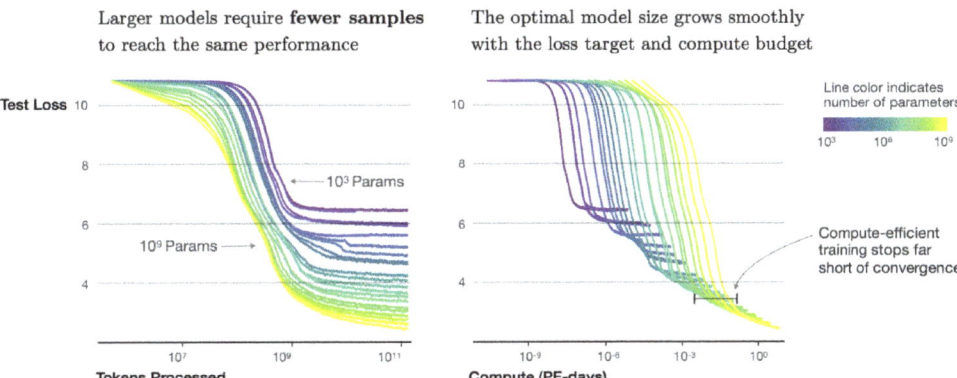

Fig. 11.9.10 Transformer language model training runs. (Figure taken from Kaplan et al. (2020).)

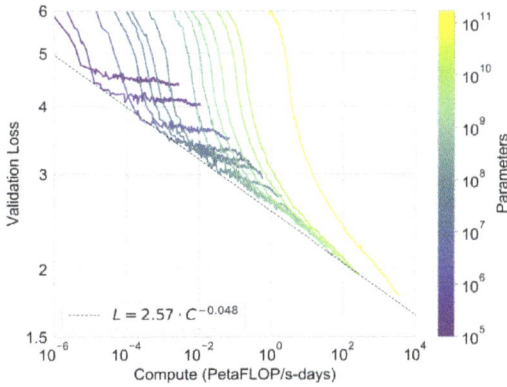

Fig. 11.9.11 GPT-3 performance (cross-entropy validation loss) follows a power-law trend with the amount of compute used for training. The power-law behavior observed in Kaplan et al. (2020) continues for an additional two orders of magnitude with only small deviations from the predicted curve. Embedding parameters are excluded from compute and parameter counts. (Caption adapted and figure taken from Brown et al. (2020).)

11.9.5 Large Language Models

The scalability of Transformers in the GPT series has inspired subsequent large language models. The GPT-2 Transformer decoder was used for training the 530-billion-parameter Megatron-Turing NLG (Smith *et al.*, 2022) with 270 billion training tokens. Following the GPT-2 design, the 280-billion-parameter Gopher (Rae *et al.*, 2021) pretrained with 300 billion tokens, performed competitively across diverse tasks. Inheriting the same architecture and using the same compute budget of Gopher, Chinchilla (Hoffmann *et al.*, 2022) is a substantially smaller (70 billion parameters) model that trains for much longer (1.4 trillion training tokens), outperforming Gopher on many tasks and with more emphasis on the number of tokens than on the number of parameters. To continue the scaling line of language modeling, PaLM (Pathway Language Model) (Chowdhery *et al.*, 2022), a 540-billion-parameter Transformer decoder with modified designs pretrained on 780 billion tokens, outperformed average human performance on the BIG-Bench benchmark (Srivastava *et al.*, 2022). Its later version, PaLM 2 (Anil *et al.*, 2023), scaled data and model roughly 1:1 and improved multilingual and reasoning capabilities. Other large language models, such as Minerva (Lewkowycz *et al.*, 2022) that further trains a generalist (PaLM) and Galactica (Taylor *et al.*, 2022) that is not trained on a general corpus, have shown promising quantitative and scientific reasoning capabilities.

Open-sourced releases, such as OPT (open pretrained Transformers) (Zhang *et al.*, 2022), BLOOM (Scao *et al.*, 2022), and FALCON (Penedo *et al.*, 2023), democratized research and use of large language models. Focusing on computational efficiency at inference time, the open-sourced Llama 1 (Touvron *et al.*, 2023a) outperformed much larger models by training on more tokens than had been typically used. The updated Llama 2 (Touvron *et al.*, 2023b) further increased the pretraining corpus by 40%, leading to product models that may match the performance of competitive close-sourced models.

Wei *et al.* (2022) discussed emergent abilities of large language models that are present in larger models, but not in smaller models. However, simply increasing model size does not inherently make models follow human instructions better. Sanh *et al.* (2021), Wei *et al.* (2021) have found that fine-tuning large language models on a range of datasets described via *instructions* can improve zero-shot performance on held-out tasks. Using *reinforcement learning from human feedback*, Ouyang *et al.* (2022) fine-tuned GPT-3 to follow a diverse set of instructions. Following the resultant InstructGPT which aligns language models with human intent via fine-tuning (Ouyang *et al.*, 2022), ChatGPT[164] can generate human-like responses (e.g., code debugging and creative writing) based on conversations with humans and can perform many natural language processing tasks zero-shot (Qin *et al.*, 2023). Bai *et al.* (2022) replaced human inputs (e.g., human-labeled data) with model outputs to partially automate the instruction tuning process, which is also known as *reinforcement learning from AI feedback*.

164

Large language models offer an exciting prospect of formulating text input to induce models to perform desired tasks via in-context learning, which is also known as *prompting*. Notably, *chain-of-thought prompting* (Wei *et al.*, 2022), an in-context learning method with few-shot "question, intermediate reasoning steps, answer" demonstrations, elicits the complex reasoning capabilities of large language models in order to solve mathematical, commonsense, and symbolic reasoning tasks. Sampling multiple reasoning paths (Wang *et al.*, 2023), diversifying few-shot demonstrations (Zhang *et al.*, 2023), and reducing complex problems to sub-problems (Zhou *et al.*, 2023) can all improve the reasoning accuracy. In fact, with simple prompts like "Let's think step by step" just before each answer, large language models can even perform *zero-shot* chain-of-thought reasoning with decent accuracy (Kojima *et al.*, 2022). Even for multimodal inputs consisting of both text and images, language models can perform multimodal chain-of-thought reasoning with higher accuracy than using text input only (Zhang *et al.*, 2023).

11.9.6 Summary and Discussion

Transformers have been pretrained as encoder-only (e.g., BERT), encoder–decoder (e.g., T5), and decoder-only (e.g., GPT series). Pretrained models may be adapted to perform different tasks with model update (e.g., fine-tuning) or not (e.g., few-shot). Scalability of

Transformers suggests that better performance benefits from larger models, more training data, and more training compute. Since Transformers were first designed and pretrained for text data, this section leans slightly towards natural language processing. Nonetheless, those models discussed above can be often found in more recent models across multiple modalities. For example, (i) Chinchilla (Hoffmann *et al.*, 2022) was further extended to Flamingo (Alayrac *et al.*, 2022), a visual language model for few-shot learning; (ii) GPT-2 (Radford *et al.*, 2019) and the vision Transformer encode text and images in CLIP (Contrastive Language-Image Pre-training) (Radford *et al.*, 2021), whose image and text embeddings were later adopted in the DALL-E 2 text-to-image system (Ramesh *et al.*, 2022). Although there have been no systematic studies on Transformer scalability in multimodal pretraining yet, an all-Transformer text-to-image model called Parti (Yu *et al.*, 2022) shows potential of scalability across modalities: a larger Parti is more capable of high-fidelity image generation and content-rich text understanding (Fig. 11.9.12).

A portrait photo of a kangaroo wearing an orange hoodie and blue sunglasses standing on the grass in front of the Sydney Opera House holding a sign on the chest that says Welcome Friends!

Fig. 11.9.12   Image examples generated from the same text by the Parti model of increasing sizes (350M, 750M, 3B, 20B). (Examples taken from Yu et al. (2022).)

11.9.7 Exercises

1. Is it possible to fine-tune T5 using a minibatch consisting of different tasks? Why or why not? How about for GPT-2?

2. Given a powerful language model, what applications can you think of?

3. Say that you are asked to fine-tune a language model to perform text classification by adding additional layers. Where will you add them? Why?

4. Consider sequence-to-sequence problems (e.g., machine translation) where the input sequence is always available throughout the target sequence prediction. What could be limitations of modeling with decoder-only Transformers? Why?

Discussions[165].

165

Tools for Deep Learning

To get the most out of *Dive into Deep Learning*, we will talk you through different tools in this appendix, such as for running and contributing to this interactive open-source book.

A.1 Using Jupyter Notebooks

This section describes how to edit and run the code in each section of this book using the Jupyter Notebook. Make sure you have installed Jupyter and downloaded the code as described in *Installation* (page xxvi). If you want to know more about Jupyter see the excellent tutorial in their documentation[166].

166

A.1.1 Editing and Running the Code Locally

Suppose that the local path of the book's code is xx/yy/d2l-en/. Use the shell to change the directory to this path (cd xx/yy/d2l-en) and run the command jupyter notebook. If your browser does not do this automatically, open http://localhost:8888 and you will see the interface of Jupyter and all the folders containing the code of the book, as shown in Fig. A.1.

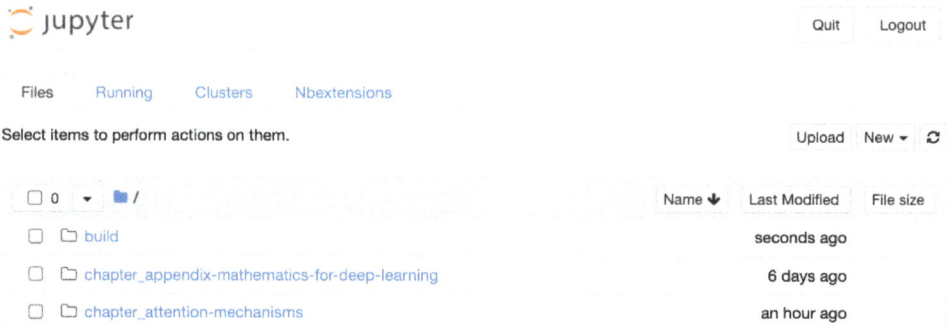

Fig. A.1 The folders containing the code of this book.

You can access the notebook files by clicking on the folder displayed on the webpage. They usually have the suffix ".ipynb". For the sake of brevity, we create a temporary "test.ipynb" file. The content displayed after you click it is shown in Fig. A.2. This notebook includes a markdown cell and a code cell. The content in the markdown cell includes "This Is a Title" and "This is text.". The code cell contains two lines of Python code.

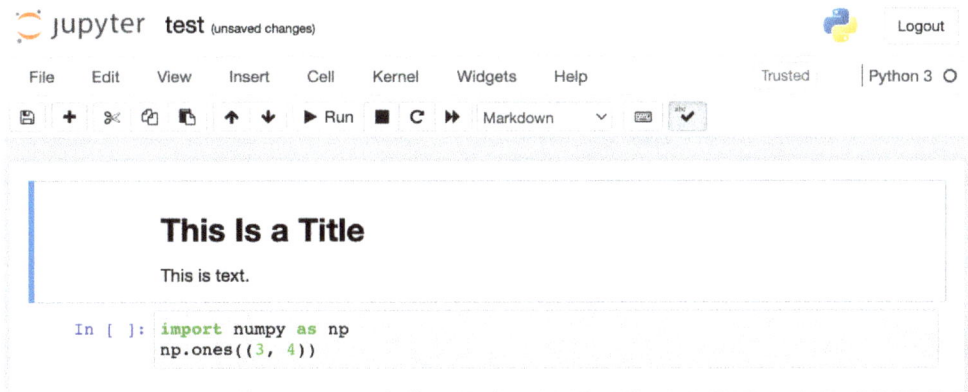

Fig. A.2   Markdown and code cells in the "text.ipynb" file.

Double click on the markdown cell to enter edit mode. Add a new text string "Hello world." at the end of the cell, as shown in Fig. A.3.

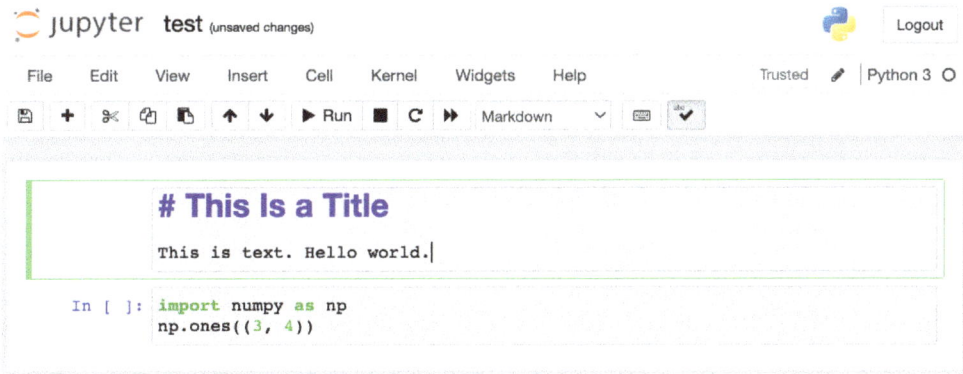

Fig. A.3   Edit the markdown cell.

As demonstrated in Fig. A.4, click "Cell" → "Run Cells" in the menu bar to run the edited cell.

After running, the markdown cell is shown in Fig. A.5.

Next, click on the code cell. Multiply the elements by 2 after the last line of code, as shown in Fig. A.6.

You can also run the cell with a shortcut ("Ctrl + Enter" by default) and obtain the output result from Fig. A.7.

When a notebook contains more cells, we can click "Kernel" → "Restart & Run All" in the menu bar to run all the cells in the entire notebook. By clicking "Help" → "Edit Keyboard Shortcuts" in the menu bar, you can edit the shortcuts according to your preferences.

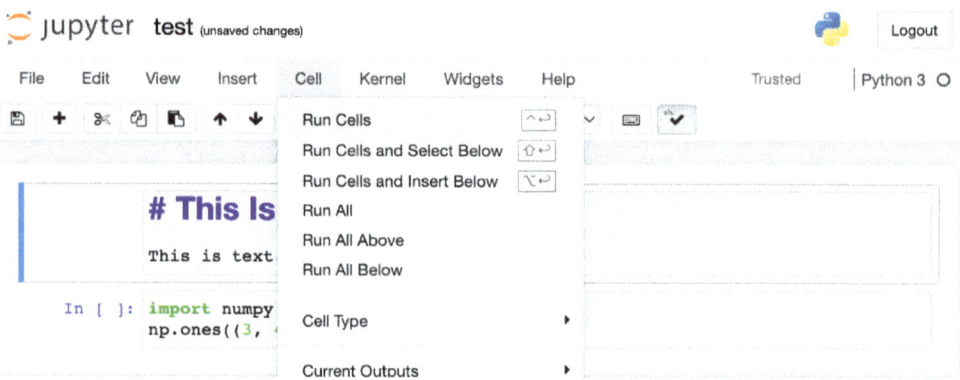

Fig. A.4 Run the cell.

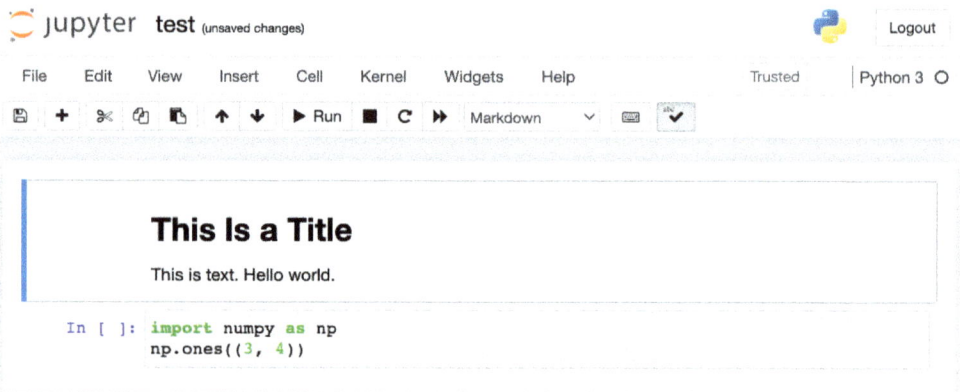

Fig. A.5 The markdown cell after running.

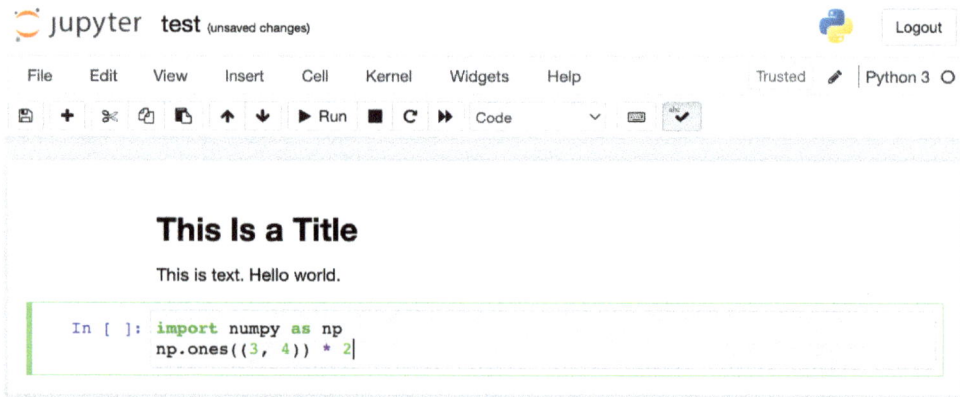

Fig. A.6 Edit the code cell.

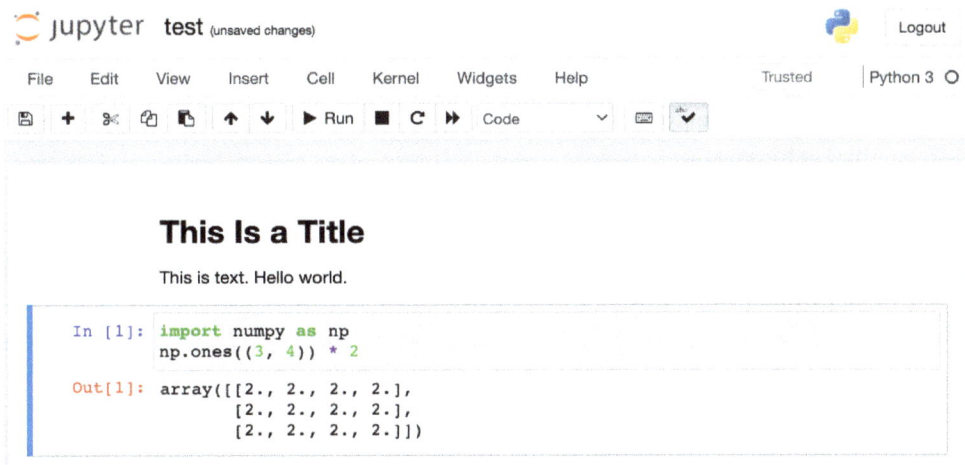

Fig. A.7 Run the code cell to obtain the output.

A.1.2 Advanced Options

Beyond local editing two things are quite important: editing the notebooks in the markdown format and running Jupyter remotely. The latter matters when we want to run the code on a faster server. The former matters since Jupyter's native ipynb format stores a lot of auxiliary data that is irrelevant to the content, mostly related to how and where the code is run. This is confusing for Git, making reviewing contributions very difficult. Fortunately there is an alternative—native editing in the markdown format.

Markdown Files in Jupyter

If you wish to contribute to the content of this book, you need to modify the source file (md file, not ipynb file) on GitHub. Using the notedown plugin we can modify notebooks in the md format directly in Jupyter.

First, install the notedown plugin, run the Jupyter Notebook, and load the plugin:

```
pip install d2l-notedown  # You may need to uninstall the original notedown.
jupyter notebook --NotebookApp.contents_manager_class='notedown.
↪NotedownContentsManager'
```

You may also turn on the notedown plugin by default whenever you run the Jupyter Notebook. First, generate a Jupyter Notebook configuration file (if it has already been generated, you can skip this step).

```
jupyter notebook --generate-config
```

Then, add the following line to the end of the Jupyter Notebook configuration file (for Linux or macOS, usually in the path ~/.jupyter/jupyter_notebook_config.py):

```
c.NotebookApp.contents_manager_class = 'notedown.NotedownContentsManager'
```

After that, you only need to run the `jupyter notebook` command to turn on the notedown plugin by default.

Running Jupyter Notebooks on a Remote Server

Sometimes, you may want to run Jupyter notebooks on a remote server and access it through a browser on your local computer. If Linux or macOS is installed on your local machine (Windows can also support this function through third-party software such as PuTTY), you can use port forwarding:

```
ssh myserver -L 8888:localhost:8888
```

The above string `myserver` is the address of the remote server. Then we can use http://localhost:8888 to access the remote server `myserver` that runs Jupyter notebooks. We will detail on how to run Jupyter notebooks on AWS instances later in this appendix.

Timing

We can use the `ExecuteTime` plugin to time the execution of each code cell in Jupyter notebooks. Use the following commands to install the plugin:

```
pip install jupyter_contrib_nbextensions
jupyter contrib nbextension install --user
jupyter nbextension enable execute_time/ExecuteTime
```

A.1.3 Summary

- Using the Jupyter Notebook tool, we can edit, run, and contribute to each section of the book.

- We can run Jupyter notebooks on remote servers using port forwarding.

A.1.4 Exercises

1. Edit and run the code in this book with the Jupyter Notebook on your local machine.

2. Edit and run the code in this book with the Jupyter Notebook *remotely* via port forwarding.

3. Compare the running time of the operations $A^\top B$ and AB for two square matrices in $\mathbb{R}^{1024 \times 1024}$. Which one is faster?

Discussions[167].

A.2 Using Amazon SageMaker

Deep learning applications may demand so much computational resource that easily goes beyond what your local machine can offer. Cloud computing services allow you to run GPU-intensive code of this book more easily using more powerful computers. This section will introduce how to use Amazon SageMaker to run the code of this book.

A.2.1 Signing Up

First, we need to sign up an account at https://aws.amazon.com/. For additional security, using two-factor authentication is encouraged. It is also a good idea to set up detailed billing and spending alerts to avoid any surprise, e.g., when forgetting to stop running instances. After logging into your AWS account, go to your console[168] and search for "Amazon Sage-Maker" (see Fig. A.8), then click it to open the SageMaker panel.

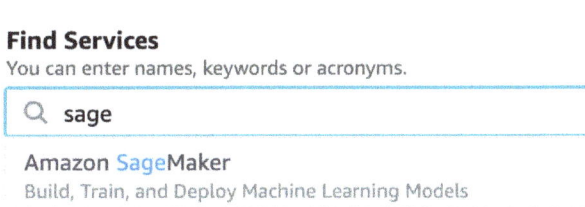

Fig. A.8   Search for and open the SageMaker panel.

A.2.2 Creating a SageMaker Instance

Next, let's create a notebook instance as described in Fig. A.9.

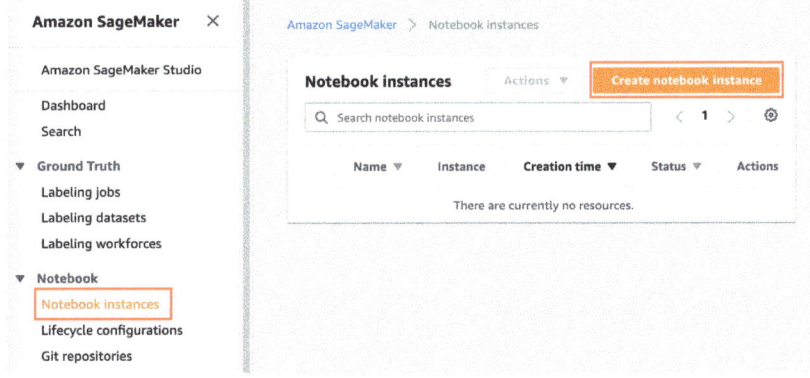

Fig. A.9   Create a SageMaker instance.

SageMaker provides multiple instance types [169] with varying computational power and prices. When creating a notebook instance, we can specify its name and type. In Fig. A.10, we choose `ml.p3.2xlarge`: with one Tesla V100 GPU and an 8-core CPU, this instance is powerful enough for most of the book.

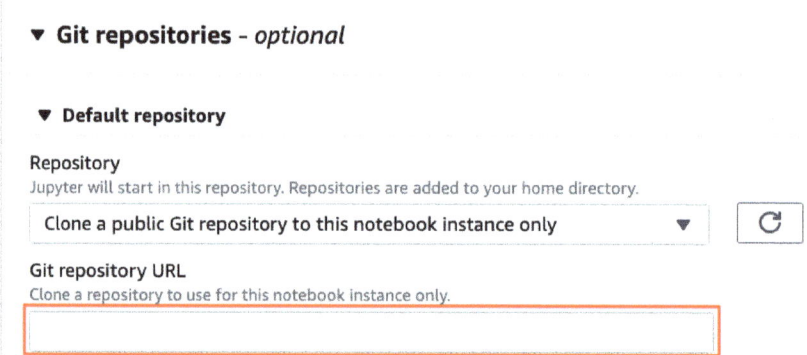

Fig. A.10   Choose the instance type.

The entire book in the ipynb format for running with SageMaker is available at https://github.com/d2l-ai/d2l-pytorch-sagemaker. We can specify this GitHub repository URL (Fig. A.11) to allow SageMaker to clone it when creating the instance.

Fig. A.11   Specify the GitHub repository.

A.2.3 Running and Stopping an Instance

Creating an instance may take a few minutes. When it is ready, click on the "Open Jupyter" link next to it (Fig. A.12) so you can edit and run all the Jupyter notebooks of this book on this instance (similar to steps in Section A.1).

Fig. A.12   Open Jupyter on the created SageMaker instance.

After finishing your work, do not forget to stop the instance to avoid being charged further (Fig. A.13).

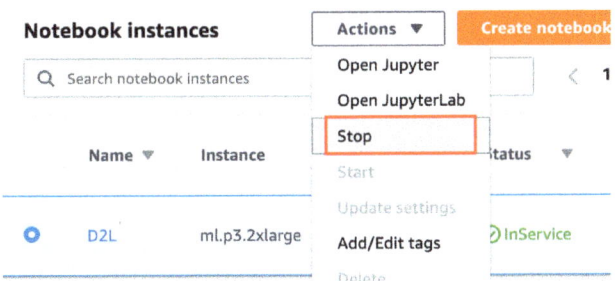

Fig. A.13   Stop a SageMaker instance.

A.2.4 Updating Notebooks

Notebooks of this open-source book will be regularly updated in the d2l-ai/d2l-pytorch-sagemaker[170] repository on GitHub. To update to the latest version, you may open a terminal on the SageMaker instance (Fig. A.14).

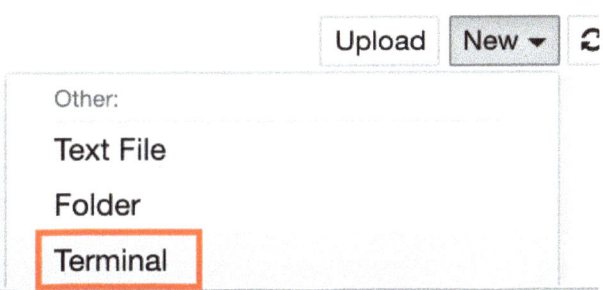

Fig. A.14   Open a terminal on the SageMaker instance.

You may wish to commit your local changes before pulling updates from the remote repository. Otherwise, simply discard all your local changes with the following commands in the terminal:

```
cd SageMaker/d2l-pytorch-sagemaker/
git reset --hard
git pull
```

A.2.5 Summary

- We can create a notebook instance using Amazon SageMaker to run GPU-intensive code of this book.

- We can update notebooks via the terminal on the Amazon SageMaker instance.

A.2.6 Exercises

1. Edit and run any section that requires a GPU using Amazon SageMaker.

2. Open a terminal to access the local directory that hosts all the notebooks of this book.

Discussions[171].

171

A.3 Using AWS EC2 Instances

In this section, we will show you how to install all libraries on a raw Linux machine. Recall that in Section A.2 we discussed how to use Amazon SageMaker, while building an instance by yourself costs less on AWS. The walkthrough includes three steps:

1. Request for a GPU Linux instance from AWS EC2.

2. Install CUDA (or use an Amazon Machine Image with preinstalled CUDA).

3. Install the deep learning framework and other libraries for running the code of the book.

This process applies to other instances (and other clouds), too, albeit with some minor modifications. Before going forward, you need to create an AWS account, see Section A.2 for more details.

A.3.1 Creating and Running an EC2 Instance

After logging into your AWS account, click "EC2" (Fig. A.15) to go to the EC2 panel.

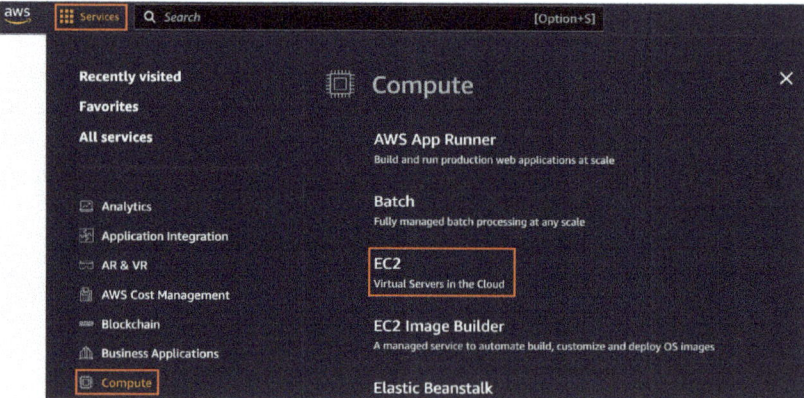

Fig. A.15   Open the EC2 console.

Fig. A.16 shows the EC2 panel.

Presetting Location

Select a nearby data center to reduce latency, e.g., "Oregon" (marked by the red box in the top-right of Fig. A.16). If you are located in China, you can select a nearby Asia Pacific

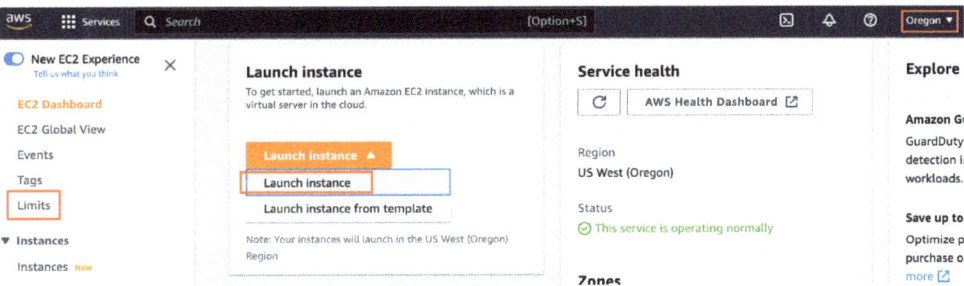

Fig. A.16   The EC2 panel.

region, such as Seoul or Tokyo. Please note that some data centers may not have GPU instances.

Increasing Limits

Before choosing an instance, check if there are quantity restrictions by clicking the "Limits" label in the bar on the left as shown in Fig. A.16. Fig. A.17 shows an example of such a limitation. The account currently cannot open "p2.xlarge" instances according to the region. If you need to open one or more instances, click on the "Request limit increase" link to apply for a higher instance quota. Generally, it takes one business day to process an application.

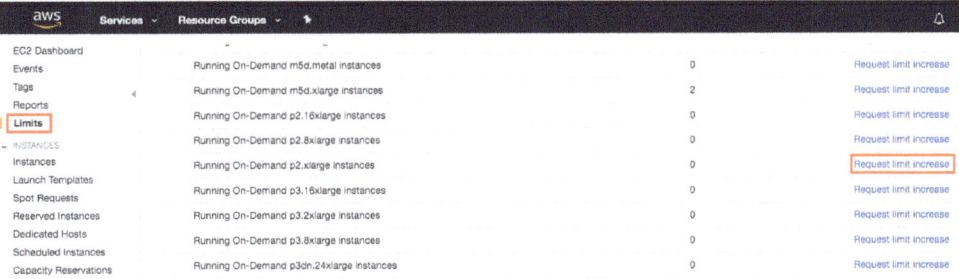

Fig. A.17   Instance quantity restrictions.

Launching an Instance

Next, click the "Launch Instance" button marked by the red box in Fig. A.16 to launch your instance.

We begin by selecting a suitable Amazon Machine Image (AMI). Select an Ubuntu instance (Fig. A.18).

EC2 provides many different instance configurations to choose from. This can sometimes feel overwhelming to a beginner. Table A.1 lists different suitable machines.

All these servers come in multiple flavors indicating the number of GPUs used. For exam-

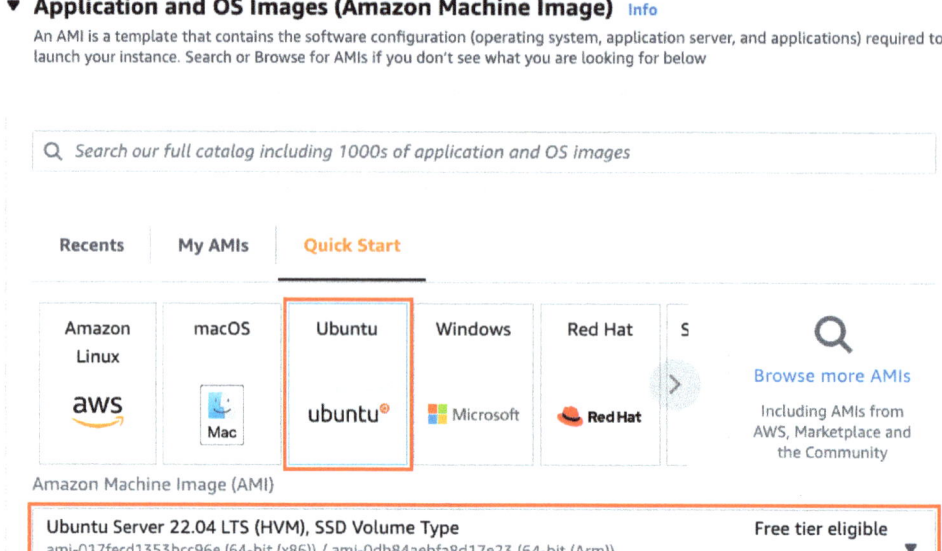

Fig. A.18 Choose an AMI.

Table A.1 Different EC2 instance types		
Name	**GPU**	**Notes**
g2	Grid K520	ancient
p2	Kepler K80	old but often cheap as spot
g3	Maxwell M60	good trade-off
p3	Volta V100	high performance for FP16
p4	Ampere A100	high performance for large-scale training
g4	Turing T4	inference optimized FP16/INT8

ple, a p2.xlarge has 1 GPU and a p2.16xlarge has 16 GPUs and more memory. For more details, see the AWS EC2 documentation[172] or a summary page[173]. For the purpose of illustration, a p2.xlarge will suffice (marked in the red box of Fig. A.19).

172

173

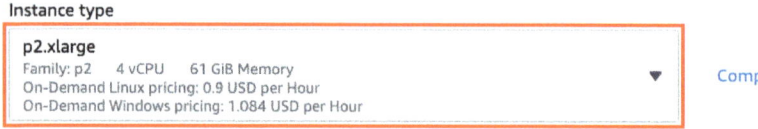

Fig. A.19 Choose an instance.

Note that you should use a GPU-enabled instance with suitable drivers and a GPU-enabled deep learning framework. Otherwise you will not see any benefit from using GPUs.

We go on to select the key pair used to access the instance. If you do not have a key pair, click "Create new key pair" in Fig. A.20 to generate a key pair. Subsequently, you can select the previously generated key pair. Make sure that you download the key pair and store it in a safe location if you generated a new one. This is your only way to SSH into the server.

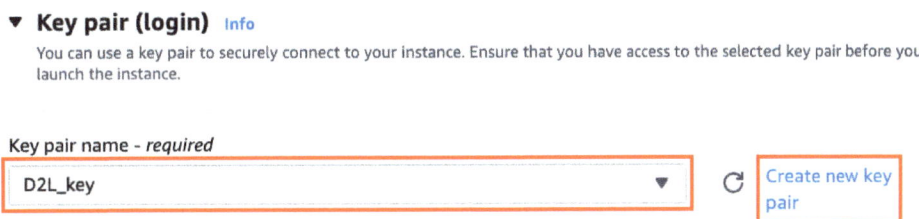

Fig. A.20   Select a key pair.

In this example, we will keep the default configurations for "Network settings" (click the "Edit" button to configure items such as the subnet and security groups). We just increase the default hard disk size to 64 GB (Fig. A.21). Note that CUDA by itself already takes up 4 GB.

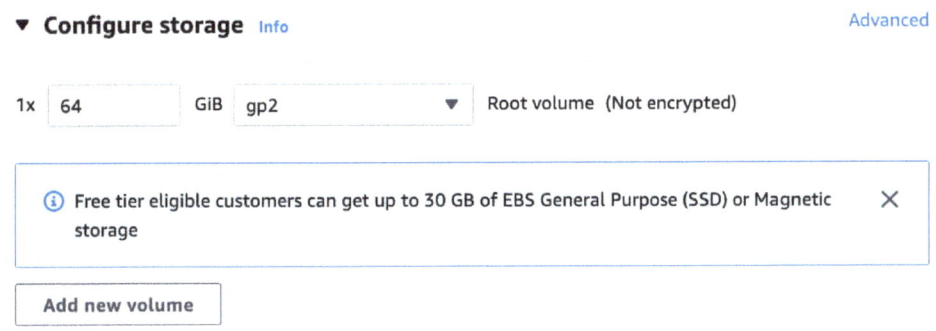

Fig. A.21   Modify the hard disk size.

Click "Launch Instance" to launch the created instance. Click the instance ID shown in Fig. A.22 to view the status of this instance.

Connecting to the Instance

As shown in Fig. A.23, after the instance state turns green, right-click the instance and select Connect to view the instance access method.

If this is a new key, it must not be publicly viewable for SSH to work. Go to the folder where you store D2L_key.pem and execute the following command to make the key not publicly viewable:

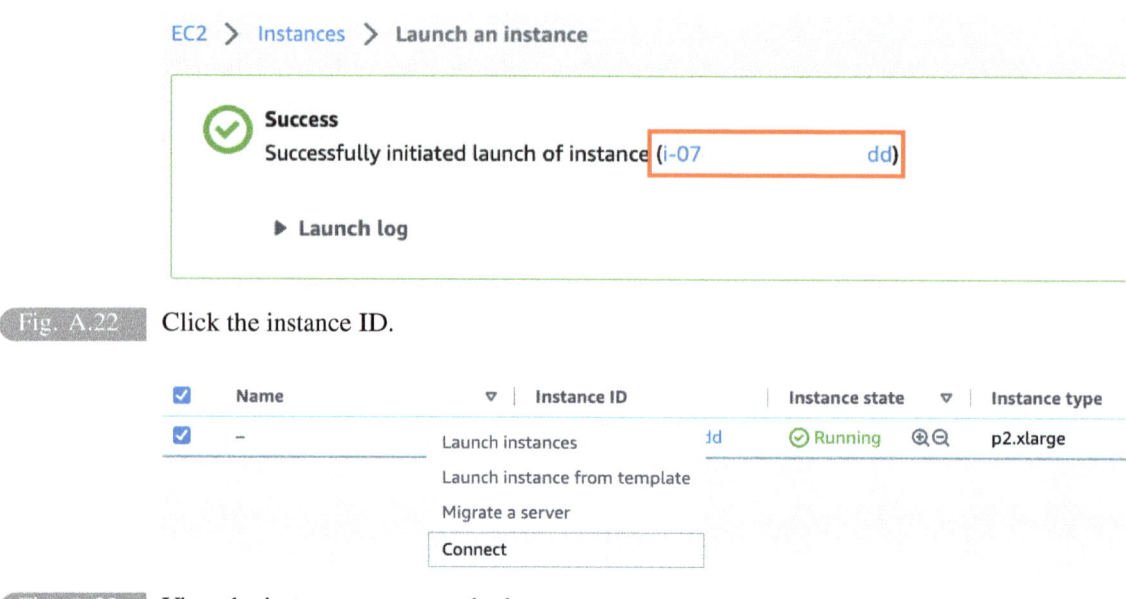

Fig. A.22   Click the instance ID.

Fig. A.23   View the instance access method.

```
chmod 400 D2L_key.pem
```

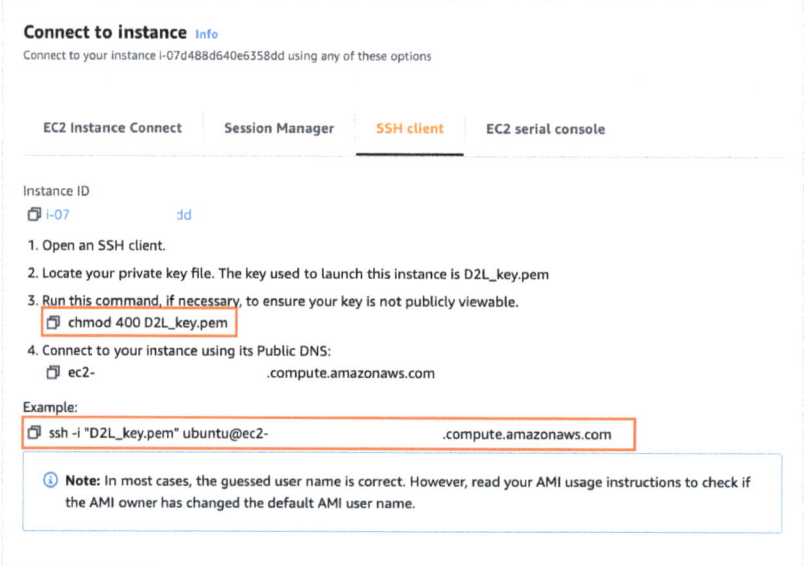

Fig. A.24   View instance access and startup method.

Now, copy the SSH command in the lower red box of Fig. A.24 and paste onto the command line:

```
ssh -i "D2L_key.pem" ubuntu@ec2-xx-xxx-xxx-xxx.y.compute.amazonaws.com
```

When the command line prompts "Are you sure you want to continue connecting (yes/no)", enter "yes" and press Enter to log into the instance.

Your server is ready now.

A.3.2 Installing CUDA

Before installing CUDA, be sure to update the instance with the latest drivers.

```
sudo apt-get update && sudo apt-get install -y build-essential git libgfortran3
```

Here we download CUDA 12.1. Visit NVIDIA's official repository[174] to find the download link as shown in Fig. A.25.

CUDA Toolkit 12.1 Downloads

Home

Select Target Platform

Click on the green buttons that describe your target platform. Only supported platforms will be shown. By downloading and using the software, you agree to fully comply with the terms and conditions of the CUDA EULA.

Operating System	Linux	Windows						
Architecture	x86_64	ppc64le	arm64-sbsa	aarch64-jetson				
Distribution	CentOS	Debian	Fedora	KylinOS	OpenSUSE	RHEL	Rocky	SLES	Ubuntu
	WSL-Ubuntu							
Version	18.04	20.04	22.04					
Installer Type	deb (local)	deb (network)	runfile (local)					

Download Installer for Linux Ubuntu 22.04 x86_64

The base installer is available for download below.

Fig. A.25   Find the CUDA 12.1 download address.

Copy the instructions and paste them onto the terminal to install CUDA 12.1.

```
# The link and file name are subject to changes
wget https://developer.download.nvidia.com/compute/cuda/repos/ubuntu2204/x86_
↪64/cuda-ubuntu2204.pin
sudo mv cuda-ubuntu2204.pin /etc/apt/preferences.d/cuda-repository-pin-600
wget https://developer.download.nvidia.com/compute/cuda/12.1.0/local_
↪installers/cuda-repo-ubuntu2204-12-1-local_12.1.0-530.30.02-1_amd64.deb
sudo dpkg -i cuda-repo-ubuntu2204-12-1-local_12.1.0-530.30.02-1_amd64.deb
```

(continues on next page)

(continued from previous page)

```
sudo cp /var/cuda-repo-ubuntu2204-12-1-local/cuda-*-keyring.gpg /usr/share/
↪keyrings/
sudo apt-get update
sudo apt-get -y install cuda
```

After installing the program, run the following command to view the GPUs:

```
nvidia-smi
```

Finally, add CUDA to the library path to help other libraries find it, such as appending the following lines to the end of ~/.bashrc.

```
export PATH="/usr/local/cuda-12.1/bin:$PATH"
export LD_LIBRARY_PATH=${LD_LIBRARY_PATH}:/usr/local/cuda-12.1/lib64
```

A.3.3 Installing Libraries for Running the Code

To run the code of this book, just follow steps in *Installation* (page xxvi) for Linux users on the EC2 instance and use the following tips for working on a remote Linux server:

- To download the bash script on the Miniconda installation page, right click the download link and select "Copy Link Address", then execute wget [copied link address].

- After running ~/miniconda3/bin/conda init, you may execute source ~/.bashrc instead of closing and reopening your current shell.

A.3.4 Running the Jupyter Notebook remotely

To run the Jupyter Notebook remotely you need to use SSH port forwarding. After all, the server in the cloud does not have a monitor or keyboard. For this, log into your server from your desktop (or laptop) as follows:

```
# This command must be run in the local command line
ssh -i "/path/to/key.pem" ubuntu@ec2-xx-xxx-xxx-xxx.y.compute.amazonaws.com -L↲
↪8889:localhost:8888
```

Next, go to the location of the downloaded code of this book on the EC2 instance, then run:

```
conda activate d2l
jupyter notebook
```

Fig. A.26 shows the possible output after you run the Jupyter Notebook. The last row is the URL for port 8888.

Since you used port forwarding to port 8889, copy the last row in the red box of Fig. A.26, replace "8888" with "8889" in the URL, and open it in your local browser.

```
(d2l)                                          $ jupyter notebook
[I 05:23:30.157 NotebookApp] Writing notebook server cookie secret to /home/ubuntu/.lo
[I 05:23:30.711 NotebookApp] Serving notebooks from local directory: /home/ubuntu/d2l-
[I 05:23:30.711 NotebookApp] Jupyter Notebook 6.4.4 is running at:
[I 05:23:30.711 NotebookApp] http://localhost:8888/?token=7ae1f41cbd5c6ca705c1ad9f5115
[I 05:23:30.711 NotebookApp]  or http://127.0.0.1:8888/?token=7ae1f41cbd5c6ca705c1ad9f
[I 05:23:30.712 NotebookApp] Use Control-C to stop this server and shut down all kerne
[W 05:23:30.717 NotebookApp] No web browser found: could not locate runnable browser.
[C 05:23:30.717 NotebookApp]

    To access the notebook, open this file in a browser:
        file:///home/ubuntu/.local/share/jupyter/runtime/nbserver-1615-open.html
    Or copy and paste one of these URLs:
        http://localhost:8888/?token=7ae1f41cbd5c6ca705c1ad9f5115931bb8929817fd507ba3
```

Fig. A.26 Output after running the Jupyter Notebook. The last row is the URL for port 8888.

A.3.5 Closing Unused Instances

As cloud services are billed by the time of use, you should close instances that are not being used. Note that there are alternatives:

- "Stopping" an instance means that you will be able to start it again. This is akin to switching off the power for your regular server. However, stopped instances will still be billed a small amount for the hard disk space retained.

- "Terminating" an instance will delete all data associated with it. This includes the disk, hence you cannot start it again. Only do this if you know that you will not need it in the future.

If you want to use the instance as a template for many more instances, right-click on the example in Fig. A.23 and select "Image" → "Create" to create an image of the instance. Once this is complete, select "Instance State" → "Terminate" to terminate the instance. The next time you want to use this instance, you can follow the steps in this section to create an instance based on the saved image. The only difference is that, in "1. Choose AMI" shown in Fig. A.18, you must use the "My AMIs" option on the left to select your saved image. The created instance will retain the information stored on the image hard disk. For example, you will not have to reinstall CUDA and other runtime environments.

A.3.6 Summary

- We can launch and stop instances on demand without having to buy and build our own computer.

- We need to install CUDA before using the GPU-enabled deep learning framework.

- We can use port forwarding to run the Jupyter Notebook on a remote server.

A.3.7 Exercises

175

1. The cloud offers convenience, but it does not come cheap. Find out how to launch spot instances[175] to see how to reduce costs.

2. Experiment with different GPU servers. How fast are they?

3. Experiment with multi-GPU servers. How well can you scale things up?

Discussions[176].

176

A.4 Using Google Colab

We introduced how to run this book on AWS in Section A.2 and Section A.3. Another option is running this book on Google Colab[177] if you have a Google account.

177

To run the code of a section on Colab, simply click the Colab button as shown in Fig. A.27.

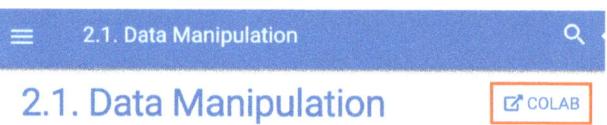

Fig. A.27   Run the code of a section on Colab.

If it is your first time to run a code cell, you will receive a warning message as shown in Fig. A.28. Just click "RUN ANYWAY" to ignore it.

Warning: This notebook was not authored …

This notebook is being loaded from **GitHub**. It may request access to your data stored with Google, or read data and credentials from other sessions. Please review the source code before executing this notebook.

CANCEL   RUN ANYWAY

Fig. A.28   Ignore the warning message by clicking "RUN ANYWAY".

Next, Colab will connect you to an instance to run the code of this section. Specifically, if a GPU is needed, Colab will be automatically requested for connecting to a GPU instance.

A.4.1 Summary

- You can use Google Colab to run each section's code in this book.

- Colab will be requested to connect to a GPU instance if a GPU is needed in any section of this book.

A.4.2 Exercises

1. Open any section of this book using Google Colab.

2. Edit and run any section that requires a GPU using Google Colab.

Discussions[178].

A.5 Selecting Servers and GPUs

Deep learning training generally requires large amounts of computation. At present GPUs are the most cost-effective hardware accelerators for deep learning. In particular, compared with CPUs, GPUs are cheaper and offer higher performance, often by over an order of magnitude. Furthermore, a single server can support multiple GPUs, up to 8 for high end servers. More typical numbers are up to 4 GPUs for an engineering workstation, since heat, cooling, and power requirements escalate quickly beyond what an office building can support. For larger deployments, cloud computing (e.g., Amazon's P3 [179] and G4 [180] instances) is a much more practical solution.

A.5.1 Selecting Servers

There is typically no need to purchase high-end CPUs with many threads since much of the computation occurs on the GPUs. That said, due to the global interpreter lock (GIL) in Python single-thread performance of a CPU can matter in situations where we have 4–8 GPUs. All things equal this suggests that CPUs with a smaller number of cores but a higher clock frequency might be a more economical choice. For example, when choosing between a 6-core 4 GHz and an 8-core 3.5 GHz CPU, the former is much preferable, even though its aggregate speed is less. An important consideration is that GPUs use lots of power and thus dissipate lots of heat. This requires very good cooling and a large enough chassis to use the GPUs. Follow the guidelines below if possible:

1. **Power Supply**. GPUs use significant amounts of power. Budget with up to 350W per device (check for the *peak demand* of the graphics card rather than typical demand, since efficient code can use lots of energy). If your power supply is not up to the demand you will find that your system becomes unstable.

2. **Chassis Size**. GPUs are large and the auxiliary power connectors often need extra space. Also, large chassis are easier to cool.

3. **GPU Cooling**. If you have a large number of GPUs you might want to invest in water cooling. Also, aim for *reference designs* even if they have fewer fans, since they are thin enough to allow for air intake between the devices. If you buy a multi-fan GPU it might be too thick to get enough air when installing multiple GPUs and you will run into thermal throttling.

4. **PCIe Slots**. Moving data to and from the GPU (and exchanging it between GPUs) requires lots of bandwidth. We recommend PCIe 3.0 slots with 16 lanes. If you mount multiple GPUs, be sure to carefully read the motherboard description to ensure that 16× bandwidth is still available when multiple GPUs are used at the same time and that you are getting PCIe 3.0 as opposed to PCIe 2.0 for the additional slots. Some motherboards downgrade to 8× or even 4× bandwidth with multiple GPUs installed. This is partly due to the number of PCIe lanes that the CPU offers.

In short, here are some recommendations for building a deep learning server:

- **Beginner**. Buy a low end GPU with low power consumption (cheap gaming GPUs suitable for deep learning use 150–200W). If you are lucky your current computer supports it.

- **1 GPU**. A low-end CPU with 4 cores will be sufficient and most motherboards suffice. Aim for at least 32 GB DRAM and invest into an SSD for local data access. A power supply with 600W should be sufficient. Buy a GPU with lots of fans.

- **2 GPUs**. A low-end CPU with 4-6 cores will suffice. Aim for 64 GB DRAM and invest into an SSD. You will need in the order of 1000W for two high-end GPUs. In terms of mainboards, make sure that they have *two* PCIe 3.0 x16 slots. If you can, get a mainboard that has two free spaces (60mm spacing) between the PCIe 3.0 x16 slots for extra air. In this case, buy two GPUs with lots of fans.

- **4 GPUs**. Make sure that you buy a CPU with relatively fast single-thread speed (i.e., high clock frequency). You will probably need a CPU with a larger number of PCIe lanes, such as an AMD Threadripper. You will likely need relatively expensive mainboards to get 4 PCIe 3.0 x16 slots since they probably need a PLX to multiplex the PCIe lanes. Buy GPUs with reference design that are narrow and let air in between the GPUs. You need a 1600–2000W power supply and the outlet in your office might not support that. This server will probably run *loud and hot*. You do not want it under your desk. 128 GB of DRAM is recommended. Get an SSD (1–2 TB NVMe) for local storage and a bunch of hard disks in RAID configuration to store your data.

- **8 GPUs**. You need to buy a dedicated multi-GPU server chassis with multiple redundant power supplies (e.g., 2+1 for 1600W per power supply). This will require dual socket server CPUs, 256 GB ECC DRAM, a fast network card (10 GBE recommended), and you will need to check whether the servers support the *physical form factor* of the GPUs. Airflow and wiring placement differ significantly between consumer and server GPUs (e.g., RTX 2080 vs. Tesla V100). This means that you might not be able to install the consumer GPU in a server due to insufficient clearance for the power cable or lack of a suitable wiring harness (as one of the coauthors painfully discovered).

A.5.2 Selecting GPUs

At present, AMD and NVIDIA are the two main manufacturers of dedicated GPUs. NVIDIA was the first to enter the deep learning field and provides better support for deep learning frameworks via CUDA. Therefore, most buyers choose NVIDIA GPUs.

NVIDIA provides two types of GPUs, targeting individual users (e.g., via the GTX and RTX series) and enterprise users (via its Tesla series). The two types of GPUs provide comparable compute power. However, the enterprise user GPUs generally use (passive) forced cooling, more memory, and ECC (error correcting) memory. These GPUs are more suitable for data centers and usually cost ten times more than consumer GPUs.

If you are a large company with 100+ servers you should consider the NVIDIA Tesla series or alternatively use GPU servers in the cloud. For a lab or a small to medium company with 10+ servers the NVIDIA RTX series is likely most cost effective. You can buy preconfigured servers with Supermicro or Asus chassis that hold 4–8 GPUs efficiently.

GPU vendors typically release a new generation every one to two years, such as the GTX 1000 (Pascal) series released in 2017 and the RTX 2000 (Turing) series released in 2019. Each series offers several different models that provide different performance levels. GPU performance is primarily a combination of the following three parameters:

1. **Compute Power**. Generally we look for 32-bit floating-point compute power. 16-bit floating point training (FP16) is also entering the mainstream. If you are only interested in prediction, you can also use 8-bit integer. The latest generation of Turing GPUs offers 4-bit acceleration. Unfortunately at the time of writing the algorithms for training low-precision networks are not yet widespread.

2. **Memory Size**. As your models become larger or the batches used during training grow bigger, you will need more GPU memory. Check for HBM2 (High Bandwidth Memory) vs. GDDR6 (Graphics DDR) memory. HBM2 is faster but much more expensive.

3. **Memory Bandwidth**. You can only get the most out of your compute power when you have sufficient memory bandwidth. Look for wide memory buses if using GDDR6.

For most users, it is enough to look at compute power. Note that many GPUs offer different types of acceleration. For example, NVIDIA's TensorCores accelerate a subset of operators by 5×. Ensure that your libraries support this. The GPU memory should be no less than 4 GB (8 GB is much better). Try to avoid using the GPU also for displaying a GUI (use the built-in graphics instead). If you cannot avoid it, add an extra 2 GB of RAM for safety.

Fig. A.29 compares the 32-bit floating-point compute power and price of the various GTX 900, GTX 1000 and RTX 2000 series models. The prices suggested are those found on Wikipedia at the time of writing.

We can see a number of things:

1. Within each series, price and performance are roughly proportional. Titan models command a significant premium for the benefit of larger amounts of GPU memory. However, the newer models offer better cost effectiveness, as can be seen by comparing the 980 Ti and 1080 Ti. The price does not appear to improve much for the RTX 2000 series. However, this is due to the fact that they offer far superior low precision performance (FP16, INT8, and INT4).

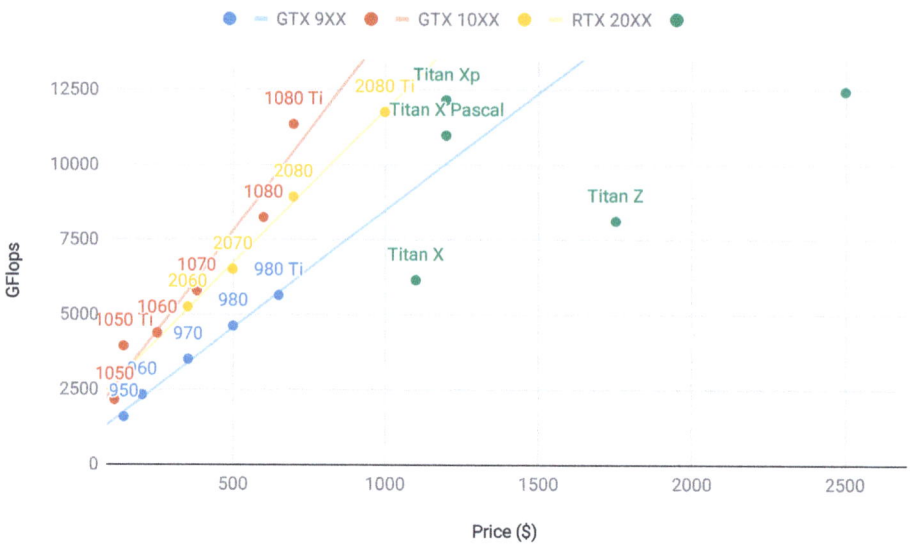

Floating-point compute power and price comparison.

2. The performance-to-cost ratio of the GTX 1000 series is about two times greater than the 900 series.

3. For the RTX 2000 series the performance (in GFLOPs) is an *affine* function of the price.

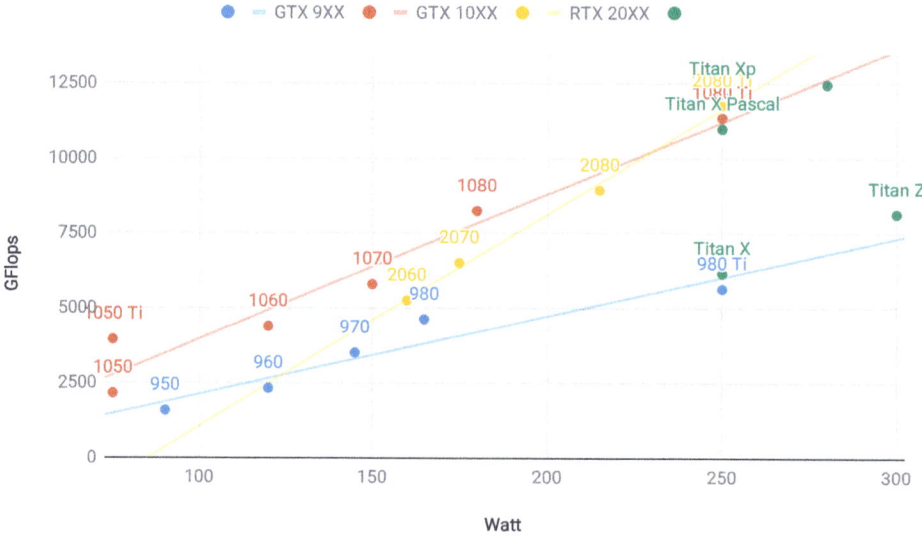

Floating-point compute power and energy consumption.

Fig. A.30 shows how energy consumption scales mostly linearly with the amount of computation. Second, later generations are more efficient. This seems to be contradicted by the graph corresponding to the RTX 2000 series. However, this is a consequence of the TensorCores that draw disproportionately much energy.

A.5.3 Summary

- Watch out for power, PCIe bus lanes, CPU single thread speed, and cooling when building a server.

- You should purchase the latest GPU generation if possible.

- Use the cloud for large deployments.

- High density servers may not be compatible with all GPUs. Check the mechanical and cooling specifications before you buy.

- Use FP16 or lower precision for high efficiency.

Discussions[181].

A.6 Contributing to This Book

Contributions by readers[182] help us improve this book. If you find a typo, an outdated link, something where you think we missed a citation, where the code does not look elegant or where an explanation is unclear, please contribute back and help us help our readers. While in regular books the delay between print runs (and thus between typo corrections) can be measured in years, it typically takes hours to days to incorporate an improvement in this  book. This is all possible due to version control and continuous integration (CI) testing. To do so you need to submit a pull request[183] to the GitHub repository. When your pull request is merged into the code repository by the authors, you will become a contributor.

A.6.1 Submitting Minor Changes

The most common contributions are editing one sentence or fixing typos. We recommend that you find the source file in the GitHub repository[184] and edit the file directly. For example, you can search the file through the Find file[185] button (Fig. A.31) to locate the source file (a markdown file). Then you click the "Edit this file" button on the upper-right corner to make your changes in the markdown file.

After you are done, fill in your change descriptions in the "Propose file change" panel on the page bottom and then click the "Propose file change" button. It will redirect you to a new page to review your changes (Fig. A.37). If everything is good, you can submit a pull request by clicking the "Create pull request" button.

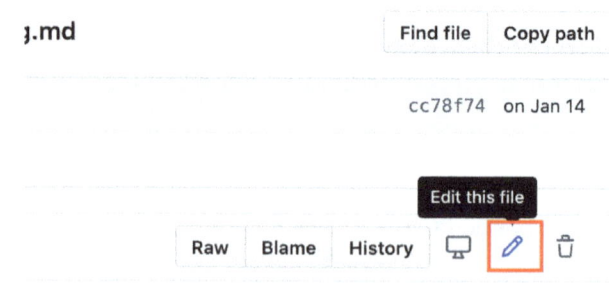

Fig. A.31   Edit the file on Github.

A.6.2 Proposing Major Changes

186

If you plan to update a large portion of text or code, then you need to know a little bit more about the format this book is using. The source file is based on the markdown format[186] with a set of extensions through the D2L-Book[187] package such as referring to equations, images, chapters, and citations. You can use any markdown editors to open these files and make your changes.

187

If you would like to change the code, we recommend that you use the Jupyter Notebook to open these markdown files as described in Section A.1, so that you can run and test your changes. Please remember to clear all outputs before submitting your changes since our CI system will execute the sections you updated to generate outputs.

Some sections may support multiple framework implementations. If you add a new code block, please use `%%tab` to mark this block on the beginning line. For example, `%%tab pytorch` for a PyTorch code block, `%%tab tensorflow` for a TensorFlow code block, or `%%tab all` a shared code block for all implementations. You may refer to the `d2lbook` package for more information.

A.6.3 Submitting Major Changes

We suggest you to use the standard Git process to submit a major change. In a nutshell the process works as described in Fig. A.32.

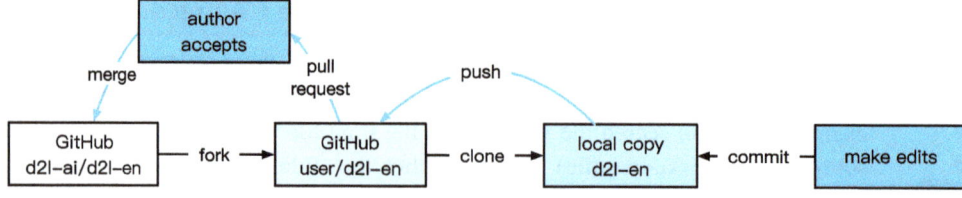

Fig. A.32   Contributing to the book.

We will walk you through the steps in detail. If you are already familiar with Git you can skip this section. For concreteness we assume that the contributor's user name is "astonzhang".

Installing Git

The Git open-source book describes how to install Git[188]. This typically works via `apt install git` on Ubuntu Linux, by installing the Xcode developer tools on macOS, or by using GitHub's desktop client[189]. If you do not have a GitHub account, you need to sign up for one.

Logging in to GitHub

Enter the address[190] of the book's code repository in your browser. Click on the Fork button in the red box at the upper-right of Fig. A.33, to make a copy of the repository of this book. This is now *your copy* and you can change it any way you want.

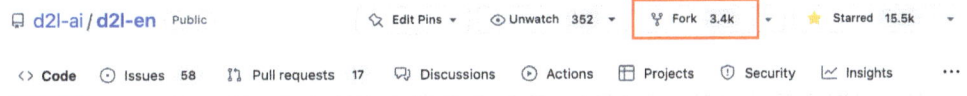

Fig. A.33   The code repository page.

Now, the code repository of this book will be forked (i.e., copied) to your username, such as `astonzhang/d2l-en` shown at the upper-left of Fig. A.34.

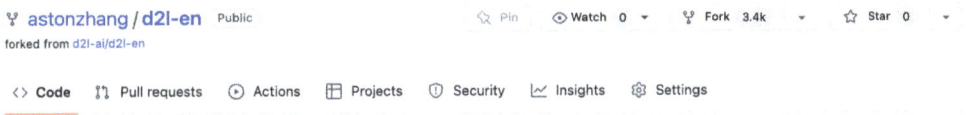

Fig. A.34   The forked code repository.

Cloning the Repository

To clone the repository (i.e., to make a local copy) we need to get its repository address. The green button in Fig. A.35 displays this. Make sure that your local copy is up to date with the main repository if you decide to keep this fork around for longer. For now simply follow the instructions in *Installation* (page xxvi) to get started. The main difference is that you are now downloading *your own fork* of the repository.

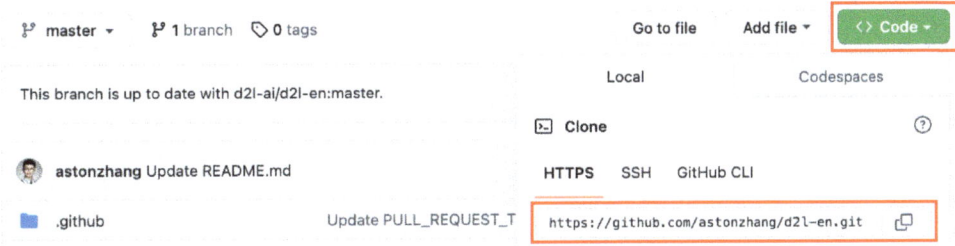

Fig. A.35   Cloning the repository.

```
# Replace your_github_username with your GitHub username
git clone https://github.com/your_github_username/d2l-en.git
```

Editing and Pushing

Now it is time to edit the book. It is best to edit it in the Jupyter Notebook following instructions in Section A.1. Make the changes and check that they are OK. Assume that we have modified a typo in the file ~/d2l-en/chapter_appendix-tools-for-deep-learning/ contributing.md. You can then check which files you have changed.

At this point Git will prompt that the chapter_appendix-tools-for-deep-learning/ contributing.md file has been modified.

```
mylaptop:d2l-en me$ git status
On branch master
Your branch is up-to-date with 'origin/master'.

Changes not staged for commit:
  (use "git add <file>..." to update what will be committed)
  (use "git checkout -- <file>..." to discard changes in working directory)

    modified:   chapter_appendix-tools-for-deep-learning/contributing.md
```

After confirming that this is what you want, execute the following command:

```
git add chapter_appendix-tools-for-deep-learning/contributing.md
git commit -m 'Fix a typo in git documentation'
git push
```

The changed code will then be in your personal fork of the repository. To request the addition of your change, you have to create a pull request for the official repository of the book.

Submitting Pull Requests

As shown in Fig. A.36, go to your fork of the repository on GitHub and select "New pull request". This will open up a screen that shows you the changes between your edits and what is current in the main repository of the book.

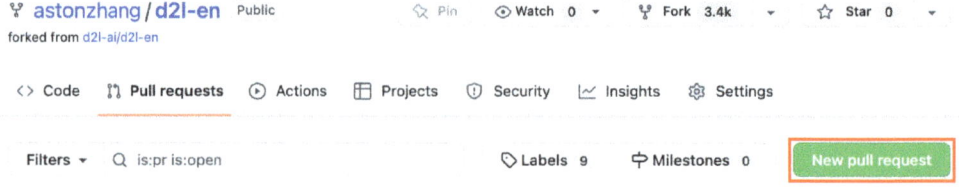

Fig. A.36   New pull request.

Finally, submit a pull request by clicking the button as shown in Fig. A.37. Make sure to describe the changes you have made in the pull request. This will make it easier for the authors to review it and to merge it with the book. Depending on the changes, this might get accepted right away, rejected, or more likely, you will get some feedback on the changes. Once you have incorporated them, you are good to go.

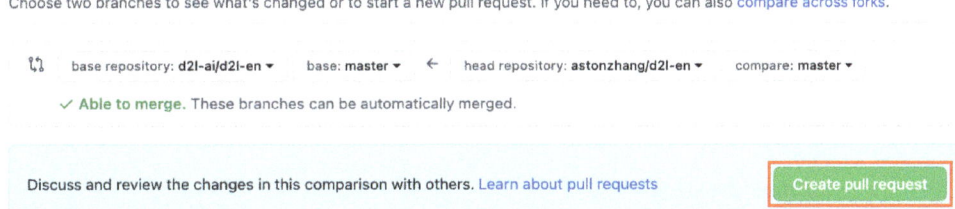

Fig. A.37   Create pull request.

A.6.4 Summary

- You can use GitHub to contribute to this book.

- You can edit the file on GitHub directly for minor changes.

- For a major change, please fork the repository, edit things locally, and only contribute back once you are ready.

- Pull requests are how contributions are being bundled up. Try not to submit huge pull requests since this makes them hard to understand and incorporate. Better send several smaller ones.

A.6.5 Exercises

191   1. Star and fork the d2l-ai/d2l-en repository.

192   2. If you spot anything that needs improvement (e.g., missing a reference), submit a pull request.

3. It is usually a better practice to create a pull request using a new branch. Learn how to do it with Git branching[191].

Discussions[192].

A.7 The d2l API Document

This section displays classes and functions (sorted alphabetically) in the d2l package, showing where they are defined in the book so you can find more detailed implementations and explanations. See also the source code on the GitHub repository [193].

193

A.7.1 Classes

class d2l.torch.AdditiveAttention(*num_hiddens*, *dropout*, ****kwargs*)

Bases: Module

Additive attention.

Defined in Section 11.3.2

forward(*queries*, *keys*, *values*, *valid_lens*)

Defines the computation performed at every call.

Should be overridden by all subclasses.

Note: Although the recipe for forward pass needs to be defined within this function, one should call the Module (page 505) instance afterwards instead of this since the former takes care of running the registered hooks while the latter silently ignores them.

---

training: bool

class d2l.torch.AddNorm(*norm_shape*, *dropout*)

Bases: Module

The residual connection followed by layer normalization.

Defined in Section 11.7.2

forward(*X*, *Y*)

Defines the computation performed at every call.

Should be overridden by all subclasses.

Note: Although the recipe for forward pass needs to be defined within this function, one should call the Module (page 505) instance afterwards instead of this since the former takes care of running the registered hooks while the latter silently ignores them.

---

training: bool

class d2l.torch.AttentionDecoder

    Bases: Decoder (page 501)

    The base attention-based decoder interface.

    Defined in Section 11.4

    property attention_weights

    training: bool

class d2l.torch.Classifier(*plot_train_per_epoch=2*, *plot_valid_per_epoch=1*)

    Bases: Module (page 505)

    The base class of classification models.

    Defined in Section 4.3

    accuracy(*Y_hat*, *Y*, *averaged=True*)

        Compute the number of correct predictions.

        Defined in Section 4.3

    layer_summary(*X_shape*)

        Defined in Section 7.6

    loss(*Y_hat*, *Y*, *averaged=True*)

        Defined in Section 4.5

    training: bool

    validation_step(*batch*)

class d2l.torch.DataModule(*root='../data'*, *num_workers=4*)

    Bases: HyperParameters (page 504)

    The base class of data.

    Defined in Section 3.2.2

    get_dataloader(*train*)

    get_tensorloader(*tensors*, *train*, *indices=slice(0, None, None)*)

        Defined in Section 3.3

    train_dataloader()

    val_dataloader()

class d2l.torch.Decoder

> Bases: Module
>
> The base decoder interface for the encoder–decoder architecture.
>
> Defined in Section 10.6
>
> forward(*X*, *state*)
>
> > Defines the computation performed at every call.
> >
> > Should be overridden by all subclasses.
> >
> > ---
> > **Note:** Although the recipe for forward pass needs to be defined within this function, one should call the Module (page 505) instance afterwards instead of this since the former takes care of running the registered hooks while the latter silently ignores them.
> > ---
>
> init_state(*enc_all_outputs*, **args*)
>
> training: bool

class d2l.torch.DotProductAttention(*dropout*)

> Bases: Module
>
> Scaled dot product attention.
>
> Defined in Section 11.3.2
>
> forward(*queries*, *keys*, *values*, *valid_lens=None*)
>
> > Defines the computation performed at every call.
> >
> > Should be overridden by all subclasses.
> >
> > ---
> > **Note:** Although the recipe for forward pass needs to be defined within this function, one should call the Module (page 505) instance afterwards instead of this since the former takes care of running the registered hooks while the latter silently ignores them.
> > ---
>
> training: bool

class d2l.torch.Encoder

> Bases: Module
>
> The base encoder interface for the encoder–decoder architecture.
>
> Defined in Section 10.6

forward(*X*, **args*)

Defines the computation performed at every call.

Should be overridden by all subclasses.

Note: Although the recipe for forward pass needs to be defined within this function, one should call the `Module` (page 505) instance afterwards instead of this since the former takes care of running the registered hooks while the latter silently ignores them.

---

training: bool

class d2l.torch.EncoderDecoder(*encoder*, *decoder*)

Bases: `Classifier` (page 501)

The base class for the encoder–decoder architecture.

Defined in Section 10.6

forward(*enc_X*, *dec_X*, **args*)

Defines the computation performed at every call.

Should be overridden by all subclasses.

Note: Although the recipe for forward pass needs to be defined within this function, one should call the `Module` (page 505) instance afterwards instead of this since the former takes care of running the registered hooks while the latter silently ignores them.

---

predict_step(*batch*, *device*, *num_steps*, *save_attention_weights=False*)

Defined in Section 10.7.6

training: bool

class d2l.torch.FashionMNIST(*batch_size=64*, *resize=(28, 28)*)

Bases: `DataModule` (page 501)

The Fashion-MNIST dataset.

Defined in Section 4.2

get_dataloader(*train*)

Defined in Section 4.2

text_labels(*indices*)

Return text labels.

Defined in Section 4.2

visualize(*batch*, *nrows=1*, *ncols=8*, *labels=[]*)

    Defined in Section 4.2

class d2l.torch.GRU(*num_inputs*, *num_hiddens*, *num_layers*, *dropout=0*)

    Bases: RNN (page 508)

    The multilayer GRU model.

    Defined in Section 10.3

    training: bool

class d2l.torch.HyperParameters

    Bases: object

    The base class of hyperparameters.

    save_hyperparameters(*ignore=[]*)

        Save function arguments into class attributes.

        Defined in sec_utils

class d2l.torch.LeNet(*lr=0.1*, *num_classes=10*)

    Bases: Classifier (page 501)

    The LeNet-5 model.

    Defined in Section 7.6

    training: bool

class d2l.torch.LinearRegression(*lr*)

    Bases: Module (page 505)

    The linear regression model implemented with high-level APIs.

    Defined in Section 3.5

    configure_optimizers()

        Defined in Section 3.5

    forward(*X*)

        Defined in Section 3.5

    get_w_b()

        Defined in Section 3.5

    loss(*y_hat*, *y*)

        Defined in Section 3.5

    training: bool

class d2l.torch.LinearRegressionScratch(*num_inputs*, *lr*, *sigma=0.01*)

    Bases: Module (page 505)

    The linear regression model implemented from scratch.

    Defined in Section 3.4

    configure_optimizers()

        Defined in Section 3.4

    forward(*X*)

        Defined in Section 3.4

    loss(*y_hat*, *y*)

        Defined in Section 3.4

    training: bool

class d2l.torch.Module(*plot_train_per_epoch=2*, *plot_valid_per_epoch=1*)

    Bases: Module, HyperParameters (page 504)

    The base class of models.

    Defined in Section 3.2

    apply_init(*inputs*, *init=None*)

        Defined in Section 6.4

    configure_optimizers()

        Defined in Section 4.3

    forward(*X*)

        Defines the computation performed at every call.

        Should be overridden by all subclasses.

Note: Although the recipe for forward pass needs to be defined within this function, one should call the Module (page 505) instance afterwards instead of this since the former takes care of running the registered hooks while the latter silently ignores them.

---

    loss(*y_hat*, *y*)

    plot(*key*, *value*, *train*)

        Plot a point in animation.

    training: bool

    training_step(*batch*)

validation_step(*batch*)

class d2l.torch.MTFraEng(*batch_size*, *num_steps=9*, *num_train=512*,
                                      num_val=128)

   Bases: DataModule (page 501)

   The English-French dataset.

   Defined in Section 10.5

   build(*src_sentences*, *tgt_sentences*)

      Defined in Section 10.5.3

   get_dataloader(*train*)

      Defined in Section 10.5.3

class d2l.torch.MultiHeadAttention(*num_hiddens*, *num_heads*, *dropout*,
                                              bias=False, ***kwargs*)

   Bases: Module (page 505)

   Multi-head attention.

   Defined in Section 11.5

   forward(*queries*, *keys*, *values*, *valid_lens*)

      Defines the computation performed at every call.

      Should be overridden by all subclasses.

   Note: Although the recipe for forward pass needs to be defined within this function, one should call the Module (page 505) instance afterwards instead of this since the former takes care of running the registered hooks while the latter silently ignores them.

   ---

   training: bool

   transpose_output(*X*)

      Reverse the operation of transpose_qkv.

      Defined in Section 11.5

   transpose_qkv(*X*)

      Transposition for parallel computation of multiple attention heads.

      Defined in Section 11.5

class d2l.torch.PositionalEncoding(*num_hiddens*, *dropout*, *max_len=1000*)
   Bases: Module

   Positional encoding.

Defined in Section 11.6

forward(*X*)

Defines the computation performed at every call.

Should be overridden by all subclasses.

Note: Although the recipe for forward pass needs to be defined within this function, one should call the Module (page 505) instance afterwards instead of this since the former takes care of running the registered hooks while the latter silently ignores them.

---

training: bool

class d2l.torch.PositionWiseFFN(*ffn_num_hiddens, ffn_num_outputs*)

Bases: Module

The positionwise feed-forward network.

Defined in Section 11.7

forward(*X*)

Defines the computation performed at every call.

Should be overridden by all subclasses.

Note: Although the recipe for forward pass needs to be defined within this function, one should call the Module (page 505) instance afterwards instead of this since the former takes care of running the registered hooks while the latter silently ignores them.

---

training: bool

class d2l.torch.ProgressBoard(*xlabel=None, ylabel=None, xlim=None,*
ylim=None, xscale='linear', yscale='linear',
ls=['-', '--', '-.', ':'], colors=['C0', 'C1', 'C2',
'C3'], fig=None, axes=None, figsize=(3.5, 2.5),
display=True)

Bases: HyperParameters (page 504)

The board that plots data points in animation.

Defined in Section 3.2

draw(*x, y, label, every_n=1*)

Defined in sec_utils

class d2l.torch.Residual(*num_channels*, *use_1x1conv=False*, *strides=1*)

Bases: `Module`

The Residual block of ResNet models.

Defined in Section 8.6

forward(*X*)

Defines the computation performed at every call.

Should be overridden by all subclasses.

Note: Although the recipe for forward pass needs to be defined within this function, one should call the `Module` (page 505) instance afterwards instead of this since the former takes care of running the registered hooks while the latter silently ignores them.

---

training: bool

class d2l.torch.ResNeXtBlock(*num_channels*, *groups*, *bot_mul*,
                              use_1x1conv=False, *strides=1*)

Bases: `Module`

The ResNeXt block.

Defined in Section 8.6.2

forward(*X*)

Defines the computation performed at every call.

Should be overridden by all subclasses.

Note: Although the recipe for forward pass needs to be defined within this function, one should call the `Module` (page 505) instance afterwards instead of this since the former takes care of running the registered hooks while the latter silently ignores them.

---

training: bool

class d2l.torch.RNN(*num_inputs*, *num_hiddens*)

Bases: `Module` (page 505)

The RNN model implemented with high-level APIs.

Defined in Section 9.6

forward(*inputs*, *H=None*)

Defines the computation performed at every call.

Should be overridden by all subclasses.

Note: Although the recipe for forward pass needs to be defined within this function, one should call the `Module` (page 505) instance afterwards instead of this since the former takes care of running the registered hooks while the latter silently ignores them.

`training: bool`

`class d2l.torch.RNNLM`(*rnn, vocab_size, lr=0.01*)

Bases: `RNNLMScratch` (page 509)

The RNN-based language model implemented with high-level APIs.

Defined in Section 9.6

`init_params()`

`output_layer`(*hiddens*)

Defined in Section 9.5

`training: bool`

`class d2l.torch.RNNLMScratch`(*rnn, vocab_size, lr=0.01*)

Bases: `Classifier` (page 501)

The RNN-based language model implemented from scratch.

Defined in Section 9.5

`forward`(*X, state=None*)

Defined in Section 9.5

`init_params()`

`one_hot`(*X*)

Defined in Section 9.5

`output_layer`(*rnn_outputs*)

Defined in Section 9.5

`predict`(*prefix, num_preds, vocab, device=None*)

Defined in Section 9.5

`training: bool`

`training_step`(*batch*)

`validation_step`(*batch*)

class d2l.torch.RNNScratch(*num_inputs*, *num_hiddens*, *sigma=0.01*)

    Bases: Module (page 505)

    The RNN model implemented from scratch.

    Defined in Section 9.5

    forward(*inputs*, *state=None*)

        Defined in Section 9.5

    training: bool

class d2l.torch.Seq2Seq(*encoder*, *decoder*, *tgt_pad*, *lr*)

    Bases: EncoderDecoder (page 503)

    The RNN encoder–decoder for sequence to sequence learning.

    Defined in Section 10.7.3

    configure_optimizers()

        Defined in Section 4.3

    training: bool

    validation_step(*batch*)

class d2l.torch.Seq2SeqEncoder(*vocab_size*, *embed_size*, *num_hiddens*, *num_layers*, *dropout=0*)

    Bases: Encoder (page 502)

    The RNN encoder for sequence-to-sequence learning.

    Defined in Section 10.7

    forward(*X*, **args*)

        Defines the computation performed at every call.

        Should be overridden by all subclasses.

    Note: Although the recipe for forward pass needs to be defined within this function, one should call the Module (page 505) instance afterwards instead of this since the former takes care of running the registered hooks while the latter silently ignores them.

---

    training: bool

class d2l.torch.SGD(*params*, *lr*)

    Bases: HyperParameters (page 504)

    Minibatch stochastic gradient descent.

    Defined in Section 3.4

```
step()
```

```
zero_grad()
```

class d2l.torch.SoftmaxRegression(*num_outputs*, *lr*)

    Bases: Classifier (page 501)

    The softmax regression model.

    Defined in Section 4.5

    forward(*X*)

        Defines the computation performed at every call.

        Should be overridden by all subclasses.

    Note: Although the recipe for forward pass needs to be defined within this function, one should call the Module (page 505) instance afterwards instead of this since the former takes care of running the registered hooks while the latter silently ignores them.

---

    training: bool

class d2l.torch.SyntheticRegressionData(*w*, *b*, *noise=0.01*, *num_train=1000*, *num_val=1000*, *batch_size=32*)

    Bases: DataModule (page 501)

    Synthetic data for linear regression.

    Defined in Section 3.3

    get_dataloader(*train*)

        Defined in Section 3.3

class d2l.torch.TimeMachine(*batch_size*, *num_steps*, *num_train=10000*, *num_val=5000*)

    Bases: DataModule (page 501)

    The Time Machine dataset.

    Defined in Section 9.2

    build(*raw_text*, *vocab=None*)

        Defined in Section 9.2

    get_dataloader(*train*)

        Defined in Section 9.3.3

class d2l.torch.Trainer(*max_epochs*, *num_gpus=0*, *gradient_clip_val=0*)

    Bases: HyperParameters (page 504)

The base class for training models with data.

Defined in Section 3.2.2

    clip_gradients(*grad_clip_val*, *model*)

        Defined in Section 9.5

    fit(*model*, *data*)

    fit_epoch()

        Defined in Section 3.4

    prepare_batch(*batch*)

        Defined in Section 6.7

    prepare_data(*data*)

    prepare_model(*model*)

        Defined in Section 6.7

class d2l.torch.TransformerEncoder(*vocab_size*, *num_hiddens*, *ffn_num_hiddens*, *num_heads*, *num_blks*, *dropout*, *use_bias=False*)

    Bases: Encoder (page 502)

The Transformer encoder.

Defined in Section 11.7.4

    forward(*X*, *valid_lens*)

        Defines the computation performed at every call.

        Should be overridden by all subclasses.

Note: Although the recipe for forward pass needs to be defined within this function, one should call the Module (page 505) instance afterwards instead of this since the former takes care of running the registered hooks while the latter silently ignores them.

---

    training: bool

class d2l.torch.TransformerEncoderBlock(*num_hiddens*, *ffn_num_hiddens*, *num_heads*, *dropout*, *use_bias=False*)

    Bases: Module

The Transformer encoder block.

Defined in Section 11.7.2

forward(*X*, *valid_lens*)

Defines the computation performed at every call.

Should be overridden by all subclasses.

Note: Although the recipe for forward pass needs to be defined within this function, one should call the Module (page 505) instance afterwards instead of this since the former takes care of running the registered hooks while the latter silently ignores them.

---

training: bool

class d2l.torch.Vocab(*tokens=[]*, *min_freq=0*, *reserved_tokens=[]*)

Bases: object

Vocabulary for text.

to_tokens(*indices*)

property unk

A.7.2 Functions

d2l.torch.add_to_class(*Class*)

Register functions as methods in created class.

Defined in Section 3.2

d2l.torch.bleu(*pred_seq*, *label_seq*, *k*)

Compute the BLEU.

Defined in Section 10.7.6

d2l.torch.check_len(*a*, *n*)

Check the length of a list.

Defined in Section 9.5

d2l.torch.check_shape(*a*, *shape*)

Check the shape of a tensor.

Defined in Section 9.5

d2l.torch.corr2d(*X*, *K*)

Compute 2D cross-correlation.

Defined in Section 7.2

`d2l.torch.cpu()`

Get the CPU device.

Defined in Section 6.7

`d2l.torch.gpu(`*i=0*`)`

Get a GPU device.

Defined in Section 6.7

`d2l.torch.init_cnn(`*module*`)`

Initialize weights for CNNs.

Defined in Section 7.6

`d2l.torch.init_seq2seq(`*module*`)`

Initialize weights for sequence-to-sequence learning.

Defined in Section 10.7

`d2l.torch.masked_softmax(`*X*, *valid_lens*`)`

Perform softmax operation by masking elements on the last axis.

Defined in Section 11.3

`d2l.torch.num_gpus()`

Get the number of available GPUs.

Defined in Section 6.7

`d2l.torch.plot(`*X*, *Y=None*, *xlabel=None*, *ylabel=None*, *legend=[ ]*, *xlim=None*, *ylim=None*, *xscale='linear'*, *yscale='linear'*, *fmts=('-', 'm--', 'g-.', 'r:')*, *figsize=(3.5, 2.5)*, *axes=None*`)`

Plot data points.

Defined in Section 2.4

`d2l.torch.set_axes(`*axes*, *xlabel*, *ylabel*, *xlim*, *ylim*, *xscale*, *yscale*, *legend*`)`
Set the axes for matplotlib.

Defined in Section 2.4

`d2l.torch.set_figsize(`*figsize=(3.5, 2.5)*`)`

Set the figure size for matplotlib.

Defined in Section 2.4

`d2l.torch.show_heatmaps(`*matrices*, *xlabel*, *ylabel*, *titles=None*, *figsize=(2.5, 2.5)*, *cmap='Reds'*`)`

Show heatmaps of matrices.

Defined in Section 11.1

`d2l.torch.show_list_len_pair_hist`(*legend*, *xlabel*, *ylabel*, *xlist*, *ylist*)

Plot the histogram for list length pairs.

Defined in Section 10.5

`d2l.torch.try_all_gpus()`

Return all available GPUs, or [cpu(),] if no GPU exists.

Defined in Section 6.7

`d2l.torch.try_gpu`(*i=0*)

Return gpu(i) if exists, otherwise return cpu().

Defined in Section 6.7

`d2l.torch.use_svg_display()`

Use the svg format to display a plot in Jupyter.

Defined in Section 2.4

References

Abadi, M., Barham, P., Chen, J., Chen, Z., Davis, A., Dean, J., ... et al. (2016). Tensor-Flow: a system for large-scale machine learning. *12th USENIX Symposium on Operating Systems Design and Implementation (OSDI 16)* (pp. 265–283).

Abdel-Hamid, O., Mohamed, A.-R., Jiang, H., Deng, L., Penn, G., & Yu, D. (2014). Convolutional neural networks for speech recognition. *IEEE/ACM Transactions on Audio, Speech, and Language Processing*, *22*(10), 1533–1545.

Ahmed, A., Aly, M., Gonzalez, J., Narayanamurthy, S., & Smola, A. J. (2012). Scalable inference in latent variable models. *Proceedings of the Fifth ACM International Conference on Web Search and Data Mining* (pp. 123–132).

Akiba, T., Sano, S., Yanase, T., Ohta, T., & Koyama, M. (2019). Optuna: a next-generation hyperparameter optimization framework. *Proceedings of the 25th ACM SIGKDD International Conference on Knowledge Discovery & Data Mining*.

Alayrac, J.-B., Donahue, J., Luc, P., Miech, A., Barr, I., Hasson, Y., ... et al. (2022). Flamingo: a visual language model for few-shot learning. *ArXiv:2204.14198*.

Alsallakh, B., Kokhlikyan, N., Miglani, V., Yuan, J., & Reblitz-Richardson, O. (2020). Mind the PAD – CNNs can develop blind spots. *ArXiv:2010.02178*.

Anil, R., Gupta, V., Koren, T., Regan, K., & Singer, Y. (2020). Scalable second-order optimization for deep learning. *ArXiv:2002.09018*.

Anil, R., Dai, A. M., Firat, O., Johnson, M., Lepikhin, D., Passos, A., ... et al. (2023). PaLM 2 Technical Report. *ArXiv:2305.10403*.

Aronszajn, N. (1950). Theory of reproducing kernels. *Transactions of the American Mathematical Society*, *68*(3), 337–404.

Ba, J. L., Kiros, J. R., & Hinton, G. E. (2016). Layer normalization. *ArXiv:1607.06450*.

Baevski, A., & Auli, M. (2018). Adaptive input representations for neural language modeling. *International Conference on Learning Representations*.

Bahdanau, D., Cho, K., & Bengio, Y. (2014). Neural machine translation by jointly learning to align and translate. *ArXiv:1409.0473*.

Bai, Y., Kadavath, S., Kundu, S., Askell, A., Kernion, J., Jones, A., ... et al. (2022). Constitutional AI: harmlessness from AI feedback. *ArXiv:2212.08073*.

Baptista, R., & Poloczek, M. (2018). Bayesian optimization of combinatorial structures. *Proceedings of the 35th International Conference on Machine Learning*.

Bardenet, R., Brendel, M., Kégl, B., & Sebag, M. (2013). Collaborative hyperparameter tuning. *Proceedings of the 30th International Conference on Machine Learning (ICML'13)*.

Bay, H., Tuytelaars, T., & Van Gool, L. (2006). SURF: Speeded up robust features. *European Conference on Computer Vision* (pp. 404–417).

Bellman, R. (1952). On the theory of dynamic programming. *Proceedings of the National Academy of Sciences*, *38*(8), 716–719.

Bellman, R. (1957a). A Markovian decision process. *Journal of Mathematics and Mechanics*, *6*(5), 679–684. URL: http://www.jstor.org/stable/24900506

Bellman, R. (1957b). *Dynamic Programming*. Dover Publications.

Bellman, R. (1966). Dynamic programming. *Science*, *153*, 34–37.

Beltagy, I., Peters, M. E., & Cohan, A. (2020). Longformer: the long-document transformer. *ArXiv:2004.05150*.

Bengio, Y., Simard, P., & Frasconi, P. (1994). Learning long-term dependencies with gradient descent is difficult. *IEEE Transactions on Neural Networks*, *5*(2), 157–166.

Bengio, Y., Ducharme, R., Vincent, P., & Jauvin, C. (2003). A neural probabilistic language model. *Journal of Machine Learning Research*, *3*(Feb), 1137–1155.

Bergstra, J., Breuleux, O., Bastien, F., Lamblin, P., Pascanu, R., Desjardins, G., ... Bengio, Y. (2010). Theano: a CPU and GPU math compiler in Python. *Proc. 9th Python in Science Conference* (pp. 3–10).

Bergstra, J., Bardenet, R., Bengio, Y., & Kégl, B. (2011). Algorithms for hyper-parameter optimization. *Advances in Neural Information Processing Systems*, *24*.

Beutel, A., Murray, K., Faloutsos, C., & Smola, A. J. (2014). CoBaFi: collaborative Bayesian filtering. *Proceedings of the 23rd International Conference on World Wide Web* (pp. 97–108).

Bishop, C. M. (1995). Training with noise is equivalent to Tikhonov regularization. *Neural Computation*, *7*(1), 108–116.

Bishop, C. M. (2006). *Pattern Recognition and Machine Learning*. Springer.

Black, F., & Scholes, M. (1973). The pricing of options and corporate liabilities. *Journal of Political Economy*, *81*, 637–654.

Bodla, N., Singh, B., Chellappa, R., & Davis, L. S. (2017). Soft-NMS-improving object detection with one line of code. *Proceedings of the IEEE International Conference on Computer Vision* (pp. 5561–5569).

Bojanowski, P., Grave, E., Joulin, A., & Mikolov, T. (2017). Enriching word vectors with subword information. *Transactions of the Association for Computational Linguistics*, *5*, 135–146.

Bollobás, B. (1999). *Linear Analysis*. Cambridge University Press.

Bommasani, R., Hudson, D. A., Adeli, E., Altman, R., Arora, S., von Arx, S., ... et al. (2021). On the opportunities and risks of foundation models. *ArXiv:2108.07258*.

Bottou, L. (2010). Large-scale machine learning with stochastic gradient descent. *Proceedings of COMPSTAT'2010* (pp. 177–186). Springer.

Bottou, L., & Le Cun, Y. (1988). SN: a simulator for connectionist models. *Proceedings of NeuroNimes 88* (pp. 371–382). Nimes, France. URL: http://leon.bottou.org/papers/bottou-lecun-88

Boucheron, S., Bousquet, O., & Lugosi, G. (2005). Theory of classification: a survey of some recent advances. *ESAIM: Probability and Statistics*, *9*, 323–375.

Bowman, S. R., Angeli, G., Potts, C., & Manning, C. D. (2015). A large annotated corpus for learning natural language inference. *ArXiv:1508.05326*.

Boyd, S., & Vandenberghe, L. (2004). *Convex Optimization*. Cambridge, England: Cambridge University Press.

Bradley, R. A., & Terry, M. E. (1952). Rank analysis of incomplete block designs: I. The method of paired comparisons. *Biometrika*, *39*(3/4), 324–345.

Brown, N., & Sandholm, T. (2017). Libratus: the superhuman AI for no-limit poker. *IJCAI* (pp. 5226–5228).

Brown, P. F., Cocke, J., Della Pietra, S. A., Della Pietra, V. J., Jelinek, F., Lafferty, J., … Roossin, P. S. (1990). A statistical approach to machine translation. *Computational Linguistics*, *16*(2), 79–85.

Brown, T., Mann, B., Ryder, N., Subbiah, M., Kaplan, J. D., Dhariwal, P., … et al. (2020). Language models are few-shot learners. *Advances in Neural Information Processing Systems*, *33*, 1877–1901.

Brown, P. F., Cocke, J., Della Pietra, S. A., Della Pietra, V. J., Jelinek, F., Mercer, R. L., & Roossin, P. (1988). A statistical approach to language translation. *COLING Budapest 1988 Volume 1: International Conference on Computational Linguistics*.

Buslaev, A., Iglovikov, V. I., Khvedchenya, E., Parinov, A., Druzhinin, M., & Kalinin, A. A. (2020). Albumentations: Fast and flexible image augmentations. *Information*, *11*(2), 125.

Campbell, M., Hoane Jr, A. J., & Hsu, F.-h. (2002). Deep blue. *Artificial Intelligence*, *134*(1-2), 57–83.

Canny, J. (1987). A computational approach to edge detection. *Readings in Computer Vision* (pp. 184–203). Elsevier.

Cer, D., Diab, M., Agirre, E., Lopez-Gazpio, I., & Specia, L. (2017). SemEval-2017 Task 1: semantic textual similarity multilingual and crosslingual focused evaluation. *Proceedings of the 11th International Workshop on Semantic Evaluation (SemEval-2017)* (pp. 1–14).

Chan, W., Jaitly, N., Le, Q. V., & Vinyals, O. (2015). Listen, attend and spell. *ArXiv:1508.01211*.

Chen, T., Li, M., Li, Y., Lin, M., Wang, N., Wang, M., … Zhang, Z. (2015). MXNET: a flexible and efficient machine learning library for heterogeneous distributed systems. *ArXiv:1512.01274*.

Chen, L., Lu, K., Rajeswaran, A., Lee, K., Grover, A., Laskin, M., … Mordatch, I. (2021). Decision transformer: reinforcement learning via sequence modeling. *Advances in Neural Information Processing Systems*, *34*, 15084–15097.

Cheng, J., Dong, L., & Lapata, M. (2016). Long short-term memory-networks for machine reading. *Proceedings of the 2016 Conference on Empirical Methods in Natural Language Processing* (pp. 551–561).

Chetlur, S., Woolley, C., Vandermersch, P., Cohen, J., Tran, J., Catanzaro, B., & Shelhamer, E. (2014). CuDNN: Efficient primitives for deep learning. *ArXiv:1410.0759*.

Cho, K., Van Merriënboer, B., Bahdanau, D., & Bengio, Y. (2014a). On the properties of neural machine translation: Encoder–decoder approaches. *ArXiv:1409.1259*.

Cho, K., Van Merriënboer, B., Gulcehre, C., Bahdanau, D., Bougares, F., Schwenk, H., & Bengio, Y. (2014b). Learning phrase representations using RNN encoder–decoder for statistical machine translation. *ArXiv:1406.1078*.

Chowdhery, A., Narang, S., Devlin, J., Bosma, M., Mishra, G., Roberts, A., … et al. (2022). PaLM: scaling language modeling with pathways. *ArXiv:2204.02311*.

Chung, J., Gulcehre, C., Cho, K., & Bengio, Y. (2014). Empirical evaluation of gated recurrent neural networks on sequence modeling. *ArXiv:1412.3555*.

Clark, K., Luong, M.-T., Le, Q. V., & Manning, C. D. (2020). ELECTRA: pre-training text encoders as discriminators rather than generators. *International Conference on Learning Representations*.

Collobert, R., Weston, J., Bottou, L., Karlen, M., Kavukcuoglu, K., & Kuksa, P. (2011). Natural language processing (almost) from scratch. *Journal of Machine Learning Research*, *12*, 2493–2537.

Cordonnier, J.-B., Loukas, A., & Jaggi, M. (2020). On the relationship between self-attention and convolutional layers. *International Conference on Learning Representations*.

Cover, T., & Thomas, J. (1999). *Elements of Information Theory*. John Wiley & Sons.

Csiszár, I. (2008). Axiomatic characterizations of information measures. *Entropy*, *10*(3), 261–273.

Cybenko, G. (1989). Approximation by superpositions of a sigmoidal function. *Mathematics of Control, Signals and Systems*, *2*(4), 303–314.

Dalal, N., & Triggs, B. (2005). Histograms of oriented gradients for human detection. *2005 IEEE Computer Society Conference on Computer Vision and Pattern Recognition (CVPR'05)* (pp. 886–893).

De Cock, D. (2011). Ames, Iowa: alternative to the Boston housing data as an end of semester regression project. *Journal of Statistics Education*, *19*(3).

Dean, J., Corrado, G. S., Monga, R., Chen, K., Devin, M., Le, Q. V., … et al. (2012). Large scale distributed deep networks. *Proceedings of the 25th International Conference on Neural Information Processing Systems, Volume 1* (pp. 1223–1231).

DeCandia, G., Hastorun, D., Jampani, M., Kakulapati, G., Lakshman, A., Pilchin, A., … Vogels, W. (2007). Dynamo: Amazon's highly available key-value store. *ACM SIGOPS Operating Systems Review* (pp. 205–220).

Deng, J., Dong, W., Socher, R., Li, L.-J., Li, K., & Fei-Fei, L. (2009). Imagenet: a large-scale hierarchical image database. *2009 IEEE Conference on Computer Vision and Pattern Recognition* (pp. 248–255).

Der Kiureghian, A., & Ditlevsen, O. (2009). Aleatory or epistemic? does it matter? *Structural Safety*, *31*(2), 105–112.

Devlin, J., Chang, M.-W., Lee, K., & Toutanova, K. (2018). BERT: Pre-training of deep bidirectional transformers for language understanding. *ArXiv:1810.04805*.

Dinh, L., Krueger, D., & Bengio, Y. (2014). NICE: non-linear independent components estimation. *ArXiv:1410.8516*.

Dinh, L., Sohl-Dickstein, J., & Bengio, S. (2017). Density estimation using real NVP. *International Conference on Learning Representations*.

Doersch, C., Gupta, A., & Efros, A. A. (2015). Unsupervised visual representation learning by context prediction. *Proceedings of the IEEE International Conference on Computer Vision* (pp. 1422–1430).

Dosovitskiy, A., Beyer, L., Kolesnikov, A., Weissenborn, D., Zhai, X., Unterthiner, T., … et al. (2021). An image is worth 16 x 16 words: transformers for image recognition at scale. *International Conference on Learning Representations*.

Duchi, J., Hazan, E., & Singer, Y. (2011). Adaptive subgradient methods for online learning and stochastic optimization. *Journal of Machine Learning Research*, *12*, 2121–2159.

Dumoulin, V., & Visin, F. (2016). A guide to convolution arithmetic for deep learning. *ArXiv:1603.07285*.

Dwivedi, V. P., & Bresson, X. (2020). A generalization of transformer networks to graphs. *ArXiv:2012.09699*.

Dwork, C., Feldman, V., Hardt, M., Pitassi, T., Reingold, O., & Roth, A. L. (2015). Preserving statistical validity in adaptive data analysis. *Proceedings of the 47th Annual ACM Symposium on Theory of Computing* (pp. 117–126).

Elman, J. L. (1990). Finding structure in time. *Cognitive Science*, *14*(2), 179–211.

Elsken, T., Metzen, J. H., & Hutter, F. (2018). Neural architecture search: a ssurvey. *ArXiv:1808.05377 [stat.ML]*.

Fechner, G. T. (1860). *Elemente der Psychophysik*. Vol. 2. Breitkopf u. Härtel.

Fedus, W., Zoph, B., & Shazeer, N. (2022). Switch transformers: scaling to trillion parameter models with simple and efficient sparsity. *Journal of Machine Learning Research*, *23*(120), 1–39.

Fernando, R. (2004). *GPU Gems: Programming Techniques, Tips, and Tricks for Real-Time Graphics*. Addison-Wesley.

Feurer, M., & Hutter, F. (2018). Hyperparameter ptimization. *Automatic Machine Learning: Methods, Systems, Challenges*. Springer.

Feurer, M., Letham, B., Hutter, F., & Bakshy, E. (2022). Practical transfer learning for Bayesian optimization. *ArXiv:1802.02219 [stat.ML]*.

Field, D. J. (1987). Relations between the statistics of natural images and the response properties of cortical cells. *JOSA A*, *4*(12), 2379–2394.

Fisher, R. A. (1925). *Statistical Methods for Research Workers*. Oliver & Boyd.

Flammarion, N., & Bach, F. (2015). From averaging to acceleration, there is only a step-size. *Conference on Learning Theory* (pp. 658–695).

Forrester, A. I., Sóbester, A., & Keane, A. J. (2007). Multi-fidelity optimization via surrogate modelling. *Proceedings of the Royal Society A: Mathematical, Physical and Engineering Sciences*, *463*(2088), 3251–3269.

Franceschi, L., Donini, M., Frasconi, P., & Pontil, M. (2017). Forward and reverse gradient-based hyperparameter optimization. *Proceedings of the 34th International Conference on Machine Learning (ICML'17)*.

Frankle, J., & Carbin, M. (2018). The lottery ticket hypothesis: finding sparse, trainable neural networks. *ArXiv:1803.03635*.

Frazier, P. I. (2018). A tutorial on Bayesian optimization. *ArXiv:1807.02811*.

Freund, Y., & Schapire, R. E. (1996). Experiments with a new boosting algorithm. *Proceedings of the International Conference on Machine Learning* (pp. 148–156).

Friedman, J. H. (1987). Exploratory projection pursuit. *Journal of the American Statistical Association*, *82*(397), 249–266.

Frostig, R., Johnson, M. J., & Leary, C. (2018). Compiling machine learning programs via high-level tracing. *Proceedings of Systems for Machine Learning*.

Fukushima, K. (1982). Neocognitron: a self-organizing neural network model for a mechanism of visual pattern recognition. *Competition and Cooperation in Neural Nets* (pp. 267–285). Springer.

Gardner, J., Pleiss, G., Weinberger, K. Q., Bindel, D., & Wilson, A. G. (2018). GPyTorch: blackbox matrix–matrix Gaussian process inference with GPU acceleration. *Advances in Neural Information Processing Systems*.

Garg, S., Balakrishnan, S., Kolter, Z., & Lipton, Z. (2021). RATT: leveraging unlabeled data to guarantee generalization. *International Conference on Machine Learning* (pp. 3598–3609).

Gatys, L. A., Ecker, A. S., & Bethge, M. (2016). Image style transfer using convolutional neural networks. *Proceedings of the IEEE Conference on Computer Vision and Pattern Recognition* (pp. 2414–2423).

Gauss, C. F. (1809). Theoria motus corporum coelestum. *Werke*. Königlich Preussische Akademie der Wissenschaften.

Gibbs, J. W. (1902). *Elementary Principles of Statistical Mhanics*. Scribner's.

Ginibre, J. (1965). Statistical ensembles of complex, quaternion, and real matrices. *Journal of Mathematical Physics*, *6*(3), 440–449.

Girshick, R. (2015). Fast R-CNN. *Proceedings of the IEEE International Conference on Computer Vision* (pp. 1440–1448).

Girshick, R., Donahue, J., Darrell, T., & Malik, J. (2014). Rich feature hierarchies for accurate object detection and semantic segmentation. *Proceedings of the IEEE Conference on Computer Vision and Pattern Recognition* (pp. 580–587).

Glorot, X., & Bengio, Y. (2010). Understanding the difficulty of training deep feedforward neural networks. *Proceedings of the 13th International Conference on Artificial Intelligence and Statistics* (pp. 249–256).

Goh, G. (2017). Why momentum really works. *Distill*. URL: http://distill.pub/2017/momentum

Goldberg, D., Nichols, D., Oki, B. M., & Terry, D. (1992). Using collaborative filtering to weave an information tapestry. *Communications of the ACM*, *35*(12), 61–71.

Golub, G. H., & Van Loan, C. F. (1996). *Matrix Computations*. Johns Hopkins University Press.

Goodfellow, I., Pouget-Abadie, J., Mirza, M., Xu, B., Warde-Farley, D., Ozair, S., … Bengio, Y. (2014). Generative adversarial nets. *Advances in Neural Information Processing Systems* (pp. 2672–2680).

Goodfellow, I., Bengio, Y., & Courville, A. (2016). *Deep Learning*. MIT Press. http://www.deeplearningbook.org.

Gotmare, A., Keskar, N. S., Xiong, C., & Socher, R. (2018). A closer look at deep learning heuristics: learning rate restarts, warmup and distillation. *ArXiv:1810.13243*.

Goyal, A., Bochkovskiy, A., Deng, J., & Koltun, V. (2021). Non-deep networks. *ArXiv:2110.07641*.

Graham, B. (2014). Fractional max-pooling. *ArXiv:1412.6071*.

Graves, A. (2013). Generating sequences with recurrent neural networks. *ArXiv:1308.0850*.

Graves, A., & Schmidhuber, J. (2005). Framewise phoneme classification with bidirectional LSTM and other neural network architectures. *Neural Networks*, *18*(5-6), 602–610.

Graves, A., Liwicki, M., Fernández, S., Bertolami, R., Bunke, H., & Schmidhuber, J. (2008). A novel connectionist system for unconstrained handwriting recognition. *IEEE Transactions on Pattern Analysis and Machine Intelligence*, *31*(5), 855–868.

Griewank, A. (1989). On automatic differentiation. *Mathematical Programming: Recent Developments and Applications* (pp. 83–107). Kluwer.

Gulati, A., Qin, J., Chiu, C.-C., Parmar, N., Zhang, Y., Yu, J., … et al. (2020). Conformer: convolution-augmented transformer for speech recognition. *Proc. Interspeech 2020*, pp. 5036–5040.

Guyon, I., Gunn, S., Nikravesh, M., & Zadeh, L. A. (2008). *Feature Extraction: Foundations and Applications*. Springer.

Hadjis, S., Zhang, C., Mitliagkas, I., Iter, D., & Ré, C. (2016). Omnivore: an optimizer for multi-device deep learning on CPUs and GPUs. *ArXiv:1606.04487*.

Hartley, R. I., & Kahl, F. (2009). Global optimization through rotation space search. *International Journal of Computer Vision, 82*(1), 64–79.

Hartley, R., & Zisserman, A. (2000). *Multiple View Geometry in Computer Vision*. Cambridge University Press.

He, K., Zhang, X., Ren, S., & Sun, J. (2015). Delving deep into rectifiers: surpassing human-level performance on ImageNet classification. *Proceedings of the IEEE International Conference on Computer Vision* (pp. 1026–1034).

He, K., Zhang, X., Ren, S., & Sun, J. (2016a). Deep residual learning for image recognition. *Proceedings of the IEEE Conference on Computer Vision and Pattern Recognition* (pp. 770–778).

He, K., Zhang, X., Ren, S., & Sun, J. (2016b). Identity mappings in deep residual networks. *European Conference on Computer Vision* (pp. 630–645).

He, K., Gkioxari, G., Dollár, P., & Girshick, R. (2017). Mask R-CNN. *Proceedings of the IEEE International Conference on Computer Vision* (pp. 2961–2969).

He, K., Chen, X., Xie, S., Li, Y., Dollár, P., & Girshick, R. (2022). Masked autoencoders are scalable vision learners. *Proceedings of the IEEE/CVF Conference on Computer Vision and Pattern Recognition* (pp. 16000–16009).

Hebb, D. O. (1949). *The Organization of Behavior*. Wiley.

Hendrycks, D., & Gimpel, K. (2016). Gaussian error linear units (GELUs). *ArXiv:1606.08415*.

Hennessy, J. L., & Patterson, D. A. (2011). *Computer Architecture: A Quantitative Approach*. Elsevier.

Ho, J., Jain, A., & Abbeel, P. (2020). Denoising diffusion probabilistic models. *Advances in Neural Information Processing Systems, 33*, 6840–6851.

Hochreiter, S., & Schmidhuber, J. (1997). Long short-term memory. *Neural Computation, 9*(8), 1735–1780.

Hochreiter, S., Bengio, Y., Frasconi, P., & Schmidhuber, J. (2001). Gradient flow in recurrent nets: the difficulty of learning long-term dependencies. *A Field Guide to Dynamical Recurrent Neural Networks*. IEEE Press.

Hoffmann, J., Borgeaud, S., Mensch, A., Buchatskaya, E., Cai, T., Rutherford, E., … et al. (2022). Training compute-optimal large language models. *ArXiv:2203.15556*.

Howard, A., Sandler, M., Chu, G., Chen, L.-C., Chen, B., Tan, M., … Adam, H. (2019). Searching for MobileNetV3. *Proceedings of the IEEE/CVF International Conference on Computer Vision* (pp. 1314–1324).

Hoyer, P. O., Janzing, D., Mooij, J. M., Peters, J., & Schölkopf, B. (2009). Nonlinear causal discovery with additive noise models. *Advances in Neural Information Processing Systems* (pp. 689–696).

Hu, Y., Koren, Y., & Volinsky, C. (2008). Collaborative filtering for implicit feedback datasets. *2008 8th IEEE International Conference on Data Mining* (pp. 263–272).

Hu, J., Shen, L., & Sun, G. (2018). Squeeze-and-excitation networks. *Proceedings of the IEEE Conference on Computer Vision and Pattern Recognition* (pp. 7132–7141).

Hu, Z., Lee, R. K.-W., Aggarwal, C. C., & Zhang, A. (2022). Text style transfer: a review and experimental evaluation. *SIGKDD Explor. Newsl.*, *24*(1). URL: https://doi.org/10.1145/3544903.3544906

Huang, Z., Xu, W., & Yu, K. (2015). Bidirectional LSTM–CRF models for sequence tagging. *ArXiv:1508.01991*.

Huang, G., Liu, Z., Van Der Maaten, L., & Weinberger, K. Q. (2017). Densely connected convolutional networks. *Proceedings of the IEEE Conference on Computer Vision and Pattern Recognition* (pp. 4700–4708).

Huang, C.-Z. A., Vaswani, A., Uszkoreit, J., Simon, I., Hawthorne, C., Shazeer, N., … Eck, D. (2018). Music transformer: generating music with long-term structure. *International Conference on Learning Representations*.

Hubel, D. H., & Wiesel, T. N. (1959). Receptive fields of single neurones in the cat's striate cortex. *Journal of Physiology*, *148*(3), 574–591.

Hubel, D. H., & Wiesel, T. N. (1962). Receptive fields, binocular interaction and functional architecture in the cat's visual cortex. *Journal of Physiology*, *160*(1), 106–154.

Hubel, D. H., & Wiesel, T. N. (1968). Receptive fields and functional architecture of monkey striate cortex. *Journal of Physiology*, *195*(1), 215–243.

Hutter, F., Hoos, H., & Leyton-Brown, K. (2011). Sequential model-based optimization for general algorithm configuration. *Proceedings of the Fifth International Conference on Learning and Intelligent Optimization (LION'11)*.

Hutter, F., Kotthoff, L., & Vanschoren, J. (Eds.) (2019). *Automated Machine Learning: Methods, Systems, Challenges*. Springer.

Ioffe, S. (2017). Batch renormalization: towards reducing minibatch dependence in batch-normalized models. *Advances in Neural Information Processing Systems* (pp. 1945–1953).

Ioffe, S., & Szegedy, C. (2015). Batch normalization: accelerating deep network training by reducing internal covariate shift. *ArXiv:1502.03167*.

Izmailov, P., Podoprikhin, D., Garipov, T., Vetrov, D., & Wilson, A. G. (2018). Averaging weights leads to wider optima and better generalization. *ArXiv:1803.05407*.

Jacot, A., Gabriel, F., & Hongler, C. (2018). Neural tangent kernel: convergence and generalization in neural networks. *Advances in Neural Information Processing Systems*.

Jaeger, H. (2002). *Tutorial on training recurrent neural networks, covering BPPT, RTRL, EKF and the "echo state network" approach*. GMD-Forschungszentrum Informationstechnik Bonn.

Jamieson, K., & Talwalkar, A. (2016). Non-stochastic best arm identification and hyperparameter optimization. *Proceedings of the 17th International Conference on Artificial Intelligence and Statistics*.

Jenatton, R., Archambeau, C., González, J., & Seeger, M. (2017). Bayesian optimization with tree-structured dependencies. *Proceedings of the 34th International Conference on Machine Learning (ICML'17)*.

Jia, Y., Shelhamer, E., Donahue, J., Karayev, S., Long, J., Girshick, R., ... Darrell, T. (2014). Caffe: convolutional architecture for fast feature embedding. *Proceedings of the 22nd ACM International Conference on Multimedia* (pp. 675–678).

Jia, X., Song, S., He, W., Wang, Y., Rong, H., Zhou, F., ... et al. (2018). Highly scalable deep learning training system with mixed-precision: training ImageNet in four minutes. *ArXiv:1807.11205*.

Joshi, M., Chen, D., Liu, Y., Weld, D. S., Zettlemoyer, L., & Levy, O. (2020). SpanBERT: improving pre-training by representing and predicting spans. *Transactions of the Association for Computational Linguistics*, *8*, 64–77.

Jouppi, N. P., Young, C., Patil, N., Patterson, D., Agrawal, G., Bajwa, R., ... et al. (2017). In-datacenter performance analysis of a tensor processing unit. *2017 ACM/IEEE 44th Annual International Symposium on Computer Architecture (ISCA)* (pp. 1–12).

Kalchbrenner, N., Grefenstette, E., & Blunsom, P. (2014). A convolutional neural network for modelling sentences. *ArXiv:1404.2188*.

Kalman, B. L., & Kwasny, S. C. (1992). Why tanh: choosing a sigmoidal function. *Proceedings of the International Joint Conference on Neural Networks (IJCNN)* (pp. 578–581).

Kaplan, J., McCandlish, S., Henighan, T., Brown, T. B., Chess, B., Child, R., ... Amodei, D. (2020). Scaling laws for neural language models. *ArXiv:2001.08361*.

Karnin, Z., Koren, T., & Somekh, O. (2013). Almost optimal exploration in multi-armed bandits. *Proceedings of the 30th International Conference on Machine Learning (ICML'13)*.

Karras, T., Aila, T., Laine, S., & Lehtinen, J. (2017). Progressive growing of GANs for improved quality, stability, and variation. *ArXiv:1710.10196*.

Kim, Y. (2014). Convolutional neural networks for sentence classification. *ArXiv:1408.5882*.

Kim, J., El-Khamy, M., & Lee, J. (2017). Residual LSTM: design of a deep recurrent architecture for distant speech recognition. *ArXiv:1701.03360*.

Kimeldorf, G. S., & Wahba, G. (1971). Some results on Tchebycheffian spline functions. *J. Math. Anal. Appl.*, *33*, 82–95.

Kingma, D. P., & Ba, J. (2014). Adam: a method for stochastic optimization. *ArXiv:1412.6980*.

Kingma, D. P., & Welling, M. (2014). Auto-encoding variational Bayes. *International Conference on Learning Representations (ICLR)*.

Kipf, T. N., & Welling, M. (2016). Semi-supervised classification with graph convolutional networks. *ArXiv:1609.02907*.

Kojima, T., Gu, S. S., Reid, M., Matsuo, Y., & Iwasawa, Y. (2022). Large language models are zero-shot reasoners. *arxiv.org/abs/2205.11916*.

Koller, D., & Friedman, N. (2009). *Probabilistic Graphical Models: Principles and Techniques*. MIT Press.

Kolmogorov, A. (1933). Sulla determinazione empirica di una legge di distribuzione. *Inst. Ital. Attuari, Giorn.*, *4*, 83–91.

Kolter, Z. (2008). Linear algebra review and reference. *Available online: http://cs229.stanford.edu/section/cs229-linalg.pdf.*

Krizhevsky, A., Sutskever, I., & Hinton, G. E. (2012). ImageNet classification with deep convolutional neural networks. *Advances in Neural Information Processing Systems* (pp. 1097–1105).

Kung, S. Y. (1988). VLSI Array Processors. *Prentice Hall.*

Kuzovkin, I., Vicente, R., Petton, M., Lachaux, J.-P., Baciu, M., Kahane, P., ... Aru, J. (2018). Activations of deep convolutional neural networks are aligned with gamma band activity of human visual cortex. *Communications Biology, 1*(1), 1–12.

Lan, Z., Chen, M., Goodman, S., Gimpel, K., Sharma, P., & Soricut, R. (2019). ALBERT: a lite BERT for self-supervised learning of language representations. *ArXiv:1909.11942.*

Lavin, A., & Gray, S. (2016). Fast algorithms for convolutional neural networks. *Proceedings of the IEEE Conference on Computer Vision and Pattern Recognition* (pp. 4013–4021).

Le, Q. V. (2013). Building high-level features using large scale unsupervised learning. *Proceedings of the IEEE International Conference on Acoustics, Speech and Signal Processing* (pp. 8595–8598).

LeCun, Y., Boser, B., Denker, J. S., Henderson, D., Howard, R. E., Hubbard, W., & Jackel, L. D. (1989). Backpropagation applied to handwritten zip code recognition. *Neural Computation, 1*(4), 541–551.

LeCun, Y., Bengio, Y., & et al. (1995a). Convolutional networks for images, speech, and time series. *The Handbook of Brain Theory and Neural Networks* (p. 3361). MIT Press.

LeCun, Y., Jackel, L., Bottou, L., Brunot, A., Cortes, C., Denker, J., ... et al. (1995b). Comparison of learning algorithms for handwritten digit recognition. *International Conference on Artificial Neural Networks* (pp. 53–60).

LeCun, Y., Bottou, L., Orr, G., & Muller, K.-R. (1998a). Efficient backprop. *Neural Networks: Tricks of the Trade.* Springer.

LeCun, Y., Bottou, L., Bengio, Y., & Haffner, P. (1998b). Gradient-based learning applied to document recognition. *Proceedings of the IEEE, 86*(11), 2278–2324.

Legendre, A. M. (1805). *Mémoire sur les Opérations Trigonométriques: dont les Résultats Dépendent de la Figure de la Terre.* F. Didot.

Lewis, M., Liu, Y., Goyal, N., Ghazvininejad, M., Mohamed, A., Levy, O., ... Zettlemoyer, L. (2019). BART: denoising sequence-to-sequence pre-training for natural language generation, translation, and comprehension. *ArXiv:1910.13461.*

Lewkowycz, A., Andreassen, A., Dohan, D., Dyer, E., Michalewski, H., Ramasesh, V., ... et al. (2022). Solving quantitative reasoning problems with language models. *ArXiv:2206.14858.*

Li, M. (2017). *Scaling Distributed Machine Learning with System and Algorithm Co-design* (Doctoral dissertation). PhD Thesis, CMU.

Li, M., Andersen, D. G., Park, J. W., Smola, A. J., Ahmed, A., Josifovski, V., ... Su, B.-Y. (2014a). Scaling distributed machine learning with the parameter server. *11th Symposium on Operating Systems Design and Implementation (OSDI 14)* (pp. 583–598).

Li, M., Zhang, T., Chen, Y., & Smola, A. J. (2014b). Efficient mini-batch training for stochastic optimization. *Proceedings of the 20th ACM SIGKDD International Conference on Knowledge Discovery and Data Mining* (pp. 661–670).

Li, L., Jamieson, K., Rostamizadeh, A., Gonina, K., Hardt, M., Recht, B., & Talwalkar, A. (2018). Massively parallel hyperparameter tuning. *ArXiv:1810.05934*.

Liaw, R., Liang, E., Nishihara, R., Moritz, P., Gonzalez, J., & Stoica, I. (2018). Tune: a research platform for distributed model selection and training. *ArXiv:1807.05118*.

Lin, Y., Lv, F., Zhu, S., Yang, M., Cour, T., Yu, K., ... others. (2010). ImageNet classification: fast descriptor coding and large-scale SVM training. *Large Scale Visual Recognition Challenge*.

Lin, M., Chen, Q., & Yan, S. (2013). Network in network. *ArXiv:1312.4400*.

Lin, T.-Y., Goyal, P., Girshick, R., He, K., & Dollár, P. (2017a). Focal loss for dense object detection. *Proceedings of the IEEE International Conference on Computer Vision* (pp. 2980–2988).

Lin, Z., Feng, M., Santos, C. N. d., Yu, M., Xiang, B., Zhou, B., & Bengio, Y. (2017b). A structured self-attentive sentence embedding. *ArXiv:1703.03130*.

Lipton, Z. C., & Steinhardt, J. (2018). Troubling trends in machine learning scholarship. *Communications of the ACM*, *17*, 45–77.

Lipton, Z. C., Berkowitz, J., & Elkan, C. (2015). A critical review of recurrent neural networks for sequence learning. *ArXiv:1506.00019*.

Lipton, Z. C., Kale, D. C., Elkan, C., & Wetzel, R. (2016). Learning to diagnose with LSTM recurrent neural networks. *International Conference on Learning Representations (ICLR)*.

Liu, D. C., & Nocedal, J. (1989). On the limited memory BFGS method for large scale optimization. *Mathematical Programming*, *45*(1), 503–528.

Liu, W., Anguelov, D., Erhan, D., Szegedy, C., Reed, S., Fu, C.-Y., & Berg, A. C. (2016). SSD: single shot multibox detector. *European Conference on Computer Vision* (pp. 21–37).

Liu, H., Simonyan, K., & Yang, Y. (2018). DARTS: differentiable architecture search. *ArXiv:1806.09055*.

Liu, Y., Ott, M., Goyal, N., Du, J., Joshi, M., Chen, D., ... Stoyanov, V. (2019). RoBERTa: a robustly optimized BERT pretraining approach. *ArXiv:1907.11692*.

Liu, Z., Lin, Y., Cao, Y., Hu, H., Wei, Y., Zhang, Z., ... Guo, B. (2021). Swin transformer: hierarchical vision transformer using shifted windows. *Proceedings of the IEEE/CVF International Conference on Computer Vision* (pp. 10012–10022).

Liu, Z., Mao, H., Wu, C.-Y., Feichtenhofer, C., Darrell, T., & Xie, S. (2022). A convNet for the 2020s. *ArXiv:2201.03545*.

Long, J., Shelhamer, E., & Darrell, T. (2015). Fully convolutional networks for semantic segmentation. *Proceedings of the IEEE Conference on Computer Vision and Pattern Recognition* (pp. 3431–3440).

Loshchilov, I., & Hutter, F. (2016). SGDR: stochastic gradient descent with warm restarts. *ArXiv:1608.03983*.

Lowe, D. G. (2004). Distinctive image features from scale-invariant keypoints. *International Journal of Computer Vision*, *60*(2), 91–110.

Luo, P., Wang, X., Shao, W., & Peng, Z. (2018). Towards understanding regularization in batch normalization. *ArXiv:1809.00846*.

Maas, A. L., Daly, R. E., Pham, P. T., Huang, D., Ng, A. Y., & Potts, C. (2011). Learning word vectors for sentiment analysis. *Proceedings of the 49th Annual Meeting of the Association for Computational Linguistics: Human Language Technologies, Volume 1* (pp. 142–150).

Mack, Y.-P., & Silverman, B. W. (1982). Weak and strong uniform consistency of kernel regression estimates. *Zeitschrift für Wahrscheinlichkeitstheorie und verwandte Gebiete, 61*(3), 405–415.

MacKay, D. J. (2003). *Information Theory, Inference and Learning Algorithms.* Cambridge University Press.

Maclaurin, D., Duvenaud, D., & Adams, R. (2015). Gradient-based hyperparameter optimization through reversible learning. *Proceedings of the 32nd International Conference on Machine Learning (ICML'15).*

Mangasarian, O. L. (1965). Linear and nonlinear separation of patterns by linear programming. *Oper. Res., 13*, 444-452.

Mangram, M. E. (2013). A simplified perspective of the Markowitz portfolio theory. *Global Journal of Business Research, 7*(1), 59–70.

Matthews, A. G. d. G., Rowland, M., Hron, J., Turner, R. E., & Ghahramani, Z. (2018). Gaussian process behaviour in wide deep neural networks. *ArXiv:1804.11271.*

McCann, B., Bradbury, J., Xiong, C., & Socher, R. (2017). Learned in translation: Contextualized word vectors. *Advances in Neural Information Processing Systems* (pp. 6294–6305).

McCulloch, W. S., & Pitts, W. (1943). A logical calculus of the ideas immanent in nervous activity. *Bulletin of Mathematical Biophysics, 5*(4), 115–133.

Mead, C. (1980). Introduction to VLSI systems. *IEE Proceedings I-Solid-State and Electron Devices, 128*(1), 18.

Merity, S., Xiong, C., Bradbury, J., & Socher, R. (2016). Pointer sentinel mixture models. *ArXiv:1609.07843.*

Micchelli, C. A. (1984). Interpolation of scattered data: distance matrices and conditionally positive definite functions. *Approximation Theory and Spline Functions* (pp. 143–145). Springer.

Mikolov, T., Chen, K., Corrado, G., & Dean, J. (2013a). Efficient estimation of word representations in vector space. *ArXiv:1301.3781.*

Mikolov, T., Sutskever, I., Chen, K., Corrado, G. S., & Dean, J. (2013b). Distributed representations of words and phrases and their compositionality. *Advances in Neural Information Processing Systems* (pp. 3111–3119).

Miller, G. A. (1995). WordNet: a lexical database for English. *Communications of the ACM, 38*(11), 39–41.

Mirhoseini, A., Pham, H., Le, Q. V., Steiner, B., Larsen, R., Zhou, Y., ... Dean, J. (2017). Device placement optimization with reinforcement learning. *Proceedings of the 34th International Conference on Machine Learning* (pp. 2430–2439).

Mnih, V., Kavukcuoglu, K., Silver, D., Graves, A., Antonoglou, I., Wierstra, D., & Riedmiller, M. (2013). Playing Atari with deep reinforcement learning. *ArXiv:1312.5602.*

Mnih, V., Heess, N., Graves, A., ... et al. (2014). Recurrent models of visual attention. *Advances in Neural Information Processing Systems* (pp. 2204–2212).

Mnih, V., Kavukcuoglu, K., Silver, D., Rusu, A. A., Veness, J., Bellemare, M. G., ... et al. (2015). Human-level control through deep reinforcement learning. *Nature*, *518*(7540), 529–533.

Moon, T., Smola, A., Chang, Y., & Zheng, Z. (2010). Intervalrank: isotonic regression with listwise and pairwise constraints. *Proceedings of the 3rd ACM International Conference on Web Search and Data Mining* (pp. 151–160).

Morey, R. D., Hoekstra, R., Rouder, J. N., Lee, M. D., & Wagenmakers, E.-J. (2016). The fallacy of placing confidence in confidence intervals. *Psychonomic Bulletin & Review*, *23*(1), 103–123.

Morozov, V. A. (1984). *Methods for Solving Incorrectly Posed Problems*. Springer.

Nadaraya, E. A. (1964). On estimating regression. *Theory of Probability & its Applications*, *9*(1), 141–142.

Nair, V., & Hinton, G. E. (2010). Rectified linear units improve restricted Boltzmann machines. *ICML*.

Nakkiran, P., Kaplun, G., Bansal, Y., Yang, T., Barak, B., & Sutskever, I. (2021). Deep double descent: where bigger models and more data hurt. *Journal of Statistical Mechanics: Theory and Experiment*, *2021*(12), 124003.

Naor, M., & Reingold, O. (1999). On the construction of pseudorandom permutations: Luby–Rackoff revisited. *Journal of Cryptology*, *12*(1), 29–66.

Neal, R. M. (1996). *Bayesian Learning for Neural Networks*. Springer.

Nesterov, Y. (2018). *Lectures on Convex Optimization*. Springer.

Nesterov, Y., & Vial, J.-P. (2000). Confidence level solutions for stochastic programming. *Automatica*, *44*(6), 1559–1568.

Neyman, J. (1937). Outline of a theory of statistical estimation based on the classical theory of probability. *Philosophical Transactions of the Royal Society of London. Series A, Mathematical and Physical Sciences*, *236*(767), 333–380.

Norelli, A., Fumero, M., Maiorca, V., Moschella, L., Rodolà, E., & Locatello, F. (2022). ASIF: coupled data turns unimodal models to multimodal without training. *ArXiv:2210.01738*.

Novak, R., Xiao, L., Lee, J., Bahri, Y., Yang, G., Hron, J., ... Sohl-Dickstein, J. (2018). Bayesian deep convolutional networks with many channels are Gaussian processes. *ArXiv:1810.05148*.

Novikoff, A. B. J. (1962). On convergence proofs on perceptrons. *Proceedings of the Symposium on the Mathematical Theory of Automata* (pp. 615–622).

Olshausen, B. A., & Field, D. J. (1996). Emergence of simple-cell receptive field properties by learning a sparse code for natural images. *Nature*, *381*(6583), 607–609.

Ong, C. S., Smola, A., & Williamson, R. (2005). Learning the kernel with hyperkernels. *Journal of Machine Learning Research*, *6*, 1043–1071.

OpenAI. (2023). GPT-4 Technical Report. *ArXiv:2303.08774*.

Ouyang, L., Wu, J., Jiang, X., Almeida, D., Wainwright, C. L., Mishkin, P., ... et al. (2022). Training language models to follow instructions with human feedback. *ArXiv:2203.02155*.

Papineni, K., Roukos, S., Ward, T., & Zhu, W.-J. (2002). BLEU: a method for automatic evaluation of machine translation. *Proceedings of the 40th Annual Meeting of the Association for Computational Linguistics* (pp. 311–318).

Parikh, A. P., Täckström, O., Das, D., & Uszkoreit, J. (2016). A decomposable attention model for natural language inference. *ArXiv:1606.01933*.

Park, T., Liu, M.-Y., Wang, T.-C., & Zhu, J.-Y. (2019). Semantic image synthesis with spatially-adaptive normalization. *Proceedings of the IEEE Conference on Computer Vision and Pattern Recognition* (pp. 2337–2346).

Parzen, E. (1957). On consistent estimates of the spectrum of a stationary time series. *Annals of Mathematical Statistics*, *28*, 329–348.

Paszke, A., Gross, S., Massa, F., Lerer, A., Bradbury, J., Chanan, G., … et al. (2019). PyTorch: an imperative style, high-performance deep learning library. *Advances in Neural Information Processing Systems*, *32*, 8026–8037.

Paulus, R., Xiong, C., & Socher, R. (2017). A deep reinforced model for abstractive summarization. *ArXiv:1705.04304*.

Penedo, G., Malartic, Q., Hesslow, D., Cojocaru, R., Cappelli, A., Alobeidli, H., … Launay, J. (2023). The RefinedWeb dataset for Falcon LLM: outperforming curated corpora with web data, and web data only. *ArXiv:2306.01116*.

Pennington, J., Socher, R., & Manning, C. (2014). GloVe: global vectors for word representation. *Proceedings of the 2014 Conference on Empirical Methods in Natural Language Processing (EMNLP)* (pp. 1532–1543).

Pennington, J., Schoenholz, S., & Ganguli, S. (2017). Resurrecting the sigmoid in deep learning through dynamical isometry: theory and practice. *Advances in Neural Information Processing Systems* (pp. 4785–4795).

Peters, J., Janzing, D., & Schölkopf, B. (2017a). *Elements of Causal Inference: Foundations and Learning Algorithms*. MIT Press.

Peters, M., Ammar, W., Bhagavatula, C., & Power, R. (2017b). Semi-supervised sequence tagging with bidirectional language models. *Proceedings of the 55th Annual Meeting of the Association for Computational Linguistics, Volume 1* (pp. 1756–1765).

Peters, M., Neumann, M., Iyyer, M., Gardner, M., Clark, C., Lee, K., & Zettlemoyer, L. (2018). Deep contextualized word representations. *Proceedings of the 2018 Conference of the North American Chapter of the Association for Computational Linguistics: Human Language Technologies, Volume 1* (pp. 2227–2237).

Petersen, K. B., & Pedersen, M. S. (2008). *The Matrix Cookbook*. Technical University of Denmark.

Pleiss, G., Chen, D., Huang, G., Li, T., Van Der Maaten, L., & Weinberger, K. Q. (2017). Memory-efficient implementation of densenets. *ArXiv:1707.06990*.

Polyak, B. T. (1964). Some methods of speeding up the convergence of iteration methods. *USSR Computational Mathematics and Mathematical Physics*, *4*(5), 1–17.

Prakash, A., Hasan, S. A., Lee, K., Datla, V., Qadir, A., Liu, J., & Farri, O. (2016). Neural paraphrase generation with stacked residual LSTM networks. *ArXiv:1610.03098*.

Qin, C., Zhang, A., Zhang, Z., Chen, J., Yasunaga, M., & Yang, D. (2023). Is ChatGPT a general-purpose natural language processing task solver? *ArXiv:2302.06476*.

Quadrana, M., Cremonesi, P., & Jannach, D. (2018). Sequence-aware recommender systems. *ACM Computing Surveys*, *51*(4), 66.

Quinlan, J. R. (1993). *C4.5: Programs for Machine Learning*. Elsevier.

Rabiner, L., & Juang, B.-H. (1993). *Fundamentals of Speech Recognition*. Prentice-Hall.

Radford, A., Metz, L., & Chintala, S. (2015). Unsupervised representation learning with deep convolutional generative adversarial networks. *ArXiv:1511.06434*.

Radford, A., Narasimhan, K., Salimans, T., & Sutskever, I. (2018). Improving language understanding by generative pre-training. *OpenAI*.

Radford, A., Wu, J., Child, R., Luan, D., Amodei, D., & Sutskever, I. (2019). Language models are unsupervised multitask learners. *OpenAI Blog*, *1*(8), 9.

Radford, A., Kim, J. W., Hallacy, C., Ramesh, A., Goh, G., Agarwal, S., ... et al. (2021). Learning transferable visual models from natural language supervision. *International Conference on Machine Learning* (pp. 8748–8763).

Radosavovic, I., Johnson, J., Xie, S., Lo, W.-Y., & Dollár, P. (2019). On network design spaces for visual recognition. *Proceedings of the IEEE/CVF International Conference on Computer Vision* (pp. 1882–1890).

Radosavovic, I., Kosaraju, R. P., Girshick, R., He, K., & Dollár, P. (2020). Designing network design spaces. *Proceedings of the IEEE/CVF Conference on Computer Vision and Pattern Recognition* (pp. 10428–10436).

Rae, J. W., Borgeaud, S., Cai, T., Millican, K., Hoffmann, J., Song, F., ... et al. (2021). Scaling language models: methods, analysis & insights from training gopher. *ArXiv:2112.11446*.

Raffel, C., Shazeer, N., Roberts, A., Lee, K., Narang, S., Matena, M., ... Liu, P. J. (2020). Exploring the limits of transfer learning with a unified text-to-text transformer. *Journal of Machine Learning Research*, *21*, 1–67.

Rajpurkar, P., Zhang, J., Lopyrev, K., & Liang, P. (2016). SQuAD: 100,000+ questions for machine comprehension of text. *ArXiv:1606.05250*.

Ramachandran, P., Zoph, B., & Le, Q. V. (2017). Searching for activation functions. *ArXiv:1710.05941*.

Ramachandran, P., Parmar, N., Vaswani, A., Bello, I., Levskaya, A., & Shlens, J. (2019). Stand-alone self-attention in vision models. *Advances in Neural Information Processing Systems*, *32*.

Ramesh, A., Dhariwal, P., Nichol, A., Chu, C., & Chen, M. (2022). Hierarchical text-conditional image generation with clip latents. *ArXiv:2204.06125*.

Ramón y Cajal, Santiago, & Azoulay, L. (1894). *Les Nouvelles Idées sur la Structure du Système Nerveux chez l'Homme et chez les Vertébrés*. Paris, C. Reinwald & Cie.

Ranzato, M.-A., Boureau, Y.-L., Chopra, S., & LeCun, Y. (2007). A unified energy-based framework for unsupervised learning. *Artificial Intelligence and Statistics* (pp. 371–379).

Rasmussen, C. E., & Williams, C. K. (2006). *Gaussian Processes for Machine Learning*. MIT Press.

Reddi, S. J., Kale, S., & Kumar, S. (2019). On the convergence of Adam and beyond. *ArXiv:1904.09237*.

Redmon, J., Divvala, S., Girshick, R., & Farhadi, A. (2016). You only look once: unified, real-time object detection. *Proceedings of the IEEE Conference on Computer Vision and Pattern Recognition* (pp. 779–788).

Redmon, J., & Farhadi, A. (2018). YOLOv3: an incremental improvement. *ArXiv:1804.02767*.

Reed, S., & De Freitas, N. (2015). Neural programmer-interpreters. *ArXiv:1511.06279*.

Reed, S., Zolna, K., Parisotto, E., Colmenarejo, S. G., Novikov, A., Barth-Maron, G., ... et al. (2022). A generalist agent. *ArXiv:2205.06175*.

Ren, S., He, K., Girshick, R., & Sun, J. (2015). Faster R-CNN: towards real-time object detection with region proposal networks. *Advances in Neural Information Processing Systems* (pp. 91–99).

Revels, J., Lubin, M., & Papamarkou, T. (2016). Forward-mode automatic differentiation in Julia. *ArXiv:1607.07892*.

Rezende, D. J., Mohamed, S., & Wierstra, D. (2014). Stochastic backpropagation and approximate inference in deep generative models. *International Conference on Machine Learning* (pp. 1278–1286).

Riesenhuber, M., & Poggio, T. (1999). Hierarchical models of object recognition in cortex. *Nature Neuroscience*, *2*(11), 1019–1025.

Rockafellar, R. T. (1970). *Convex Analysis*. Princeton University Press.

Rolnick, D., Veit, A., Belongie, S., & Shavit, N. (2017). Deep learning is robust to massive label noise. *ArXiv:1705.10694*.

Rudin, W. (1973). *Functional Analysis*. McGraw-Hill.

Rumelhart, D. E., Hinton, G. E., & Williams, R. J. (1988). Learning representations by back-propagating errors. *Cognitive Modeling*, *5*(3), 1.

Russakovsky, O., Deng, J., Huang, Z., Berg, A. C., & Fei-Fei, L. (2013). Detecting avocados to zucchinis: what have we done, and where are we going? *International Conference on Computer Vision (ICCV)*.

Russakovsky, O., Deng, J., Su, H., Krause, J., Satheesh, S., Ma, S., ... et al. (2015). ImageNet large scale visual recognition challenge. *International Journal of Computer Vision*, *115*(3), 211–252.

Russell, S. J., & Norvig, P. (2016). *Artificial Intelligence: A Modern Approach*. Pearson Education Limited.

Saharia, C., Chan, W., Saxena, S., Li, L., Whang, J., Denton, E., ... et al. (2022). Photorealistic text-to-image diffusion models with deep language understanding. *ArXiv:2205.11487*.

Salinas, D., Seeger, M., Klein, A., Perrone, V., Wistuba, M., & Archambeau, C. (2022). Syne Tune: a library for large scale hyperparameter tuning and reproducible research. *First Conference on Automated Machine Learning*.

Sanh, V., Debut, L., Chaumond, J., & Wolf, T. (2019). DistilBERT, a distilled version of BERT: smaller, faster, cheaper and lighter. *ArXiv:1910.01108*.

Sanh, V., Webson, A., Raffel, C., Bach, S. H., Sutawika, L., Alyafeai, Z., ... et al. (2021). Multitask prompted training enables zero-shot task generalization. *ArXiv:2110.08207*.

Santurkar, S., Tsipras, D., Ilyas, A., & Madry, A. (2018). How does batch normalization help optimization? *Advances in Neural Information Processing Systems* (pp. 2483–2493).

Sarwar, B. M., Karypis, G., Konstan, J. A., & Riedl, J. (2001). Item-based collaborative filtering recommendation algorithms. *Proceedings of 10th International Conference on World Wide Web* (pp. 285–295).

Scao, T. L., Fan, A., Akiki, C., Pavlick, E., Ilić, S., Hesslow, D., ... et al. (2022). BLOOM: a 176B-parameter open-access multilingual language model. *ArXiv:2211.05100*.

Schein, A. I., Popescul, A., Ungar, L. H., & Pennock, D. M. (2002). Methods and metrics for cold-start recommendations. *Proceedings of the 25th Annual International ACM SIGIR Conference on Research and Development in Information Retrieval* (pp. 253–260).

Schölkopf, B., & Smola, A. J. (2002). *Learning with Kernels: Support Vector Machines, Regularization, Optimization, and Beyond.* MIT Press.

Schölkopf, B., Burges, C., & Vapnik, V. (1996). Incorporating invariances in support vector learning machines. *International Conference on Artificial Neural Networks* (pp. 47–52).

Schölkopf, B., Herbrich, R., & Smola, A. J. (2001). Helmbold, D. P., & Williamson, B. (Eds.). A generalized representer theorem. *Proceedings of the Annual Conference on Computational Learning Theory* (pp. 416–426). Springer-Verlag.

Schuhmann, C., Beaumont, R., Vencu, R., Gordon, C., Wightman, R., Cherti, M., … et al. (2022). LAION-5B: an open large-scale dataset for training next generation image-text models. *ArXiv:2210.08402.*

Schuster, M., & Paliwal, K. K. (1997). Bidirectional recurrent neural networks. *IEEE Transactions on Signal Processing*, *45*(11), 2673–2681.

Sennrich, R., Haddow, B., & Birch, A. (2015). Neural machine translation of rare words with subword units. *ArXiv:1508.07909.*

Sergeev, A., & Del Balso, M. (2018). Horovod: fast and easy distributed deep learning in TensorFlow. *ArXiv:1802.05799.*

Shannon, C. E. (1948). A mathematical theory of communication. *The Bell System Technical Journal*, *27*(3), 379–423.

Shao, H., Yao, S., Sun, D., Zhang, A., Liu, S., Liu, D., … Abdelzaher, T. (2020). ControlVAE: controllable variational autoencoder. *Proceedings of the 37th International Conference on Machine Learning.*

Shaw, P., Uszkoreit, J., & Vaswani, A. (2018). Self-attention with relative position representations. *ArXiv:1803.02155.*

Shoeybi, M., Patwary, M., Puri, R., LeGresley, P., Casper, J., & Catanzaro, B. (2019). Megatron-LM: training multi-billion parameter language models using model parallelism. *ArXiv:1909.08053.*

Silver, D., Huang, A., Maddison, C. J., Guez, A., Sifre, L., Van Den Driessche, G., … et al. (2016). Mastering the game of Go with deep neural networks and tree search. *Nature*, *529*(7587), 484.

Silverman, B. W. (1986). *Density Estimation for Statistical and Data Analysis.* Chapman and Hall.

Simard, P. Y., LeCun, Y. A., Denker, J. S., & Victorri, B. (1998). Transformation invariance in pattern recognition – tangent distance and tangent propagation. *Neural Networks: Tricks of the Trade* (pp. 239–274). Springer.

Simonyan, K., & Zisserman, A. (2014). Very deep convolutional networks for large-scale image recognition. *ArXiv:1409.1556.*

Sindhwani, V., Sainath, T. N., & Kumar, S. (2015). Structured transforms for small-footprint deep learning. *ArXiv:1510.01722.*

Sivic, J., & Zisserman, A. (2003). Video Google: a text retrieval approach to object matching in videos. *Proceedings of the IEEE International Conference on Computer Vision* (pp. 1470–1470).

Smith, S., Patwary, M., Norick, B., LeGresley, P., Rajbhandari, S., Casper, J., ... et al. (2022). Using DeepSpeed and Megatron to train Megatron-Turing NLG 530B, a large-scale generative language model. *ArXiv:2201.11990*.

Smola, A., & Narayanamurthy, S. (2010). An architecture for parallel topic models. *Proceedings of the VLDB Endowment, 3*(1-2), 703–710.

Snoek, J., Larochelle, H., & Adams, R. (2012). Practical Bayesian optimization of machine learning algorithms. *Advances in Neural Information Processing Systems 25* (pp. 2951–2959).

Sohl-Dickstein, J., Weiss, E., Maheswaranathan, N., & Ganguli, S. (2015). Deep unsupervised learning using nonequilibrium thermodynamics. *International Conference on Machine Learning* (pp. 2256–2265).

Song, Y., & Ermon, S. (2019). Generative modeling by estimating gradients of the data distribution. *Advances in Neural Information Processing Systems, 32*.

Song, Y., Sohl-Dickstein, J., Kingma, D. P., Kumar, A., Ermon, S., & Poole, B. (2021). Score-based generative modeling through stochastic differential equations. *International Conference on Learning Representations*.

Speelpenning, B. (1980). *Compiling fast partial derivatives of functions given by algorithms* (Doctoral dissertation). University of Illinois at Urbana-Champaign.

Srivastava, N., Hinton, G., Krizhevsky, A., Sutskever, I., & Salakhutdinov, R. (2014). Dropout: a simple way to prevent neural networks from overfitting. *Journal of Machine Learning Research, 15*(1), 1929–1958.

Srivastava, R.K., Greff, K., & Schmidhuber, J. (2015). Highway networks. *ArXiv:1505.00387*.

Srivastava, A., Rastogi, A., Rao, A., Shoeb, A.A.M., Abid, A., Fisch, A., ... et al. (2022). Beyond the imitation game: quantifying and extrapolating the capabilities of language models. *ArXiv:2206.04615*.

Strang, G. (1993). *Introduction to Linear Algebra*. Wellesley–Cambridge Press.

Su, X., & Khoshgoftaar, T. M. (2009). A survey of collaborative filtering techniques. *Advances in Artificial Intelligence, 2009*.

Sukhbaatar, S., Weston, J., & Fergus, R. (2015). End-to-end memory networks. *Advances in Neural Information Processing Systems* (pp. 2440–2448).

Sutskever, I., Martens, J., Dahl, G., & Hinton, G. (2013). On the importance of initialization and momentum in deep learning. *International Conference on Machine Learning* (pp. 1139–1147).

Sutskever, I., Vinyals, O., & Le, Q. V. (2014). Sequence to sequence learning with neural networks. *Advances in Neural Information Processing Systems* (pp. 3104–3112).

Szegedy, C., Liu, W., Jia, Y., Sermanet, P., Reed, S., Anguelov, D., ... Rabinovich, A. (2015). Going deeper with convolutions. *Proceedings of the IEEE Conference on Computer Vision and Pattern Recognition* (pp. 1–9).

Szegedy, C., Vanhoucke, V., Ioffe, S., Shlens, J., & Wojna, Z. (2016). Rethinking the Inception architecture for computer vision. *Proceedings of the IEEE Conference on Computer Vision and Pattern Recognition* (pp. 2818–2826).

Szegedy, C., Ioffe, S., Vanhoucke, V., & Alemi, A. A. (2017). Inception-v4, Inception-ResNet and the impact of residual connections on learning. *31st AAAI Conference on Artificial Intelligence*.

Tallec, C., & Ollivier, Y. (2017). Unbiasing truncated backpropagation through time. *ArXiv:1705.08209*.

Tan, M., & Le, Q. (2019). EfficientNet: rethinking model scaling for convolutional neural networks. *International Conference on Machine Learning* (pp. 6105–6114).

Taskar, B., Guestrin, C., & Koller, D. (2004). Max-margin Markov networks. *Advances in Neural Information Processing Systems*, *16*, 25.

Tay, Y., Dehghani, M., Bahri, D., & Metzler, D. (2020). Efficient transformers: a survey. *ArXiv:2009.06732*.

Taylor, R., Kardas, M., Cucurull, G., Scialom, T., Hartshorn, A., Saravia, E., ... Stojnic, R. (2022). Galactica: a large language model for science. *ArXiv:2211.09085*.

Teye, M., Azizpour, H., & Smith, K. (2018). Bayesian uncertainty estimation for batch normalized deep networks. *ArXiv:1802.06455*.

Thomee, B., Shamma, D. A., Friedland, G., Elizalde, B., Ni, K., Poland, D., ... Li, L.-J. (2016). Yfcc100m: the new data in multimedia research. *Communications of the ACM*, *59*(2), 64–73.

Tieleman, T., & Hinton, G. (2012). Divide the gradient by a running average of its recent magnitude. *COURSERA: Neural Networks for Machine Learning, Lecture 6.5-rmsprop*.

Tikhonov, A. N., & Arsenin, V. Y. (1977). *Solutions of Ill-Posed Problems*. W.H. Winston.

Tolstikhin, I. O., Houlsby, N., Kolesnikov, A., Beyer, L., Zhai, X., Unterthiner, T., ... et al. (2021). MLP-mixer: an all-MLP architecture for vision. *Advances in Neural Information Processing Systems*, *34*.

Torralba, A., Fergus, R., & Freeman, W. T. (2008). 80 million tiny images: a large data set for nonparametric object and scene recognition. *IEEE Transactions on Pattern Analysis and Machine Intelligence*, *30*(11), 1958–1970.

Touvron, H., Cord, M., Douze, M., Massa, F., Sablayrolles, A., & Jégou, H. (2021). Training data-efficient image transformers & distillation through attention. *International Conference on Machine Learning* (pp. 10347–10357).

Touvron, H., Lavril, T., Izacard, G., Martinet, X., Lachaux, M.-A., Lacroix, T., ... et al. (2023a). LLaMA: open and efficient foundation language models. *ArXiv:2302.13971*.

Touvron, H., Martin, L., Stone, K., Albert, P., Almahairi, A., Babaei, Y., ... et al. (2023b). LLaMA 2: open foundation and fine-tuned chat models. *ArXiv:2307.09288*.

Tsoumakas, G., & Katakis, I. (2007). Multi-label classification: an overview. *International Journal of Data Warehousing and Mining*, *3*(3), 1–13.

Turing, A. (1950). Computing machinery and intelligence. *Mind*, *59*(236), 433.

Uijlings, J. R., Van De Sande, K. E., Gevers, T., & Smeulders, A. W. (2013). Selective search for object recognition. *International Journal of Computer Vision*, *104*(2), 154–171.

Vapnik, V. (1995). *The Nature of Statistical Learning Theory*. New York: Springer.

Vapnik, V. (1992). Principles of risk minimization for learning theory. *Advances in Neural Information Processing Systems* (pp. 831–838).

Vapnik, V. (1998). *Statistical Learning Theory*. New York: John Wiley and Sons.

Vapnik, V., & Chervonenkis, A. (1964). A note on one class of perceptrons. *Automation and Remote Control*, *25*.

Vapnik, V., & Chervonenkis, A. (1968). Uniform convergence of frequencies of occurence of events to their probabilities. *Dokl. Akad. Nauk SSSR*, *181*, 915-918.

Vapnik, V., & Chervonenkis, A. (1971). On the uniform convergence of relative frequencies of events to their probabilities. *Theory Probab. Appl.*, *16*(2), 264-281.

Vapnik, V. N., & Chervonenkis, A. Y. (1974). Ordered risk minimization. *Automation and Remote Control*, *35*, 1226–1235, 1403–1412.

Vapnik, V., & Chervonenkis, A. (1981). The necessary and sufficient conditions for the uniform convergence of averages to their expected values. *Teoriya Veroyatnostei i Ee Primeneniya*, *26*(3), 543-564.

Vapnik, V., & Chervonenkis, A. (1991). The necessary and sufficient conditions for consistency in the empirical risk minimization method. *Pattern Recognition and Image Analysis*, *1*(3), 283-305.

Vapnik, V., Levin, E., & Le Cun, Y. (1994). Measuring the VC-dimension of a learning machine. *Neural Computation*, *6*(5), 851–876.

Vaswani, A., Shazeer, N., Parmar, N., Uszkoreit, J., Jones, L., Gomez, A. N., ... Polosukhin, I. (2017). Attention is all you need. *Advances in Neural Information Processing Systems* (pp. 5998–6008).

Wahba, G. (1990). *Spline Models for Observational Data*. SIAM.

Waibel, A., Hanazawa, T., Hinton, G., Shikano, K., & Lang, K. J. (1989). Phoneme recognition using time-delay neural networks. *IEEE Transactions on Acoustics, Speech, and Signal Processing*, *37*(3), 328–339.

Wang, Y., Davidson, A., Pan, Y., Wu, Y., Riffel, A., & Owens, J. D. (2016). Gunrock: a high-performance graph processing library on the GPU. *ACM SIGPLAN Notices* (p. 11).

Wang, L., Li, M., Liberty, E., & Smola, A. J. (2018). Optimal message scheduling for aggregation. *Networks*, *2*(3), 2–3.

Wang, Q., Li, B., Xiao, T., Zhu, J., Li, C., Wong, D. F., & Chao, L. S. (2019). Learning deep transformer models for machine translation. *Proceedings of the 57th Annual Meeting of the Association for Computational Linguistics* (pp. 1810–1822).

Wang, H., Zhang, A., Zheng, S., Shi, X., Li, M., & Wang, Z. (2022). Removing batch normalization boosts adversarial training. *International Conference on Machine Learning* (pp. 23433–23445).

Wang, X., Wei, J., Schuurmans, D., Le, Q., Chi, E., & Zhou, D. (2023). Self-consistency improves chain of thought reasoning in language models. *International Conference on Learning Representations*.

Warstadt, A., Singh, A., & Bowman, S. R. (2019). Neural network acceptability judgments. *Transactions of the Association for Computational Linguistics*, *7*, 625–641.

Wasserman, L. (2013). *All of Statistics: A Concise Course in Statistical Inference*. Springer.

Watkins, C. J., & Dayan, P. (1992). Q-learning. *Machine Learning*, *8*(3–4), 279–292.

Watson, G. S. (1964). Smooth regression analysis. *Sankhyā: The Indian Journal of Statistics, Series A*, pp. 359–372.

Wei, J., Bosma, M., Zhao, V. Y., Guu, K., Yu, A. W., Lester, B., ... Le, Q. V. (2021). Finetuned language models are zero-shot learners. *ArXiv:2109.01652*.

Wei, J., Tay, Y., Bommasani, R., Raffel, C., Zoph, B., Borgeaud, S., ... et al. (2022a). Emergent abilities of large language models. *ArXiv:2206.07682*.

Wei, J., Wang, X., Schuurmans, D., Bosma, M., Chi, E., Le, Q., & Zhou, D. (2022b). Chain of thought prompting elicits reasoning in large language models. *ArXiv:2201.11903*.

Welling, M., & Teh, Y. W. (2011). Bayesian learning via stochastic gradient Langevin dynamics. *Proceedings of the 28th International Conference on Machine Learning (ICML-11)* (pp. 681–688).

Wengert, R. E. (1964). A simple automatic derivative evaluation program. *Communications of the ACM*, *7*(8), 463–464.

Werbos, P. J. (1990). Backpropagation through time: what it does and how to do it. *Proceedings of the IEEE*, *78*(10), 1550–1560.

Wigner, E. P. (1958). On the distribution of the roots of certain symmetric matrices. *Ann. Math.* (pp. 325–327).

Wilson, A. G., & Izmailov, P. (2020). Bayesian deep learning and a probabilistic perspective of generalization. *Advances in Neural Information Processing Systems*, *33*, 4697–4708.

Wistuba, M., Schilling, N., & Schmidt-Thieme, L. (2018). Scalable Gaussian process-based transfer surrogates for hyperparameter optimization. *Machine Learning*, *108*, 43–78.

Wistuba, M., Rawat, A., & Pedapati, T. (2019). A survey on neural architecture search. *ArXiv:1905.01392 [cs.LG]*.

Wolpert, D. H., & Macready, W. G. (1995). *No free lunch theorems for search*. Technical Report SFI-TR-95-02-010, Santa Fe Institute.

Wood, F., Gasthaus, J., Archambeau, C., James, L., & Teh, Y. W. (2011). The sequence memoizer. *Communications of the ACM*, *54*(2), 91–98.

Wu, Y., Schuster, M., Chen, Z., Le, Q. V., Norouzi, M., Macherey, W., ... et al. (2016). Google's neural machine translation system: bridging the gap between human and machine translation. *ArXiv:1609.08144*.

Wu, B., Wan, A., Yue, X., Jin, P., Zhao, S., Golmant, N., ... Keutzer, K. (2018). Shift: a zero flop, zero parameter alternative to spatial convolutions. *Proceedings of the IEEE Conference on Computer Vision and Pattern Recognition* (pp. 9127–9135).

Xiao, H., Rasul, K., & Vollgraf, R. (2017). Fashion-MNIST: a novel image dataset for benchmarking machine learning algorithms. *ArXiv:1708.07747*.

Xiao, L., Bahri, Y., Sohl-Dickstein, J., Schoenholz, S., & Pennington, J. (2018). Dynamical isometry and a mean field theory of CNNs: how to train 10,000-layer vanilla convolutional neural networks. *International Conference on Machine Learning* (pp. 5393–5402).

Xie, S., Girshick, R., Dollár, P., Tu, Z., & He, K. (2017). Aggregated residual transformations for deep neural networks. *Proceedings of the IEEE Conference on Computer Vision and Pattern Recognition* (pp. 1492–1500).

Xiong, W., Wu, L., Alleva, F., Droppo, J., Huang, X., & Stolcke, A. (2018). The Microsoft 2017 conversational speech recognition system. *2018 IEEE International Conference on Acoustics, Speech and Signal Processing (ICASSP)* (pp. 5934–5938).

Xiong, R., Yang, Y., He, D., Zheng, K., Zheng, S., Xing, C., ... Liu, T. (2020). On layer normalization in the transformer architecture. *International Conference on Machine Learning* (pp. 10524–10533).

Yamaguchi, K., Sakamoto, K., Akabane, T., & Fujimoto, Y. (1990). A neural network for speaker-independent isolated word recognition. *First International Conference on Spoken Language Processing*.

Yang, Z., Moczulski, M., Denil, M., De Freitas, N., Smola, A., Song, L., & Wang, Z. (2015). Deep fried convnets. *Proceedings of the IEEE International Conference on Computer Vision* (pp. 1476–1483).

Yang, Z., Hu, Z., Deng, Y., Dyer, C., & Smola, A. (2016). Neural machine translation with recurrent attention modeling. *ArXiv:1607.05108*.

Ye, M., Yin, P., Lee, W.-C., & Lee, D.-L. (2011). Exploiting geographical influence for collaborative point-of-interest recommendation. *Proceedings of the 34th International ACM SIGIR Conference on Research and Development in Information Retrieval* (pp. 325–334).

You, Y., Gitman, I., & Ginsburg, B. (2017). Large batch training of convolutional networks. *ArXiv:1708.03888*.

Yu, J., Xu, Y., Koh, J. Y., Luong, T., Baid, G., Wang, Z., ... Wu, Y. (2022). Scaling autoregressive models for content-rich text-to-image generation. *ArXiv:2206.10789*.

Zaheer, M., Reddi, S., Sachan, D., Kale, S., & Kumar, S. (2018). Adaptive methods for nonconvex optimization. *Advances in Neural Information Processing Systems* (pp. 9793–9803).

Zeiler, M. D. (2012). ADADELTA: an adaptive learning rate method. *ArXiv:1212.5701*.

Zeiler, M. D., & Fergus, R. (2013). Stochastic pooling for regularization of deep convolutional neural networks. *ArXiv:1301.3557*.

Zhang, W., Tanida, J., Itoh, K., & Ichioka, Y. (1988). Shift-invariant pattern recognition neural network and its optical architecture. *Proceedings of Annual Conference of the Japan Society of Applied Physics*.

Zhang, S., Yao, L., Sun, A., & Tay, Y. (2019). Deep learning based recommender system: a survey and new perspectives. *ACM Computing Surveys*, *52*(1), 5.

Zhang, A., Tay, Y., Zhang, S., Chan, A., Luu, A. T., Hui, S. C., & Fu, J. (2021a). Beyond fully-connected layers with quaternions: parameterization of hypercomplex multiplications with 1/n parameters. *International Conference on Learning Representations*.

Zhang, C., Bengio, S., Hardt, M., Recht, B., & Vinyals, O. (2021b). Understanding deep learning (still) requires rethinking generalization. *Communications of the ACM*, *64*(3), 107–115.

Zhang, Y., Sun, P., Jiang, Y., Yu, D., Yuan, Z., Luo, P., ... Wang, X. (2021). ByteTrack: multi-object tracking by associating every detection box. *ArXiv:2110.06864*.

Zhang, S., Roller, S., Goyal, N., Artetxe, M., Chen, M., Chen, S., ... et al. (2022c). OPT: open pre-trained transformer language models. *ArXiv:2205.01068*.

Zhang, Z., Zhang, A., Li, M., & Smola, A. (2023a). Automatic chain of thought prompting in large language models. *International Conference on Learning Representations*.

Zhang, Z., Zhang, A., Li, M., Zhao, H., Karypis, G., & Smola, A. (2023b). Multimodal chain-of-thought reasoning in language models. *ArXiv:2302.00923*.

Zhao, Z.-Q., Zheng, P., Xu, S.-t., & Wu, X. (2019). Object detection with deep learning: a review. *IEEE Transactions on Neural Networks and Learning Systems*, *30*(11), 3212–3232.

Zhou, D., Schärli, N., Hou, L., Wei, J., Scales, N., Wang, X., ... Chi, E. (2023). Least-to-most prompting enables complex reasoning in large language models. *International Conference on Learning Representations*.

Zhu, Y., Kiros, R., Zemel, R., Salakhutdinov, R., Urtasun, R., Torralba, A., & Fidler, S. (2015). Aligning books and movies: towards story-like visual explanations by watching movies and reading books. *Proceedings of the IEEE International Conference on Computer Vision* (pp. 19–27).

Zhu, J.-Y., Park, T., Isola, P., & Efros, A. A. (2017). Unpaired image-to-image translation using cycle-consistent adversarial networks. *Proceedings of the IEEE International Conference on Computer Vision* (pp. 2223–2232).

Zoph, B., & Le, Q. V. (2016). Neural architecture search with reinforcement learning. *ArXiv:1611.01578.*

Index